Thidiazuron: From Urea Derivative to Plant Growth Regulator

Naseem Ahmad • Mohammad Faisal
Editors

Thidiazuron: From Urea Derivative to Plant Growth Regulator

Springer

Editors
Naseem Ahmad
Department of Botany
Aligarh Muslim University
Aligarh, UP, India

Mohammad Faisal
Department of Botany & Microbiology,
College of Science
King Saud University
Riyadh, Saudi Arabia

ISBN 978-981-10-8003-6 ISBN 978-981-10-8004-3 (eBook)
https://doi.org/10.1007/978-981-10-8004-3

Library of Congress Control Number: 2018934324

Printed on acid-free paper

This Springer imprint is published by the registered company Springer Nature Singapore Pte Ltd. part of Springer Nature.
The registered company address is: 152 Beach Road, #21-01/04 Gateway East, Singapore 189721, Singapore

Preface

Plant biotechnology is a most interesting field for academicians, researchers, and students, and nowadays it becomes a very useful tool in agriculture and medicine and regarded as a popular area of research especially in biological sciences because of its integral use in biochemistry, molecular biology, and engineering sciences in order to achieve technological application of cultured tissues, cells, and microbes. Plant tissue culture (PTC), a technique refers to cultivation of plant cells and other parts on artificial nutrient medium in controlled environment under aseptic conditions, is essential to all plant biotechnologies and is an exciting area of research for basic and applied sciences. PTC requires various nutrients, pH, carbon source, gelling agent, temperature, photoperiod, humidity, etc., and most importantly the judicious use of plant growth regulators. Various natural, adenine and phenyl urea derivatives were employed for the induction and proliferation of different types of explants. Several phenyl urea derivatives were evaluated, and it was observed that thidiazuron (n-phenyl-N"-1,2,3-thidiazol-5-ulurea) was found to be the most effective among the plant growth regulators. Thidiazuron (TDZ) was initially developed as a cotton defoliant and showed high cytokinin like activity. In some examples, its activity was 100 times more than BA in tobacco callus assay and produces more number of shoots in cultures than Zeatin and 2iP. TDZ also showed major breakthrough in tissue culture of various recalcitrant legumes and woody plant species. For the last two decades, a number of laboratories have been working on TDZ with different aspect, and a number of publications have come out. To the best of our knowledge, there is no comprehensive edited volume on this particular topic. Hence, the edited volume is a deed to consolidate the scattered information on the role of TDZ in plant tissue culture and genetic manipulations that would hopefully prove informative to various researches.

Thidiazuron: From Urea Derivative to Plant Growth Regulator is a compilation of review and research papers dealing with various aspects of TDZ in plant tissue culture with profitable implications.

The book will provide basic and applied information for academicians, scientists, and researchers who want to initiate work in this fascinating area of research. The book contains 28 chapters compiled by international dignitaries, which give much weightage to this edited volume. All the chapters have been organized in a way to provide a crisp information on role of TDZ in plant tissue culture. Special attention has been given to explore the mechanism and mode of actions of TDZ which can lead to organogenesis in vitro.

We are extremely thankful to all the contributors who agreed to our proposal, thus helped in this endeavor.

Aligarh, India
September 17, 2017

Naseem Ahmad
Mohammad Faisal

Acknowledgments

We, the editors, are thankful and acknowledge the generous support established from the contributors who spared their precious time in scripting the chapters for this edited volume and also the publisher for making this book a reality. We thank the reviewers for their comments and suggestions, which have been very helpful for the improvement of the manuscript. We would also like to thank Professor M. Anis, our mentor at the Department of Botany, Aligarh Muslim University, for his assistance and support in the editing of the manuscript. We express our special thanks to the Aligarh Muslim University, King Saud University, and National Plan for Science, Technology and Innovation (MAARIFAH), King Abdulaziz City for Science and Technology, Kingdom of Saudi Arabia, for their support.

Dr. Naseem Ahmad
Dr. Mohammad Faisal

Contents

About the Editors

Naseem Ahmad is working as assistant professor in the Department of Botany, Aligarh Muslim University, Aligarh (UP) India. He has also worked as Young Scientist, has more than 8 years of experience in plant tissue culture and molecular biology, has published about 39 research papers, and has edited 1 book entitled *Plant Tissue Culture: Propagation, Conservation and Crop Improvement* (Springer; 2016). He holds a lifetime membership in many learned/scientific societies and is a fellow of Indian Botanical Society (IBS). He is recipient of "Yuva Vaigiyanik Samman (2009), CST, UP; Young Scientist (2010) by NESA, N. Delhi; Rashtriya Gaurav Award (2015) by IIFS, N. Delhi; and Young Scientist (2015) by IFEE, Kolkata. He was also a senior research fellow (SRF) of CSIR, N. Delhi.

Mohammad Faisal has over 10 years of research experience in plant biotechnology and is currently working as associate professor in the Department of Botany and Microbiology, King Saud University, Riyadh, Saudi Arabia. He received his PhD degree in Botany from the Aligarh Muslim University, India. He has been the recipient of Plant Biotechnologist Award (2017) by SESR, India, Scientist of the Year Award (2015) by NESA, and several national and international fellowships. He holds membership of several academic bodies/societies, is fellow of the SESR, and has published over 64 research articles, 3 book chapters, and 2 books, and his articles have been cited over 1395 times with Google h-index of 23. He is included the editorial board of several journals.

Thidiazuron: Modulator of Morphogenesis In Vitro

1

Elham Tavakouli Dinani, Mukund R. Shukla, Christina E. Turi, J. A. Sullivan, and Praveen K. Saxena

Abstract

Thidiazuron (TDZ) is a substituted phenylurea first synthesized in 1967 by the Schering Corporation in Germany. Compared to other plant growth regulators, TDZ is a powerful and potent synthetic growth regulator, leading to a wide array of in vitro and in vivo applications in plants including prevention of leaf yellowing, enhanced photosynthetic activity, breaking of bud dormancy, fruit ripening, as well as proliferation of adventitious shoots, callus production, and induction of somatic embryogenesis. Despite the diversity of effects attributed to TDZ, its application and mode of action for induction of in vitro morphogenesis in plants are not well understood. Thus, this review aims to summarize current understandings for TDZ during in vitro morphogenesis in order to better understand the potential applications of TDZ for induction of in vitro morphogenesis and organogenesis.

Keywords

Thidiazuron · TDZ · Morphogenesis · Auxin · Cytokinin · Plant growth regulation

1.1 Introduction

Thidiazuron (TDZ) is a substituted phenylurea first synthesized in 1967 by the Schering Corporation in Germany, originally being used as a cotton defoliant and eventually becoming registered in the USA in 1982 (Arndt et al. 1976; Pavlista and

E. T. Dinani · M. R. Shukla · C. E. Turi · J. A. Sullivan · P. K. Saxena (✉)
Department of Plant Agriculture, Gosling Research Institute for Plant Preservation, University of Guelph, Guelph, Canada
e-mail: psaxena@uoguelph.ca

N. Ahmad, M. Faisal (eds.), *Thidiazuron: From Urea Derivative to Plant Growth Regulator*, https://doi.org/10.1007/978-981-10-8004-3_1

Fig. 1.1 Summary of the physiological effect of TDZ on apple plant organs which include the stem, root, leaf, flower, and fruit

Gall 2011). Compared to other plant growth regulators (PGRs), TDZ is a powerful and potent synthetic growth regulator exhibiting both auxin- and cytokinin (CK)-like effects in plants, leading to a wide array of in vitro and in vivo applications including prevention of leaf yellowing, enhanced photosynthetic activity, breaking of bud dormancy, fruit ripening, as well as proliferation of adventitious shoots, callus production, and induction of somatic embryogenesis (Fig. 1.1). Despite this unique and dual effect, TDZ's action is often overgeneralized and referred to as a cytokinin. It is therefore important to note that although TDZ can mimic the effects of auxins and CKs, structurally it differs from both of these PGR groups, possessing both phenyl and thiadiazole functional groups, with both groups required for biological activity (Mok et al. 1987).

Compared with other PGRs, TDZ can be used for regeneration at much lower concentrations (10–1000 times lower) making it a valuable commercial agrochemical (Fig. 1.1; Guo et al. 2011). For instance, TDZ's ability to inhibit leaf yellowing, delay leaf senescence, maintain chlorophyll (Chl) concentrations, inhibit carotenoid degradation, inhibit abscisic acid (ABA) biosynthesis, and decrease ethylene sensitivity in cut flowers (Uthairatanakij et al. 2007; Ferrante et al. 2004) has led to its application in the horticultural industry for the purpose of increasing the longevity of cut flowers such alstroemeria (*Alstroemeria aurea* Graham), lilies (*Lilium* spp.), tulips (*Tulipa* spp.), and chrysanthemum (*Chrysanthemum* spp.) (Ferrante et al. 2002a, b; Sankhla et al. 2003). In addition to the above, TDZ's ability to increase fruit size without affecting seed number, through the promotion of cell division in the cortex layer of fruits (Stern et al. 2003), has led to its application for improving fruit size in a number of crops including pear (*Pyrus communis* L.), grape (*Vitis vinifera* L.), persimmon (*Diospyros virginiana* L.), cucumber (*Cucumis sativus* L.)

and kiwifruit (*Actinidia deliciosa* (A. Chev.) C.F. Liang and A.R. Ferguson) (Amarante et al. 2003; Stern et al. 2003). In stone fruits and cut flowers, TDZ has also been used to stimulate bud growth and opening and to accelerate bud breaking (Erez et al. 2006; Wang et al. 1986).

Despite the diversity of effects attributed to TDZ, its application and mode of action for induction of in vitro morphogenesis in plants is not well understood. This notion largely stems from TDZ's ability to display both CK- and auxin-like activities individually or simultaneously during in vitro regeneration. To complicate matters further, TDZ's ability to induce a defensive response in plant tissues can also initiate the up- or downregulation of other PGRs (i.e., ABA, ethylene, melatonin, serotonin) and secondary metabolites (i.e., polyamines) while also modulating the influx/efflux of specific cations (i.e., calcium) across biological membranes (Murch et al. 1997; Murch and Saxena 1997; Murthy et al. 1995; Proctor et al. 1996). In order to better understand potential applications of TDZ for induction of in vitro morphogenesis and organogenesis, the current review aims to summarize the current uses of this multipurpose synthetic PGR in plant tissue culture processes.

1.2 Application of TDZ During Plant Morphogenesis

Although shoot production and plant development reportedly vary in response to TDZ concentration, plant material, and species (Liu et al. 1998), generally, TDZ is more biologically active than BAP, kinetin, or zeatin (Capelle et al. 1983). For example, Lu (1993) observed that TDZ is more effective at lower concentrations compared to classical CKs during shoot regeneration of woody species. In addition to the above, TDZ's ability to exhibit its effects in explants well after the initial treatment (subsequently transferred to media without TDZ) indicates that some explants only require limited exposure (Matand and Prakash 2007). Short exposure time and low concentrations of TDZ have, in fact, been found to be highly effective in stimulation of shoot regeneration across diverse species (Mihaljević and Vršek 2009). TDZ's unique property of high efficacy at low doses and/or short exposure times may be explained by TDZ's ability to resist enzymatic degradation in vivo (Murthy et al. 1998; Kumar and Reddy 2012) which in turn enables TDZ levels to remain stable over time (Dey et al. 2012). For example, in bean callus incubated with radiolabeled TDZ for 33 days, TDZ was found to remain largely intact, with only a small fraction being glycosylated (Mok and Mok 1985). Tracer studies by Benezet and Knowles (1982) have also observed limited degradation (oxidation) of the TDZ molecule within etiolated hypocotyls by 13 species of microorganisms, as evidenced by limited evolution of $14CO_2$, which is one of the principle degradation products of TDZ. This indicates TDZ molecules were not undergoing significant degradation and likely remained within plant tissues over the duration of the experiment, up to a 28-day incubation period. Furthermore, through the use of 14C-TDZ and fractionation experiments, Murch and Saxena (2001) noted that TDZ may in fact exist in several forms, i.e., TDZ-free molecules, sequestered TDZ molecules, and conjugated forms associated with proteins or cell wall components within plant tissues.

1.3 Shoot Bud Induction

In plants, the induction of shoot buds is dependent upon a balance between auxin and CK levels, whereby an increased presence of auxin and CKs can inhibit or initiate bud formation, respectively (Wang et al. 1986). TDZ appears to promote shoot bud initiation by stimulating cell division and multiplication in the apical meristem while also reprogramming cells to the appropriate developmental stage for initiation of shoot differentiation (Dey et al. 2012; Vu et al. 2006). As in other processes affected by TDZ, diverse factors may affect the ability of TDZ to induce shoot bud initiation and growth including: concentration of TDZ, type and source of explant, age or phase of growth, cultivar, presence of other PGRs, particularly auxin, in the medium, balance of endogenous growth regulators, and presence of light (Sanikhani et al. 2006; Visser et al. 1992; Table 1.1).

In general, low concentrations (≥2.5 μM) of TDZ enhance axillary bud formation on cultured shoot tip meristems, while moderate concentrations of TDZ (5–10 μM) can result in somatic embryo formation. At higher concentrations, morphological abnormalities like hyperhydricity have been reported (Lu 1993; Mithila et al. 2003). Not surprisingly, TDZ is typically applied at low concentrations to a wide range of explant types in order to induce bud growth (Murashige 1974; Jiang et al. 2008); however, the concentration required varies with explant type. For instance, direct shoot bud formation occurred only on cotyledonary nodes when TDZ was applied at rates of 0.9–5.4 μM during in vitro regeneration of soybean (*Glycine max* (L.) Merr.) seeds. On the other hand, 10 μM TDZ was optimal for induction of shoot buds in leaf explants of apple (*Malus domestica* Borkh.) (Fasolo et al. 1989), while low concentrations of TDZ (0.02–0.56 μM) induced bud/shoot regeneration in excised roots (*Albizia julibrissin* Durazz.) (Sankhla et al. 1996). TDZ (10 μM) has also been found to induce bud formation and regeneration in thin cell layer (TCL) system from the common bean *Phaseolus vulgaris* L., where pretreatment significantly increased bud regeneration. Optimal bud induction and further development of the formed buds were observed in 2-week cultures of TCLs on 10 μM TDZ later reduced to 1 μM TDZ (Cruz De Carvalho et al. 2000). The length of time the explants are exposed to TDZ can also impact the ability of TDZ to induce bud formation. In *Curculigo orchioides* Gaertn., pretreatment with 15 μM TDZ for 24 h significantly stimulated adventitious shoot regeneration from leaves, while in *Tecomella undulata* (Sm.) Seem., exposure to a concentration of 0.7 μM for a duration of 1–3 weeks was most efficient for shoot regeneration (Varshney and Anis 2012). Interestingly, duration and level of exposure of explants to light during TDZ treatment can also influence shoot organogenesis. For example, de novo shoot bud formation in strawberry (*Fragaria* x *ananassa* Duch) was achieved using leaf disks cultured in the dark and on MS medium containing 9.08 μM TDZ (Husaini and Abdin 2007). Although it is not yet fully understood how light affects TDZ action, it is believed that TDZ's ability to induce shoot bud production in the dark is triggered by calcium stress, which in turn affects the production of ethylene (Mundhara and Rashid

Table 1.1 Summary of TDZ used during induction of shoot buds

Plant response	Optimal TDZ concentration	Explant used	Species	Family	Exposure time	Other PGRs used	Exposure to light/dark	References
Adventitious bud regeneration	0.01, 0.03, 0.1, 0.3, 1.0, 3.0 μM	Normal and hairy root cultures	*Centaurium erythraea* Rafn	Gentianaceae	30 days	Kinetin, BA, 6-γ,γ-dimethylallylaminopurine (2iP), N-(2-chloro-4-pyridyl)-N′-phenylurea (CPPU), 6-[4-hydroxy-3-methyl-but-2-enylamino] purine (ZEA)	16-h photoperiod	Subotić et al. (2009)
Shoot differentiation	0.1, 1, 10 μM	Leaf explants	*Rhododendron* spp.	Ericaceae	2 weeks	IBA, 2iP	16-h photoperiod	Preece and Imel (1991)
Plantlet regeneration	50 μM	Cell suspensions derived from leaf callus	*Actinidia chinensis* Planch	Actinidiaceae	1 month	Zeatin, NAA, BA, 2,4-D	Dark/16-h photoperiod	Suezawa et al. (1988)
Plant regeneration	1–50 μM	Protoplast-derived pea callus	*Pisum sativum* L.	Fabaceae	16 weeks	–	16-h photoperiod	Böhmer et al. (1995)
Adventitious bud formation	0.01–0.1 mg dm-3	Cotyledonary nodes	*Medicago sativa* L.	Fabaceae	28 days	NAA	14-h photoperiod	Li et al. (2009)
Bud induction	22.7 μM	Cotyledon explants	*Capsicum annuum* L.	Solanaceae	2–4 weeks	BA, IAA	16-h photoperiod	Hyde and Phillips (1996)
Shoot regeneration	5 μM	Root explant	*Hypericum perforatum* L.	Hypericaceae	28 days	–	16-h photoperiod	Zobayed and Saxena (2003)
Bud regeneration	1, 2.5, and 5 μM	Leaves	*Pyrus communis* L.	Rosaceae	20 days	BA, IBA	Dark	Leblay et al. (1991)

(continued)

Table 1.1 (continued)

Plant response	Optimal TDZ concentration	Explant used	Species	Family	Exposure time	Other PGRs used	Exposure to light/dark	References
Bud/shoot regeneration	0.02, 0.11, 0.56, 2.8, 14.1 μM	Root segment	*Populus alba* L.	Salicaceae	7 weeks	–	Dark/16-h photoperiod	Tsvetkov et al. (2007)
Shoot bud development	0.01–1 mg/l	Seed-derived rhizomes	*Cymbidium sinense* (Jacks.) Willd	Orchidaceae	4 months	NAA	16-h photoperiod	Chang and Chang (2000)
Shoot bud differentiation	0.90–22.72 μM	Leaf explants	*Jatropha curcas* L.	Euphorbiaceae	6 weeks	BAP, IBA	16-h photoperiod	Khurana-Kaul et al. (2010)
Shoot bud formation	0.1, 1.0 μM	Root segments, entire roots	*Albizia julibrissin* Durazz.	Fabaceae	30 days	6-BAP, TDZ	Light/dark	Hosseini-Nasr and Rashid (2002)
Shoot bud formation	0.05–0.2 mg/l	Flower clusters	*Vitis vinifera* L.	Vitaceae	3–6 weeks	2,4-D	16-h photoperiod	Oláh et al. (2003)
Shoot bud formation	4.54, 9.08, 13.12, 18.16 μM	Seedling	*Tamarindus indica* L.	Fabaceae	6 weeks	–	Dark/light	Mehta et al. (2004)
Shoot bud formation	0, 0.45, 2.27, 4.54 μM	Roots of 15-day-old seedlings	*Solanum melongena* L.	Solanaceae	28 days	BA, NAA	16-h photoperiod	Franklin et al. (2004)
Shoot bud induction	0.05, 0.1, 0.2, 0.4 μM	Leaves and cotyledons	*Solanum melongena* L.	Solanaceae	30 days	–	16-h photoperiod	Magioli et al. (1998)
Shoot bud induction	0.5–1 μM	Nodal and internodal segment	*Ephedra gerardiana* Wall. ex Stapf	Ephedraceae	10–15 days	–	Light	Sharma et al. (2013)
Shoot bud induction	0.01–0.1.0 mg/l	Axillary buds	*Prunus dulcis* (Mill.) D.A. Webb	Rosaceae	30 days	BAP, IAA, IBA, NAA	16-h photoperiod	Choudhary et al. (2015)

Shoot bud differentiation	0.1, 1, 10 μM	Leaf explants	*Rhododendron* spp.	Ericaceae	2 weeks	IBA, 2iP	16-h photoperiod	Preece and Imel (1991)
Plantlet regeneration	50 μM	Cell suspensions derived from leaf callus	*Actinidia chinensis* Planch	Actinidiaceae	1 month	Zeatin, NAA, BA, 2,4-D	Dark/16-h photoperiod	Suezawa et al. (1988)
Shoot bud regeneration	0.04, 0.45, 4.54, 13.62 μM	1-month-old root seedling	*Melia azedarach* L.	Meliaceae	3 weeks	BA, kinetin, adenine sulfate (AD)	14-h photoperiod or darkness	Vila et al. (2005)
Shoot regeneration	0.5–0.1 mg/l	Nodal explants	*Medicago scutellata* (L.) Mill., *M. rigidula* (L.) All.	Fabaceae	4 weeks	NAA, BAP	16-h photoperiod	Band et al. (2011)

2002). Given the above, future research is greatly needed to explore the interaction between light and TDZ as it will open new avenues for discovery in terms of its mechanism of action.

1.4 Shoot Growth, Elongation, and Multiplication

TDZ's CK-like activity has also shown to be useful for the development of shoot buds and shoot proliferation/multiplication in plants (Table 1.2) (Mok et al. 1982; Thomas and Katterman 1986; Fiola et al. 1990; Malik and Saxena 1992; Huetteman and Preece 1993; Murch et al. 1997; Faisal et al. 2014; Singh and Dwivedi 2014; Parveen and Shahzad 2011; Jones et al. 2015). TDZ's CK-like activity is believed to be largely responsible for its ability to release lateral buds from dormancy or induce bud regeneration in vitro (Mok et al. 2005; Singh and Dwivedi 2014). Still it is important to note that TDZ likely modulates levels of other PGRs, including auxin, to achieve shoot bud regeneration by evoking regenerative responses, i.e., dedifferentiation and redifferentiation of tissue cells (Malik and Saxena 1992; Guo et al. 2011; Visser et al. 1992). For example, treatment of geranium hypocotyl explants with TDZ in combination with auxin increased shoot regeneration (Hutchinson et al. 1996). With respect to shoot proliferation, a wide spectrum of factors can influence TDZ's effects in vitro including: plant PGR perception and transduction, dedifferentiation and subsequent redifferentiation of cells, genotype, wounding of explants, donor plant condition (e.g. explant age), and duration of exposure to TDZ (Lazzeri and Dunwell 1984; Kumar and Reddy 2012; Magyar-Tábori et al. 2010; Sharifi et al. 2010). Furthermore, TDZ's ability to influence shoot proliferation has shown to be concentration and species specific. At low concentrations, between 1 and 10 μM TDZ can be used to enhance axillary shoot proliferation (Husain et al. 2007), while at much higher concentrations, shoot elongation can be either inhibited (Kumar and Reddy 2012) or stimulated to produce adventitious shoots (Feng et al. 2012; Guo et al. 2012). This trend has been observed for several spp. including "Gala" apples (*M. domestica*), where shoot production was found to decrease with increasing concentrations of TDZ (from 1 to 10 μM) (Liu et al. 1998), while TDZ concentrations greater than 22.7 μM inhibited shoot regeneration (Montecelli et al. 1999).

In addition to concentration, other factors can impact shoot organogenesis including the presence of other PGRs. For example, in vitro shoot multiplication of *Capsicum annuum* L. from cotyledonary node explants excised from seedlings was optimized on MS medium supplemented with 1.5 μM TDZ and 0.5 μM IAA. Compared to purine-type CKs, TDZ is superior at inducing shoot proliferation (Lu 1993) while also working synergistically with other PGRs to induce a response. The synergistic effect of TDZ with other CKs may be due to differences in uptake, recognition by the cells, or mechanisms of action of these different compounds (Huetteman and Preece 1993). For instance, the effect of TDZ on axillary meristem and shoot production was found to be 5–10 times greater compared to CKs (i.e., BA) in species such as soybean (*G. max*), peanut (*Arachis hypogaea* L.), and saffron

Table 1.2 Summary of TDZ used during shoot proliferation

Plant response	Optimal TDZ concentration	Explant used	Species	Family	Exposure time	Other PGRs used	Exposure to light/dark	References
Adventitious shoot regeneration	1, 2, 4, 8, 16 μM	Hypocotyl, cotyledon, and stem	*Hyoscyamus niger* L.	Solanaceae	3 weeks	BAP	16-h photoperiod	Uranbey (2005)
Adventitious shoot formation	2.0 mg/L	Leaves	*Malus domestica* Borkh.	Rosaceae	8 weeks	NAA, BAP, 2,4-D	16-h photoperiod	Bacha et al. (2009)
Adventitious shoot formation	10 μM	Leaves	*Malus domestica* Borkh.	Rosaceae	3 months	BA	Dark/light	Fasiola et al. (1989)
Adventitious shoot induction	22.7 or 27.3 μM	Leaf	*Paulownia tomentosa* Steud.	Paulowniaceae	2 weeks	IAA, BA	16-h photoperiod	Corredoira et al. (2008)
Adventitious shoot regeneration	0–20 mg/L	Root explants	*Gentiana triflora* Pall.	Gentianaceae	8 weeks	NAA	16-h photoperiod	Hosokawa et al. (1996)
Axillary shoot formation	2.2 μM	Seeds	*Malus domestica* Borkh.	Rosaceae	Not mentioned	BA	16-h photoperiod	Sharma et al. (2004)
Callus formation, shoot regeneration	0.05 μM	Leaf	*Populus tremula* L., *Populus alba* L.	Salicaceae	6 weeks	2,4-D	16-h photoperiod	Chupeau et al. (1993)
Direct organogenesis	0, 0.2, 0.5, 1, 2 mg/L	Shoot meristem	*Oryza sativa* L.	Poaceae	3–4 weeks	2,4-D, NAA, BA	16-h photoperiod	Chakrabarty et al. (2010)
Direct organogenesis	0.1 mg/L	Stem node	*Teucrium polium* L.	Lamiaceae	4 weeks	BAP, kinetin, NAA	16-h photoperiod	Rad et al. (2014)
Direct shoot regeneration	0.0, 0.005, 0.01, 0.02, 0.04, 0.08 mM	Excised root tips	*Populus alba* L., *P. tremula* L., *P. tremuloides* Michx.	Salicaceae	8 weeks	–	16-h photoperiod	Sherif and Khattab (2011)

(continued)

Table 1.2 (continued)

Plant response	Optimal TDZ concentration	Explant used	Species	Family	Exposure time	Other PGRs used	Exposure to light/dark	References
Direct shoot regeneration and callus-mediated protocorm-like body (PLB)	0.5–9.0 μM	Root tips	× *Doritaenopsis*	Orchidaceae	8 weeks	BA, zeatin	16-h photoperiod	Park et al. (2003)
Shoot formation	0.22 μM	Nodal explants	*Manihot esculenta* Crantz	Euphorbiaceae	6–8 weeks	BA, GA3	16-h photoperiod	Bhagwat et al. (1996)
Shoot formation	0.22 and 2.20 ppm	Root	*Panax quinquefolius* L.	Araliaceae	20 weeks	–	78% shade	Proctor et al. (1996)
Shoot formation	0.05 μM	Root segments from 15- to 20-day-old seedlings	*Albizia julibrissin* Durazz.	Fabaceae	15–30 days	IAA, IBA, NAA, BA, zeatin, 2iP	16-h photoperiod	Sankhla et al. (1996)
Shoot formation	0.1–30 μM TDZ	Root	*Pelargonium domesticum* L.H. Bailey	Geraniaceae	18 days	–	16-h photoperiod	Murch et al. (1997)
Shoot formation	1pM–100 nM	Node	*Camellia sinensis* (L.) Kuntze	Myrtaceae	2 months	BAP, NAA, IBA	16-h photoperiod	Mondal et al. (1998)
Shoot formation	0.046–4.6 μM	Cotyledonary nodes	*Dalbergia sissoo* DC.	Fabaceae	30 days	BA, kinetin, 2iP	16-h photoperiod	Pradhan et al. (1998)
Shoot formation	0, 5, 10, 15, 20 mM	Hypocotyls	*Hypericum perforatum* L.	Clusiaceae	3, 6, 9, and 12 days	BAP, NAA, IAA, and 2,4-D	16-h photoperiod	Murch et al. (2000)

Shoot formation	0.05, 0.1, 0.3 mg/L	Leaf, cotyledons, and stem	*Oenothera* spp.	Agavaceae	3 weeks	IBA, IAA, 2,4-D	16-h photoperiod	De Gyves et al. (2001)
Shoot formation	0.1 μM	Hypocotyl	*Linum usitatissimum* L.	Linaceae	5–7 days	NAA	Continues light	Jain and Rashid (2001)
Shoot formation	0.001–1 mg/L	Hypocotyl	*Cuminum cyminum* L.	Apiaceae	30 days	–	1S-h photoperiod	Gupta and Bhargava (2001)
Shoot formation	50–10,000 nM	Nodal shoot segments	*Wrightia arborea* (Dennst.) Mabb.	Apocynaceae	3 weeks	BAP, GA3, kinetin	16-h photoperiod	Purohit et al. (2004)
Shoot formation	0.5 and 5.0 μM	Petal, leave	*Dianthus* spp.	Caryophyllaceae	30 days	IAA, NAA	16-h photoperiod	Casanova et al. (2004)
Shoot formation	0.05, 0.1, 0.5, 1.0, 5.0 μM	Leaf	*Hydrangea quercifolia* W. Bartram	Hydrangeaceae	16 weeks	IBA	16-h photoperiod	Ledbetter and Preece (2004)
Shoot formation	4.5 μM	Cotyledon, hypocotyls	*Arnebia euchroma* (Royle)	Boraginaceae	12 days	–	16-h photoperiod	Jiang et al. (2005)
Shoot formation	2–10 μM	Leaf	*Tylophora indica* (Burm. f.) Merr.	Apocynaceae	45 days	2,4-D, BA, kinetin	16-h photoperiod	Thomas and Philip (2005)
Shoot formation	18.16–72.64 μM	Bud	*Curcuma longa* L.	Zingiberaceae	1 week	BA	16-h photoperiod	Prathanturarug et al. (2005)
Shoot formation	0, 0.4, 1, 2.5 μM	Shoot tip	*Musa* spp.	Musaceae	3 weeks	BAP, IAA	16-h photoperiod	Gubbuk and Pekmezci (2006)
Shoot formation	0.05, 0.10, 0.25, 0.50, 1.0 mg/L	Cotyledonary leaves	*Astragalus cicer* L.	Fabaceae	3–5 weeks	–	16-h photoperiod	Basalma et al. (2008)
Shoot formation	1.0–15 μM	Seedling	*Firmiana simplex* (L.) W. Wight	Malvaceae	8 weeks	BA, 2-iP, zeatin, kinetin, GA3	16-h photoperiod	Hussain et al. (2008)

(continued)

Table 1.2 (continued)

Plant response	Optimal TDZ concentration	Explant used	Species	Family	Exposure time	Other PGRs used	Exposure to light/dark	References
Shoot formation	0.5, 0.75, 1, 1.25 mg/L	Seedling	*Hypericum triquetrifolium* Turra	Hypericaceae	8 weeks	IAA, BAP	Dark	Oluk and Orhan (2009)
Shoot formation	0.5–9.0 μM	Nodal	*Cannabis sativa* L.	Cannabaceae	28 days	–	16-h photoperiod	Lata et al. (2009)
Shoot formation	20, 40, 80 μM	Immature seed	*Epimedium alpinum* L.	Berberidaceae	7 days	CPPU, 2,4-D	Dark	Mihaljević and Vršek (2009)
Shoot formation	0.22–1.8 μM	Shoot tip	*Hoya wightii* Hook.f.	Apocynaceae	8 weeks	KN, BA, 2-iP, IBA, IAA, NAA	16-h photoperiod	Lakshmi et al. (2010)
Shoot formation	0, 0.5, 1.0, 3 mg/L	Immature embryos	*Sandersonia aurantiaca* Hook	Colchicaceae	4–6 weeks	NAA	16-h photoperiod	Deroles et al. (2010)
Shoot formation	1 mg/L	Male inflorescence	*Musa* sp.	Musaceae	4 weeks	BAP, kinetin, 2iP, zeatin	16-h photoperiod	Darvari et al. (2010)
Shoot formation	0.5, 1, 2, 3 μM	Hypocotyls and cotyledon	*Lycopersicon esculentum* Mill.	Solanaceae	2–8 weeks	BAP	16-h photoperiod	Osman et al. (2010)
Shoot formation	5–100 μM	Nodal segments	*Nyctanthes arbor-tristis* L.	Oleaceae	4, 8, 12, and 16 days	IBA	16-h photoperiod	Jahan et al. (2011)
Shoot formation	0.5, 1.0, 2.5, 5.0, 7.5, or 10.0 μM	Nodal explants	*Vitex trifolia* L.	Lamiaceae	4 weeks	BA, NAA	16-h photoperiod	Ahmed and Anis (2012)
Shoot formation	2.0 mg/L	Meristem-derived callus of leaf protoplasts	*Malus domestica* Borkh.	Rosaceae	2–3 months	2,4-D, BA, IAA, ABA	16-h photoperiod	Saito and Suzuki (1999)

Shoot formation and development	0.36 mM	Peach palm plants	*Bactris gasipaes* Kunth	Arecaceae	84 days	NAA, BAP	16-h photoperiod	Graner et al. (2013)
Shoot formation and somatic embryogenesis	0.01, 0.1, 1.0, 10.0 μM	Immature, mature nonstratified seed	*Fraxinus americana* L.	Oleaceae	4 weeks	BA, 2iP, 2,4-D	16-h photoperiod	Bates et al. (1992)
Shoot formation and somatic embryogenesis	0, 0.5, 1.0, 1.5, 2.0, 2.5, 3.0, 5.0, 10 μM	Leaf and petiole	*Saintpaulia ionantha* H.Wendl.	Gesneriaceae	3, 6, or 9 days	CPPU, 2,4-D, NAA, BA	16-h photoperiod	Mithila et al. (2003)
Shoot induction	0.1–10 μM	Cotyledonary nodes	*Pterocarpus marsupium* Roxb. i	Fabaceae	6 weeks	BA	16-h photoperiod	Husain et al. (2007)
Shoot organogenesis	0, 0.25, 0.5, 0.75, 1.0, 2.5, 5.0 μmol/L	Leaf	*Artemisia judaica* L.	Asteraceae	20 days	–	Dark/light	Liu et al. (2003)
Shoot organogenesis	9.08, 13.62, 18.17, 22.7 μM	Leaf	*Morus* spp.	Moraceae	30 days	2,4-D, BAP, kinetin	16-h photoperiod	Chitra and Padmaja (2005)
Shoot organogenesis	0, 2.5, 5.0, 7.5, 10.0, 20.0 μM	Intact seedlings, etiolated hypocotyl, sterile stem	*Scutellaria baicalensis* Georgi	Lamiaceae	14–20 days	–	16-h photoperiod	Li et al. (2000)
Shoot organogenesis	0.5–9 mM	Nodal segment	*Kigelia africana* (Lam.) Benth.	Bignoniaceae	4 weeks	2,4-D, NAA	16-h photoperiod	Thomas and Puthur (2004)
Shoot organogenesis	0.45, 1.20, 2.30, 4.54, 22.70 μM	Leaf explants	*Rhododendron* spp.	Ericaceae	12 weeks	IAA	16-h photoperiod	Pavingerová (2009)
Shoot production	0.5–40 mg/L	Leaf, petiole, hypocotyl, internode, cotyledon, root	*Arachis hypogaea*	Legumes	15 days	BA	16-h photoperiod	Kanyand et al. (1994)

(continued)

Table 1.2 (continued)

Plant response	Optimal TDZ concentration	Explant used	Species	Family	Exposure time	Other PGRs used	Exposure to light/dark	References
Shoot regeneration	0.88–3.3 mg/L	Excised cotyledons and rooted hypocotyls	*Capsicum annuum* L.	Solanaceae	3 weeks	BAP, IAA	16-h photoperiod	Szász et al. (1995)
Shoot regeneration	1.0 or 2.0 mg/L	Cotyledons, leaves, cotyledonary nodes and shoot tip	*Capsicum annuum* L.	Solanaceae	30 days	–	16-h photoperiod	Dabauza and Pena (2001)
Shoot regeneration	0.1, 1, 5, and 10 μM	Untransformed and transformed roots	*Malus domestica* Borkh.	Rosaceae	–	NAA	16-h photoperiod	Pawlicki-Jullian et al. (2002)
Shoot regeneration	2–9 μM	Cotyledonary explants	*Morus alba* L.	Moraceae	7, 14, 21 days	BAP	16-h photoperiod	Thomas (2003)
Shoot regeneration	0.5, 1.0 μM	Root	*Fragaria* × *ananassa* (Duchesne ex Weston) Duchesne ex Rozier	Rosaceae	3–4 weeks	BA, 2,4-D, NAA	16-h photoperiod	Passey et al. (2003)
Shoot regeneration	0–12 μM	Cotyledon, hypocotyls	*Pinus strobus* L.	Pinaceae	6 weeks	2,4-D, NAA, IAA, BA, 2iP	16-h photoperiod	Tang and Newton (2005)
Shoot regeneration	0, 0.5, 2, 4 μM	Young expanding sepals	*Fragaria* × *ananassa* (Duchesne ex Weston) Duchesne ex Rozier	Rosaceae	4–5 weeks	Zeatin	Dark/light	Debnath (2005)

Shoot regeneration	0.1, 0.5, 1.0, 2.5, 5.0 µM	Root explant from 30-day-old seedlings	*Senna alexandrina* Mill.	Fabaceae	6 weeks	BA, kinetin, NAA	16-h photoperiod	Parveen and Shahzad (2011)
Shoot regeneration	0.22, 0.90, 2.27, 4.54, 9.08 µM	Cotyledonary petiole	*Jatropha curcas* L.	Euphorbiaceae	6 weeks	Kinetin, BAP, NAA, IAA	16-h photoperiod	Kumar and Reddy (2012)
Shoot regeneration	15 mg/L	Young leaf	*Prunus persica* (L.) Batsch	Rosaceae	5 weeks	NAA, IBA	Dark/light	Soliman (2013)
Shoot regeneration	0, 2.5, 5, 7.5 µM	Nodal explants, leave	*Pyrus communis* L.	Rosaceae	4 weeks	NAA, BAP	Dark/16-h photoperiod	Yousefiara et al. (2014)
Shoot regeneration	0.45–22.7 µM	Apical and axillary buds, leaf, and internode	*Guizotia abyssinica* (L.f.) Cass	Asteraceae	3–4 months	–	16-h photoperiod	Baghel and Bansal (2014)
Shoot regeneration	4.54, 6.81, 9.08 µM	Leaf explant	*Prunus persica* (L.) Batsch	Rosaceae	4–6 weeks	NAA	Dark/16-h photoperiod	San et al. (2015)
Shoot regeneration	1.0 µM	Leaf disks	*Mentha × piperita* L.	Lamiaceae	6 weeks	NAA, BA	16-h photoperiod	Faure et al. (1998)
Shoot and root regeneration	1.0 mg/L	Aerial stem explants	*Zingiber officinale* Roscoe	Zingiberaceae	3–4 months	IBA, BAP, NAA	16-h photoperiod	Lincy and Sasikumar (2010)
Shoot induction	4.5, 9.1 µM	Mature embryos		Poaceae	4 weeks	BAP	16-h photoperiod	Ganeshan et al. (2006)
Shoot multiplication	0.5, 1, 2, 3, 4, 5 µM	Nodal	*Cullen corylifolium* (L.) Medik.	Fabaceae	4 weeks	–	16-h photoperiod	Faisal et al. (2006)
Shoot regeneration	1.0 µM	Leaf disks	*Mentha x piperita* L.	Lamiaceae	6 weeks	NAA, BA	16-h photoperiod	Faure et al. (1998)

(continued)

Table 1.2 (continued)

Plant response	Optimal TDZ concentration	Explant used	Species	Family	Exposure time	Other PGRs used	Exposure to light/dark	References
Shoot regeneration	0.45–45.41 µM	Rhizome sections	*Cheilocostus speciosus* (J.Koenig) C.D.Specht	Costaceae	6 weeks	–	16-h photoperiod	Malabadi et al. (2004)
Shoot regeneration	0.045, 0.23, 1.14, 2.27, 4.54 µM	Cotyledons excised	*Boehmeria nivea* (L.) Gaudich.	Urticaceae	4 days	2,4-D, IAA, IBA, NAA	16-h photoperiod	Wang et al. (2007)
Shoot regeneration	2 mg/L	Root, cotyledon, hypocotyl, and intact seedlings	*Brassica oleracea* L.	Brassicaceae	6 weeks	BA, zeatin	16-h photoperiod	Ċosiċ et al. (2015)

(*Crocus sativus* L.) (Victor et al. 1999; Radhakrishnan et al. 2009; Sharifi et al. 2010). Furthermore, TDZ alone or in combination with other auxins/CKs (e.g., BA) can induce shoot bud formation and multiplication especially after transfer of shoots to TDZ-free medium (Singh and Dwivedi 2014). Consequently, the transfer of explants from enriched TDZ medium to a secondary medium without growth regulators has been successfully applied in plant regeneration systems for a variety of species (Malik and Saxena 1992; Victor et al. 1999).

1.5 Somatic Embryogenesis

TDZ is a substitute for the auxin/CK requirement that is needed during somatic embryogenesis, thereby increasing the number of formed somatic embryos (Visser et al. 1992). Somatic embryogenesis changes somatic cells to embryonic cells in a physiological sequence that is tightly regulated by a delicate balance of PGRs (Murthy et al. 1998). Induction and development of somatic embryogenesis are associated with endogenous PGRs including auxins and CKs; not surprisingly, TDZ promotes somatic embryogenesis, alone or in combination with other PGRs, for a wide range of recalcitrant species (Durkovic and Misalova 2008; Nhut et al. 2006) as well as a variety of commercial crops including tobacco (*Nicotiana tabacum* L.), peanut (*A. hypogaea*), geranium (*Pelargonium* spp.), African violet (*Saintpaulia* spp.; Mithila et al. 2003; Shukla et al. 2013), and chickpea (*Cicer arietinum* L.) (Visser et al. 1992; Saxena et al. 1992; Gill and Saxena 1993; Murthy et al. 1995) (Table 1.3).

Different types of tissues can be selected for induction, for instance, TDZ has been described to induce somatic embryos on hypocotyl, epicotyl, cotyledonary node, cotyledon, and leaves of intact seedlings of *Azadirachta indica* A. Juss. (Gairi and Rashid 2004; Saxena et al. 1992; Iantcheva et al. 1999). In peanut, induction of direct somatic embryogenesis occurs by culturing mature intact seeds on a medium supplemented with 0.5–10 μM TDZ or N-(2-chloro-4-pyridyl)-N′-phenylurea (CPPU). Explants with no cotyledons, and thus no embryogenic potential, did not respond to increasing levels of TDZ. In contrast, retention of one or both cotyledons resulted in increased response to TDZ (Saxena et al. 1992; Murthy and Saxena 1994; Murthy et al. 1995). Exposure time can impact the effectiveness of TDZ. For example, application of TDZ on plant tissues alone or in combination with other PGRs for short periods of time at low concentration (10 μM) has been found to induce embryogenic responses (Hutchinson et al. 1997; Malik and Saxena 1992; Murthy et al. 1998), while exposure to TDZ for longer than 3–4 weeks (10 μM) led to a reduced induction of roots (Malik et al. 1993). This is interesting, as it reflects patterns also observed in induction of somatic embryogenesis by the synthetic auxin and pesticide, 2,4-dichlorophenoxyacetic acid (2,4-D). Similar to TDZ, short exposure to 2,4-D followed by explant transfer to growth regulator-free medium allows for first an accumulation of 2,4-D in tissues followed by a gradual decrease over time with somatic embryos developing with these falling concentrations (Zee 1981; Fujimura and Komamine 1980; Feher et al. 2002). The similarity between this

Table 1.3 Summary of TDZ used during somatic embryo formation

Plant response	Optimal TDZ concentration	Explant used	Species	Family	Exposure time	Other PGRs used	Exposure to light/dark	References
Somatic embryogenesis	10 μM	Hypocotyl explants	*Pelargonium* × *hortorum* Bailey	Geraniaceae	3 days	Gas	16-h photoperiod	Hutchinson et al. (1997)
Somatic embryogenesis	0. 045, 0.45, 4.54 μM	Immature embryo		Fagacea	2, 4, and 8 weeks	–	Dark	Tsvetkov (1999)
Somatic embryogenesis	0.01–5.0 μM	Embryonic axes	*Quercus robur* L.	Fagacea	4 weeks	BA, NAA	16-h photoperiod	Martínez et al. (2008)
Somatic embryogenesis	2.5, 5, 7.5 mg/L	Nucellar tissue	*Mangifera indica* L.	Anacardiaceae	1–2 months	BAP, 2iP, IAA	15-h photoperiod	Kidwai et al. (2009)
Somatic embryogenesis	5.0 or 10.0 μM	Leaf disk	*Vitis vinifera* L.	Vitaceae	40 days	2,4-D, BA, KIN, ZEA, 2iP	16-h photoperiod	Matsuta and Hirabayashi (1989)
Somatic embryogenesis	0.455 μM	Nodal and internodal tissues	*Bambusa odashimae* Hatus. ex D.Z.Li & Stapleton	Poaceae	3–4 weeks	Kinetin, 2,4-D	16-h photoperiod	Lin et al. (2004)
Somatic embryogenesis	0.1–2.5 μM	Stem segments and shoot tips	*Capsicum annuum* L.	Solanaceae	4 weeks	–	16-h photoperiod	Khan et al. (2006)
Somatic embryos	1 μM	Leaf	*Echinacea purpurea* (L.) Moench	Asteraceae	4 weeks	–	Dark	Jones et al. (2007)
Somatic embryos	1,5, 10, 20, 40 μM	Seedling	*Arachis hypogaea* L.	Fabaceae	4–6 weeks	–	16-h photoperiod	Saxena et al. (1992)

well-documented process and the pattern observed in TDZ treatment supports a strong auxin-like role for TDZ in this mechanism. Further, it is likely that the inherent stability of TDZ in living tissues is a strong contributing factor in establishing this function.

1.6 Intact Seedling Development

TDZ enhances seed germination via improvement of shoot regeneration, with positive effects being reported in soybean (*G. max*), pea (*Pisum sativum* L.), common bean (*P. vulgaris*), chickpea (*C. arietinum*), and lentil (*Lens culinaris* Medik) (Radhakrishnan et al. 2009; Malik and Saxena 1992) (Table 1.4). In contrast, the intact seedling regeneration system is a unique morphogenetic system which involves the direct development of multiple shoots on the germinating seedling. For the first time, Malik (1993) reported a direct seed culture method for de novo differentiation of shoots from intact seedling without explanting. The number of shoots regenerated from intact seedling of *Lathyrus sativus* L., *L. cicera* L., and *L. ochrus* L. DC. was significantly higher than that observed with explants. These results indicated that excision of explant is not always necessary for induction of morphogenesis and also that the morphological integrity of intact seedlings plays a critical role in the induction of organogenesis/somatic embryogenesis (Malik 1993). TDZ induction of shoot production in the intact seedling system effectively depends on the applied concentration. For example, intact seedlings of silk tree (*A. julibrissin*) grown on MS medium containing 0.1–10 μM TDZ produced shoots indirectly through callus. Interestingly, at higher TDZ concentrations (2.5–10 μM), shoots were produced, but did not form callus (Mok et al. 1987). Sankhla et al. (1994) also reported high efficiency of TDZ in inducing shoot formation from roots of intact seedling of *A. julibrissin* at 0.1–1.0 μM TDZ. Regeneration of multiple shoots from intact seedlings of switch grass (*Panicum virgatum* L.) was induced on MS medium supplemented with 4.5 μM2, 4-dichlorophenoxyacetic acid (2,4-D), and 18.2 μM TDZ (Gupta and Conger 1998). An in vitro propagation system for *Artemisia judaica* L., a medicinal plant, induced shoot organogenesis by culturing intact seedlings on medium supplemented with 1 μM TDZ for 20 days (Liu et al. 2003). In a study with seeds of *Firmiana simplex* (L.) W. Wight, induction of shoot proliferation was assessed on MS medium supplemented with 5.0 μM TDZ + 1.5 M GA_3 + 0.1% ascorbic acid compared to various levels (1.0–15 M) of several different cytokinins (BA, 2-iP, zeatin, and kinetin). Shoots formed within 8 weeks of culture and the shoot-forming capacity of seeds were found to be influenced by the type and concentration of CKs, with TDZ showing up to 13% greater regeneration rates than other cytokinins tested (Hussain et al. 2008). Induction of shoot organogenesis for felty germander (*Teucrium polium* L.), an endangered medicinal plant, was obtained using intact seedlings cultivated in MS medium supplemented with 22.72 μM TDZ (Rad et al. 2014). Regeneration ability of kohlrabi (*Brassica oleracea var. gongylodes*) cultivars Vienna Purple (VP) and Vienna White (VW) has also been tested. Intact seedlings were cultivated on MS media

Table 1.4 Summary of TDZ used during seedling development

Plant response	Optimal TDZ concentration	Explant used	Species	Family	Exposure time	Other PGRs used	Exposure to light/dark	References
Somatic embryogenesis	0.1–1.0 μM	Hypocotyl, epicotyl, cotyledon, leaves	*Azadirachta indica* A. Juss.	Meliaceae	1 week	–	24-h light	Gairi and Rashid (2004)
Direct somatic embryogenesis	0, 0.91, 2.27, 4.54, 45.41 mM	Hypocotyls, cotyledons, petioles, nodal stem	*Medicago* spp.	Fabaceae	20 days	BAP	16-h photoperiod	Iantcheva et al. (1999)
Shoot formation	10 μM	Mature seeds and growing seedlings	*Phaseolus vulgaris* L.	Fabaceae	7 days to 4 weeks	BAP	Dark/light	Malik and Saxena (1992)
Somatic embryogenesis	0.5–10 μM	Seedling	*Arachis hypogaea* L.	Fabaceae	5 weeks	–	Dark/light	Murthy et al. (1995)
Direct embryogenesis	1, 5, 10, 20, 40 μM	Seedling	*Arachis hypogaea* L.	Fabaceae	6–9 weeks	–	16-h photoperiod	Saxena et al. (1992)
Somatic embryogenesis	0.05, 1, 5, 10, 20 μM	Seeds	*Cajanus cajan* (L.) Millsp.	Fabaceae	24 h to 14 days	–	16-h photoperiod	Singh et al. (2003)
Shoot formation	1–100 μM	Cotyledonary region of the seedlings	*Cicer arietinum* L.	Fabaceae	2–3 weeks	NAA, BAP	Dark/light	Murthy et al. (1996)
Somatic embryogenesis and regeneration	1, 5, 10, 15, 20, 25 μM	Seedling explants	*Arachis hypogaea* L.	Fabaceae	2 weeks	BAP	Dark/light	Gill and Saxena (1992)
Shoot formation	1 μM	Seedling	*Linum usitatissimum* L.	Linaceae	4 days	BAP	24-h light	Mundhara and Rashid (2006)
Bud induction	10, 1 μM	Seedlings	*Phaseolus vulgaris* L.	Fabaceae	2 weeks	BAP	Dark or light	de Carvalho et al. (2000)
Shoot formation	0.88, 1.1, 1.76, 2.2, 2.75, 3.3 mg/L	Seedling	*Capsicum annuum* L.	Solanaceae	1 week	BAP, IAA	16-h photoperiod	Szasz et al. (1995)

supplemented with BA, TDZ, and trans- or cis-zeatin. All tested CKs induced shoot regeneration with 47.5–60% shoot regeneration frequency from hypocotyl explants and intact seedlings (Ċosiċ et al. 2015).

1.7 Mechanisms of TDZ Activity

1.7.1 Cytokinin-Related Effects of TDZ

TDZ was first reported to have CK activity in 1982 by Mok et al. and later confirmed by Visser et al. (1992). TDZ exhibits a considerably higher degree of biological activity when compared with traditional CKs for inducing regeneration in plant species (Mok et al. 1987; Van Nieuwkerk et al. 1985; Escalettes and Dosba 1993), stimulating organogenesis and somatic embryogenesis, and retarding senescence or leaf yellowing in plants (Mehrotra et al. 2015). For example, callus tissue of *Phaseolus lunatus* L. which cannot grow without CKs is able to grow after exposure to TDZ (Murthy et al. 1998). Similarly, lower concentrations of TDZ are needed to initiate shoot differentiation and regeneration responses compared to levels required for CKs (Baker and Bhatia 1993). TDZ's CK-like activity is believed to stem from its ability to modulate pathways responsible for CK biosynthesis in plants (Mok et al. 1987) by acting on endogenous adenine-based CK metabolism (Capelle et al. 1983). To date it is unclear whether TDZ causes CK responses by interacting directly with CK receptors or indirectly either by stimulating the conversion of CK nucleotides to active ribonucleosides or by inducing the accumulation of endogenous adenine-based CKs.

It has been proposed that TDZ promotes the conversion of CK ribonucleotides (inactive CKs) to active forms of CKs (i.e., ribonucleosides and free bases) by encouraging the synthesis of endogenous purine CKs while also inhibiting their degradation (Capelle et al. 1983; Lu 1993; Murthy et al. 1995; Mok and Mok 1985). On the other hand, TDZ has demonstrated binding affinity for CK receptors such as CRE1 as well as CRE1/AHK4, AHK2, and AHK3 (de Melo Ferreira et al. 2006; Susan 1996; Rolli et al. 2012). It is interesting to note that both purine- and urea-type CKs have demonstrated binding affinities for cytokinin-specific binding proteins (CSBPs). A stronger association has, however, been demonstrated for compounds containing phenylurea derivatives (Murthy et al. 1998); this could help to explain TDZ's ability to modulate plant morphogenesis at lower concentrations. In addition to the above, TDZ can also increase endogenous levels of CKs by reducing catabolism, increasing synthesis, and changing non-active CK molecules to active forms (Kefford et al. 1968; Murthy et al. 1995), possibly through inactivation of CK oxidase/dehydrogenase (CKX) (an enzyme responsible for CK inactivation through cleavage of the unsaturated N6 side chain of most isoprenoid CKs) (Nikolić et al. 2006). TDZ can also modify CK biosynthesis pathways by decreasing endogenous pools of the CK 2iP and by increasing the concentration of purine-based CKs (Zhang et al. 2005).

In general, reduced rooting capacity and inhibition of shoot elongation are attributable to the high CK activity of TDZ. Medium concentrations (approx. 10–20 μM) of TDZ may result in both axillary and adventitious shoot organogenesis, and high concentrations tend to stimulate callus formation. Concentrations of TDZ much smaller than most CKs often stimulate higher shoot proliferation. Combinations of TDZ with other CKs result in better shoot proliferation due to differences in uptake, recognition by the cells and receptors, or mechanisms of action of different compounds (Huetteman and Preece 1993). TDZ facilitates efficient multiplication of apical meristem cells and their reprogramming to appropriate developmental stages for shoot differentiation (Dey et al. 2012).

1.7.2 Auxin-Related Activity of TDZ

The auxin-like activity of TDZ was first assessed by Suttle (1984). Following this work, TDZ's ability to modulate auxin levels in plants was reported by Yip and Yang (1986) who found that TDZ stimulated auxin concentrations in mung bean (*Vigna radiata* (L.) R.Wilczek) hypocotyl tissue. Similarly, results by Visser et al. (1992) suggested that auxin(s) were involved during the induction and/or expression of TDZ-induced morphogenic differentiation.

To date TDZ's auxin-like activity is believed to act through the modulation of metabolism and transport for endogenous hormones including auxins, cytokinins, ethylene, abscisic acid, and gibberellins (Feng et al. 2012; Murch and Saxena 2001). While a significant amount of work has been performed to understand TDZ's cytokinin-like effects in plants, far less is understood in terms of its relationship to auxin. Currently, two concepts have been proposed: (1) TDZ directly promotes growth due to its own biological activity, and (2) TDZ may modulate the synthesis and accumulation of endogenous auxins or auxin-like bioregulators in synergism with CKs (Capelle et al. 1983; Mok and Mok 1985).

Auxins including natural (IAA) and synthetic auxins (e.g., naphthaleneacetic acid (NAA) and 2,4-D) are responsible for cell proliferation and development of callus (a mass of dedifferentiated cells), which are the first part of the morphogenetic process. They are also strongly associated with regeneration and somatic embryogenesis (Murthy et al. 1998). TDZ via auxin-like activity has been shown to induce callus formation on the graft and bud cutting of grape and leaf disks of cotton (Lin et al. 1988; Kartomysheva et al. 1983), increasing proliferation and growth rate of callus 30 times more than the common auxins. Tracer studies by Murch and Saxena (2001) noted that the translocation of auxin is essential for TDZ-induced morphogenesis through the observation that radiolabeled IAA accumulated in the hypocotyl of geraniums and was translocated over a great distance within the tissues. TDZ may also mimic an auxin response by modifying endogenous auxin metabolism, for example, TDZ had a stimulating effect on auxin synthesis when peanut seedlings were treated with TDZ, causing an increase in cytosolic auxin followed by induction of somatic embryogenesis (Murthy et al. 1995).

The relationship between TDZ and auxin metabolism has also been confirmed through inhibitor studies. Suppression of TDZ-induced regeneration by inhibitors of auxin action and transport has been employed in several studies to better understand the relationship between auxin and TDZ across several different regeneration studies (Hutchinson et al. 1996; Murch and Saxena 2001; Murch et al. 2002). For example, application of 2-(ρ-chlorophenoxy)-2-methylpropionic acid (PCIB, an auxin biosynthesis inhibitor) in peanut and geranium demonstrated an increasing effect of TDZ during somatic embryogenesis (Murthy et al. 1998). Although use of 2,3,5-triiodobenzoic acid (TIBA, an inhibitor of polar auxin transport) in samples treated with TDZ did not change auxin levels, a decrease in the rate of somatic embryogenesis was observed (Hutchinson et al. 1996). Reduced rate of embryogenesis in TDZ-exposed tissues treated with TIBA and PCIB suggests TDZ may modulate auxin metabolism during developmental processes such as embryogenesis (Hutchinson et al. 1996). Furthermore, in TDZ-exposed leaf tissue of *Echinacea purpurea* L., inclusion of TIBA and PCIB decreased TDZ-induced morphogenesis (shoot organogenesis and somatic embryogenesis) but increased concentrations of auxin and endogenous indoleamines (i.e., melatonin and serotonin) (Jones et al. 2007). The above examples indicate that TDZ-induced regeneration is correlated with a metabolic cascade, i.e., accumulation and transport of endogenous signals auxin and melatonin, and the activation of a stress response.

Endogenous and exogenous auxin levels are closely associated with somatic embryogenesis in plants, and TDZ plays a crucial role in modulating the interaction among different hormones. It is important to note that TDZ's ability to induce somatic embryogenesis is not solely dependent upon its auxin-like properties, as CKs have also been implicated. For example, embryogenesis was repressed in TDZ-treated geranium tissues by applying diaminopurine (DAP, an inhibitor of a purine-based CK) (Hutchinson and Saxena 1996). Unlike purine-based CKs, TDZ alone can induce somatic embryogenesis (Murthy et al. 1998), which in turn highlights the ability of TDZ to act as both an auxin and cytokinin. In addition to somatic embryogenesis, TDZ's auxin-like activity has also been shown to be beneficial during callus formation by increasing proliferation and growth rate of callus (Lin et al. 1988). Synthetic auxins such as NAA and 2,4-D are responsible for stimulation, multiplication, and differentiation of cells into somatic embryos and callus development (Murthy et al. 1998). The regulatory role of TDZ appears to be partially mediated through inactivation of genes responsible for auxin and CK biosynthesis, which in turn causes changes in developmental patterns in plants (Malik 1993).

In general, TDZ inhibits root meristem activity effectively by acting as an auxin antagonist (Rolli et al. 2012). Auxin-like activity of TDZ is also strongly associated with regeneration, somatic embryogenesis, organogenesis, and development of adventitious shoots in many plant species (Huetteman and Preece 1993; Lu 1993; Feng et al. 2012; Guo et al. 2012). A low concentration of TDZ induces organogenesis of axillary buds on cultured shoot tip meristem by reducing apical dominance (Lu 1993). However, it is important to note that auxin-like properties of TDZ are dependent on a multitude of factors including the basal medium used, type of cultivar, source of the explant, developmental stage of explant, and age of the donor plant (Radhakrishnan et al. 2009). TDZ seems to act via reprogramming the fate of cells, developmental pathway, and interaction between endogenous hormones (Malik 1993).

1.7.3 Calcium Signaling

TDZ is believed to modulate plant morphogenesis through its ability to influence inter- and intracellular calcium (Ca^{2+}) concentrations and signaling cascades (Trewaves 1999). Plant cells and tissues react to different hormones due to changes in concentration of external Ca^{2+} (Guo et al. 2011), and the balance of cytosolic Ca^{2+} may relate to the TDZ induction. Ca^{2+} is an important secondary messenger and signaling molecule in plants, facilitating different morphological responses in plant cells and tissues through modulation of PGR levels (Guo et al. 2011; Allen and Schroeder 2001). In response to TDZ, Ca^{2+} channels will open, leading to changes in plant cytosolic Ca^{2+} levels; intermittent signals are then sent across the cell initiating a cascade of metabolic events (White and Broadley 2003). Several studies have confirmed the above noted theory. Hosseini-Nasr and Rashid (2002) reported that addition of Ca^{2+} uptake inhibitors (lanthanum, calmodulin, trifluoperazine (TFP), chlorpromazine (CPZ)) to culture medium supplemented with TDZ led to decreased levels of shoot production, while Jones et al. (2007) applied a Ca^{2+} channel activator, (S)-Bay K8644, in TDZ-treated explants of *E. purpurea* and noted changes in cell polarity, increased auxin concentration, callus induction, and regeneration. Murch et al. (2003) found that treatment with the calcium channel antagonist (S)-Bay K8644 increased influx of Ca^{2+}, leading to a change in the pattern of somatic embryogenesis. Increases in cytosolic Ca^{2+} for a long period, however, can also lead to apoptosis and cell death (White and Broadley 2003).

1.7.4 Relationship to Other PGRs and Stress Signaling Molecules

Plants interpret TDZ as stress, and it has been suggested that TDZ's ability to initiate stress in plants helps to induce morphogenesis through modulation of PGRs as well as other metabolites and ions. For instance, proline is considered a marker of stress as it enables plants to produce more $NADP^+$/NADPH. Proline levels have been found to increase in tissues which have been treated with TDZ and which show a capacity to switch from shoot formation to somatic embryogenesis (Hare and Cress 1997). In addition to proline, accumulation of mineral ions in TDZ-treated tissues may also act as a trigger factor for induction of somatic embryogenesis and regeneration in carrot (*Daucus carota* subsp. *sativus* (Hoffm.) Arcang) (Guo et al. 2011). Stress-related metabolites 4-aminobutyrate, ABA, proline, and mineral ions increased in the TDZ-treated root tissues of geranium (*Pelargonium domesticum* L.H. Bailey) (Murch et al. 1997, Murch and Saxena 1997).

TDZ treatment has also been found to significantly improve accumulation of endogenous hormones (IAA, zeatin, GA3, and ABA) during shoot organogenesis. For instance, in leaf explants of *E. purpurea*, the levels for auxin, melatonin, and serotonin were found to increase after exposure to TDZ during regeneration. Furthermore, TDZ exposure stimulates ethylene production concurrent to accumulations of ABA, auxin, proline, and Ca^{2+} (Jones et al. 2007). Inhibition of rooting and hypocotyl elongation, swelling at the base of hypocotyl, and tightening of cotyledons toward the apex induced by TDZ are characteristics of an ethylene action (Mundhara and Rashid 2006); not surprisingly, TDZ is more effective than CK for inducing "stress ethylene" production in plants (Yip and Yang 1986). Some negative effects of TDZ on growth parameters like rooting can be related to the stimulatory effect of TDZ on endogenous ethylene production (Pourebad et al. 2015). An increase in ethylene production following TDZ treatment results in an inhibition of auxin transport in many dicots (Radhakrishnan et al. 2009), which in turn further highlights the complex relationship between TDZ's auxin-like activity in terms of downstream effects with other PGRs.

1.7.5 Morphological Abnormalities Resulting from TDZ Use

Genetic evaluation of TDZ-induced explants using flow cytometry, inter-simple sequence repeat (ISSR), molecular markers, and directed amplification of minisatellite-region DNA (DAMD) has shown uniformity and stability in genome size and consistent ploidy level (Faisal et al. 2014). Still unfavorable side effects involving TDZ have been reported including hyperhydricity, dwarfing, uncontrolled callusing, abnormal shoot growth, and difficulty in rooting of shoots. The above side effects are manageable by transferring samples to TDZ-free medium and altering concentration and exposure time (Huetteman and Preece 1993; Mok et al. 2005; Singh and Dwivedi 2014; Magyar-Tábori et al. 2010). In addition to the concentration, undesired side effects associated with the use of TDZ will increase over time as a result of overexposure to TDZ (Manjula et al. 2014; Zhihui et al. 2009; Franklin et al. 2004).

Observed abnormalities demonstrated by cultured tissues exposed to TDZ are likely to be specific to plant organ and species; still certain trends have been observed including: enlarged dark-green cotyledons and leaves (Lu 1993; Murch et al. 1999), short, compact shoots and shoot buds, inhibited shoot elongation, deformation and hyperhydricity of seedlings (Franklin et al. 2004; Hosokawa et al. 1996; Varshney and Anis 2012; Zaytseva et al. 2016; Zhihui et al. 2009; Hare and Van Staden 1994; Dobránszki and da Silva 2010; Lu 1993), inhibited rooting, stunted and thickened root systems (Lu 1993; Murch et al. 1999; Proctor et al. 1996; Dobránszki and da Silva 2010), and necrosis and browning of tissue in seedlings (Zhihui et al. 2009). Morphological effects caused through exposure to TDZ can also be species specific, for instance, morphological abnormalities have been observed in *C. annuum* and *Malus* spp. along with positive effects including increased bud production. TDZ promotes abnormal regenerated shoots from roots of *Bixa orellana* L. (da Cruz et al.

2014). Seedlings developed in the presence of TDZ exhibited reduced root, epicotyl, and hypocotyl elongation. Sankhla et al. (1994) found that *A. julibrissin* roots developed under the influence of TDZ were very thick and short and the development of secondary roots was inhibited. On the other hand, no abnormalities (i.e., including fasciated shoots, hyperhydricity, and inhibited shoot elongation) were reported for other species including white pine and *Dendrocalamus strictus* (Roxb.) Nees (Mihaljevic and Vrsek 2009; Huetteman and Preece 1993; Tang and Newton 2005; Singh and Dwivedi 2014).

Abnormalities caused by TDZ can be overcome. For instance, vitrification can be reduced by using unsealed petri dishes during shoot bud initiation, vented caps for jars during shoot elongation, and a higher concentration of gelling agent. Also, transferring regenerated shoots induced by TDZ to a second medium containing different CKs BA, 2iP, or IBA but lacking TDZ, can lead to regenerated shoots with normal growth and development (Lu 1993; Husain et al. 2007). Another solution to reduce the frequency of shoot fasciation is subculturing induced shoots to medium without TDZ which results in elongated shoots and normal leaves (Huetteman and Preece 1993; Varshney and Anis 2012). Furthermore, the type and combination of other CKs with TDZ significantly influence the occurrence of morphologically abnormal plants (Manjula et al. 2014). Generally, most morphological abnormalities associated with applying TDZ can be overcome by reducing TDZ concentrations and exposure time (Lu 1993). Additionally, TDZ-induced abnormal variations may be overcome by testing various concentrations and times of TDZ application in balance with other phytohormones. Application of TDZ in root-based regeneration systems may also be useful as roots are considered to be genetically more stable in regeneration responses. Regardless of these shortcomings of TDZ in inducing regeneration, it still remains a very useful tool to achieve the designed goals in a range of short-term and long-term micropropagation projects.

1.8 Conclusion

Although TDZ was discovered half a century ago, many questions still remain with respect to its mode of action and function during morphogenesis and organogenesis, which in turn provides interesting opportunities for researchers to explore. For instance, it is still largely unclear as to how plants metabolize TDZ upon exposure and how the mode of action of TDZ may contribute toward its ability to induce morphogenesis in plants even after being removed from growth media. Similarly, the observation that a relationship between TDZ and photoperiod exists suggests that additional mechanisms of action may exist for TDZ, potentially via downstream interactions with phytochrome. While a wealth of attention has been given to TDZ's auxin- and CK-like properties, there is growing information in the literature to suggest that its mode of action is far more complicated than once initially thought, with PGRs and regulatory signals likely playing a greater role than once imagined. The diversity of mechanisms with which TDZ is thought to act is reflected across the wide spectrum of morphological responses that has been observed for TDZ in plants. For example, specific responses including bud development, shoot proliferation, somatic embryogenesis, and seedling development are known to vary significantly across species, explant, concentration, exposure time, and photoperiod, as well as in the presence or absence of other PGRs. As greater efforts are put forth to understand TDZ's multifaceted role in vivo, new ways for utilizing this intriguing PGR will undoubtedly be realized as researchers will be better equipped to predict plant growth and developmental responses when inducing morphogenesis in vitro.

References

Ahmed MR, Anis M (2012) Role of TDZ in the quick regeneration of multiple shoots from nodal explant of *Vitex trifolia* L. an important medicinal plant. App Biochem Biotechnol 168(5):957–966. https://doi.org/10.1007/s12010-012-9799-0

Allen G, Schroeder J (2001) Combining genetics and cell biology to crack the code of plant cell calcium signaling. Sci STKE 102:1–7

Amarante C, Megguer C, Blum L (2003) Effect of preharvest spraying with thidiazuron on fruit quality and maturity of apples. Rev Bras Frutic 25(1):59–62

Arndt F, Rusch R, Stillfried H (1976) SN 49537, a new cotton defoliant. Plant Physiol 57:S-99

Bacha N, Darkazanli K, Abdul-Kader A (2009) Direct organogenesis and plantlet multiplication from leaf explants of in vitro-grown shoots of apple (*Malus domestica* Borkh.) cv.'golden delicious' and 'MM111' rootstock. Fruit Veg Cereal Sci Biotechnol 3(1):28–34

Baghel S, Bansal Y (2014) Thidiazuron promotes in vitro plant regeneration and phytochemical screening of *Guizotia abyssinica* Cass. A multipurpose oil crop. 1193–1217

Baker SB, Bhatia SK (1993) Factors effecting adventitious shoot regeneration from leaf explants of quince (*Cydonia oblonga*). Plant Cell Tissue Organ Cult 35(3):273–277

Band S, Ghadimzadeh M, Jafari M et al (2011) Direct shoot regeneration from stem nodal explants of two wild 'Medicago' species- '*Medicago Scutellata*' and '*Medicago Rigidula*'. Aust J Crop Sci 5(6):668

Basalma D, Uranbey S, Gürlek D et al (2008) TDZ-induced plant regeneration in *Astragalus cicer* L. Afr J of Biotech 7(8):955–959

Bates S, Preece J, Navarrete N et al (1992) Thidiazuron stimulates shoot organogenesis and somatic embryogenesis in white ash (*Fraxinus americana* L.) Plant Cell Tissue Organ Cult 31(1):21–29

Benezet H, Knowles C (1982) Microbial degradation of thidiazuron and its photoproduct. Arch Environ Contam Toxicol 11(1):107–110. https://doi.org/10.1007/BF01055195

Bhagwat B, Vieiral L, Erickson L (1996) Stimulation of in vitro shoot proliferation from nodal explants of cassava by thidiazuron, benzyladenine and gibberellic acid. Plant Cell Tissue Organ Cult 46(1):1–7. https://doi.org/10.1007/BF00039690

Böhmer P, Meyer B, Jacobsen H (1995) Thidiazuron-induced high frequency of shoot induction and plant regeneration in protoplast derived pea callus. Plant Cell Rep 15(1):26–29. https://doi.org/10.1007/BF01690247

Capelle S, Mok D, Kirchner S et al (1983) Effects of thidiazuron on cytokinin autonomy and the metabolism of N6-(Δ^2-isopentenyl)[8-^{14}C] adenosine in callus tissues of *Phaseolus lunatus* L. Plant Physiol 73(3):796–802

Casanova E, Valdés A, Fernández B et al (2004) Levels and immunolocalization of endogenous cytokinins in thidiazuron-induced shoot organogenesis in carnation. J Plant Physiol 161(1):95–104

Chakrabarty D, Trivedi P, Shri M et al (2010) Differential transcriptional expression following thidiazuron induced callus differentiation developmental shifts in rice. Plant Biol 12(1):46–59. https://doi.org/10.1111/j.1438-8677.2009.00213

Chang C, Chang W (2000) Effect of thidiazuron on bud development of *Cymbidium sinense* Willd in vitro. Plant Growth Regul 30(2):171–175. https://doi.org/10.1023/A:1006341300416

Chitra D, Padmaja G (2005) Shoot regeneration via direct organogenesis from in vitro derived leaves of mulberry using thidiazuron and 6-benzylaminopurine. Sci Hortic 106(4):593–602

Choudhary R, Chaudhury R, Malik S et al (2015) An efficient regeneration and rapid micropropagation protocol for Almond using dormant axillary buds as explants. Indian J Exp Biol 53(7):462–467

Chupeau M, Lemoine M, Chupeau Y (1993) Requirement of thidiazuron for healthy protoplast development to efficient tree regeneration of a hybrid poplar (*Populus tremula* x *P. alba*). J Plant Physiol 141(5):601–609

Corredoira E, Ballester A, Vieitez A (2008) Thidiazuron-induced high frequency plant regeneration from leaf explants of *Paulownia tomentosa* mature trees. Plant Cell Tissue Organ Cult 95:197–208. https://doi.org/10.1007/s11240-008-9433-6

Ċosiċ T, Motyka V, Raspor M et al (2015) In vitro shoot organogenesis and comparative analysis of endogenous phytohormones in kohlrabi (*Brassica oleracea* var. gongylodes): effects of genotype, explant type and applied cytokinins. Plant Cell Tissue Organ Cult 121(3):741–760. https://doi.org/10.1007/s11240-008-9433-6

Da Cruz A, Rocha D, Iarema L et al (2014) In vitro organogenesis from root culture segments of *Bixa orellana* L. (Bixaceae). In Vitro Cell Dev Biol-Plant 50:76–83. https://doi.org/10.1007/s11627-013-9580-2

Dabauza M, Pena L (2001) High efficiency organogenesis in sweet pepper (*Capsicum annuum* L.) tissues from different seedling explants. Plant Growth Regul 33(3):221–229. https://doi.org/10.1023/A:1017585407870

Darvari F, Sariah M, Puad M et al (2010) Micropropagation of some Malaysian banana and plantain (*Musa* sp.) cultivars using male flowers. Afr J Biotechnol 9(16):2360–2366

de Carvalho MHC, Van Le B, Zuily-Fodil Y, Thi ATP, Van KTT (2000) Efficient whole plant regeneration of common bean (*Phaseolus vulgaris* L.) using thin-cell-layer culture and silver nitrate. Plant Sci 159(2):223–232

De Gyves EM, Sparks CA, Fieldsend AF et al (2001) High frequency of adventitious shoot regeneration from commercial cultivars of evening primrose (*Oenothera* spp.) using thidiazuron. Ann Appl Biol 138(3):329–332. https://doi.org/10.1111/j.1744-7348.2001.tb00117.x

Debnath SC (2005) Strawberry sepal: another explant for thidiazuron-induced adventitious shoot regeneration. In Vitro Cell Dev Biol-Plant 41(5):671–676

Deroles SC, Seelye JF, Javellana J, Mullan AC (2010) In vitro propagation of *Sandersonia aurantiaca* Hook using thidiazuron. Plant Cell Tissue Organ Cult 102(1):115–119. https://doi.org/10.1007/s11240-010-9705-9

Dey M, Bakshi S, Galiba G et al (2012) Development of a genotype independent and transformation amenable regeneration system from shoot apex in rice (*Oryza sativa* spp. indica) using TDZ. 3 Biotech 2(3):233–240. https://doi.org/10.1007/s13205-012-0051

Dobránszki J, da Silva J (2010) Micropropagation of apple—a review. Biotechnol Adv 28(4):462–488

Ďurkovič J, Mišalová A (2008) Micropropagation of temperate noble hardwoods: an overview. Funct Plant Sci Biotechnol 2:1–19

El Sherif F, Khattab S (2011) Direct shoot regeneration from leaf, root and stem internode segments of male poplar trees and the molecular analysis of variant regenerated plants. J Am Sci 7(8):200–206

Erez A, Yablowitz Z, Aronovitz A et al (2006) Dormancy breaking chemicals; efficiency with reduced phytotoxicity. In: XXVII international horticultural congress-IHC2006: international symposium on enhancing economic and environmental 772, pp 105–112

Escalettes V, Dosba F (1993) In vitro adventitious shoot regeneration from leaves of *Prunus* spp. Plant Sci 90(2):201–209

Faisal M, Siddique I, Anis M (2006) In vitro rapid regeneration of plantlets from nodal explants of *Mucuna pruriens* – a valuable medicinal plant. Ann Appl Biol 148(1):1–6. https://doi.org/10.1111/j.1744-7348.2005.00034

Faisal M, Alatar A, Hegazy A et al (2014) Thidiazuron induced in vitro multiplication of *Mentha arvensis* and evaluation of genetic stability by flow cytometry and molecular markers. Ind Crop Prod 62:100–106

Fasolo F, Zimmerman R, Fordham I (1989) Adventitious shoot formation on excised leaves of in vitro grown shoots of apple cultivars. Plant Cell Tissue Organ Cult 16(2):75–87. https://doi.org/10.1007/BF00036516

Faure O, Diemer F, Moja S et al (1998) Mannitol and thidiazuron improve in vitro shoot regeneration from spearmint and peppermint leaf disks. Plant Cell Tissue Organ Cult 52(3):209–212. https://doi.org/10.1023/A:1006029123437

Fehér A, Pasternak T, Otvos K et al (2002) Induction of embryogenic competence in somatic plant cells: a review. Biol Sect Bot 51(1):5–12

Feng B, Wu B, Zhang C et al (2012) Cloning and expression of 1-aminocyclopropane-1-carboxylate oxidase cDNA induced by thidiazuron during somatic embryogenesis of alfalfa (*Medicago sativa*). J Pant Physiol 169(2):176–182

Ferrante A, Hunter D, Hackett W et al (2002a) Thidiazuron—a potent inhibitor of leaf senescence in *Alstroemeria*. Postharvest Biol Tecnol 25(3):333–338

Ferrante A, Tognoni F, Mensuali-Sodi A (2002b) Treatment with thidiazuron for preventing leaf yellowing in cut tulips and chrysanthemum. In: XXVI international horticultural congress: elegant science in floriculture 624, pp 357–363

Ferrante A, Vernieri P, Serra G et al (2004) Changes in abscisic acid during leaf yellowing of cut stock flowers. Plant Growth Regul 43(2):127–134. https://doi.org/10.1023/B:GROW.0000040119.27627.b2

Fiola J, Hassan M, Swartz H et al (1990) Effect of thidiazuron, light fluence rates and kanamycin on in vitro shoot organogenesis from excised *Rubus* cotyledons and leaves. Plant Cell Tissue Organ Cult 20(3):223–228. https://doi.org/10.1007/BF00041885

Franklin G, Sheeba C, Sita G (2004) Regeneration of eggplant (*Solanum melongena* L.) from root explants. In Vitro Cell Dev Biol-Plant 40(2):188–191. https://doi.org/10.1079/IVP2003491

Fujimura T, Komamine A (1980) Mode of action of 2, 4-D and zeatin on somatic embryogenesis in a carrot cell suspension culture. Z Pflanzenphysiol 99(1):1–8

Gairi A, Rashid A (2004) Direct differentiation of somatic embryos on different regions of intact seedlings of *Azadirachta* in response to thidiazuron. J Plant Physiol 161(9):1073–1077

Ganeshan S, Chodaparambil S, Båga M et al (2006) In vitro regeneration of cereals based on multiple shoot induction from mature embryos in response to thidiazuron. Plant Cell Tissue Organ Cult 85(1):63–73. https://doi.org/10.1007/s11240-005-9049-z

Gill R, Saxena PK (1992) Direct somatic embryogenesis and regeneration of plants from seedling explants of peanut (*Arachis hypogaea*): promotive role of thidiazuron. Can J Bot 70(6):1186–1192

Gill R, Saxena PK (1993) Somatic embryogenesis in *Nicotiana tabacum* L.: induction by thidiazuron of direct embryo differentiation from cultured leaf discs. Plant Cell Rep 12(3):154–159. https://doi.org/10.1007/BF00239097

Graner É, Oberschelp G, Brondani G et al (2013) TDZ pulsing evaluation on the in vitro morphogenesis of peach palm. Physiol Mol Biol Plant 19(2):283–288. https://doi.org/10.1007/s12298-012-0160-4

Gubbuk H, Pekmezci M (2006) In vitro propagation of banana (*Musa* spp.) using thidiazuron and activated charcoal. Acta Agric Scand Sect B-Soil Plant Sci 56(1):65–69

Guo B, Abbasi B, Zeb A et al (2011) Thidiazuron: a multi-dimensional plant growth regulator. Afr J Biotechnol 10(45):8984–9000

Guo B, Stiles A, Liu C (2012) Thidiazuron enhances shoot organogenesis from leaf explants of *Saussurea involucrata* Kar. et Kir. In Vitro Cell Dev Biol-Plant 48(6):609–612. https://doi.org/10.1007/s11627-012-9468-6

Gupta D, Bhargava S (2001) Thidiazuron induced regeneration in *Cuminum cyminum* L. J Plant Biochem Biotechnol 10(1):61–62. https://doi.org/10.1007/BF03263109

Gupta S, Conger B (1998) In vitro differentiation of multiple shoot clumps from intact seedlings of switchgrass. In Vitro Cell Dev Biol-Plant 34(3):196–202. https://doi.org/10.1007/BF02822708

Hare P, Cress W (1997) Metabolic implications of stress-induced proline accumulation in plants. Plant Growth Regul 21(2):79–102. https://doi.org/10.1023/A:1005703923347
Hare P, Van Staden J (1994) Inhibitory effect of thidiazuron on the activity of cytokinin oxidase isolated from soybean callus. Plant Cell Physiol 35(8):1121–1125
Hosokawa K, Nakano M, Oikawa Y et al (1996) Adventitious shoot regeneration from leaf, stem and root explants of commercial cultivars of *Gentiana*. Plant Cell Rep 15(8):578–581. https://doi.org/10.1007/BF00232456
Hosseini-Nasr M, Rashid A (2002) Thidiazuron-induced shoot-bud formation on root segments of *Albizia julibrissin* is an apex-controlled, light-independent and calcium-mediated response. Plant Growth Regul 36(1):81–85. https://doi.org/10.1023/A:1014771130101
Huetteman C, Preece J (1993) Thidiazuron: a potent cytokinin for woody plant tissue culture. Plant Cell Tissue Organ Cult 33(2):105–119. https://doi.org/10.1007/BF01983223
Husain M, Anis M, Shahzad A (2007) In vitro propagation of Indian Kino (*Pterocarpus marsupium* Roxb.) using thidiazuron. In Vitro Cell Dev Biol-Plant 43(1):59–64
Husaini A, Abdin M (2007) Interactive effect of light, temperature and TDZ on the regeneration potential of leaf discs of *Fragaria x ananassa* Duch. In Vitro Cell Dev Biol-Plant 43(6):576–584. https://doi.org/10.1007/s11627-007-9048-3
Hussain TM, Chandrasekhar T, Gopal GR (2008) Micropropagation of *Sterculia urens* Roxb., an endangered tree species from intact seedlings. Afr J Biotechnol 7(2):95–101
Hutchinson M, Saxena PK (1996) Role of purine metabolism in thidiazuron-induced somatic embryogenesis of geranium (*Pelargonium*× *hortorum*) hypocotyl cultures. Physiol Plant 98(3):517–522. https://doi.org/10.1111/j.1399-3054.1996.tb05706.x
Hutchinson M, Murch S, Saxena PK (1996) Morphoregulatory role of thidiazuron: evidence of the involvement of endogenous auxin in thidiazuron-induced somatic embryogenesis of geranium (*Pelargonium*× *hortorum* Bailey). J Plant Physiol 149(5):573–579
Hutchinson M, Murr D, Krishnaraj S et al (1997) Does ethylene play a role in thidiazuron-regulated somatic embryogenesis of geranium (*Pelargonium*× *hortorum* bailey) hypocotyl cultures. In Vitro Cell Dev Biol-Plant 33(2):136–141. https://doi.org/10.1007/s11627-997-0012-z
Hyde C, Phillips G (1996) Silver nitrate promotes shoot development and plant regeneration of chile pepper (*Capsicum annuum* L.) via organogenesis. In Vitro Cell Dev Biol-Plant 32(2):72–80. https://doi.org/10.1007/BF02823134
Iantcheva A, Vlahova M, Bakalova E et al (1999) Regeneration of diploid annual medics via direct somatic embryogenesis promoted by thidiazuron and benzylaminopurine. Plant Cell Rep 18(11):904–910. https://doi.org/10.1007/s002990050682
Jahan A, Anis M, Aref I (2011) Preconditioning of axillary buds in thidiazuron-supplemented liquid media improves in vitro shoot multiplication in *Nyctanthes arbor-tristis* L. Appl Biochem Biotechnol 163(7):851–859. https://doi.org/10.1007/s12010-010-9089-7
Jain P, Rashid A (2001) Stimulation of shoot regeneration on *Linum* hypocotyl segments by thidiazuron and its response to light and calcium. Biol Plant 44(4):611–613. https://doi.org/10.1023/A:1013767426219
Jiang B, Yang Y, Guo Y et al (2005) Thidiazuron-induced in vitro shoot organogenesis of the medicinal plant *Arnebia euchroma* (Royle) Johnst. In Vitro Cell Deve Biol-Plant 41(5):677–681
Jiang C, Wu L, Macnish A et al (2008) Thidiazuron, a non-metabolized cytokinin, shows promise in extending the life of potted plants. In: IX international symposium on postharvest quality of ornamental plants 847, p 59–66
Jones M, Yi Z, Murch S et al (2007) Thidiazuron-induced regeneration of *Echinacea purpurea* L.: micropropagation in solid and liquid culture systems. Plant Cell Rep 26(1):13–19. https://doi.org/10.1007/s00299-006-0209-3
Jones M, Shukla M, Biswas G et al (2015) Protoplast-to-plant regeneration of American elm (*Ulmus americana*). Protoplasma 252:925–931
Kanyand M, Dessai A, Prakash C (1994) Thidiazuron promotes high frequency regeneration of peanut (*Arachis hypogaea*) plants in vitro. Plant Cell Rep 14(1):1–5. https://doi.org/10.1007/BF00233288

Kartomysheva O, Volkova T, Nikitenko S (1983) Dropp, a new promising stimulant of callus formation. Sintez, Biologicheskaya Aktivnost'i Primenenie Pestitsidov:131–135
Kefford N, Zwar A, Bruce M (1968) Antagonism of purine and urea cytokinin activities by derivatives of benzylurea. Biochemistry and physiology of plant growth substances. Runge Press, Ottawa, pp 61–69
Khan H, Siddique I, Anis M (2006) Thidiazuron induced somatic embryogenesis and plant regeneration in *Capsicum annuum*. Biol Plant 50(4):789–792. https://doi.org/10.1007/s10535-006-0133-y
Khurana-Kaul V, Kachhwaha S, Kothari S (2010) Direct shoot regeneration from leaf explants of *Jatropha curcas* in response to thidiazuron and high copper contents in the medium. Biol Plant 54(2):369–372. https://doi.org/10.1007/s10535-010-0066-3
Kidwai N, Jain M, Chaturvedi H (2009) Role of thidiazuron in in vitro induction of embryogenesis in nucellar tissue of *Mangifera indica* L. var. Dashehari, leading to plantlets. Curr Sci 96:1119–1124
Kumar N, Reddy M (2012) Thidiazuron (TDZ) induced plant regeneration from cotyledonary petiole explants of elite genotypes of *Jatropha curcas*: a candidate biodiesel plant. Ind Crop Prod 39:62–68
Lakshmi S, Benjamin J, Kumar T et al (2010) In vitro propagation of *Hoya wightii* ssp. palniensis KT Mathew, a highly vulnerable and endemic species of Western Ghats of Tamil Nadu, India. Afr J Biotechnol 9(5):620–627
Lata H, Chandra S, Khan I et al (2009) Thidiazuron-induced high-frequency direct shoot organogenesis of *Cannabis sativa* L. In Vitro Cell Dev Biol-Plant 45(1):12–19. https://doi.org/10.1007/s11627-008-9167-5
Lazzeri P, Dunwell J (1984) Establishment of isolated root cultures of Brassica species and regeneration from cultured-root segments of *Brassica oleracea* var (ed). italica. Ann Bot 54(3):351–361
Leblay C, Chevreau E, Raboin L (1991) Adventitious shoot regeneration from in vitro leaves of several pear cultivars (*Pyrus communis* L.) Plant Cell Tissue Organ Cult 25(2):99–105. https://doi.org/10.1007/BF00042180
Ledbetter D, Preece J (2004) Thidiazuron stimulates adventitious shoot production from *Hydrangea quercifolia* leaf explants. Sci Hortic 101:121–126
Li H, Murch S, Saxena PK (2000) Thidiazuron-induced de novo shoot organogenesis on seedlings, etiolated hypocotyls and stem segments of Huang-qin. Plant Cell Tissue Organ Cult 62(3):169–173. https://doi.org/10.1023/A:1006491408762
Li J, Wu Y, Wang T et al (2009) In vitro direct organogenesis and regeneration of *Medicago sativa*. Biol Plant 53(2):325–328. https://doi.org/10.1007/s10535-009-0059-2
Lin C, Wang R, Jauh G (1988) Enhancement of callus formation on grape single bud cuttings by thidiazuron. In: VI international symposium on growth regulators in fruit production 239, pp 129–132
Lin C, Lin C, Chang W (2004) Effect of thidiazuron on vegetative tissue-derived somatic embryogenesis and flowering of bamboo *Bambusa edulis*. Plant Cell Tissue Organ Cult 76(1):75–82. https://doi.org/10.1023/A:1025848016557
Lincy A, Sasikumar B (2010) Enhanced adventitious shoot regeneration from aerial stem explants of ginger using TDZ and its histological studies. Tur J Bot 34(1):21–29
Liu Q, Salih S, Hammerschlag F (1998) Etiolation of Royal Gala'apple (*Malus*× *domestica* Borkh.) shoots promotes high-frequency shoot organogenesis and enhanced,-glucuronidase expression from stem internodes. Plant Cell Rep 18(1):32–36. https://doi.org/10.1007/s002990050527
Liu C, Murch S, El-Demerdash M et al (2003) Regeneration of the Egyptian medicinal plant *Artemisia judaica* L. Plant Cell Rep 21(6):525–530. https://doi.org/10.1007/s00299-002-0561-x
Lu C (1993) The use of thidiazuron in tissue culture. In Vitro Cell Dev Biol-Plant 29(2):92–96. https://doi.org/10.1007/BF02632259
Magioli C, Rocha A, De Oliveira D et al (1998) Efficient shoot organogenesis of eggplant (*Solanum melongena* L.) induced by thidiazuron. Plant Cell Rep 17(8):661–663. https://doi.org/10.1007/s002990050461

Magyar-Tábori K, Dobránszki J, da Silva J et al (2010) The role of cytokinins in shoot organogenesis in apple. Plant Cell Tissue Organ Cult 101(3):251–267. https://doi.org/10.1007/s11240-010-9696-6

Malabadi R, Mulgund G, Nataraja K et al (2004) Thidiazuron induced shoot regeneration of *Costus speciosus* (Koen.) Sm using thin rhizome sections. South Afr J Bot 70(2):255–258

Malik K, Saxena PK (1992) Regeneration in *Phaseolus vulgaris* L.: high-frequency induction of direct shoot formation in intact seedlings by N 6-benzylaminopurine and thidiazuron. Planta 186(3):384–389. https://doi.org/10.1007/BF00195319

Malik K, Ali-Khan S, Saxena PK (1993) High-frequency organogenesis from direct seed culture in *Lathyrus*. Ann Bot 72(6):629–637

Manjula R, Jholgiker P, Subbaiah K et al (2014) Morphological abnormality among hardened shoots of Banana cv. Rajapuri (AAB) after in vitro multiplication with TDZ and BAP from excised shoot tips. Int J Agric Env Biotechnol 7(3):465. https://doi.org/10.5958/2230-732X.2014.01350.3

Martínez T, Corredoira E, Valladares S et al (2008) Germination and conversion of somatic embryos derived from mature *Quercus robur* trees: the effects of cold storage and thidiazuron. Plant Cell Tissue Organ Cult 95(3):341–351. https://doi.org/10.1007/s11240-008-9448-z

Matand K, Prakash C (2007) Evaluation of peanut genotypes for in vitro plant regeneration using thidiazuron. J Biotechnol 130(2):202–207

Matsuta N, Hirabayashi T (1989) Embryogenic cell lines from somatic embryos of grape (*Vitis vinifera* L.) Plant Cell Rep 7(8):684–687

Mehrotra S, Goel M, Srivastava V et al (2015) Hairy root biotechnology of *Rauwolfia serpentina*: a potent approach for the production of pharmaceutically important terpenoid indole alkaloids. Biotechnol Lett 37(2):253–263. https://doi.org/10.1007/s10529-014-1695-y

Mehta U, Barreto S, Hazra S (2004) Effect of thidiazuron in germinating tamarind seedlings. In Vitro Cell Dev Biol-Plant 40(3):279–283

de Melo FW, Kerbauy G, Kraus J et al (2006) Thidiazuron influences the endogenous levels of cytokinins and IAA during the flowering of isolated shoots of *Dendrobium*. J Plant Physiol 163(11):1126–1134

Mihaljevic S, Vrsek I (2009) In vitro shoot regeneration from immature seeds of *Epimedium alpinum* induced by thidiazuron and CPPU. Sci Hortic 120:406–410

Mithila J, Hall J, Victor J et al (2003) Thidiazuron induces shoot organogenesis at low concentrations and somatic embryogenesis at high concentrations on leaf and petiole explants of African violet (*Saintpaulia ionantha* Wendl.) Plant Cell Rep 21(5):408–414. https://doi.org/10.1007/s00299-002-0544-y

Mok M, Mok D (1985) The metabolism of [^{14}C]-tMdiaziiroii in callus tissues of *Phaseolus lunatus*. Physiol Plant 65(4):427–432

Mok M, Mok D, Armstrong D et al (1982) Cytokinin activity of N-phenyl-N′-1, 2, 3-thiadiazol-5-ylurea (thidiazuron). Phytochemistry 21(7):1509–1511

Mok M, Mok D, Turner J, Mujer C (1987) Biological and biochemical effects of cytokinin-active phenylurea derivatives in tissue culture systems. Hort Sci 22(6):1194–1197

Mok M, Martin R, Dobrev P et al (2005) Topolins and hydroxylated thidiazuron derivatives are substrates of cytokinin O-glucosyltransferase with position specificity related to receptor recognition. Plant Physiol 137(3):1057–1066

Mondal TK, Bhattacharya A, Sood A et al (1998) Micropropagation of tea (*Camellia sinensis* (L.) O. Kuntze) using Thidiazuron. Plant Growth Regul 26(1):57–61. https://doi.org/10.1023/A:1006019206264

Montecelli S, Gentile A, Damiano C (1999) In vitro shoot regeneration of apple cultivar gala. In: International symposium on methods and markers for quality assurance in micropropagation. Acta Hortic 530:219–224

Mundhara R, Rashid A (2002) Stimulation of shoot-bud regeneration on hypocotyl of Linum seedlings, on a transient withdrawal of calcium: effect of calcium, cytokinin and thidiazuron. Plant Sci 162(2):211–214

Mundhara R, Rashid A (2006) TDZ-induced triple-response and shoot formation on intact seedlings of Linum, putative role of ethylene in regeneration. Plant Sci 170(2):185–190

Murashige T (1974) Plant propagation through tissue cultures. Ann Rev Plant Physiol 25(1):135–166

Murch S, Saxena PK (1997) Modulation of mineral and fatty acid profiles during thidiazuron mediated somatic embryogenesis in peanuts (*Arachis hypogaea* L.) J Plant Physiol 151:358–361

Murch S, Saxena PK (2001) Molecular fate of thidiazuron and its effects on auxin transport in hypocotyls tissues of *Pelargonium*× *hortorum* Bailey. Plant Growth Regul 35(3):269–275. https://doi.org/10.1023/A:1014468905953

Murch S, KrishnaRaj S, Saxena PK (1997) Thidiazuron-induced regeneration: a potential stress response. Plant Physiol 114(3):177–177

Murch S, Victor J, Krishnaraj S et al (1999) The role of proline in thidiazuron-induced somatic embryogenesis of peanut. In Vitro Cell Dev Biol-Plant 35(1):102–105. https://doi.org/10.1007/s11627-999-0018-9

Murch S, Choffe K, Victor J et al (2000) Thidiazuron-induced plant regeneration from hypocotyl cultures of St. John's wort (*Hypericum perforatum*. cv'Anthos'). Plant Cell Rep 19(6):576–581. https://doi.org/10.1007/s002990050776

Murch S, Victor J, Saxena PK (2002) Auxin, calcium and sodium in somatic embryogenesis of African violet (*Saintpaulia ionantha* Wendl. Cv. Benjamin). In: XXVI international horticultural congress: biotechnology in horticultural crop improvement: achievements, opportunities and 625, pp 201–209

Murthy BNS, Saxena PK (1994) Somatic embryogenesis in peanut (*Arachis hypogaea* L.): stimulation of direct differentiation of somatic embryos by forchlorfenuron (CPPU). Plant Cell Rep 14(2–3):145–150

Murthy B, Murch S, Saxena PK (1995) Thidiazuron-induced somatic embryogenesis in intact seedlings of peanut (*Arachis hypogaea*): endogenous growth regulator levels and significance of cotyledons. Physiol Plant 94(2):268–276

Murthy B, Victor J, Singh R et al (1996) In vitro regeneration of chickpea (*Cicer arietinum* L.): stimulation of direct organogenesis and somatic embryogenesis by thidiazuron. Plant Growth Regul 19(3):233–240. https://doi.org/10.1007/BF00037796

Murthy B, Murch S, Saxena PK (1998) Thidiazuron: a potent regulator of in vitro plant morphogenesis. In Vitro Cell Dev Biol-Plant 34(4):267. https://doi.org/10.1007/BF02822732

Nhut DT, Hanh N, Tuan P et al (2006) Liquid culture as a positive condition to induce and enhance quality and quantity of somatic embryogenesis of *Lilium longiflorum*. Sci Hortic 110(1):93–97

Nikolić R, Mitić N, Miletić R et al (2006) Effects of cytokinins on in vitro seed germination and early seedling morphogenesis in *Lotus corniculatus* L. J Plant Growth Regul 25(3):187–194. https://doi.org/10.1007/s00344-005-0129-4

Olah R, Szegedi E, Ruthner S, Korbuly J (2003) Thidiazuron-induced regeneration and genetic transformation of grapevine rootstock varieties. Vitis 42(4):207–207

Oluk E, Orhan S (2009) Thidiazuron induced micropropagation of *Hypericum triquetrifolium* Turra. Afr J Biotechnol 8(15):3506–3510

Osman M, Elhadi E, Khalafalla M (2010) Callus formation and organogenesis of tomato (*Lycopersicon esculentum* Mill, CV Omdurman) induced by thidiazuron. Afr J Biotechnol 9(28):4407–4413

Park S, Murthy H, Paek K (2003) Protocorm-like body induction and subsequent plant regeneration from root tip cultures of *Doritaenops*. Plant Sci 164(6):919–923

Parveen S, Shahzad A (2011) A micropropagation protocol for Cassia angustifolia Vahl. from root explants. Acta Physiol Plant 33(3):789–796. https://doi.org/10.1007/s11738-010-0603-x

Passey A, Barrett K, James D (2003) Adventitious shoot regeneration from seven commercial strawberry cultivars (*Fragaria*× *ananassa* Duch.) using a range of explant types. Plant Cell Rep 21(5):397–401. https://doi.org/10.1007/s00299-002-0530-4

Pavingerová D (2009) The influence of thidiazuron on shoot regeneration from leaf explants of fifteen cultivars of Rhododendron. Biol Plant 53(4):797–799. https://doi.org/10.1007/s10535-009-0147-3

Pavlista A, Gall C (2011) Delaying early blight onset in potato with thidiazuron. Am J Potato Res 88(2):114–120. https://doi.org/10.1007/s12230-010-9168-x

Pawlicki-Jullian N, Sedira M, Welander M (2002) The use of Agrobacterium rhizogenes transformed roots to obtain transgenic shoots of the apple rootstock Jork 9. Plant Cell Tissue Organ Cult 70(2):163–171. https://doi.org/10.1023/A:1016387004712

Pourebad N, Motafakkerazad R, Kosari-Nasab M et al (2015) The influence of TDZ concentrations on in vitro growth and production of secondary metabolites by the shoot and callus culture of *Lallemantia iberica*. Plant Cell Tissue Organ Cult 122(2):331–339. https://doi.org/10.1007/s11240-015-0769-4

Pradhan C, Kar S, Pattnaik S et al (1998) Propagation of *Dalbergia sissoo* Roxb. through in vitro shoot proliferation from cotyledonary nodes. Plant Cell Rep 18(1–2):122–126. https://doi.org/10.1007/s002990050543

Prathanturarug S, Soonthornchareonnon N, Chuakul W et al (2005) Rapid micropropagation of *Curcuma longa* using bud explants pre-cultured in thidiazuron-supplemented liquid medium. Plant Cell Tissue Organ Cult 80(3):347–351. https://doi.org/10.1007/s11240-004-1020-x

Preece J, Imel M (1991) Plant regeneration from leaf explants of Rhododendron 'PJM Hybrids'. Sci Hortic 48(1–2):159–170

Proctor J, Slimmon T, Saxena PK (1996) Modulation of root growth and organogenesis in thidiazuron-treated ginseng (*Panax quinquefolium* L.) Plant Growth Regul 20(3):201–208. https://doi.org/10.1007/BF00043309

Purohit S, Joshi P, Tak K et al (2004) Development of high efficiency micropropagation protocol of an adult tree—*Wrightia tomentosa*. Plant Biotechnol Mol Markers 3:217–227

Rad F, Jafari M, Khezrinejad N et al (2014) An efficient plant regeneration system via direct organogenesis with in vitro flavonoid accumulation and analysis of genetic fidelity among regenerants of *Teucrium polium* L. Hort Env Biotechnol 55(6):568–577. https://doi.org/10.1007/s13580-014-0611-7

Radhakrishnan R, Ramachandran A, Kumari BR (2009) Rooting and shooting: dual function of thidiazuron in in vitro regeneration of soybean (*Glycine max* L.) Acta Physiol Plant 31(6):1213–1217. https://doi.org/10.1007/s11738-009-0356-6

Rolli E, Incerti M, Brunoni F et al (2012) Structure–activity relationships of N-phenyl-N′-benzothiazol-6-ylurea synthetic derivatives: cytokinin-like activity and adventitious rooting enhancement. Phytochemistry 74:159–165

Saito A, Suzuki M (1999) Plant regeneration from meristem-derived callus protoplasts of apple (*Malus domestica* cv.Fuji'). Plant Cell Rep 18(7):549–553. https://doi.org/10.1007/s002990050620

San B, Li Z, Hu Q et al (2015) Adventitious shoot regeneration from in vitro cultured leaf explants of peach rootstock Guardian® is significantly enhanced by silver thiosulfate. Plant Cell Tissue Organ Cult 120(2):757–765. https://doi.org/10.1007/s11240-014-0645-7

Sanikhani M, Frello S, Serek M (2006) TDZ induces shoot regeneration in various *Kalanchoe blossfeldiana* Poelln. Cultivars in the absence of auxin. Plant Cell Tissue Organ Cult 85(1):75–82. https://doi.org/10.1007/s11240-005-9050-6

Sankhla D, Davis T, Sankhla N (1994) Thidiazuron-induced in vitro shoot formation from roots of intact seedlings of *Albizzia julibrissin*. Plant Growth Regul 14(3):267–272. https://doi.org/10.1007/BF00024802

Sankhla D, Davis T, Sankhla N (1996) In vitro regeneration of silktree (*Albizzia julibrissin*) from excised roots. Plant Cell Tissue Organ Cult 44(1):83–86. https://doi.org/10.1007/BF00045917

Sankhla N, Mackay W, Davis T (2003) Effect of thidiazuron on senescence of flowers in cut inflorescences of *Lupinus densiflorus* Benth. In: VIII international symposium on postharvest physiology of ornamental plants 669, pp 239–244

Saxena PK, Malik K, Gill R (1992) Induction by thidiazuron of somatic embryogenesis in intact seedlings of peanut. Planta 187(3):421–424. https://doi.org/10.1007/BF00195667

Sharifi G, Ebrahimzadeh H, Ghareyazie B et al (2010) Globular embryo-like structures and highly efficient thidiazuron-induced multiple shoot formation in saffron (*Crocus sativus* L.) In Vitro Cell Dev Biol-Plant 46(3):274–280. https://doi.org/10.1007/s11627-009-9264-0

Sharma RK, Barna K, Wakhlu A (2004) Effect of Thidiazuron on micropropagation of *Malus pumila* Mill. Cv. Ambri. Oriental Sci 9:31–36

Sharma V, Gupta S, Dhiman M (2013) Regeneration of plants from nodal and internodal segment cultures of *Ephedra gerardiana* using thidiazuron. Plant Tiss Cult Biotechnol 22(2):53–161

Shukla M, Sullivan A, Jain S et al (2013) Micropropagation of African violet (*Saintpaulia ionantha* Wendl.) In: Maurizio L et al (eds) Protocols for micropropagation of selected economically-important horticultural plants, methods in molecular biology, vol 994. Springer, New York, pp 279–288. https://doi.org/10.1007/978-1-62703-074-8_22

Singh. P., Dwivedi, P. (2014). Two-stage culture procedure using thidiazuron for efficient micro-propagation of Stevia rebaudiana, an anti-diabetic medicinal herb. 3 Biotechnology 4(4), 431–437

Singh N, Sahoo L, Sarin N et al (2003) The effect of TDZ on organogenesis and somatic embryo-genesis in pigeonpea (*Cajanus cajan* L. Millsp). Plant Sci 164(3):341–347

Soliman H (2013) In vitro regeneration and genetic transformation of peach (*Prunus Persica* L.) plants. Life Sci J 10(2):487–496

Stern R, Shargal A, Flaishman M (2003) Thidiazuron increases fruit size of 'Spadona'and 'Coscia'pear (*Pyrus communis* L.) J Horticult Sci Biotechnol 78(1):51–55

Subotić A, Jevremović S, Grubišić D (2009) Influence of cytokinins on in vitro morphogenesis in root cultures of *Centaurium erythraea*—valuable medicinal plant. Sci Hortic 120(3):386–390

Suezawa K, Matsuta N, Omura M et al (1988) Plantlet formation from cell suspensions of kiwifruit (*Actinidia chinensis* Planch. var. chinensis). Sci Hortic 37(2):123–128

Susan J (1996) Morphoregulatory role of thidiazuron: effect on metabolic process during regenera-tion. Dissertation, The University of Guelph

Suttle J (1984) Effect of the defoliant thidiazuron on ethylene evolution from mung bean hypocotyl segments. Plant Physiol 75(4):902–907

Szász A, Nervo G, Fári M (1995) Screening for in vitro shoot-forming capacity of seedling explants in bell pepper (*Capsicum annuum* L.) genotypes and efficient plant regeneration using thidiazuron. Plant Cell Rep 14(10):666–669

Tang W, Newton R (2005) Plant regeneration from callus cultures derived from mature zygotic embryos in white pine (*Pinus strobus* L.) Plant Cell Rep 24(1):1–9. https://doi.org/10.1007/s00299-005-0914-3

Thengane S, Kulkarni D, Shrikhande V et al (2001) Effect of thidiazuron on adventitious shoot regeneration from seedling explants of *Nothapodytes foetida*. In Vitro Cell Dev Biol-Plant 37(2):206–210. https://doi.org/10.1007/s11627-001-0036-8

Thomas T (2003) Thidiazuron induced multiple shoot induction and plant regeneration from cotyledonary explants of mulberry. Biol Plant 46(4):529–533. https://doi.org/10.1023/A:1024807426591

Thomas J, Katterman F (1986) Cytokinin activity induced by thidiazuron. Plant Physiol 81(2):681–683

Thomas TD, Philip B (2005) Thidiazuron-induced high-frequency shoot organogenesis from leaf-derived callus of a medicinal climber, *Tylophora indica* (Burm. F.) Merrill. In Vitro Cell Dev Bio Plant 41(2):124–128

Thomas T, Puthur J (2004) Thidiazuron induced high frequency shoot organogenesis in callus from *Kigelia pinnata* L. Bot Bull Acad Sin 45:307–313

Trewavas A (1999) Le calcium, c'est la vie: calcium makes waves. Plant Physiol 120(1):1–6

Tsvetkov I (1999) Thidiazuron-induced somatic embryogenesis in common oak (*Q. Robur* L.) Biotechnol Equip 13(1):44–46

Tsvetkov I, Husman J, Jouve L (2007) Thidiazuron-induced regeneration in root segments of white poplar (*P. alba* L.) Bulgarian J Agric Sci 13(5):623

Twyford C, Mantell S (1996) Production of somatic embryos and plantlets from root cells of the Greater Yam. Plant Cell Tissue Organ Cult 46(1):17–26. https://doi.org/10.1007/BF00039692

Uranbey S (2005) Thidiazuron induced adventitious shoot regeneration in *Hyoscyamus niger*. Biol Plant 49(3):427–430. https://doi.org/10.1007/s10535-005-0021-x

Uthairatanakij A, Jeenbuntug J, Buanong M, Kanlayanarat S (2007) Effect of Thidiazuron puls-ing on physiological changes of cut tuberose flower (*Polianthes tuberose* L.). In: International conference on quality management in supply chains of ornamentals 755, pp 477–481

Van Nieuwkerk J, Zimmerman R, Fordham I (1985) Response of apple cultivars in vitro to thidiazuron. Hort Sci 20:523

Varshney A, Anis M (2012) Improvement of shoot morphogenesis in vitro and assessment of changes of the activity of antioxidant enzymes during acclimation of micropropagated plants of Desert Teak. Acta Physiol Plant 34(3):859–867. https://doi.org/10.1007/s11738-011-0883-9

Victor J, Murthy B, Murch S et al (1999) Role of endogenous purine metabolism in thidiazuron-induced somatic embryogenesis of peanut (*Arachis hypogaea* L.) Plant Growth Regul 28(1):41–47. https://doi.org/10.1023/A:1006251531319

Vila S, Gonzalez A, Rey H et al (2005) Plant regeneration, origin, and development of shoot buds from root segments of *Melia azedarach* L. (Meliaceae) seedlings. In Vitro Cell Dev Biol-Plant 41(6):746–751

Visser C, Qureshi J, Gill R et al (1992) Morphoregulatory role of thidiazuron substitution of auxin and cytokinin requirement for the induction of somatic embryogenesis in geranium hypocotyl cultures. Plant Physiol 99(4):1704–1707

Vu N, Anh P, Nhut D (2006) The role of sucrose and different cytokinins in the in vitro floral morphogenesis of rose (hybrid tea) cv."First Prize". Plant Cell Tissue Organ Cult 87(3):315–320. https://doi.org/10.1007/s11240-006-9089-z

Wang S, Steffens G, Faust M (1986) Breaking bud dormancy in apple with a plant bioregulator, thidiazuron. Phytochemistry 25(2):311–317

Wang B, Peng D, Liu L et al (2007) An efficient adventitious shoot regeneration system for ramie (*Boehmeria nivea* Gaud) using thidiazuron. Bot Stud 48(2):173–180

Wang Q, Zheng L, Yuan H et al (2013) Propagation of Salvia miltiorrhiza from hairy root explants via somatic embryogenesis and tanshinone content in obtained plants. Ind Crop Prod 50:648–653

White P, Broadley M (2003) Calcium in plants. Ann Bot 92(4):487–511

Yip W, Yang S (1986) Effect of thidiazuron, a cytokinin-active urea derivative, in cytokinin-dependent ethylene production systems. Plant Physiol 80(2):515–519

Yousefiara M, Kermani M, Bagheri A et al (2014) Induction of direct adventitious shoot regeneration in pear (*Pyrus communis* L.) Plant Tiss Cult Biotechnol 24(1):87–92

Zaytseva Y, Poluboyarova T, Novikova T (2016) Effects of thidiazuron on in vitro morphogenic response of *Rhododendron sichotense* Pojark and *Rhododendron catawbiense* cv. Grandiflorum leaf explants. In Vitro Cell Dev Biol-Plant 52(1):56–63. https://doi.org/10.1007/s11627-015-9737-2

Zee S (1981) Studies on adventive embryo formation in the petiole expiants of coriander (*Coriandrum sativum*). Protoplasma 107(1):21–26. https://doi.org/10.1007/BF01275604

Zhang CG, Li W, Mao YF et al (2005) Endogenous hormonal levels in *Scutellaria baicalensis* calli induced by thidiazuron. Russ J Plant Physiol 52(3):345–351. https://doi.org/10.1007/s11183-005-0052-3

Zhihui S, Tzitzikas M, Raemakers K et al (2009) Effect of TDZ on plant regeneration from mature seeds in pea (*Pisum sativum*). In Vitro Cell Dev Biol-Plant 45(6):776–782. https://doi.org/10.1007/s11627-009-9212-z

Zobayed S, Saxena PK (2003) In vitro-grown roots: a superior explants for prolific shoot regeneration of St. John's wort (*Hypericum perforatum* L. cv. "New Stem") in a temporary immersion bioreactor. Plant Sci 165:463–470

2 TDZ: Mode of Action, Use and Potential in Agriculture

Jaroslav Nisler

Abstract

Strong cytokinin effects of thidiazuron (TDZ) in many plant species have been observed since its discovery in the 1970s. Several of these effects, such as cell division stimulatory activity, anti-senescence, anti-stress activity and ethylene production stimulation, have been adopted by agriculturalists and horticulturalists for a wide range of use. TDZ has been shown to promote the growth of various fruits, delay senescence of cut and potted flowers, increase stress tolerance and yield of several crops and cause defoliation of cotton. In this chapter, the mechanisms of how TDZ affects the desired traits are described, and the literature provides evidences reviewed. The information given here should convince everyone that TDZ is not a mysterious substance but that it triggers classical cytokinin responses in plants as successfully as natural cytokinins, no matter whether directly or indirectly. A direct TDZ effect is mediated through the activation of all the cytokinin receptors in plants and their downstream associated signalling pathways. The indirect effect of TDZ is considered to be its ability to inhibit the enzyme cytokinin oxidase/dehydrogenase which degrades cytokinins. This should lead to the elevation of endogenous cytokinin levels; however, it is not possible to distinguish whether the cytokinin effect was the effect of TDZ or the effect of endogenous cytokinins, since both share the same binding site in the proteins and the mechanism of action.

Keywords

Thidiazuron · *Arabidopsis* · Cytokinin oxidase/dehydrogenase · Horticulture · Agriculture · Senescence · Stress tolerance · Defoliation · Cotton · Cut flower · Ethylene

J. Nisler (✉)
Laboratory of Growth Regulators, Centre of the Region Hana for Biotechnological and Agricultural Research, Institute of Experimental Botany, AS CR & Palacký University, Slechtitel u 27, Olomouc, 783 71, Czech Republic
e-mail: jaroslav.nisler@upol.cz

N. Ahmad, M. Faisal (eds.), *Thidiazuron: From Urea Derivative to Plant Growth Regulator*, https://doi.org/10.1007/978-981-10-8004-3_2

2.1 TDZ: Cytokinin Activity in *Arabidopsis*

TDZ belongs to bis-substituted urea derivatives, which showed very strong cytokinin activity. Among them 1-(2-chloro-pyridin-4-yl)-3-phenyl-urea (CPPU) was the first compound proved to bind specifically to the same site at cytokinin-binding protein from mung bean (*Vigna radiata*) as adenine-type cytokinins (Arima et al. 1995). Six years later, the identification of cytokinin receptor in *Arabidopsis* was reported, and it was shown that the receptor is activated also by TDZ (Inoue et al. 2001). The protein called cytokinin response 1 (CRE1) is a sensor histidine kinase which autophosphorylates after the activation by a cytokinin molecule (Inoue et al. 2001). Interestingly, CRE1 was found to be identical to *Arabidopsis* histidine kinase 4 (AHK4) (Ueguchi et al. 2001) and to Wooden leg receptor protein, which is required for root development in *Arabidopsis* (Mähönen et al. 2000). Later, Spíchal et al. (2004) clearly demonstrated that TDZ activates also the other *Arabidopsis* cytokinin receptor AHK3 in bacterial test system (Suzuki et al. 2001; Yamada et al. 2001). These authors quantified the response of CRE1/AHK4 and AHK3 to TDZ and compared it with other naturally occurring cytokinins. TDZ strongly activated both receptors and, interestingly, activated the receptor AHK3 more effectively than *trans*-zeatin (*t*Z), being the best activator of this receptor from all the tested cytokinins.

Using the same bacteria, expressing cytokinin receptors AHK3 or CRE1/AHK4, in a competition assay, it was demonstrated that TDZ binds to the same binding sites in both receptors as *t*Z (Romanov et al. 2006). Finally, these data were completed by Stoltz et al. (2011) who showed that TDZ possesses also a very high affinity to the ligand-binding (called CHASE) domain of the last *Arabidopsis* cytokinin receptor AHK2. Actually, the K_D of AHK2-CHASE domain and TDZ (2.5 nM) was lower than the K_D of AHK2-CHASE domain and *t*Z (4.0 nM). TDZ showed again the highest affinity to this receptor from all cytokinins tested. More accurate data were obtained when AHK2, AHK3 and ZmHK1 (the cytokinin receptor from *Zea mays*) were expressed in full length in a plant assay system (Lomin et al. 2015). In this assays, TDZ exhibited higher K_D with cytokinin receptors than *t*Z, but only to a minor extent, having still a very high affinity to all cytokinin receptors, including ZmHK1, which is a maize orthologue of AHK4.

Last doubts about whether the urea- and adenine-derived cytokinins bind to the same binding site were refuted by Hothorn et al. (2011) who achieved to co-crystalize TDZ with CRE1/AHK4. The phenyl moiety of TDZ binds to the same pocket as isoprenoid tail of *t*Z, and the two nitrogens of thiadiazolyl group mimic the polar interactions of N^9 and N^3 of adenine ring. The urea moiety of TDZ establishes polar interactions which are very similar to those observed for N^7 and N^6 in the adenine-type complexes (for more details, see Hothorn et al. 2011). This co-crystal thus rationalizes why the urea derivatives are capable to bind to the cytokinin binding site in the receptor, which also determines their cytokinin activity.

In *Arabidopsis*, it was shown that the activation of CRE1/AHK4 is responsible for the inhibition of root growth associated with the use of classical cytokinins (Riefler et al. 2006). Accordingly, we have observed that TDZ strongly inhibited the root

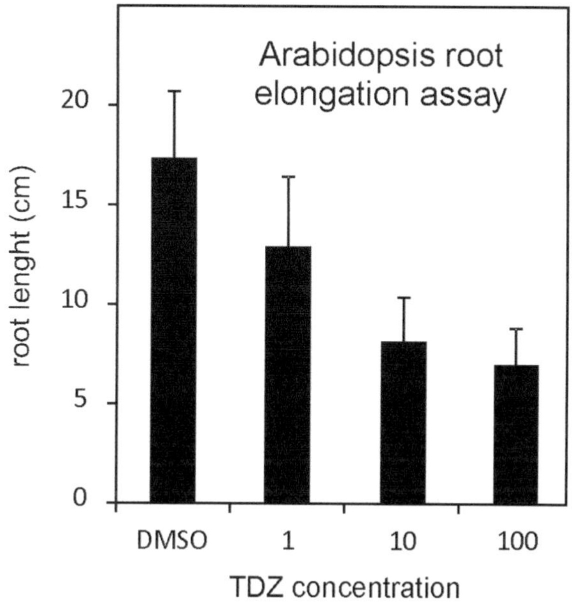

Fig. 2.1 The effect of TDZ on root growth of *Arabidopsis* (14 days old) grown on half-strength MS medium (supplemented with 0.1% sucrose and 6 g/L Phytagel) containing TDZ or 0.01% DMSO (22 °C, 16/8 light/dark, 130 μM m^{-2} s^{-1}) ($n = 40$)

growth and development of *Arabidopsis* seedlings from 1 nanomolar concentration (Fig. 2.1). TDZ also retards the dark-induced senescence in detached *Arabidopsis* leaves (from 1 micromolar concentration, our unpublished data), which has been shown to be mediated mainly by the activation of AHK3 and through specific phosphorylation of a cytokinin response regulator *ARR2* (Kim et al. 2006). Our group reported that TDZ activates the expression of *ARR5* gene in *Arabidopsis* seedlings (Spíchal et al. 2004; Nisler et al. 2016). The *ARR5* is a primary response regulator with a cytokinin-dependent promoter (D'Agostino et al. 2000), the activation of which integrates the responses of several putative cytokinin signalling pathways and therefore does not distinguish between the contributions of individual cytokinin receptors. Notably, TDZ was the most active cytokinin in this assay, exceeding again the activity of cytokinins *tZ* and N^6-benzyladenine (BA) (Spíchal et al. 2004; Nisler et al. 2016). It was also shown earlier that TDZ (and CPPU) induced the expression of *IBC6* and *IBC7* (induced by cytokinin) genes, now known as type-A response regulatory genes *ARR5* and *ARR4*. In *Arabidopsis* genome, 23 genes encoding cytokinin response regulators (*ARRs*) – divided into type A and type B – were found (e.g. Schaller et al. 2008). The transcription of all members of the type-A family, but not the type-B family, is rapidly induced by cytokinins (Brandstatter and Kieber 1998; Taniguchi et al. 1998; D'Agostino et al. 2000). Some of type-A ARRs negatively regulate cytokinin signalling and thus participate in a negative feedback loop (Kiba et al. 2003; To et al. 2004; Leibfried et al. 2005; To et al. 2007). When type-B ARRs are phosphorylated, they act as transcription factors which bind to target DNA sequences and activate transcription of target genes. The same role is also fulfilled by cytokinin response factors (Rashotte et al. 2006). Modern genetic methods, such as microarrays, revealed in *Arabidopsis* more than 3000 genes in total which are regulated by cytokinins (reviewed in Brenner et al. 2012).

To conclude, and as reviewed also in this chapter, there are numerous proofs that at least in *Arabidopsis*, TDZ functions as a regular cytokinin whose effect is mediated through the activation of cytokinin receptors and downstream-associated signalling pathways. TDZ strongly exhibits properties common to all cytokinins; it is therefore not a mystery, as many authors state, how TDZ functions in plants. If there are some doubts or uncertainties regarding the TDZ mode of action and function in plants, those are not related to TDZ but to cytokinins in general.

2.2 Inhibitory Strength of TDZ Towards Cytokinin Oxidase/Dehydrogenase

Naturally occurring cytokinins are adenine derivatives with a distinct N^6 side chain. In plants cytokinin levels are regulated by the enzyme cytokinin oxidase/dehydrogenase (abbreviated as CKO or CKX, EC 1.5.99.12). CKX catalyses the irreversible oxidative breakdown of cytokinins to form adenine/adenosine and the corresponding aldehyde (Whitty and Hall 1974; Brownlee et al. 1975; Chatfield and Armstrong 1986). Model plant *Arabidopsis thaliana* contains seven CKX isoforms, of which AtCKX2 is the most active and well-studied one (Galuszka et al. 2007).

The negative effect of TDZ on CKX activity was firstly observed by Chatfield and Armstrong (1986) in callus tissue of *Phaseolus vulgaris*. TDZ inhibited breakdown of N^6-(Δ^2-isopentenyl) adenine-8-^{14}C to the same extent as the excess of unlabelled N^6-(Δ^2-isopentenyl) adenine. The authors hypothesized that TDZ could be the substrate of CKX and therefore compete with cytokinin in the enzyme active site, but other mechanisms of inhibition were not excluded. Hare and Van Staden (1994) studied the mechanism of how TDZ inhibits the activity of CKX from soybean callus and suggested that it is a mixed and predominantly uncompetitive mechanism. K_i value for the inhibition was 168 μM. More recent studies have shown that phenyl urea derivatives, including TDZ, are competitive inhibitors of CKX (Bilyeu et al. 2001; Kopečný et al. 2010, Nisler et al. 2016). Monocotyledonous *Zea mays* contains 13 CKX isoforms, of which ZmCKX1 is the best studied isoform and which has been shown to play a crucial role in cytokinin degradation in maize (Houba-Hérin et al. 1999; Morris et al. 1999). Bilyeu et al. (2001) provided the molecular and biochemical characterization of ZmCKX1 and determined the K_i values for several urea derivatives with this enzyme. Compound F-PU (K_i = 0.1 μM, Table 2.1) which is CPPU with fluorine in *meta*-position of phenyl ring and CPPU (K_i = 3 μM, Table 2.1) were better inhibitors than TDZ (K_i = 8 μM). Kopečný et al. (2010) further identified several new potent inhibitors of ZmCKX1 derived from CPPU and showed, as well as the previous work, that TDZ is a rather weaker inhibitor of ZmCKX1 when compared to CPPU and its derivatives. The most recent work showed that specific substitutions on phenyl ring of TDZ lead to the improvement of the CKX inhibitory activity of TDZ molecule towards AtCKX2, ZmCKX1 and ZmCKX4a (Nisler et al. 2016). Specifically, methyl hydroxy-moiety (compound 8)

Table. 2.1 Structures and designation of substances and reference

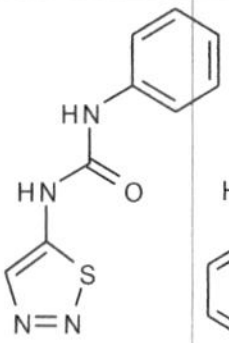	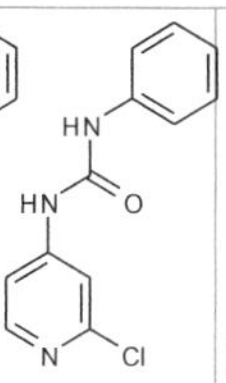	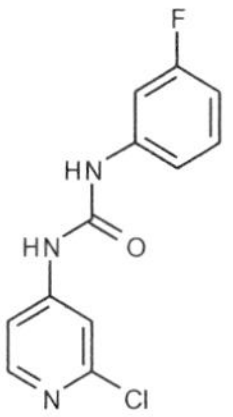	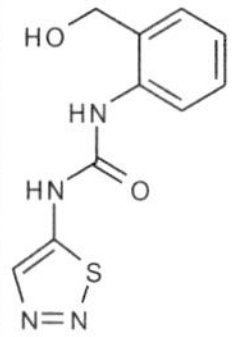	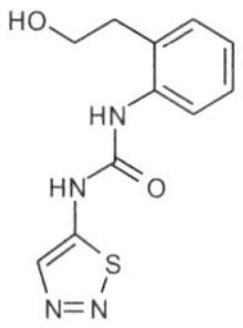	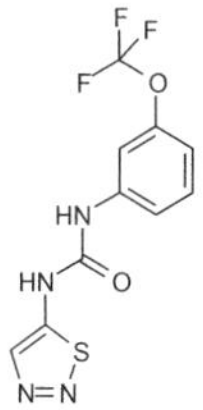
TDZ	CPPU	F-PU	Comp. 8	HETDZ	3FMTDZ
		Bilyeu et al.	Nisler et al.	Nisler et al.	Nisler et al.

and ethyl hydroxy-moiety (compound HETDZ) in *ortho*-position and trifluoromethoxy group in *meta*-position of phenyl ring (compound 3FMTDZ) caused the most evident increase in CKX inhibition (structures in Table 2.1). On the other hand, but in accordance with previous literature (Bruce and Zwar 1966; Mok et al. 2005), any substitutions on the TDZ phenyl ring lowered the cytokinin activity of the molecule when compared to TDZ (Nisler et al. 2016).

To conclude, TDZ in addition to a direct cytokinin effect at the receptor level can increase endogenous cytokinin concentrations in plants, which together could result in an enhanced plant response to cytokinin signal. However, TDZ is apparently a weak inhibitor of CKX enzymes, when compared to the other urea-type compounds. In addition to that, it was reported that application of exogenous cytokinin induces strong expression of CKX genes (Mik et al. 2011). Therefore, it is very possible that the level of endogenous cytokinins decreases in plants after the TDZ treatment and that the plant response to cytokinin is mainly the response to TDZ. Indeed, strong cytokinin effect of TDZ is most probably associated with its high ability to activate cytokinin receptors and the inability of CKX to degrade it. This is the most obvious difference between natural and synthetic urea-based cytokinins and the probable reason why the synthetic cytokinins exhibit stronger effects in plants than the natural ones.

2.3 TDZ: Mode of Action, Uses and Potential in Agriculture

2.3.1 Fruit Crop Improvement

A defining characteristic of cytokinins is their ability to promote plant cell division (cytokinesis) (Miller et al. 1955). Most recent works revealed that cytokinins are implicated in controlling the cell cycle and specifically in both the G1/S and G2/M transitions. In G1/S transition, cytokinins were shown to induce the expression of the three D-type cyclin genes (Riou-Khamlichi et al. 1999; Scofield et al. 2013), which control the type of cycle, favouring the mitotic cell cycle over endo reduplication (Dewitte et al. 2007). It was suggested that D-type cyclins are the key targets of cytokinins in G1/S transition (Dewitte et al. 2007). Despite this, it is assumed that a major role of cytokinins in the cell cycle may be in regulating the G2/M transition (Schaller et al.

2014). For example, it was shown that zeatin is indispensable for the G2/M transition in synchronized tobacco BY-2 cells (Laureys et al. 1998) and that the G2/M transition is blocked in these cells when treated with lovastatin, an inhibitor of cytokinin biosynthesis (Redig et al. 1996). In the G2/M transition, the cytokinins seem to play a crucial role, although the mechanism of this regulation at molecular level is not yet clear.

The cell division promoting activity of cytokinins is widely used in horticulture and is very well documented. The most efficient and best known of all compounds tested so far is CPPU, known as forchlorfenuron or KT-30 (Jo et al. 2003). CPPU is the active ingredient of several currently marketed formulations used for the improvement of various fruit crops. Shortly after the discovery of CPPU in 1978, it was shown by Mok et al. (1982) that the TDZ activity in *Phaseolus* callus assay was a little higher than the activity of CPPU, whose activity was similar to that of *t*Z. Since that time both compounds were tested for a fruit yield improvement. The effect of TDZ and CPPU on several parameters of wine grapes (*Vitis vinifera*) was firstly tested and compared in Canada (Reynolds et al. 1992). Both plant growth regulators when applied on wine clusters by dipping them in the compound solutions linearly with the concentration increased the cluster and berry weight in either fresh or stored clusters of all four tested varieties (Reynolds et al. 1992). The number of berries per cluster was not affected. The grapes exhibited generally delayed ripening, which was reflected by lower sugar and anthocyanin contents and higher content of titratable acids. The effect of TDZ appeared to be equal to, or better than, that of CPPU. A similar research was conducted with kiwifruits (*Actinidia deliciosa*), which were also dipped in TDZ or CPPU (Famiani et al. 1999). It was found that TDZ remarkably increased fruit growth soon after the application, and at harvest, the fruits were by 50–60% heavier than untreated ones. Contrary to the previously mentioned study, the fruits exhibited faster ripening, which was demonstrated by a higher content of soluble solids and sugars and lowered total titratable acidity. CPPU had the same effect as TDZ. The positive effect of TDZ on yield forming parameters was also observed on cucumbers (*Cucumis sativus*) (Yang et al. 1992), persimmon fruit (*Diospyros kaki L.*) (Itai et al. 1995), pears (*Pyrus communis*) (Petri et al. 2001; Stern et al. 2003) and apples (Greene 1995; Petri et al. 2001). In all cases a significant enlargement of fruits was observed. The effect of TDZ on the yield of pear trees was examined during five consecutive years (1998–2002) in Israel. The increase in fruit size was accompanied by a slight thinning, but the fruit shape, or return yield in the following year, was not affected (Stern et al. 2003). It is known that the chemical thinning of post-bloom apples improves the fruit size and increases return bloom (e.g. Forshey 1987). A strong thinning effect of TDZ was observed on apple trees, which however in consequence of that had lower yield (Elfwing and Cline 1993). Interestingly, this effect was stronger than that of BA and was probably caused by a too high concentration of TDZ. Other authors reported that TDZ during seven growing seasons, every year, significantly increased the fruit set and fruit weight on apple trees. These apples had reduced the seed number and the calcium content per fruit, but increased the fruit firmness (Petri et al. 2001).

Abad et al. (2004) published a comprehensive study on the effect of 17 phenyl-fluorinated analogues of TDZ on the growth and quality of kiwifruits. Two different concentrations (10 and 25 ppm) were tested. TDZ-treated fruits were by 31% (10 ppm) and 38% (25 ppm) heavier than the untreated ones. Interestingly, most of the fluorinated TDZ analogues enhanced the fruit size to the similar extent as TDZ. The highest increase in weight was achieved by the application of 2,3,5,6-tetrafluorinated TDZ derivative at 25 ppm (at harvest, fruits were by 58% heavier than the untreated ones) followed by a 3,5-difluorinated TDZ derivative at 10 ppm (weight increase by 50%). In general, the treatments did not significantly affect the fruit shape or the number of seeds and consistently with previous results accelerated the fruit ripening; however, some of the more active derivatives delayed postharvest maturation (for more details see Abad et al. 2004).

In summary, it is reported that TDZ as well as CPPU increases the fruit size by stimulating cell division and expansion in the early stages of fruit development and thus have a positive effect on overall yield. TDZ can be a powerful tool for improving fruit cropping, fruit quality and storage potential in a wide range of fruit species. However, it seems that TDZ and CPPU exhibit very similar effects, but CPPU is historically older and commercially cheaper, which is probably the reason why CPPU has been preferred at the market.

2.3.2 Improvement of the Vase Life of Cut Flowers

Postharvest senescence of cut flowers is a major limitation of their marketing; thus, considerable effort has been devoted to retard this process. Cutting the flowering stem from the mother plant causes stress-induced senescence in the cut flower, and it is accompanied and accelerated by an increased production of ethylene by tissue and/or its sensitivity to this phytohormone. Ethylene is a positive regulator of senescence in plants (Abeles et al. 1988; Reid 1995), and its endogenous production by plant organs is a highly regulated and age-dependent process (Jing et al. 2005). Increasing ethylene production occurs in senescing tissues and as a response to stress (El-Beltagy and Hall 1974), including cutting (references herein). Therefore, most of the efforts to improve postharvest quality of plant products were focused on diminishing the ethylene evolution or its sensing. The enhanced level of endogenous cytokinins as well as exogenously applied cytokinins was shown to delay the senescence (Gan and Amasino 1995; Rivero et al. 2007; Richmond and Lang 1957; Ferrante et al. 2002a, b) and therefore also the production of ethylene. Some of the mechanisms how cytokinins function in senescence have been described (Weaver et al. 1998; Jordi et al. 2000; Balibrea Lara et al. 2004; Kim et al. 2006). Several works provided evidence that there is a strong correlation between decreased cytokinin content in the leaves and the onset and the progression of their senescence (Van Staden 1973; Singh et al. 1992; Gan and Amasino 1997; Masferrer et al. 2002); therefore, supplying cytokinins to cut flowers to solve their postharvest senescence is the right strategy.

Historically, Richmond and Lang (1957) were the first who reported that cytokinin kinetin reduced protein and chlorophyll loss in detached *Xanthium* leaves. Later, several urea derivatives with cytokinin activity were found to delay chlorophyll degradation in radish (*Raphunus sutivus*) leaf discs in one micromolar concentration as well as kinetin (Kefford et al. 1973). The preventing effect of TDZ on leaf senescence was firstly reported quite late by Ferrante et al. (2001), although its high cytokinin activity had already been known for 20 years (Mok et al. 1982). Ferrante focused his attention on cut *Alstroemeria* flowers and found that TDZ is by far the most effective substance in retarding leaf yellowing (Ferrante et al. 2002a). While the leaves of control plants turned yellow after 6 days, 24 h pulse treatment with 10 μM TDZ kept the leaves green for more than 60 days. In contrast the second most effective hormone, applied by the same way in 100 μM concentration, was GA_3 which was effective approximately 30 days. However, Ferrante compared the effect of TDZ with GA_3 which was shown to have much lower activity in delaying yellowing of *Alstroemeria* leaves than gibberellins GA_4 and GA_7 (Jordi et al. 1995). In Ferrante's report, BA, CPPU and DPU were almost ineffective even in 100 μM concentration in preventing senescence of *Alstroemeria* leaves. Unfortunately, none of the tested compounds, including TDZ, prevented petal abscission of *Alstroemeria* flower, which is known to be ethylene insensitive (Ferrante et al. 2002a). In cut flowers of *Matthiola incana*, TDZ prolonged the fresh appearance of both the leaves and flowers, although the effect on leaf longevity was much more apparent than the effect on flower vitality (Ferrante et al. 2009). In the same plant, TDZ prevented also the degradation of carotenoids and the production of abscisic acid (Ferrante et al. 2004). In addition to ethylene, abscisic acid accumulation also accelerates senescence (Mayak and Halevy 1972). Abscisic acid occurs as a response to stress and is considered to be a marker of stress-induced senescence (Cutler et al. 2010). It was therefore demonstrated that TDZ reduces stress responses in the stems of cut flowers (Ferrante et al. 2004). In another work Ferrante et al. (2009) showed that the co-application of TDZ and GA_3 did not have any beneficial effects in *Matthiola incana*, although GA_3 alone slightly delayed leaves and petal senescence. Interestingly, combined application of TDZ (5 μM) and GA_3 (500 μM) induced higher production of ethylene by flowers and leaves and had a negative effect on leaf colour (Ferrante et al. 2009). From these results, it seems that high doses of GA_3 antagonized the anti-senescence effect of TDZ in *Matthiola incana*. Indeed it was shown that gibberellins inhibits cytokinin responses in *Arabidopsis*, and it was suggested that SPINDLY proteins mediate this hormone interaction (Greenboim-Wainberg et al. 2004). Yet, many commercial preparations for cut flowers contain a combination of cytokinins and gibberellins. This combination is probably advantageous when both phytohormones are used in appropriate concentrations.

On the other hand, there is clear evidence showing that the effect of TDZ can be enhanced by other factors or chemicals which prevent the progress of senescence. It was found that TDZ in combination with light significantly increased chlorophyll content in cut stock (*Matthiola incana L.*) flowers, cut chrysanthemum, tulips and lilies and also cut eucalyptus foliage when compared to the initial values measured at the harvesting time (Ferrante et al. 2002a, b, 2003, 2004, 2005, 2009, 2011). The effect of TDZ on dark-stored flowers was weaker. These results are not as surprising if we take into account that light inhibits leaf senescence (Wingler et al. 1998), and it is required for the synthesis of chlorophyll (Kraepiel and Miginiac 1997). It was further shown that cytokinins in combination with light accelerate biosynthesis of the photosynthetic complexes in developing chloroplasts (Yaronskaya et al. 2006). TDZ and light therefore exhibited synergistic effect in delaying senescence and moreover together stimulated biosynthesis of new chlorophyll in leaves of cut stems. There was also observed a synergistic effect of TDZ with the effect of sodium nitroprusside, a nitric oxide donor (Mortazavi et al. 2011). Nitric oxide is known to inhibit senescence by decreasing ethylene production (Leshem and Wills 1998). Both substances TDZ and nitric oxide when applied separately significantly decreased ethylene production in cut roses (*Rosa* sp.) – in concentration-dependent manner – and prolonged their shelf life. When applied together, no further decrease in ethylene production was observed; however, the longevity of flower was prolonged (Mortazavi et al. 2011). Very similar results were also reported for *Lilium* cut flowers (Kaviani and Mortazavi 2013). Sankhla et al. (2005) further reported that TDZ alone and sucrose alone can counteract the accelerating effect of ethylene on senescence of cut inflorescences of *Lupinus densiflorus* and that TDZ and sucrose have synergistic effect. Sucrose is a transportable form of carbohydrates in plants and provides energy supply for biological processes associated with active growth, metabolism and development. It was shown that radioactively labelled nutrients are preferentially transported and accumulated in cytokinin-treated tissue (Mothes and Engelbrecht 1963); hence, TDZ most probably facilitated the mobilization of sucrose in sink tissues, such as in senescent leaves of cut flowers. Later, the mechanism was clarified when it was shown that cytokinins upregulate the expression of extracellular cell wall invertase, an enzyme which has a crucial role in source-sink regulation (Balibrea Lara et al. 2004). Thus, in cut flowers, TDZ-induced sucrose mobilization exhibited an additional senescence delaying effect to the other effects of TDZ itself. It is in accordance with reports showing the suppressive effect of sucrose on ethylene production and on ethylene sensitivity (Mayak and Dilley 1976; Ichimura and Hiraya 1999). Sucrose is also a common part of commercial preparations for the extension of flower longevity. Sankhla et al. (2003) further reported that TDZ significantly enhanced the postharvest quality of perennial phlox flowers (*Phlox paniculata*) and induced the opening of additional flower buds during vase life. In this case, TDZ counteracted the negative effect of abscisic acid on flower abscission and leaf senescence (Sankhla et al. 2003).

Contrary to the above evidence, it was shown that TDZ can induce ethylene production in cut flowers. In experiments with Dutch iris cut flowers (*Iris* × *hollandica*), TDZ increased the evolution of ethylene by opening flower but also prolonged its vase life, stimulated the growth of the stem and the opening of second flower and delayed stem yellowing (Macnish et al. 2010). In *Pelargonium* cuttings, TDZ significantly increased ethylene evolution, but simultaneously induced strong expression of *PhETR1* (a negative acting ethylene receptor gene), thus decreasing the sensitivity of *Pelargonium* leaves to ethylene and having delaying effect on the onset of senescence (Mutui et al. 2007). Similar results were obtained with petunia (*Petunia x hybrida*) transformed with P_{SAG12}-IPT, which overproduce endogenous cytokinins in senescence period. The transformed flowers produced the same amount of ethylene as wild-type flowers but were less sensitive to exogenous ethylene than wild-type flowers and exhibited a much longer lifetime (Chang et al. 2003). It demonstrates that both exogenous cytokinin (TDZ) and endogenous cytokinins lowered the sensitivity of plants to ethylene and extended their lifetime. It is very possible that TDZ in addition to delaying ethylene production lowers the plant sensitivity to ethylene also in other species. The evidence given here further demonstrates that even in cases when TDZ enhanced ethylene production, the TDZ anti-senescence effect was prevailing over the senescence effect of TDZ-induced ethylene. This suggests that the action of cytokinins is superior to the action of ethylene in the regulation of senescence. In addition to the works reviewed in this chapter, several other authors had reported positive effect of TDZ on marketable traits of other cut and potted species. The complete overview is presented in Table 2.2.

Table 2.2 Overview of studies reporting on the effects of TDZ on extending leaf and flower longevity in various cut and potted species. Ordered alphabetically according to the first author

Cut flowers/potted species	Effect of TDZ/other (co)-applications/ hormone interaction	References
Alstroemeria hybrida	Retardation of leaf and flower senescence	Bagheri and Sedaghathour (2013)
	NAA, ethylene	
Rosa hybrida	Extension of flower vase life in one out of seven cultivars	Chamani et al. (2006)
	Induction of lateral shoot development	
Lilium sp.	Retardation of leaf and flower senescence	Dhiman et al. (2015)
	GA_3, BA	
Alstroemeria sp.	Retardation of leaf senescence	Ferrante et al. (2001)
	BA; GA_3	
Alstroemeria sp.	Retardation of leaf senescence	Ferrante et al. (2002a)
	4-CPPU; 1,3-DPU; BA; GA_3	
Eucalyptus parvifolia cut branches	Retardation of leaf senescence	Ferrante et al. (2002b)
	BA, ACC, AOA, CoCl2	
Tulipa sp. and *Chrysanthemum* sp.	Retardation of leaf senescence	Ferrante et al. (2003)
	Inhibition of rooting	
	Promotion of lateral shoot elongation	
Matthiola incana	Retardation of leaf senescence	Ferrante et al. (2004)
	Inhibition of carotenoid degradation	
	Inhibition of ABA production	
	8-HQS, ABA	
Chrysanthemum sp.	Retardation of leaf senescence	Ferrante et al. (2005)
	BA	
Matthiola incana	Retardation of leaf senescence and petal wilting	Ferrante et al. (2009)
	Light, ethylene, GA_3	
Lilium sp.	Retardation of leaf and flower senescence	Ferrante et al. (2011)
	BA	
Matthiola incana	Retardation of leaf and flower senescence	Ferrante et al. (2012)
	1-MCP, ethylene	
Pelargonium zonale	Retardation of leaf senescence	Hatami et al. (2013)
Potted plants	Enhancement of antioxidant enzyme activity	

(continued)

Table 2.2 (continued)

Cut flowers/potted species	Effect of TDZ/other (co)-applications/ hormone interaction	References
Alstroemeria hybrid	Retardation of leaf and flower senescence	Hatamzadeh et al. (2012)
	BA	
Freesia sp., *Ornithogalum* sp., *Euphorbia fulgens* potted *Geranium* sp.	Retardation of leaf senescence	Jiang et al. (2009)
	Delayed leaf abscission	
	Extended flower life and increased flower number	
Chrysanthemum morifolium	Retardation of leaf and flower senescence	Kaur and Singh (2015)
	Enhancement of antioxidant enzyme activity	
	Ethrel (contains CEPA), STS, BA	
Lilium sp.	Increased flower life and flower diameter	Kaviani and Mortazavi (2013)
	Nitric oxide	
Lilium sp.	Delayed leaf chlorosis	Leonard and Nell (2004)
	Chrysal BVB, fascination	
Iris × *hollandica*	Extension of flowers vase life	Macnish et al. (2010)
	Enhancement of flower opening GA_3	
Rosa hybrida	Extended flower life	Mortazavi et al. (2011)
	Nitric oxide, SNP	
Pelargonium cuttings	Retardation of leaf senescence	Mutui et al. (2003)
	Delayed onset of leaf senescence	
Potted *Pelargonium*	Delayed onset of leaf senescence	Mutui et al. (2005)
	Reduced root formation	
	Ethylene, ABA, darkness, IBA	
Pelargonium cuttings	Retardation of leaf senescence	Mutui et al. (2007)
	Reduction of sensitivity to ethylene by the induction of higher expression of the negative acting ethylene receptor – PhETR1 ethylene, ABA, darkness	
Phlox paniculata cut inflorescences	Retardation of leaf and flower senescence	Sankhla et al. (2003)
	Stimulation of opening of flower buds	
	Counteracted ABA effect	
Lupinus spp. cut inflorescences	Delayed onset of flower senescence	Sankhla et al. (2005)
	Sucrose, STS, 1-MCP, CEPA, ethylene	
Alstroemeria hybrida	Extension of the flowers vase life SA	Tirtashi et al. (2014)
Rosa sp.	Retardation of leaf and flower senescence	Talebi et al. (2013)
	SNP (nitric oxide donor)	
Polianthes tuberosa	Extension of flowers vase life	Uthairatanakij et al. (2007)
	Stimulation of opening of flower buds ethylene, sucrose	

1-MCP 1-methylcyclopropene, *AOA* aminooxyacetic acid, $CoCl_2$, *STS* silver thiosulfate (ethylene action inhibitors), *8-HQS* 8-hydroxyquinoline sulphate (biocide inhibiting growth of microorganism), *ABA* abscisic acid, *ACC* 1-aminocyclopropane-1-carboxylicacid, *CEPA* 2-chloroethylphosphonic acid (ethylene source), *BA* N^6-benzyladenine (cytokinin), *Ethrel* plant growth regulator containing CEPA, *GA_3* gibberellic acid, *IBA* indole-3-butyric acid (auxin), *NAA* naphthalene acetic acid (auxin), *SA* salicylic acid

The evidence reported herein shows that TDZ functions by the same way as natural cytokinins in delaying senescence. By supplying the cytokinin signal in plants, TDZ delays the production of ethylene and/or lowers the sensitivity of plant tissue to ethylene, which retards the senescence process. In light TDZ even stimulates chlorophyll biosynthesis which has an enormous positive additional effect. It has been hypothesized that the remarkable effect of TDZ, exceeding often the activity of adenine cytokinins, is most probably a combination of its high cytokinin activity, CKX inhibitory activity and high in vivo chemical stability (Ferrante et al. 2002a, b). The latter was supported by Mok and Mok (1985) who showed that [^{14}C] thidiazuron was not broken down in *Phaseolus lunatus* callus tissue over a 1-month period. Further work also provided evidence that the TDZ molecule remains stable in both free and conjugated forms within the plant tissues (Murch and Saxena 2001). To conclude, the evidence reported herein shows that TDZ is extraordinarily effective in improving the postharvest quality of many cut flowers and foliage plants. Since it is already registered as an agricultural chemical and it is available on the market, it has a great commercial potential as a senescence retardant in many species.

2.4 Improvement of Stress Tolerance and Yield of Main Crops

Chloroplasts are among the main targets of cytokinin action in the plant cell. Cytokinins promote synthesis and prevent the degradation of many components, which are implicated in chloroplast biosynthesis; therefore, they are crucial for photosynthesis and plant development (e.g. Flores and Tobin 1988; Chory et al. 1994; Kusnetsov et al. 1994, Yaronskaya et al. 2006; Chernyad'ev 2009). By this action, cytokinins enhance photosynthetic rate (P_N) and efficiency (Debata and Murty 1981; Yoon and Richter 1990) which can improve the production and yield in crop plants (reviewed by Chernyad'ev 2009). There are more than 50 scientific works providing evidence that cytokinins such as kinetin and BA improved the grain yield of main agricultural crops, such as wheat, barley, rice, maize, etc. at optimal or stress conditions (reviewed by Koprna et al. 2016). Most of the works reported the cytokinin induced delayed senescence, enhanced photosynthetic activity and stress tolerance, which in most cases lead to the yield improvement. Unexpectedly, there are only a few similar reports regarding the effect of TDZ on main crop species. In the first study, the effect of TDZ on grain yield of spring wheat (*Triticum aestivum*) grown in pots in optimal or under salinity stress conditions was evaluated (Beckett and van Staden 1992). TDZ applied during floret initiation reduced the yield of unstressed plants, had little effect on plants receiving 50 mM NaCl and increased the yield of plants receiving 70 mM NaCl. TDZ generally reduced the number of spikelets and grains per ear, but increased the mean weight of seeds preferably in stressed plants. Chernyad'ev (1994) studied the effect of TDZ and BA on P_N and photosynthetic enzyme activities (e.g. RuBisCO) in sugar beet (*Beta vulgaris*), Pea (*Pisum sativum*), meadow fescue (*Festuca pratensis*) and reed fescue (*Festuca arundinacea*). In all tested plants both cytokinins enhanced the activity of all studied enzymes and increased the P_N, with BA having always at least a little better effect than TDZ (Chernyad'ev 1994). The effect on the overall yield was not determined.

In the most recent work, the effect of combined application of TDZ, paclobutrazol and ascorbic acid on physiological traits of four wheat genotypes under optimal and water deficit conditions was evaluated (Dwivedi et al. 2017). The results indicate that the application of the mixture enhanced P_N and had positive effect on yield parameters in wheat under both conditions. Unfortunately, the authors did not evaluate the effects of individual components; therefore, the positive contribution of TDZ cannot be estimated. A very similar research was conducted by Nagar et al. (2015), who showed that BA increased the grain yield in drought-tolerant and drought-susceptible wheat cultivars under both optimal and water deficit conditions. It has been shown that the positive result is related to the higher assimilation of nitrogen, because BA enhanced the activity of nitrate reductase and glutamine synthetase, the two major enzymes of nitrogen assimilation pathway. The positive regulatory role of cytokinins in nitrogen assimilation has been described (Taniguchi et al. 1998; Sakakibara 2003; Brenner et al. 2005); hence, it can be assumed that TDZ could have the same positive effect on yield parameters as other cytokinins. Interestingly, there is also a work showing that TDZ delayed the progress of leaf necrosis induced by early blight (*Alternaria solani*), a disease in potato (*Solanum tuberosum*), and exhibited a positive synergistic effect with some fungicides on the inhibition of the necrosis progress (Pavlista 2003). Apparently, TDZ protects chlorophyll in plant leaves regardless of the origin of the senescence initiator.

Although the benefits associated with the use of cytokinins were proven, they do not have a stable place among commercialized plant growth regulators, possibly due to the complexity of their effects. Cytokinins inhibit root development (Werner et al. 2003), which is an undesirable side effect. However, this can be overcome by the application of cytokinins in later developmental stages. It can be concluded that the potential of cytokinin utilization in typical agriculture attracted a lot of attention and that TDZ is in this regard much neglected. Given that TDZ showed much higher cytokinin activity than BA or kinetin (Mok et al. 1982; Ferrante et al. 2002a; Spíchal et al. 2004, Nisler et al. 2016), I am convinced that there is an extraordinary potential of TDZ in this field, which is unfortunately still waiting to be discovered. In my opinion, TDZ could also represent a useful tool in ameliorating the negative effects of biotic stresses, such as plant diseases which disrupt the plant photosynthetic activity. However, this remains to be evaluated by future studies.

2.5 Defoliation of Cotton

It is thought that TDZ has been invented for the defoliation of cotton, which is one of the most important steps prior to its mechanical harvesting. However, the first report on TDZ molecule came from a patent describing nine 1,2,3-thiadiazole-5-yl urea derivatives as growth retardants for grasses, beans and peanuts (Schulz and Arndt 1973). Another work showed that TDZ, as well as some of the effective herbicides, uncouples ATP formation in isolated chloroplasts or mitochondria of spinach (*Spinacia oleracea*) and therefore inhibits energy conservation in respiration and photosynthesis (Hauska et al. 1975). Only a year later Arndt et al. (1976) for the first time presented a cotton defoliating activity of TDZ at an annual meeting of the American society of plant physiologists in New Orleans. It has been known for a long time that abscission inducing or accelerating substances function through promoting ethylene evolution (Abeles 1966). The same mechanism of action had been proposed for TDZ by Suttle (1983), who later showed that TDZ induced ethylene evolution in mung bean (*Vigna radiata*) hypocotyls and that the ethylene is synthesized from methionine *via*1-aminocyclopropane-1-carboxylic acid (ACC) in methionine pathway (Suttle 1984). In cotton it was demonstrated that TDZ induced a massive production of ACC and its oxidation to ethylene (Suttle 1986). In another work Suttle (1985) demonstrated that TDZ induced a large production of endogenous ethylene by leaf blades and abscission zone explants. The ethylene then activated the abscission zone tissue between the plant stem and the petioles of the leaves which caused a leaf drop. Inhibitors of ethylene action (silver thiosulfate, hypobaric pressure) or ethylene synthesis (aminoethoxyvinylglycine) resulted in an inhibition of TDZ-induced defoliation of cotton seedlings. TDZ stimulated an abscission of the younger leaves of cotton from a concentration of 10 μM, while the oldest true leaves required a higher concentration (100 μM) and exhibited a lower rate of total abscission. Interestingly, the leaves abscised green, suggesting that senescence was not the physiological basis for ethylene evolution and leaf abscission (Suttle 1985). It was suggested that previously registered defoliants with herbicidal activity are toxic for plants, therefore causing stress-induced overproduction of ethylene and consequently leaf senescence and abscission, while TDZ stimulates ethylene evolution as a result of hormone-like activity of TDZ itself (Suttle 1984, 1985, 1986). It was supported by the fact that herbicides are applied in much higher amounts (1–3 kg/ha) than TDZ (50–250 g) to achieve the same result. Other proofs came shortly after. It had been known that TDZ exhibited cytokinin-like activity, the same as adenine-type cytokinins; therefore, their effect on cotton seedlings was also examined (Suttle 1986; Grossmann 1991). It was found that all cytokinins tested (4 PU-30, TDZ, BA, tZ, iP, Kin) stimulated the production of ethylene in cotton seedlings and concluded that the induction of leaf abscission in cotton is a common effect of urea- and adenine-type cytokinins (Suttle 1986; Grossmann 1991). Two other works also reported on stimulation of ethylene production in mung bean seedlings by common cytokinins (Yu et al. 1981; Yip and Yang 1986), further supporting this conclusion. Later, more mechanistic explanation of how cytokinins are involved in ethylene biosynthesis was described. It was suggested that low concentrations of cytokinin post-transcriptionally increase the activity of aminocyclopropane-1-carboxylic acid synthase 4 and 5 by relieving the negative effect of the carboxy-terminus of the protein (Vogel et al. 1998; Woeste et al. 1999).

Later works were focused on the effect of TDZ on cotton yield and fibre quality. It was reported that TDZ-induced leaf abscission facilitates light penetration which in return accelerates the maturity and opening of bolls (Malik et al. 1991), while the yield and quality of fibre were maintained (Malik and Din 1997; Malik et al. 2002). It was found that the plant developmental stage and weather conditions at the time of application are the major factors that limit the efficiency of a defoliation process. The best conditions are when water turgor in leaves is high as well as temperature and humidity are high (Cathey 1986). Malike et al. (2002) further specified that the atmospheric minimum and maximum temperatures of 16 and 30 °C are the most suitable for the defoliation process. In fact, the use of TDZ for the defoliation of cotton is the TDZ only currently commercial utilization.

2.6 TDZ, Cytokinins and Ethylene in Plant Development

It had been shown that cytokinins enhanced ethylene production in mung bean and cotton seedlings while inhibiting its production in cut flowers. This might seem a little confusing; thus, in this part I will explain this contradiction.

All plant tissues can produce ethylene, although the production rate is normally very low (Yang and Hoffman 1984). During the life of a plant, ethylene production is induced during germination, fruit ripening, leaf abscission, flower opening and senescence. Ethylene production can also be induced by wounding, environmental stresses and other hormones including cytokinin. Cary et al. (1995) showed that BA-induced inhibition of root and hypocotyl elongation in *Arabidopsis* seedlings is mediated by ethylene. Subsequently, it was demonstrated that exogenously applied cytokinin to etiolated *Arabidopsis* seedlings led to an increased ethylene production, and it was suggested that cytokinins increase the activity of the ethylene biosynthesis enzymes (Vogel et al. 1998; Woeste et al. 1999). By this mechanism TDZ most probably induces the evolution of ethylene in cotton seedlings and leaf drop. Interestingly, it was shown that the progress of cotton defoliation is not uniform, because young leaves abscise more easily then the mature ones upon TDZ or ethylene treatment (Suttle 1985). These works provide evidence that ethylene modulates some of the cytokinin effects in the early stages of plant development and suggest that seedlings and young leaves have higher sensitivity to ethylene. Later, it was proposed that the ethylene-induced faster abscission of young over mature cotton leaves is primarily a result of the limited abscission-inhibiting effects of auxin in the abscission zone of the younger leaves (Suttle and Hultstrand 1991). In other words, higher auxin level in mature leaves of cotton inhibited their abscission. The same hormonal balance was found in leaves during senescence; the auxin concentration declined and tissue sensitivity to ethylene increased (Brown 1997). In addition to that, it seems that also different plant species possess different sensitivity to TDZ and/or ethylene. It was shown that plants from the *Malvaceae* family (to which cotton belongs to) are more sensitive to TDZ than, e.g. maize (*Zea mays*) and soybean (*Glycine max*) (Hodgson and Snyder 1988).

It can be stated that in the early stages of plant development cytokinin-induced leaf abscission is mediated by ethylene to which young leaves are more sensitive than the mature ones, because of their lower content of auxin. This applies at least to cotton seedlings; however, the situation in other plant species can be different. Nevertheless, the statement is consistent with observations showing that the effect of ethylene on leaf senescence is under the direct influence of age-related changes

(Hensel et al. 1993; Grbić and Bleecker 1995; Jing et al. 2005). Most of the components that control age-related changes in leaves are still unclear. Parameters such as the photosynthetic capacity of the leaf (Thomas and Howarth 2000) and balanced level of cytokinins (Gan and Amasino 1997) seem to play a dominant role in determining the time to senescence. It was shown that both TDZ and adenine-type cytokinins lower the plant sensitivity to ethylene in mature leaves (Chang et al. 2003; Mutui et al. 2007). In *Pelargonium* TDZ induced the expression of *PhETR1* (a negative acting ethylene receptor gene) (Mutui et al. 2007). These results suggest that cytokinins and ethylene possess antagonistic effects in the later stages of plant development, including senescence. During leaf senescence, the level of auxins and cytokinins is decreasing (Brown 1997; Gan and Amasino 1995) alongside with the increasing production of ethylene (Brown 1997; Hunter et al. 1999). This is in accordance with Schippers et al. (2008), who suggested that, while cytokinins are increasing the sink strength of young tissue, ethylene promotes the source strength of old leaves. This is also the reason why the application of sucrose lowers ethylene production in cut flowers and why the co-application of TDZ and sucrose has synergistic effect. Another aspect of this is the fact that in senescence, TDZ and cytokinins actually do not inhibit the ethylene production in itself but, by delaying the onset of senescence, the ethylene production is delayed. In senescence, the primary stimulus of ethylene production is not a cytokinin but rather age-related changes preceding the senescence. Even though some authors observed an elevated ethylene evolution after a TDZ treatment, they always reported that the anti-senescence effect of TDZ was prevailing (Macnish et al. 2010; Mutui et al. 2007). These results demonstrate that the ethylene role is inferior to the role of cytokinins in senescence.

In summary, while ethylene mediates the plant response to cytokinin in seedlings, these two phytohormones have opposite roles in mature and senescent plant organs. It should be noted that the plant responses to ethylene vary considerably during plant development and between species.

2.7 Conclusion

The aim of this chapter was to review and discuss the current use and potential of TDZ in agriculture and the physiological mechanisms behind it and to couple these mechanisms with the action of cytokinins. This is because I am strongly convinced that the TDZ effects observed so far in many plant species are associated with the ability of TDZ to trigger the cytokinin responses through cytokinin signalling components. This is apparent from the evidence given in part 1. From other parts of this chapter, it is clear that every TDZ effect has been observed also with the use of adenine cytokinins. It was further suggested that TDZ owes its high cytokinin activity to its high chemical stability in plants. This is the reason why TDZ exhibits strong and long-lasting effects, when compared to degradable natural cytokinins. These unique characteristics were shown to be useful in promoting fruit growth, improving life of cut flowers and enhancing stress tolerance and yield of important crops. On the other hand, I have to say that it is striking how much effort has been devoted to TDZ utilization in in vitro cultures and the other herein mentioned fields in comparison with the lack of studies regarding the use of TDZ on our main agricultural crops. The latter represents a large gap in the present knowledge, which should be covered in the near future to the benefit of all.

Acknowledgements I would like to thank professor Miroslav Strnad and Lukáš Spíchal for the valuable comments, which contributed to the final version of this chapter. I wish to thank my wife Tereza Nislerová for the English corrections.

References

Abad A, Agulloä C, Cunat AC et al (2004) Preparation and promotion of fruit growth in kiwifruit of fluorinated N-phenyl-N′-1,2,3-thiadiazol-5-yl ureas. J Agric Food Chem 52:4675–4683

Abeles FB (1966) Mechanism of action of abscission accelerators. Physiol Plant 20:442–454

Abeles FB, Dunn LJ, Morgens P et al (1988) Induction of 33-kD and 60-kD peroxidases during ethylene-induced senescence of cucumber cotyledons. Plant Physiol 87:609–615

Arima Y, Oshima K, Shudo K (1995) Evolution of a novel urea-type cytokinin: horticultural uses of forchlorfenuron. Acta Hortic 394:75–84

Arndt F, Rusch R, Stilfried HV (1976) SN 49537, a new cotton defoliant. Plant Physiol 57:S–99

Bagheri H, Sedaghathour S (2013) Effect of thidiazuron and naphthalene acetic acid (NAA) on the vase life and quality of cut *Alestroemeria hybrida*. J Ornamental Hort Plants 3:111–116

Balibrea Lara ME, Gonzalez Garcia MC, Fatima T et al (2004) Extracellular invertase is an essential component of cytokinin-mediated delay of senescence. Plant Cell 16:1276–1287

Beckett RP, van Staden J (1992) The effect of thidiazuron on the yield of salinity stressed wheat. Ann Bot 70:47–51

Bilyeu KD, Cole JL, Laskey JG et al (2001) Molecular and biochemical characterization of a cytokinin oxidase from maize. Plant Physiol 125:378–386

Brandstatter I, Kieber JJ (1998) Two genes with similarity to bacterial response regulators are rapidly and specifically induced by cytokinin in Arabidopsis. Plant Cell 10:1009–1019

Brenner WG, Ramireddy E, Heyl A et al (2012) Gene regulation by cytokinin in *Arabidopsis*. Front Plant Sci 3:8. https://doi.org/10.3389/fpls.2012.00008

Brenner WG, Romanov GA, Kollmer I et al (2005) Immediate early and delayed cytokinin response genes of *Arabidopsis thaliana* identified by genome wide expression profiling reveal novel cytokinin sensitive processes and suggest cytokinin action through transcriptional cascades. Plant Mol Biol 44:314–333

Brown KM (1997) Ethylene and abscission. Physiol Plant 100:567–576

Brownlee BG, Hall RH, Whitty CD (1975) 3-Methyl-2-butenal: an enzymatic degradation product of the cytokinin, N-6-(delta-2 isopentenyl) adenine. Can J Biochem 53:37–41

Bruce MI, Zwar JA (1966) Cytokinin activity of some substituted ureas and thioureas. Proc R Soc Lond Ser B 165:245–265

Cary AJ, Liu W, Howell SH (1995) Cytokinin action is coupled to ethylene in its effects on the inhibition of root and hypocotyl elongation in *Arabidopsis thaliana* seedlings. Plant Physiol 107:1075–1082

Cathey GW (1986) Physiology of defoliation in cotton production. In: Mauney JR, Stewart JM (eds) Cotton physiology. The Cotton Foundation, Memphis, pp 143–154

Chamani E, Irving DE, Joyce DC et al (2006) Studies with thidiazuron on the vase life of cut rose flowers. J Appl Hortic 8:42–44

Chang H, Jones ML, Banowetz GM et al (2003) Overproduction of cytokinins in petunia flowers transformed with P_{SAG12}-IPT delays corolla senescence and decreases sensitivity to ethylene. Plant Physiol 132:2174–2183

Chatfield JM, Armstrong DJ (1986) Regulation of cytokinin oxidase activity in callus tissues of *Phaseolus vulgaris* L. cv Great Northern. Plant Physiol 80:493–499

Chernyad'ev II (1994) Effect of 6-benzylaminopurine and thidiazuron on photosynthesis in crop plants. Photosynthetica 30:287–292

Chernyad'ev II (2009) The protective action of cytokinins on the photosynthetic machinery and productivity of plants under stress (review). Appl Biochem Microbiol 45:351–362

Chory J, Reinecke D, Sim S et al (1994) A role for cytokinins in de-etiolation in *Arabidopsis* (det mutants have an altered response to cytokinins). Plant Physiol 104:339–347

Cutler SR, Rodriguez PL, Finkelstein RR et al (2010) Abscisic acid: emergence of a core signaling network. Annu Rev Plant Biol 61:651–679

D'Agostino IB, Deruere J, Kieber JJ (2000) Characterization of the response of the *Arabidopsis* response regulator gene family to cytokinin. Plant Physiol 124:1706–1717

Debata A, Murty KS (1981) Relation between leaf and panicle senescence in rice. Indian J Exp Biol 19:1183–1184

Dewitte W, Scofield S, Alcasabas AA et al (2007) *Arabidopsis* CYCD3 D-type cyclins link cell proliferation and endocycles and are rate-limiting for cytokinin responses. PNAS USA 104:14537–14542

Dhiman MR, Guleria MS, Parkash C et al (2015) Effect of different chemical compounds on leaf chlorophyll content and postharvest quality of *Lilium*. Int J Hort 5:1–6. https://doi.org/10.5376/ijh.2015.05.0018

Dwivedi SK, Arora A, Singh VP et al (2017) Induction of water deficit tolerance in wheat due to exogenous application of plant growth regulators: membrane stability, water relations and photosynthesis. Photosynthetica. https://doi.org/10.1007/s11099-017-0695-2

El-Beltagy AS, Hall MA (1974) Effect of water stress upon endogenous ethylene levels in *Vicia faba*. New Phytol 73:47–59

Elfving DC, Cline RA (1993) Benzyladenine and other chemicals for thinning 'Empire' apple trees. J Am Soc Hort Sci 118:593–598

Famiani F, Battistelli A, Moscatello S et al (1999) Thidiazuron affects growth, ripening and quality of *Actinidia deliciosa*. J Hort Sci Biotech 74:375–380

Ferrante A, Hunter DA, Hackett WP (2002b) Thidiazuron – a potent inhibitor of leaf senescence in *Alstroemeria*. Postharvest Biol Technol 25:333–338

Ferrante A, Hunter D, Hackett W et al (2001) TDZ: a novel tool for preventing leaf yellowing in *Alstroemeria* flowers. Hortic Sci 36.: Poster:599

Ferrante A, Mensuali-Sodi A, Serra G (2009) Effect of thidiazuron and gibberellic acid on leaf yellowing of cut stock flowers. Cent Eur J Biol 4:61–468

Ferrante A, Mensuali-Sodi A, Serra G et al (2002a) Effects of ethylene and cytokinins on vase life of cut *Eucalyptus parvifolia* Cambage branches. Plant Growth Regul 38:119–125

Ferrante A, Mensuali-Sodi A, Serra G et al (2003) Treatment with thidiazuron for preventing leaf yellowing in cut tulips, and chrysanthemum. Acta Hortic 624:357–363

Ferrante A, Mensuali-sodi A, Tognoni F et al (2005) Postharvest studies on leaf yellowing of chrysanthemum cut flowers. Adv Hortic Sci 19:81–82

Ferrante A, Trivellini A, Mensuali-Sodi A (2012) Interaction of 1-methylcyclopropene and thidiazuron on cut stock flowers vase life. Open Hort J 5:1–5

Ferrante A, Trivellini A, Serra A (2011) Benzyladenine and thidiazuron postharvest treatments for preserving cut lily flowers. Acta Hortic 900:301–307

Ferrante A, Vernieri P, Serra G et al (2004) Changes in abscisic acid during leaf yellowing of cut stock flowers. Plant Growth Regul 43:127–134

Flores S, Tobin EM (1988) Cytokinin modulation of LHCP mRNA levels: the involvement of post-transcriptional regulation. Plant Mol Biol 11:409–415

Forshey CG (1987) A review of chemical fruit thinning. Proc NE Fruit Meet 93:68–73

Galuszka P, Popelková H, Werner T et al (2007) Biochemical characterization and histochemical localization of cytokinin oxidases/dehydrogenases from *Arabidopsis thaliana* expressed in *Nicotiana tabaccum* L. J Plant Growth Regul 26:255–267

Gan S, Amasino RM (1995) Inhibition of leaf senescence by autoregulated production of cytokinin. Science 270:1986–1988

Gan S, Amasino RM (1997) Making sense of senescence (molecular genetic regulation and manipulation of leaf senescence). Plant Physiol 113:313–319

Grbić V, Bleecker AB (1995) Ethylene regulates the timing of leaf senescence in *Arabidopsis*. Plant J 8:595–602

Greenboim-Wainberg Y, Maymon I, Borochov R et al (2004) Cross talk between gibberellin and cytokinin: the *Arabidopsis* GA response inhibitor SPINDLY plays a positive role in cytokinin signaling. Plant Cell 17:92–102

Greene DW (1995) Thidiazuron effects on fruit set, fruit quality, and return bloom of apples. Hortic Sci 30:1238–1240

Grossmann K (1991) Induction of leaf abscission in cotton is a common effect of urea- and adenine-type cytokinins. Plant Physiol 95:234–237

Hare PD, Van Staden J (1994) Inhibitory effect of thidiazuron on the activity of cytokinin oxidase isolated from soybean callus. Plant Cell Physiol 35:1121–1125

Hatami M, Hatamzadeh A, Ghasemnezhad M et al (2013) Antioxidant enzymatic protection during *Pelargonium* plant leaf senescence is mediated by thidazuron. Trakia J Sci 11:152–157

Hatamzadeh A, Rezvanypour S, Asil MH (2012) Postharvest life of *Alstroemeria* cut flowers is extended by thidiazuron and benzyladenine. South West J Hort Biol Env 3:41–53

Hauska G, Trebst A, Koetter C et al (1975) 1,2,3-Thiadiazolyl-phenyl-ureas, new inhibitors of photosynthetic and respiratory energy conservation. Z Naturforsch C J Biosci 30:505–510

Hensel LL, Grbić V, Baumgarten DA et al (1993) Developmental and age-related processes that influence the longevity and senescence of photosynthetic tissues in *Arabidopsis*. Plant Cell 5:553–564

Hodgson RH, Snyder RH (1988) Thidiazuron effects on *Malvaceae*; corn, (*Zea mays*); and soybean, (*Glycine max*). Weed Technol 2:342–349

Hothorn M, Dabi T, Chory J (2011) Structural basis for cytokinin recognition by *Arabidopsis thaliana* histidine kinase 4. Nat Chem Biol 7:766–768

Houba-Hérin N, Pethe C, d'Alayer J et al (1999) Cytokinin oxidase from *Zea mays*: purification, cDNA cloning and expression in moss protoplasts. Plant J 17:615–626

Hunter DA, Yoo SD, Butcher SM et al (1999) Expression of 1-aminocyclopropane-1-carboxylate oxidase during leaf ontogeny in white clover. Plant Physiol 120:131–142

Ichimura K, Hiraya T (1999) Effect of silver thiosulfate complex (STS) in combination with sucrose on the vase life of cut sweet pea flowers. J Jpn Soc Hort Sci 68:23–27

Inoue T, Higuchi M, Hashimoto Y et al (2001) Identification of CRE1 as a cytokinin receptor from *Arabidopsis*. Nature 409:1060–1063

Itai A, Tanabe K, Tamura F et al (1995) Synthetic cytokinins control persimmon fruit shape, size and quality. J Hort Sci 70:867–873

Jiang CZ, Wu L, Macnish AJ et al (2009) Thidiazuron, a non-metabolized cytokinin, shows promise in extending the life of potted plants. Acta Hortic 847:59–66

Jing HC, Schippers JH, Hille J et al (2005) Ethylene-induced leaf senescence depends on age-related changes and OLD genes in *Arabidopsis*. J Exp Bot 56:2915–2923

Jo YS, Cho HS, Park MY et al (2003) Comparison of CPPU effects on fruit development in several actinidia species. Acta Hortic 610:539–543

Jordi W, Schapendonk A, Davelaar E et al (2000) Increased cytokinin levels in transgenic PSAG12-IPT tobacco plants have large direct and indirect effects on leaf senescence, photosynthesis and N partitioning. Plant Cell Environ 23:279–289

Jordi W, Stoopen GM, Kelepouris K et al (1995) Gibberellin-induced delay of leaf senescence of Alstroemeria cut flowering stems is not caused by an increase in the endogenous cytokinin content. J Plant Growth Regul 14:121–127

Kaur P, Singh K (2015) Influence of growth regulators on physiology and senescence of cut stems of Chrysanthemum (*Chrysanthemum morifolium* Ramat) Var. Thai Ching Queen IJAPRR 2:31–41

Kaviani M, Mortazavi SN (2013) Effect of nitric oxide and thidiazuron on *Lilium* cut flowers during postharvest. Int J Agron Plant Prod 4:664–669

Kefford NP, Bruce MI, Zwar JA (1973) Retardation of leaf senescence by urea cytokinins in *Raphanus sativus*. Phytochemistry 12:995–1003

Kiba T, Yamada H, Sato S et al (2003) The type-A response regulator, *ARR15*, acts as a negative regulator in the cytokinin-mediated signal transduction in *Arabidopsis thaliana*. Plant Cell Physiol 44:868–874

Kim HJ, Ryu H, Hong SH et al (2006) Cytokinin-mediated control of leaf longevity by AHK3 through phosphorylation of ARR2 in Arabidopsis. PNAS USA 103:814–819

Kopečný D, Briozzo P, Popelková H et al (2010) Phenyl- and benzylurea cytokinins as competitive inhibitors of cytokinin oxidase/dehydrogenase: a structural study. Biochimie 92:1052–1062

Koprna R, De Diego N, Dundálková L et al (2016) Use of cytokinins as agrochemicals. Bioorg Med Chem 24:484–492

Kraepiel Y, Miginiac E (1997) Photomorphogenesis and phytohormones. Plant Cell Environ 20:807–812

Kusnetsov VV, Oelmüller R, Sarwat MI et al (1994) Cytokinins, abscisic acid and light affect accumulation of chloroplast proteins in *Lupinus luteus* cotyledons without notable effect on steady-state mRNA levels. Planta 194:318–327

Laureys F, Dewitte W, Witters E et al (1998) Zeatin is indispensable for the G2-M transition in tobacco BY-2 cells. FEBS Lett 426:29–32

Leibfried A, To JPC, Stehling S et al (2005) WUSCHEL controls meristem size by direct transcriptional regulation of cytokinin inducible response regulators. Nature 438:1172–1175

Leonard RT, Nell TA (2004) Short-term pulsing improves postharvest leaf quality of cut oriental lilies. Hort Tech 14:405–411

Leshem YY, Wills RBH (1998) Harnessing senescence delaying gases nitric oxide and nitrous oxide: a navel approach to postharvest control of fresh horticultural produce. Biol Plant 41:1–100

Lomin SN, Krivosheev DM, Steklov MY et al (2015) Plant membrane assays with cytokinin receptors underpin the unique role of free cytokinin bases as biologically active ligands. J Exp Bot 66:1851–1863

Macnish AJ, Jiang CZ, Reid MS (2010) Treatment with thidiazuron improves opening and vase life of iris flowers. Postharvest Biol Technol 56:77–84

Mähönen AP, Bonke M, Kaupinnen L et al (2000) A novel two-component hybrid molecule regulates vascular morphogenesis of the Arabidopsis root. Genes Dev 14:2938–2943

Malik MN, Din S (1997) Efficacy of thidiazuron defoliant in cotton cultivars varying in maturity. Pak Cottons 41:36–42

Malik MN, Din S, Makhdum MI (1991) Accelerated boll dehiscence with thidiazuron. Trop Agric 68:149–150

Malik MN, Din S, Makhdum MI et al (2002) Use of thidiazuron as harvest-aid in early and late planted cotton. Int J Agric Biol 4:71–73

Masferrer A, Arro M, Manzano D et al (2002) Overexpression of *Arabidopsis thaliana* farnesyl diphosphate synthase (FPS1S) in transgenic Arabidopsis induces a cell death/senescence-like response and reduced cytokinin levels. Plant J 30:123–132

Mayak S, Dilley DR (1976) Regulation of senescence in carnation (*Dianthus caryophylus*). Plant Physiol 58:663–665

Mayak S, Halevy AH (1972) Interrelationships of ethylene and abscisic acid in the control of rose petal senescence. Plant Physiol 50:341–346

Mik V, Szüčová L, Šmehilová M et al (2011) N9-substituted derivatives of kinetin: effective anti-senescence agents. Phytochemistry 72:821–831

Miller CO, Skoog F, Von Saltza MH et al (1955) Kinetin, a cell division factor from deoxyribonucleic acid. J Amer Chem Soc 77:1392

Mok MC, Martin RC, Dobrev PI et al (2005) Topolins and hydroxylated thidiazuron derivatives are substrates of cytokinin o-glucosyltransferase with position specificity related to receptor recognition. Plant Physiol 137:1057–1066

Mok MC, Mok DWS (1985) The metabolism of [^{14}C] thidiazuron in callus tissues of *Phaseolus lunatus*. Physiol Plant 65:427–432

Mok MC, Mok DWS, Amstrong DJ et al (1982) Cytokinin activity of N-phenyl-N-1,2,3-thidiazol-5-ylurea (thidiazuron). Phytochemistry 21:1509–1511

Morris RO, Bilyeu KD, Laskey JG et al (1999) Isolation of a gene encoding a glycosylated cytokinin oxidase from maize. Biochem Biophys Res Commun 255:328–333

Mortazavi SN, Talebi SF, Naderi RA et al (2011) Regulation of ethylene biosynthesis by nitric oxide and thidiazuron during postharvest of rose. J Med Plant Res 5:5177–5183

Mothes K, Engelbrecht L (1963) On the activity of a kinetin-like root factor. Life Sci 11:852–857

Murch SJ, Saxena PK (2001) Molecular fate of thidiazuron and its effects on auxin transport in hypocotyls tissues of *Pelargonium x hortorum* Bailey. Plant Growth Regul 35:269–275

Mutui TM, Emongor VN, Hutchinson MJ (2003) Effect of benzyladenine on the vase life and keeping quality of *Alstroemeria* cut flowers. J Agric Sci Technol 5:91–105

Mutui TM, Mibus H, Serek M (2005) Effects of thidiazuron, ethylene, abscisic acid and dark storage on leaf yellowing and rooting of *Pelargonium* cuttings. J Hortic Sci Biotechnol 80:543–550

Mutui TM, Mibus H, Serek M (2007) Influence of thidiazuron, ethylene, abscisic acid and dark storage on the expression levels of ethylene receptors (ETR) and ACC synthase (ACS) genes in *Pelargonium*. Plant Growth Regul 53:87–96

Nagar S, Arora A, Singh VP et al (2015) Effect of cytokinin analogues on cytokinin metabolism and stress responsive genes under osmotic stress in wheat. Bioscan 10:67–72

Nisler J, Kopečný D, Končitíková R et al (2016) Novel thidiazuron-derived inhibitors of cytokinin oxidase/dehydrogenase. Plant Mol Biol 92:235–248

Pavlista AD (2003) Thidiazuron, a cytokinin-like compound, enhances fungicidal activity against early blight in potato. Acta Hortic 619:145–152

Petri JL, Schuck E, Leite GB (2001) Effects of thidiazuron (tdz) on fruiting of temperate tree fruits. Rev Bras Frutic 23:513–517

Rashotte AM, Mason MG, Hutchison CE et al (2006) A subset of *Arabidopsis* AP2 transcription factors mediates cytokinin responses in concert with a two-component pathway. PNAS USA 103:11081–11085

Redig P, Shaul O, Inze D et al (1996) Levels of endogenous cytokinins, indole-3-acetic acid and abscisic acid during the cell cycle of synchronized tobacco BY-2 cells. FEBS Lett 391:175–180

Reid MS (1995) Ethylene in plant growth, development, and senescence. In: Davis PJ (ed) Plant hormones. Springer, Dordrecht. https://doi.org/10.1007/978-94-011-0473-9_23

Reynolds AG, Wardle DA, Zurowski C et al (1992) Phenylureas CPPU and thidiazuron affect yield components, fruit composition, and storage potential of four seedless grape selections. J Am Soc Hort Sci 117:85–89

Richmond AE, Lang A (1957) Effect of kinetin on protein content and survival of detached *Xanthium* leaves. Sci NY 125:650–651

Riefler M, Novak O, Strnad M et al (2006) Arabidopsis cytokinin receptor mutants reveal functions in shoot growth, leaf senescence, seed size, germination, root development, and cytokinin metabolism. Plant Cell 18:40–54

Riou-Khamlichi C, Huntley R, Jacqmard A et al (1999) Cytokinin activation of *Arabidopsis* cell division through a D-type cyclin. Science 283:1541–1544

Rivero RM, Kojima M, Gepstein A et al (2007) Delayed leaf senescence induces extreme drought tolerance in a flowering plant. PNAS USA 104:19631–19636

Romanov GA, Lomin SN, Schmülling T (2006) Biochemical characteristics and ligand-binding properties of *Arabidopsis* cytokinin receptor AHK3 compared to CRE1/AHK4 as revealed by a direct binding assay. J Exp Bot 57:4051–4058

Sakakibara H (2003) Nitrate-specific and cytokinin-mediated nitrogen signaling pathways in plants. J Plant Res 116:253–257

Sankhla N, Mackay WA, Davis TD (2003) Reduction of flower abscission and leaf senescence in cut *Phlox* inflorescences by thidiazuron. Acta Hort 628:837–841

Sankhla N, Mackay WA, Davis TD (2005) Effect of thidiazuron on senescence of flowers in cut inflorescences of *Lupinus densiflorus* benth. Acta Hortic 669:239–244

Schaller GE, Kieber JJ, Shiu S (2008) Two-component signalling elements and histidyl-aspartyl phosphorelays. In: Somerville C, Meyerowitz EM (eds) The *Arabidopsis* book. American Society of Plant Biologists, Rockville, pp 1–12

Schaller GE, Street IH, Kieber JJ (2014) Cytokinin and the cell cycle. Curr Opin Plant Biol 21:7–15

Schippers JHM, Breeze E, Buchanan-Wollaston V (2008). A role for cytokinin in the onset of leaf senescence by ethylene in *Arabidopsis*. Molecular aspects of ageing and the onset of leaf senescence. Schippers JHMn.d. s.n. 164 p. Doctoral thesis

Schulz H, Arndt F (1973) 1,2,3-Thiadiazole plant growth retardants. From Ger. Offen.DE 2214632 A1 19731004

Scofield S, Dewitte W, Nieuwland J et al (2013) The Arabidopsis homeobox gene SHOOT MERISTEMLESS has cellular and meristem-organisational roles with differential requirements for cytokinin and CYCD3 activity. Plant J 75:53–66

Singh S, Letham DS, Palni LMS (1992) Cytokinin biochemistry in relation to leaf senescence. 7. Endogenous cytokinin levels and exogenous applications of cytokinins in relation to sequential leaf senescence of tobacco. Physiol Plant 86:388–397

Spíchal L, Rakova NY, Riefler M et al (2004) Two cytokinin receptors of *Arabidopsis thaliana*, CRE1/AHK4 and AHK3, differ in their ligand specificity in a bacterial assay. Plant Cell Physiol 45:1299–1305

Stern R, Shargal A, Flaishman M (2003) Thidiazuron increases fruit size of 'Spadona' and 'Coscia' pear (*Pyrus communis* L.) J Hortic Sci Biotechnol 78:51–55

Stolz A, Riefler M, Lomin SN et al (2011) The specificity of cytokinin signalling in *Arabidopsis thaliana* is mediated by differing ligand affinities and expression profiles of the receptors. Plant J 67:157–168

Suttle JC (1983) Effect of the defoliant thidiazuron on ethylene production. Plant Physiol 72:S-121

Suttle JC (1984) Effect of the defoliant thidiazuron on ethylene evolution from mung bean hypocotyl segments. Plant Physiol 75:902–907

Suttle JC (1985) Involvement of ethylene in the action of the cotton defoliant thidiazuron. Plant Physiol 78:272–276

Suttle JC (1986) Cytokinin-induced ethylene biosynthesis in non senescing cotton leaves. Plant Physiol 82:930–935

Suttle JC, Hultstrand JF (1991) Ethylene-induced leaf abscission in cotton seedlings. The physiological bases for age-dependent differences in sensitivity. Plant Physiol 95:29–33

Suzuki T, Miwa K, Ishikawa K et al (2001) The *Arabidopsis* sensor His-kinase, AHK4, can respond to cytokinins. Plant Cell Physiol 42:107–113

Talebi SF, Mortazavi SN, Naderi RA et al (2013) Role of nitric oxide and Thidiazuron on changes of pigments during postharvest in *Rosa* (Cv. 'Sensiro'). Int J Agron Plant Prod 4:121–126

Taniguchi M, Kiba T, Sakakidara H et al (1998) Expression of *Arabidopsis* response regulator homologues is induced by cytokinins and nitrate. FEBS Lett 429:259–262

Thomas H, Howarth CJ (2000) Five ways to stay green. J Exp Bot 51:329–337

Tirtashi ZB, Hashemabadi D, Kaviani B et al (2014) Effect of thidiazuron and salicylic acid on the vase life and quality of Alstroemeria (*Alstroemeria hybrida* L .cv. 'Modena') cut flower. J Ornamental Plants 4:163–168

To JPC, Deruere J, Maxwell BB et al (2007) Cytokinin regulates type-A *Arabidopsis* response regulator activity and protein stability via two-component phosphorelay. Plant Cell 19:3901–3914

To JPC, Haberer G, Ferreira FJ et al (2004) Type-A *Arabidopsis* response regulators are partially redundant negative regulators of cytokinin signalling. Plant Cell 16:658–671

Ueguchi C, Sato S, Kato T et al (2001) The AHK4 gene involved in the cytokinin-signalling pathway as a direct receptor molecule in *Arabidopsis thaliana*. Plant Cell Physiol 42:751–755

Uthairatanakij A, Jeenbuntug J, Buanong M et al (2007) Effect of thidiazuron pulsing on physiological changes of cut tuberose flower (*Polianthes tuberosa* L.) Acta Hortic 755:477–480

Van Staden J (1973) Changes in endogenous cytokinin levels during abscission and senescence of streptocarpus leaves. J Exp Bot 24:667–671

Vogel JP, Woeste KE, Theologis A et al (1998) Recessive and dominant mutations in the ethylene biosynthetic gene ACS5 of *Arabidopsis* confer cytokinin insensitivity and ethylene overproduction, respectively. PNAS USA 95:4766–4771

Weaver LM, Gan S, Quirino B et al (1998) A comparison of the expression patterns of several senescence-associated genes in response to stress and hormone treatment. Plant Mol Biol 37:455–469

Werner T, Motyka V, Laucou V et al (2003) Cytokinin-deficient transgenic *Arabidopsis* plants show multiple developmental alterations indicating opposite functions of cytokinins in the regulation of shoot and root meristem activity. Plant Cell 15:2532–2550

Whitty CD, Hall RH (1974) A cytokinin oxidase in *Zea mays*. Can J Biochem 52:789–799

Wingler A, von Schaewen A, Leegood RC et al (1998) Regulation of leaf senescence by cytokinin, sugars, and light. Effect on NADH-dependent hydroxypyruvate reductase. Plant Physiol 116:329–335

Woeste KE, Vogel JP, Kieber JJ (1999) Factors regulating ethylene biosynthesis in etiolated *Arabidopsis thaliana* seedlings. Physiol Plant 105:478–484

Yamada H, Suzuki T, Terada K et al (2001) The Arabidopsis AHK4 histidine kinase is a cytokinin-binding receptor that transduces cytokinin signals across the membrane. Plant Cell Physiol 42:1017–1023

Yang SF, Hoffman NE (1984) Ethylene biosynthesis and its regulation in higher plants. Annu Rev Plant Physiol 35:155–189

Yang YZ, Lin DC, Guo ZY (1992) Promotion of fruit development in cucumber (*Cucumis sativus*) by thidiazuron. Sci Hortic 50:47–51

Yaronskaya E, Vershilovskaya I, Poers Y et al (2006) Cytokinin effects on tetrapyrrole biosynthesis and photosynthetic activity in barley seedlings. Planta 224:700–709

Yip WK, Yang SF (1986) Effect of thidiazuron, a cytokinin active urea derivative, in cytokinin-dependent ethylene production systems. Plant Physiol 80:515–519

Yoon TM, Richter H (1990) Seasonal changes in stomatal responses of sweet cherry and plum to water status in detached leaves. Physiol Plant 80:520–526

Yu Y, Yang SF, Corse J et al (1981) Structures of cytokinins influence synergistic production of ethylene. Phytochemistry 20:1191–1195

TDZ-Induced Morphogenesis Pathways in Woody Plant Culture

3

Tatyana I. Novikova and Yulianna G. Zaytseva

Abstract

Thidiazuron (TDZ) possesses a unique property to stimulate both the auxin- and cytokinin-like activities to induce in vitro morphogenesis pathways in the different explants of many species. An additional advantage of TDZ in low concentrations for many recalcitrant woody species in comparison to common amino purine cytokinins is a higher efficiency in overcoming monopodial growth habits by stimulating the axillary shoot development in in vitro culture. The application of TDZ for in vitro woody tissue culture inducts the wide range of morphological reactions including somatic embryogenesis and shoot organogenesis followed by root organogenesis, which occurs directly or through callus formation. These responses suggested that TDZ-induced regeneration systems could be used as the models for studying the fundamental aspects of plant biology and better understanding the developmental pathways. Despite the progress of recent biochemical, physiological, and molecular researches of TDZ effect on plant regeneration through organogenesis and somatic embryogenesis, the data on chronological sequences of morphological events are still limited. The abnormality formation of de novo structures is known to be undesirable side effect of TDZ. Therefore, the morpho-histological approach based on the observation of developmental route under TDZ at microscopic level and detailed histological analysis is necessary to reveal the type of morphogenic response (organogenic or embryogenic), improve the regeneration efficiency, and create systems of large-scale propagation for woody species. The chapter discusses the data on TDZ-induced regeneration systems in various woody tissue cultures analyzed by morpho-histological approach as a valuable tool to perform morphogenesis studies.

T. I. Novikova · Y. G. Zaytseva (✉)
Biotechnology Laboratory, Central Siberian Botanical Garden, Siberian Branch of the Russian Academy of Sciences, Novosibirsk, Russian Federation
e-mail: zaytseva@csbg.nsc.ru

N. Ahmad, M. Faisal (eds.), *Thidiazuron: From Urea Derivative to Plant Growth Regulator*, https://doi.org/10.1007/978-981-10-8004-3_3

Keywords
Thidiazuron · Shoot organogenesis · Somatic embryogenesis · Morpho-histological analysis

3.1 Introduction

Woody plants are extremely important for the world economy as well as for maintaining and conservation of natural ecosystems. Despite the in vitro propagation techniques that have become a reliable and routine approach for large-scale rapid herbaceous plant multiplication, many woody species still remain recalcitrant to micropropagation due to inherent slow-growing nature and low regeneration potential. However, mass clonal propagation of elite varieties along with the accelerated tree improvement programs using genetic transformation approach is necessary to meet growing demands for forest products and reforestation as well as for application in tree breeding (Tzfira et al. 1998).

Breakthrough innovations in the woody plant micropropagation have been made by using thidiazuron (TDZ) as a plant growth regulator (PGR) in culture media (Huetteman and Preece 1993). As synthetic cytokinin, TDZ is known to induce higher regeneration rates in the comparison to purine-based cytokinins and also demonstrate the ability of fulfilling both the cytokinin and auxin requirements of regeneration responses in a number of woody plants (Jones et al. 2007). To understand the events involved in vitro wide range morphogenic responses in the presence of TDZ, detailed studies using light and electron microscopy are required. Although, morpho-histological observations have been carried out for evaluating the effect of both cytokinins and auxins on the cellular development patterns including somatic embryogenesis (SE) and de novo organogenesis, it is worth noting that not all findings can be applied to TDZ due to the complicated mechanism of its action. Furthermore, physiological disorders in regenerants induced by plant growth regulators including TDZ are better revealed using the microscopic systems (Bairu and Kane 2011).

This review aims to analyze of wide range TDZ-induced morphogenic responses in various woody plant cultures on the base of microscopic investigations providing deep insights into in vitro plant development.

3.2 In Vitro Morphogenesis of Woody Plants

3.2.1 Fundamental Aspects and Theories

Morphogenesis is an array of processes of cell differentiation during the development of multicellular organism which is regulated at the cellular, tissue, and organism levels (Butenko 1999). The study of in vitro morphogenetic pathways is of a particular interest, since the absence of control at the organism level and the use of growth regulators allow obtaining a wide range of morphogenetic reactions necessary for modeling of in vivo plant development processes (Zhuravlev and Omelko 2008). Although in vitro culture techniques are widely used for many woody species, propagating adult woody plants characterized by long life cycle remains difficult since only juvenile material is employed for tissue culture of the most tree species (Ahuja 1993). The development of the accessible model systems in vitro also plays an important role in understanding some mechanisms of woody plant morphogenesis that are promising for both practical applications as the base for large-scale trees propagation and the basic developmental biology.

Among various in vitro morphogenic responses, only SE and de novo shoot organogenesis followed by root organogenesis result in regeneration of whole plant (Phillips 2004). De novo organogenesis is the formation of unipolar structures, shoot or root apical meristems resulting in regeneration of adventitious shoot or root with vascular connection with maternal explants. Whereas, SE is the multistep process forming the bipolar structures that are similar to zygotic embryos. The bipolar structures are developed through the characteristic embryological stages from a non-zygotic cell without a vascular connection with the original tissue (Hicks 1994; von Arnold et al. 2002; Salaj et al. 2005). These major regeneration pathways can be realized directly without an intermediate unorganized callus formation or indirectly through a callus stage (Gahan and George 2008).

Using the conception of plant stem cell, some researchers have clarified that pluripotency of cells underlies organogenesis, whereas cell totipotency is the basis for SE (Verdeil et al. 2007; Atta et al. 2009). In vitro cell cultures can be initiated from the explants of either meristematic origin or from the differentiated organs. When explants contain cells that have retained the capacity to divide, they reenter the cell cycle, and direct initiation occurs. Whereas such predetermined cells are not present, explants have to dedifferentiate through cell division before they can form meristems (Atta et al. 2009). Consequently, to start a new biological program that gives rise to somatic embryos or de novo shoots/roots, the somatic cells undergo dedifferentiation or redifferentiation.

According to the developmental model proposed by Christianson and Warnick (1985), de novo organogenesis is conceptually divided into three distinct phases with different temporal requirements for exogenous phytohormones: (I) somatic cells in the explants are dedifferentiated to acquire a competence in response to external signals (plant growth regulators or stresses); (II) the competent cells are determined under the influence of exogenous plant growth regulators; and (III) morphogenesis occurs independently on the plant growth regulators (Christianson and Warnick 1985; Sugiyama 1999).

The sequence of events for SE as a morphogenic processes is frequently described as distinct phases as well. Nevertheless, there is no agreement among the authors regarding the number of phases and the terminology for each of them. Von Arnold et al. (2002) considered five steps in the plant regeneration via SE: (I) initiation of embryogenic cultures by culturing the primary explant on the medium supplemented with PGRs, mainly auxin but often also cytokinin, (II) proliferation of embryogenic cultures on the medium supplemented with PGRs, (III) prematuration of somatic embryos on the medium lacking PGRs, (IV) maturation of somatic embryos by culturing on the medium supplemented with ABA and/or reduced osmotic potential, and (V) plant development on the medium without PGRs (von Arnold et al. 2002). Jiménez (2001) proposed that SE consists of two major distinct phases which are independent of each other and are controlled by different factors: (I) the initial stage is induction which involves the somatic cells attaining the embryogenic state usually by the exogenous application of PGRs; and (II) the following stage is expression, where the newly differentiated embryonic cells develop into an embryo without any further exogenous signals. Since in vitro and in vivo embryogenesis share high similarities at almost all developmental stages, SE has become an attractive model system to study embryogenesis at the molecular, cellular, and tissue levels (Willemsen and Scheres 2004).

Initiation of both de novo organogenesis and SE is connected with the "competent" state of somatic cells resulting in their reprogramming followed by realizing the de novo developmental fate (Meng et al. 2005). The competence of the cells greatly depends on the genetic background, the developmental and physiological state of explant, the culture medium, and conditions (Christianson and Warnick 1985; Yancheva et al. 2003; Vasil 2008). However, the choice of PGRs and their concentration is of the most profound effect on reprogramming the somatic cells toward organogenic or embryogenic pathways.

3.2.2 TDZ as Bioregulator with Dual Hormonal Activities

The classical experiments of Skoog and Miller carried out 60 years ago have clearly demonstrated that the ratio of auxin and cytokinin is the critical factor in triggering the in vitro developmental events: an excess of cytokinins over auxins promotes shoot formation in callus; auxin excess induces root formation; when the same level of auxin and cytokinin is added into the medium, the explant cells proliferate to form callus (Skoog and Miller 1957). This discovery is still widely exploited for plant micropropagation; however, a new synthetic PGR – TDZ – can be used as a substitute for both auxin and cytokinin (Thomas and Katterman 1986; Visser et al. 1992; Casanova et al. 2004). This TDZ peculiarity is displayed in inducing a wide array of in vitro morphogenic reactions including callus formation, initiation of SE, adventitious shoot regeneration, and axillary shoot proliferation. Moreover, TDZ is able to alter the pathway of morphogenesis from shoot organogenesis to SE in the same explant through a change in TDZ concentration (Mithila et al. 2003; Ma et al. 2011).

TDZ was shown to induce shoot organogenesis on leaf and petiole explants of *Saintpaulia ionantha* at concentrations lower than 2.5 μM, whereas at higher doses (5–10 μM) somatic embryos were formed (Mithila et al. 2003). Similar results were achieved in cultures of pigeon pea (Dolendro et al. 2003). The redirection of the development from shoot organogenesis to SE at higher TDZ (2.5–15 μM) doses may have occurred due to the optimum phytohormone balance within the tissue or as a result of increased stress imposed by high concentrations of TDZ. However, in tissue culture of woody plants, the difference between morphogenic responses to low and high TDZ concentration is not so unambiguous. The concentration-dependent redirection of the development from shoot organogenesis to SE has been observed in the studies of regeneration from leaf and shoot explants of *Ochna integerrima*, a medicinal and ornamental tree (Ma et al. 2011). The low concentrations of TDZ (5.0 μM) could only induce adventitious shoots, whereas a higher concentration of thidiazuron (10.0–15.0 μM TDZ) could induce both SE and adventitious shoot formation. At the same time, it was recorded in the development of three different morphogenic responses – direct organogenesis, direct embryogenesis, and indirect embryogenesis – in the cotyledon explants of a medicinal forest tree *Semecarpus anacardium* cultured on woody plant medium (WPM) supplemented with the same (9.08 μM) TDZ concentration (Panda and Hazra 2012). Consequently, organogenic and direct/indirect embryogenic morphogenic pathways can occur simultaneously in the same culture conditions. Thus, TDZ depending on the concentration in the induction medium promotes dedifferentiation of explant cells and determines the cell fate, which can be realized through organogenesis or SE.

3.2.3 Physiological and Biochemical Basis of TDZ-Mediated Responses

The physiological basis of unique TDZ efficiency in the induction of a wide array of morphogenic responses either alone or in combination with the other plant growth regulators was discussed in many articles and reviews (Huetmann and Preece 1993; Murthy et al. 1998; Murch and Saxena 2001). TDZ application in vitro culture results in modulation of endogenous hormone levels especially auxin/cytokinin ratio and exerts significant effects on the morphogenetic pathways (Mok et al. 1987). TDZ as a biostimulator with high cytokinin-like activity is known to induce the endogenous cytokinin biosynthesis and accumulation (Ruzic and Vujovic 2008). Recent studies have indicated that TDZ induces metabolic cascade modifying a primary signaling event, accumulation and transfer of various endogenous plant signals including the secondary messengers and a concurrent stress response (Elhiti et al. 2013).

It was confirmed that the action mechanism of this PGR could be closely linked to the biosynthesis and transportation of indole-3-acetic acid (IAA) (Chhabra et al. 2008). Besides these activities, TDZ can act by modulating the metabolism and transport of the other PGRs (ethylene, abscisic acid, and gibberellin) (Murch and Saxena 2001). The low TDZ levels (<0.005 μM) induce accumulation of zeatin, whereas high concentration (0.5 μM) stimulates isopentenyl adenine activity. This fact can clarify the concentration-dependent mode of TDZ. The other mechanisms include the modification of cell membranes, energy levels, nutrient uptake, or nutrient assimilation (Casanova et al. 2004). The efficiency of this phenylurea derivative may be attributed to its ability to induce a metabolite stress connected with increased accumulation of mineral ions, which predispose the plant to stress (Murch et al. 1997). TDZ has been shown recently to act through regulation of oxidative stresses by changing superoxide dismutase and catalase enzyme activities in plant cells, especially during shoot regeneration or embryo formation (Guo et al. 2017).

Although some progress has been made in the past decades to understand the physiological-biochemical basis of TDZ effectiveness in in vitro culture, the chronological sequences of morphological events underlying de novo shoot regeneration or somatic embryo development in the TDZ presence are not always clear. To reveal the developmental events during morphogenesis, detailed morpho-histological analyses are still needed. Such studies allow to regulate and improve each stage of these processes efficiently and to use the most appropriate strategies for plant regeneration as well as for the selection and validation of the techniques used in genetic transformation programs.

3.3 Shoot Production

Most of studies on in vitro multiplication report that cytokinins play a crucial role in shoot production. Among cytokinins, TDZ has been reported to be very efficient for axillary shoot proliferation and adventitious shoot formation in a wide array of woody plants including the recalcitrant species (Huetmann and Preece 1993; Murthy et al. 1998; Yancheva et al. 2003).

3.3.1 Axillary Shoot Proliferation

Clonal micropropagation of plants from axillary buds or shoots is the most appropriate method of true-to-type in vitro multiplication (Varshney and Anis 2014). This approach is based on the natural ontogenetic pathway for plant development from preexisting meristems and does not involve the dedifferentiation of cells. Since meristems are more resistant to genetic changes than disorganized callus tissues, genetic stability is usually observed in the regenerated plantlets, although epigenetic alterations may occur (Baránek et al. 2015).

Shoot explants of some woody species are known to demonstrate the naturally strong monopodial growth without branching under aminopurine cytokinin treatments (Huetteman and Preece 1993). TDZ application allows to overcome these habits by releasing the lateral buds from dormancy (Wang et al. 1986). As a result much higher frequency of shoot proliferation was observed in comparison to commonly used cytokinins (Thomas and Katterman 1986). Successful shoot proliferation from axillary or apical meristems of different woody explants including seedlings, shoot apexes, cotyledonary nodes, and nodal segments was recorded under the treatment of TDZ only or in combination with the other PGRs (Table 3.1). TDZ has been shown to have a potent cytokinin-like activity at 50–100 times lower concentrations than BA (Genkov and Iordanka 1995). TDZ ability to effectively remove an apical dominance is very important in the case of mature genotypes and/or recalcitrant species exhibiting a strong influence of ontogenetic aging (Durkovic and Mišalová 2008). Moreover, this peculiarity is of prime importance for

Table 3.1 TDZ-induced axillary shoot proliferation in woody plant culture

		Step 1		Step 2		
		TDZ-induced axillary shoot proliferation		Shoot elongation		
Plant species or cultivars	Explant type	Nutrient medium and PGRs	TDZ exposure	Nutrient medium and PGRs	Time of culturing	References
Acacia sinuata Lour.	Shoot segments with single node	MS + 8.9 μM BA + 2.5 μM TDZ	60 days	MS + 1.8 μM GA_3	25–30 days	Vengadesan et al. (2003)
Camellia sinensis L.	Nodal segments	Liquid MS 5.0 μM TDZ	6–8 weeks	–	–	Sandal et al. (2001)
Cassia sophera L.	Cotyledonary node	MS + 2.5 μM TDZ	6 weeks	MS + BA 1.0 μM	6 weeks	Parveen and Shahzad (2010)
Cercis canadensis L.	Cotyledonary node[a]	DKW + 10 or 15 μM BA + 0.5 or 1.0 μM TDZ	20 days	DKW	9 weeks	Distabanjong and Geneve (1997)
Cotoneaster wilsonii Nakai	Node	MS + 0.5 mg L^{-1} TDZ + 0.1 mg L^{-1} NAA	5 weeks	MS	3 weeks	Sivanesan et al. (2011)
Embelia ribes Burm.f.	Nodal shoot segments	MS + 1.13 μM TDZ + 0.49 μM IBA	4 weeks	MS + 11.10 μM BAP	8 weeks	Dhavala and Rathore (2010)
Ficus religiosa L.	Nodal segments	WPM + 0.5 TDZ mg/l	60 days	WPM + 0.5 mg/l BAP + 0.5 mg/l IAA + 0.5 mg/l GA_3	4 weeks	Siwach and Gill (2011)
Fraxinus pennsylvanica Marsh	Axillary shoots	MSB5 + 10 μM TDZ	5 weeks	–	–	Kim et al. (1997b)
Leucaena leucocephala (Lam.) de Wit	Immature zygotic embryos	MS + 0.26 μM TDZ	3 weeks	MS	3 weeks	Pal et al. (2012)
	Cotyledonary nodes	MS + 0.23 μM TDZ	30 days	–	–	Shaik et al. (2009)
Medusagyne oppositifolia Baker	Nodal segments	MS + 1 μM TDZ + 25 μM silver nitrate	8 weeks	½ MS	3 weeks	Marriott and Sarasan (2010)
Pongamia pinnata L.	Mature-tree-derived axillary meristems	MS + 0.45 μM TDZ	4–12 weeks	MS	8 weeks	Sujatha and Hazra (2007)
Pterocarpus marsupium Roxb.	Cotyledonary nodes	MS + 0.4 μM TDZ	6 weeks	MS + 5 μM BA	6 weeks	Husain et al. (2007)

(continued)

Table 3.1 (continued)

Plant species or cultivars	Explant type	Step 1: TDZ-induced axillary shoot proliferation: Nutrient medium and PGRs	Step 1: TDZ exposure	Step 2: Shoot elongation: Nutrient medium and PGRs	Step 2: Time of culturing	References
Quercus rubra L.	Cotyledonary node[a]	MS + 4.4 μM BA + 0.45 μM TDZ	3 weeks	WPM +0.44 μM BA + 0.29 μM GA_3	3 weeks	Vengadesa and Pijut (2009)
Rauvolfia tetraphylla L.	Nodal explants	1. MS + 5 μM TDZ	4 weeks	MS	8 weeks	Faisal et al. (2005)
Robinia pseudoacacia L.	Seed	TDZ. 1.0 μM	4 weeks	–	–	Hosseine-Nasr and Rashid (2003)
Rhododendron schlippenbachii Maxim.	Seedlings without roots	AM + 1.0 μM TDZ	8 weeks	AM	8 weeks	Zaytseva and Novikova (2015)
Salix tetrasperma Roxb.	Nodal segments	WPM + 2.5 μM TDZ	4 weeks	WPM + 1.0 μM BA + 0.5 μM NAA	12 weeks	Khan and Anis (2012)
Sesbania drummondii Rydb.	Nodal segments	MS + 2.2 or 4.5 μM TDZ	4–5 weeks	Elongation on rooting medium MS + 1.14 μM IAA	3 weeks	Cheepala et al. (2004)
Tamaríndus índica L.	Seeds[a]	MS + 9.08–18.16 μM μM TDZ	6 weeks	–	–	Menhta et al. (2005)
Theobroma cacao L.	Microcuttings	DKW + 0.01 μM TDZ	5 days	DKW	6 weeks	Traore et al. (2003)
Vitex negundo L.	Nodal segments	MS + 1.8 μM TDZ	4 weeks	MS + 2.4 μM GA_3	12 weeks	Rani and Nair (2006)
	Nodal segments with axillary buds	MS + 1.0 μM TDZ	4 weeks	MS + of 1.0 μM BA + 0.5 μM NAA	8 weeks	Ahmad and Anis (2007)
Vitex trifolia L.	Nodal explants	MS medium fortified with 5.0 μM TDZ	7 days	MS + 1.0 μM BA + 0.5 μM NAA		Ahmed and Anis (2012)

AM Anderson's medium, *MS* Murashige and Skoog medium, *WPM* woody plant medium, *DKW* Driver-Kuniyuki walnut medium
[a]Histological analysis was performed

long-term cultures displaying characteristics of a recalcitrant nature since the loss of regenerative ability in serially subcultivation is usual for woody plant (Gaspar et al. 2000). Marriott and Sarasan (2010) have shown that the effect of TDZ on triggering axillary shoot proliferation of recalcitrant species *Medusagyne oppositifolia* was stronger than 2-isopentenyladenine (2-iP) and 6-benzylaminopurine (BA). *Leucaena leucocephala* is considered to be highly recalcitrant with low rate of shoot multiplication in in vitro culture (Rastogi et al. 2008). However, Pal et al. (2012) substantially increased the axillary shoot production of this species using very low concentration of TDZ (0.26 μM).

The most obvious advantages of TDZ application in woody plant multiplication were demonstrated when cytokinins of purine type did not induce shoot proliferation (Lu 1993). Since the shoots might be susceptible to fasciation and hyperhydricity under high TDZ doses, extremely low ranges of TDZ (1 nM–10 μM), whereas the most of other cytokinins are not effective, were recommended for axillary shoot proliferation by Huetteman and Preece (1993). Ahmed and Anis (2012) recorded a linear correlation between the increase in the concentration of TDZ to an optimal dose (5.0 μM) and the number of shoots per explant developed from nodal explant of *Vitex trifolia*. At the same time, the increase of the optimal concentration resulted in decrease of regeneration and often was accompanied with undesirable basal callus and shoots fasciation (Ahmed and Anis 2012). Besides, the increase of TDZ level in a nutrient medium entails simultaneous development of axillary and adventitious buds (Kim et al. 1997b; Zaytseva and Novikova 2015). Thus, even when culturing explants having the preexisting meristems, a control of the morphogenesis process can be realized by varying TDZ concentration inducing either callus formation, axillary, or adventitious shoot production. This peculiarity undoubtedly increases the productivity of such protocols; however, if a clonal fidelity is required, the potential somaclonal variants derived from adventitious shoots should be avoided by using low TDZ levels. Therefore, for clarifying the origin of shoots induced by TDZ, the morpho-histological analysis is needed.

In spite of a large number of these reports, the sufficient histological data are not available on morphogenic peculiarities of shoot proliferation (Table 3.1). Morpho-histological examination of shoot production from nodal segments of *Disanthus cercidifolius* revealed more active and early development of multiple axillary bud primordia after the exposition to 1.34 μM TDZ for 14 days (Fig. 3.1a) in comparison to the same treatment duration with 2.2 μM BA (Fig. 3.1c). Thus, the choice of optimum TDZ exposition for *D. cercidifolius* was based on the histological analysis. Almost each of the TDZ-induced axillary buds gave rise to more shoots after transferring to MS medium supplemented with 2.2 μM BA for 28 days than under BA treatment in the same exposition time (Novikova and Poluboyarova 2013). Moreover, the morphology of the shoots was rather different. The shortening and fasciation of TDZ-induced shoots of *D. cercidifolius* were observed (Fig. 3.1b), whereas BA-induced shoots were of the usual structure for in vitro conditions (Fig. 3.1d). TDZ is known often to cause the abnormality in shoots structure. This aspect will be considered in more detail in Sect. 3.3.

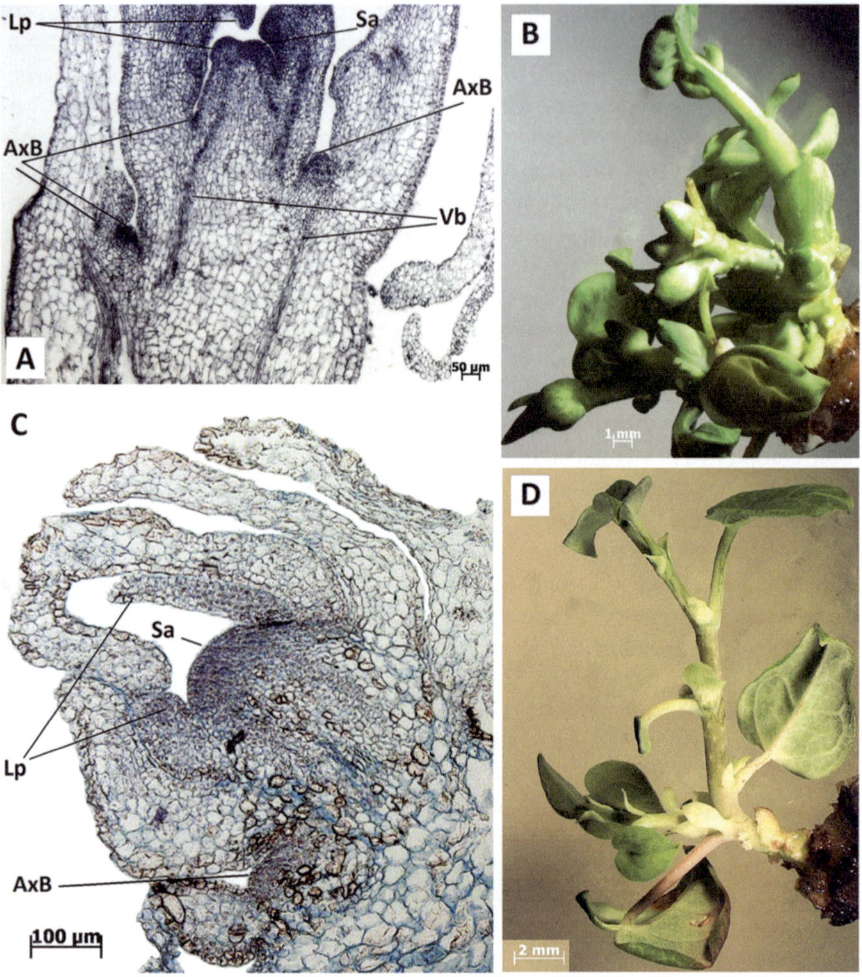

Fig. 3.1 Axillary shoot proliferation induced by TDZ and BA in *Disanthus cercidifolius* nodal explants (Novikova and Poluboyarova 2013): (**a**) longitudinal section of forming buds on MS, supplemented with 1.1 μM TDZ at 14 days; (**b**) axillary shoot induction at 28 days after transferring to MS, supplemented with 2.2 μM BA; (**c**) longitudinal section of forming buds on MS supplemented with 2.2 μM BA at 14 days; (**d**) axillary shoot induction at 42 days on the same medium. *AxB* axillary bud proliferation, *Lp* leaf primordia, *Sa* shoot apex, *Vp* vascular bundle

Histological studies indicated de novo shoot origin from actively dividing cells located in the axillary bud region of cotyledonary node explants of *Cercis canadensis* under the combination of TDZ and BA (Distabanjong and Geneve 1997). At the same time Menhta et al. (2005) found an interesting pattern of morphogenic developments around the existing meristem of *Tamaríndus índica* seedlings. In response to 9.08 mM TDZ, the protrusions were initiated from the nodal meristems, which extended laterally in circles toward the neighboring non-meristemic regions. Subsequently, besides axillary meristem proliferation, de novo shoot organogenesis was triggered.

3.3.2 Adventitious Shoot Organogenesis

Since de novo shoot organogenesis may occur both through direct and indirect route, the morpho-histological analysis of regeneration processes is of great demand. For propagation of rare or elite woody plant genotypes, the direct shoot organogenesis is more preferable than regeneration through callus to avoid somaclonal variation and conserve genetic fidelity of the progeny. From the other point of view, indirect organogenesis is a major pathway used for genetic transformation of woody plants. Since the shoots originating on the explant surface are of a better chance to be transformed, the origin of the cells involved in the organogenesis process should be detected and identified for development of successful transformation protocols (Jose et al. 2014).

De novo shoot organogenesis through direct or indirect ways from leaf, root, stem, and other explants culturing on different basal media supplemented with TDZ alone or in combinations with other PGRs has been well demonstrated for various woody species (Table 3.2). According to these data, the most efficient concentrations of TDZ to induce adventitious shoot formation are varied in a wide range from 0.5 to 10 μM. Furthermore, the TDZ efficiency can be strengthened using combination with the other cytokinins and/or auxins. For instance, Sriskandarajah and Lundquist (2009) observed synergistic effect of TDZ and BA on direct regeneration from leaf and root explants of *Hippophae rhamnoides*, whereas Kim et al. (1997a) found that TDZ efficiency was greatly enhanced in the presence of 2,4-dichlorophenoxyacetic acid (2,4-D) in the induction medium. Indole-3-butyric acid (IBA) addition reinforced the TDZ-induced de novo regeneration in leaf explants of *Viburnum dentatum* (Dai et al. 2011). At the same time, exceeding the optimal TDZ concentrations resulted in the callus formation and triggered the indirect shoot organogenesis (Feyissa et al. 2005).

Among various explants the leaf is the most frequently and successfully used for de novo shoot regeneration (Table 3.2) and further genetic transformation compared to other explants. Leaf cultures are known to serve as a model for studying in vitro morphogenesis, because the lack of apical meristems in leaves provides the opportunity to induce a broad range of morphogenic responses (Lo et al. 1997; Woo and Wetzstein 2008; Zaytseva et al. 2016). Exogenous PGRs, in particular the auxins, are known to play a principal role in the regulation of morphogenesis from the leaf tissues. Since various concentrations of TDZ can influence the endogenous level of phytohormones, the same type of explant cells under TDZ can become multi-, pluri-, or totipotent, realizing different morphogenetic developmental programs, respectively.

Thus, it was shown that the same levels of TDZ could simultaneously stimulate direct and/or indirect organogenesis and/or SE (Table 3.2). Various morphogenic

Table 3.2 TDZ-induced adventitious shoot regeneration in woody plants

		Regeneration of adventitious shoot		Step 1 TDZ-induced regeneration		Step 2	
Plant species or cultivars	Explant type	Pathway	Competent cells	Nutrient medium and PGRs	TDZ exposure	Shoot elongation	References
Acacia mangium Willd.	Embryo axes leaf, petiole, stems of seedlings	Indirect	Not found	1. MS + 9.05 μM 2,4-D + 13.95 μM KT	40 days	MS + 0.045 μM TDZ + 7.22 μM GA_3	Xie and Hong (2001)
				2. MS + 4.55 μM TDZ + 1.43 μM IAA			
Acer pseudoplatanus L.	Plumule	Direct	No HO	MS + 0.04 μM TDZ + 1.0 μM BA	5 weeks	–	Wilhelm (1999)
	Hypocotyl	Indirect					
Berberis aristata	Leaf-derived callus	Indirect	No HO	WPM + 0.5 μM TDZ	50 days	WPM + 2.89 μM GA_3 + 4.44 μM BA	Brijwal et al. (2015)
Cydonia oblonga Mill.	Leaf	Direct org. and SE	No HO	MS + 4.5 μM TDZ + 0.5 μM NAA	20–25 days	MS + 4.5 μM BA + 0.5 μM NAA	D'onofrio and Morini (2005)
Elliottia racemosa Muhl.	Leaf	Direct	Subepidermal cell layers	GB5 + 10 μM TDZ + 5 μM IAA	–	–	Woo and Wetzstein (2008)
Embelia ribes Burm. f.	Leaf	Direct	No HO	MS + 0.272 μM TDZ	2–3 weeks	MS0	Raghu et al. (2006)
Eucalyptus grandis × *E. urophylla* AEC 224	Leaf	Indirect	Tissue of petiole and apex of the leaf, but competent cells not found	1. WPM + 0.25 μM TDZ + 0.1 μM NAA	30 days	Shoots elongated in the rooting medium MS + 2.46 μM IBA	De Oliveira et al. (2017)
				2. WPM + 5.0 μM BA + 0.5 μM NAA	30 days		
Eugenia myrtifolia Sims.	Immature seeds	Indirect	No HO	MS + 2.5 μM TDZ	90 days	MS 4.4 μM BA + 0.05 μM NAA or hormone-free ½ MS	Blando et al. (2013)
Fagus sylvatica L.	Internode segments	Indirect	No HO	WPM + 4.5 μM TDZ + 2.9 μM IAA	8 weeks	MS + 2.2 μM BA + 9.1 μM zeatin + 2.9 μM IAA	Cuenca et al. (2000)

Ficus carica L.	Leaf segments	Indirect	No HO	MS + 2 mg L^{-1} IBA + 0.5 or 1.0 mg L^{-1} TDZ	4–6 weeks	–	Kim et al. (2007)
Fraxinus profunda Bush.	Hypocotyl	Indirect	No HO	MS + 22.2 μM BA and 4.5 μM TDZ	4 weeks	MS + 13.3 μM BA + 1 μM IBA + 0.29 μM GA_3	Stevens and Pijut (2012)
Gaultheria fragrantissima Wall.	Leaf, internode, root	Direct	No HO	AM 1–2 mg/l TDZ	6 weeks	1/2 AM + 5 mg/l 2iP	Ranyaphia et al. (2011)
Gymnocladus dioicus L.	Root and petiole	Direct	No HO	WPM + 0.5 or 1.0 μM TDZ + 5.0 or 10 μM BA	12 weeks	–	Geneve (2005)
Hagenia abyssinica Bruce.	Leaf[a]	Direct	Subepidermal and inner parenchyma cells	MS + 0.1 μM TDZ + 0.01 μM NAA	9–12 weeks	–	Feyissa et al. (2005)
		Indirect		MS + 1.0–10.0 μM TDZ			
Hippophae rhamnoides L.	Leaf, roots	Direct	No HO	MS + 4.5 or 2.25 μM TDZ + 2.2 μM BA + 0.53 μM NAA	5 weeks	WPM + 4.4 μM BA + 0.29 μM GA_3 + 57.0 μM IAA	Sriskandarajah and Lundquist (2009)
Jatropha curcas L.	Leaf, cotyledo, hypocoty, radicle	Direct	No HO	MS + 1.0 mg/L TDZ + 0.5 mg/L Kn + 0.5 mg/L GA_3	30 days	MS + 0.3 mg/L BA + 0.01 mg/L IBA	Zhang et al. (2013)
Liquidambar formosana L.	Leaf	Direct	No HO	WPM + 1.14 μM TDZ + 0.27 μM NAA	7 weeks	WPM + 0.54 μM NAA, 2.22 μM BA + 1.44 μM GA_3	Xu et al. (2007)
Liquidambar styraciflua L.	Hypocotyl segments	Direct	No HO	Sommer and Brown + 0.1 mg/l TDZ + 0.01 mg/l 2,4-D	7 weeks	Sommer and Brown + 0.01 mg/l NAA + 0.5 mg/l BA	Kim et al. (1997a)
Macadamia tetraphylla L.	Immature cotyledon	Indirect	Parenchyma tissue below the cut surfaces	MS + 10 or 15 μM TDZ	12 weeks	MS + 8.8 μM BA	Mulwa and Bhalla (2006)

(continued)

Table 3.2 (continued)

Plant species or cultivars	Explant type	Regeneration of adventitious shoot		Step 1		Step 2	References
				TDZ-induced regeneration			
		Pathway	Competent cells	Nutrient medium and PGRs	TDZ exposure	Shoot elongation	
Malus domestica "Gala" "Fuji"	Leaf segments	Indirect	No HO	MS + 2–3 mg L^{-1} TDZ, + 0.5 mg L^{-1} IBA	11 weeks	–	Li et al. (2014)
Nothapodytes foetida Wight	Hypocotyl segments	Direct	No HO	MS + 2.2 μM TDZ	8 weeks	MS + 2.2 μM BA	Rai (2002)
Paulownia tomentosa Thunb.	Leaf	Direct and callus formation	No HO	MS + 22.7 or 27.3 μM TDZ + 2.9 μM IAA	2 weeks	MS + 0.44 μM BA	Corredoira et al. (2008)
	Leaf[a]	Direct and indirect	Subepidermal and inner parenchyma cells	MS + 22.7 μM TDZ + 2.9 μM IAA	4 weeks	–	José et al. (2014)
Phellodendron amurense Rupr.	Leaf	Indirect	No HO	1. MS + 2.0 μM TDZ + 4.0 μM 2,4-D or 4.0 μM NAA	3 weeks	–	Azad et al. (2005)
				2. MS + 1.5 μM BAP + 1.0 μM NAA	4 weeks		
Pongamia pinnata L.	Cotyledon	Direct	No HO	MS + 11.35 μM TDZ	20 days	MS	Sujatha et al. (2008)
Populus interamericana × *Populus euramericana*	Stem cuttings	Indirect	No HO	1. ½ MS or WPM + 10 μM NAA + 5 μM 2-iP	25 days	WPM	Noël et al. (2002)
				2. WPM + 0.25 μM TDZ	30 days		
Populus alba × *P. berolinensis*	Stem	Direct	No HO	MS + TDZ 0.1 mgL^{-1} + 0.02-mg L^{-1} NAA	4 weeks	–	Wang et al. (2008)

Populus tomentosa Carriere	Petioles	Indirect	No HO	1. 1/2MS + 0.5 mg/L ZT + 1.0 mg/l NAA	3–4 weeks	–	Wei et al. (2017)
				2. 1/2MS + 0.25 mg/L BA + 0.25 mg/L ZT + 0.25 mg/L NAA + 0.01 mg/L TDZ + 0.5 mg/L KT	6 weeks		
Populus tremula L.	Roots	Direct	Epidermal and subepidermal	MS liquid $4*10^{-2}$ μM TDZ	6–8 weeks	–	Vinocur et al. (2000)
Prunus avium "Lapins" "Sweetheart"	Leaf	Indirect	No HO	MS + 2.27 or 4.54 μM TDZ + 0.27 μM NAA	6 weeks	MS + 3 μM BA + 1 μM IBA	Bhagwat and Lane (2004)
Prunus dulcis "Ne Plus Ultra" 'Nonpareil'	Leaf	Indirect	No HO	AP + 9.1 μM TDZ + 9.8 μM IBA	14–21 days	AP + 2.3–4.5 μM TDZ or TDZ + 22.2 μM BA	Ainsley et al. (2000)
Prunus persica × *P. davidiana* "Nemaguard"	Non-expanded leaves	Indirect	No HO	½ MS + 9.08 μM TDZ + 0.54 μM IBA	9 weeks	–	Zhou et al. (2010)
Prunus serotina Ehrh.	Leaf	Indirect	No HO	WPM + 9.08 μM TDZ + 0.54 or 1.07 μM NAA	5 weeks	MS + 8.88 μM BA, 0.49 μM IBA, + 0.29 μM GA_3	Liu and Pijut (2008)
Robinia pseudoacacia L.	Hypocotyl segments	Direct	No HO	N6 + TDZ 0.1 μM	–	–	Hosseine-Nasr and Rashid (2003)
Rose cvs	Leaf	Direct	No HO	MS + 0.05 or 1.0 mg l^{-1} TDZ	4 weeks	–	Kucharska and Orlikowska (2009)
Semecarpus anacardium L.	Cotyledon	Direct	Not found	WPM + 9.08 μM TDZ	4 weeks	WPM	Panda and Hazra (2012)

(continued)

Table 3.2 (continued)

				Step 1		Step 2	
		Regeneration of adventitious shoot		TDZ-induced regeneration			
Plant species or cultivars	Explant type	Pathway	Competent cells	Nutrient medium and PGRs	TDZ exposure	Shoot elongation	References
Vaccinium angustifolium Ait.	Leaf	Indirect and direct	No HO	Cranberry medium [1] + 2.3–4.5 μM TDZ	6 weeks	2.3–4.6 μM zeatin	Debnath (2009)
Vaccinium vitis-idaea L.	Hypocotyl segments	Indirect	No HO	MS + 5–10 μM TDZ	8 weeks	MS + 1–2 μM zeatin	Debnath (2003)
Viburnum dentatum L.	Leaf	Indirect	No HO	1. WPM + 40 μM BA + 8 μM TDZ	8 weeks	–	Dai et al. (2011)
Vitex negundo L.	Leaf and stem segments	Indirect	No HO	1. MS + 0.5–2.15 μM TDZ + 1.7 μM IAA	4 weeks	MS + 2.4 μM GA_3	Rani and Nair (2006)
				2. MS + 2.7 μM TDZ			
Zizyphus jujuba Mill.	Leaf	Direct	No HO	WPM + 4.54 μM TDZ + 2.85 μM IAA	20 days	MS + 0.89 μM BA + 5.77 μM GA_3	Gu and Zhang (2005)
Rhododendron sichotense Pojark.	Leaf	Direct	Epidermal cells	AM + 0.5 or 1.0 μM TDZ 16 weeks	16 weeks	AM	Zaytseva et al. (2016)
Rhododendron catawbiense 'Grandiflorum'							
Malus domestica "Topred"	Leaf	Direct	Epidermal and subepidermal cell layers	1. MS + 5 mg l^{-1} TDZ + 1 mg l^{-1} 2,4-D	6 days	–	Yancheva et al. (2003)
				2. MS + 5 mg l^{-1} TDZ + 1 mg l^{-1} 2 IBA	33 days		

HO histological analysis, *AP* Almehdi and Parfitt's medium, *AM* Anderson's medium, *GB5* Gamborg medium, *MS* Murashige and Skoog medium, *WPM* woody plant medium responses were successfully demonstrated in leaf cultures of *Vaccinium angustifolium* Ait. (Debnath 2009), *Paulownia tomentosa* (José et al. 2014), *Cydonia oblonga* Mill. (D'onofrio and Morini 2005). For better understanding the different scenarios occurred during in vitro plant regeneration and revealing the origin of de novo structures, the detailed morpho-histological studies were carried out.

There are few examples of deep histological and morphological examination of TDZ-induced de novo organogenesis (Table 3.2). For instance, the study of apple shoot organogenesis from leaf explants of cv. Topred was conducted with the help of light and scanning electron microscopies (SEM) on the basis of the current concept of tissue competence, determination, and differentiation (Yancheva et al. 2003). The authors revealed that cell determination occurred after only 1 day of exposure to TDZ with IBA or TDZ with 2,4-D, prior to transfer to hormone-free medium. Histological analysis showed that the early shoot developing, appearing on the surface of the explant within 14 days, originated from the epidermal and subepidermal cell layers, whereas adventitious shoots arising later developed from de novo internally formed meristematic centers (Yancheva et al. 2003). SEM observations allowed to reveal the developmental patterns of apple regeneration from the first cell divisions to adventitious shoots or globular embryo-like structure in dependence on TDZ combinations with IBA or /2,4-D (Yancheva et al. 2003).

Vieitez and San-Jose (1996) observed two pathways of adventitious shoot formation on the transversally divided expanding leaves of *Fagus sylvatica* in the presence of 4.5 μM TDZ with 2.9 μM IAA. It was found that a part of adventitious buds originated directly from a small protuberance of parenchymatic tissue produced by the subepidermis and bundle sheath cells located in the adaxial part of the main vein. However, most frequently shoot bud primordia arose from the meristematic centers formed in the callus produced at the cut end of the petiole by parenchymatic cells and numerous tracheids (Vieitez and San-Jose 1996).

TDZ-induced in vitro morphogenesis in *Rhododendron* leaf cultures demonstrated species-specific characteristic (Pavingerova 2009; Zaytseva et al. 2016). In order to realize outstanding property of TDZ to induce organogenesis and SE in the same type of explant with different TDZ concentrations, the experiments were conducted in leaf cultures of *R. sichotense* and *R. catawbiense* cv. Grandiflorum (Zaytseva et al. 2016). Histological examination provided detailed analysis of the differences in regeneration pathways of the genotypes tested, but de novo structures formed under TDZ exposure were epidermal-derived in both genotypes. The adventitious shoot buds of *R. sichotense* developed from protuberances (Fig. 3.2) whereas shoot organogenesis of *R. catawbiense* cv. Grandiflorum occurred from embryo-like protrusions (Fig. 3.3a, b) at the same TDZ concentration. Consequently, the shoot regeneration via direct organogenesis for both genotypes was confirmed.

Epidermal cells of both TDZ-treated leaf and root explants more frequently acquired the competence to organogenesis than other tissue cells (Table 3.2). In the study of Vinocur et al. (2000) carried out on *Populus tremula* roots treated with TDZ, the high regeneration capacity of epidermal and subepidermal cells was observed. In contrast, in BA or in growth regulator-free media, only the cells in close proximity to lateral root primordia redifferentiated to form buds (Vinocur et al. 2000).

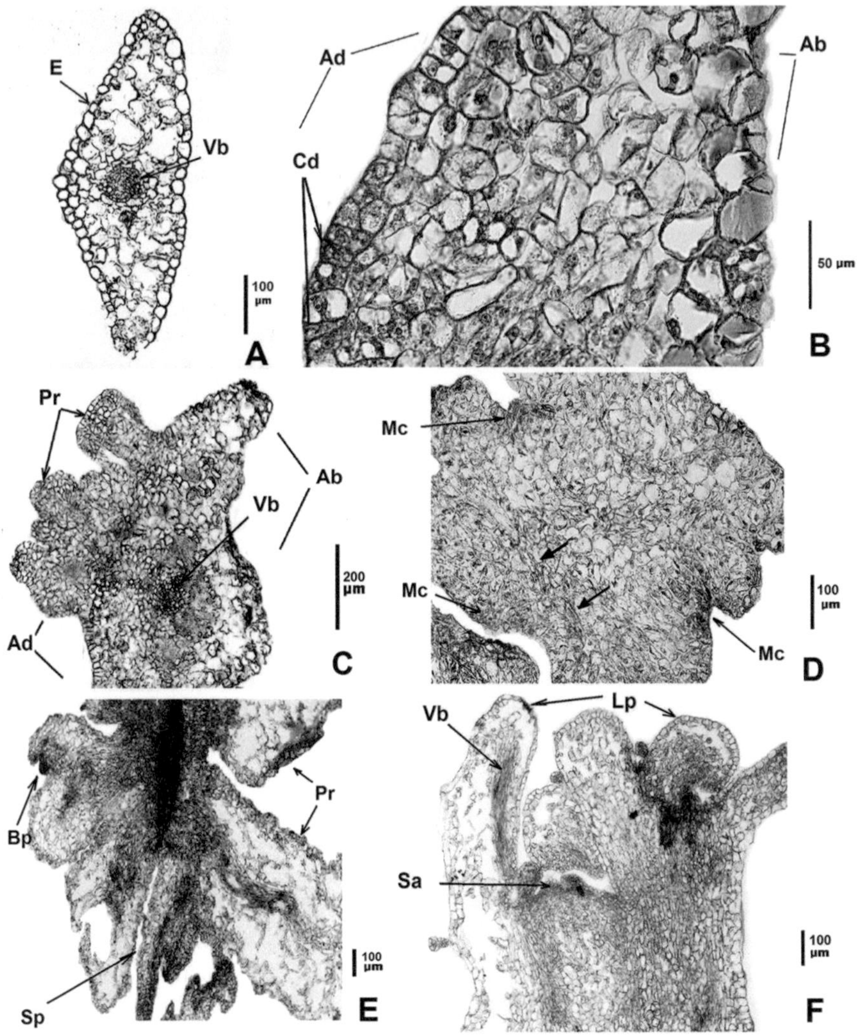

Fig. 3.2 Histological observations of adventitious buds formation from *R. sichotense* leaf explants under 1.0 µM TDZ exposure (Zaytseva et al. 2016). (**a**) Cross section through leaf explant at 0 day. (**b**) First cell divisions in adaxial epidermis of leaf explant after 10 days of culture. (**c**) Protuberance formation at the adaxial side of leaf explant at 14 days. (**d**) Formation of meristematic center and vascular system at the protuberance (arrow). (**e**) Longitudinal section of differentiated bud with leaf primordia at 35 days. (**f**) Formation of multiple leaf primordia at 8 weeks of culture. *Ab* abaxial surface, *Ad* adaxial surface, *Bp* bud primordium, *Cd* cell divisions, *E* epidermis, *Lp* leaf primordium, *Mc* meristematic center, *Pr* protuberance, *Sa* shoot apex, *Sp* shoot primordium, *Vb* vascular bundle

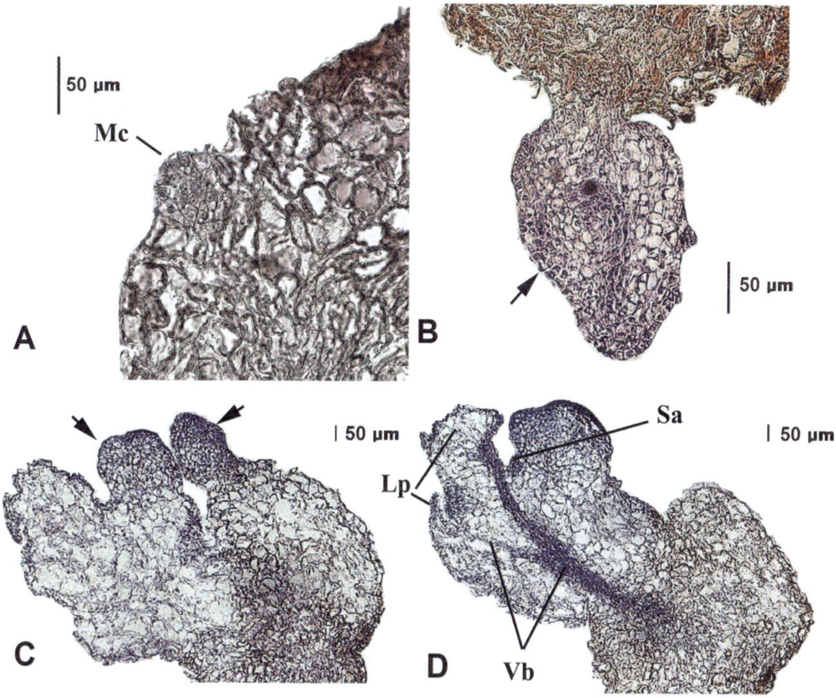

Fig. 3.3 Light microscopy observations of *R. catawbiense* cv. Grandiflorum leaf culture under TDZ exposure (Zaytseva et al. 2016). (**a**) Meristematic center in adaxial epidermis after 12 days exposure to 0.5 µM TDZ. (**b**) Embryo-like structure formation (arrow) on AM supplemented with 0.5 µM TDZ at 35 days of culture. (**c**) Embryo-like structure (arrows) developed under 5.0 µM TDZ. (**d**) Bud development in tissues exposed to 5.0 µM TDZ after 15 weeks of culture. *Lp* leaf primordium, *Mc* meristematic center, *Sa* shoot apex, *Vb* vascular bundle

Nevertheless, the TDZ-treated parenchymal cells of leaf and cotyledon explants have been found to be of competence to organogenesis. The epidermal cell layers of *Hagenia abyssinica* leaf explants were intact and the cell divisions appeared to be from the inner cells (Feyissa et al. 2005). At the same time, José et al. (2014) observed a simultaneous dedifferentiation of some subepidermal and inner parenchyma cells of *Paulownia tomentosa* leaf explants. Histological examinations in the culture of immature cotyledon of *Macadamia tetraphylla* revealed that shoot regeneration was primarily by organogenesis originating from cell division in the parenchyma tissue of explants (Mulwa and Bhalla 2006). Thus, based on the recent studies on histological observation during organogenesis, it can assume that the most competent cells occur in epidermal and subepidermal layers of leaf or cotyledon explants, but in some cases, parenchymal cells of leaf and cotyledon explants have been found to be of competence to organogenesis. Consequently, in each experiment, the histological examinations of target cell localization are a prerequisite for further genetic transformation.

3.3.3 Some Approaches to Overcome TDZ-Induced Abnormalities

The reverse side of TDZ high efficiency in the induction of various morphogenetic responses in woody plant tissue culture is the development of different undesired anomalies or disorders at cell, tissue, and organ levels. Several physiological and developmental aberrations may occur with the plants during the in vitro clonal propagation due to an artificial environment of culture growth conditions of high humidity, low light levels, and hetero- or mixotrophic nutrition (Hazarica 2006). Some of the culture-induced abnormalities can become genetic inherited by clonal progenies while others are temporary and can be corrected at a later stage (Isah 2015). The morphophysiological response to TDZ treatment varies with genotype, explant type, ontogeny, and recalcitrance of woody plants. However, TDZ-induced phenotype has characteristic features similar to various species of different systematic groups (Figs. 3.1b and 3.4e). Such anomalies as the hyperhydricity (Sandal et al. 2001; Kadota and Niimi 2003; Siwach and Gill 2011; Ahmed and Anis 2012), abnormal leaf morphology (Kim et al. 1997a; Sandal et al. 2001), fasciation, and reduced elongation (Rai 2002; Gu and Zhang 2005; Parveen and Shahzad 2010) as well as inhibition of root development (Lu 1993) occur most often.

Morpho-histological studies allow to reveal the developmental basis of some anomalies observed. For instance, some authors consider the fasciation as fusion of organs due to the deviation from normal meristematic processes while others suggest that it is the result of transformation of a single growing point into a line (Clark et al. 1993). Histological study of *Elliottia racemosa* morphogenesis detected the formation of bud-like clusters, elongated embryo-like protrusions, and shoot-like structures produced from the leaf explants under 10 μM TDZ treatment in combination with 5 μM IAA (Woo and Wetzstein 2008). Some of these structures were incapable of new plant regeneration due to such developmental aberrations in the structure as a lack of shoot apical meristem or root apex. Panda and Hazra (2012) showed that numerous buds were regenerated on the hypocotyls surface of *Semecarpus anacardium*; however, only a few of them differentiated to shoots on TDZ-containing nutrient medium. The formation of embryo-like and fused shoot-like structures was observed in the leaf culture of *Rhododendron catawbiense* "Grandiflorum" in the presence of TDZ (Zaytseva et al. 2016). Most of these disorders may be connected with nondegradable nature of this urea-based compound due to the absence of specific oxidase enzymes in cells that distinguishes TDZ from the natural phytohormones. Since physiological and developmental problems arising during in vitro propagation may have severe impact on survival of plantlets, the optimization of TDZ application protocols is needed to decrease the observed anomalies.

There are some approaches to overcome TDZ-induced disorders in in vitro culture of woody plants (Tables 3.2 and 3.3):

Fig. 3.4 Morphogenesis of *R. catawbiense* cv. Grandiflorum in vitro from leaf explants (Zaytseva et al. 2016). (**a**) Formation of embryo-like structures and stunted abnormal shoots at the base of leaf after 35 days on AM supplemented with 0.5 μM TDZ. (**b**) Clusters of short shoots at 45 days of culture on AM0 after pulse treatment with 30 μM TDZ. (**c**) Embryo-like structures (arrow) at 35 days on AM supplemented with 5.0 μM TDZ. (**d**) Embryo-like structures (arrow) on brown leaf explant after 15 weeks of culture with 5.0 μM TDZ. (**e**) Shoot clusters after 15 weeks of culture on AM with 0.5 μM TDZ. (**f**) Elongated shoot clusters after 8 weeks culture on AM0. *As* abnormal shoot, *El* embryo-like structure

Table 3.3 TDZ-induced somatic embryogenesis in woody plants

Plant species or cultivars	Explant type	SE pathway		Culture conditions			References
		Type	Competent cells	Step	PGRs	Time of treating	
Melia azedarach L.	Immature zygotic embryos	Direct	Epidermal and subepidermal cells	I+M	MS + 13.62 μM TDZ;	6 weeks	Vila et al. (2003)
				G	MS or1/4 MS	4 weeks	
	Immature zygotic embryos	Direct	No HO	I	MS + 4.54 μM TDZ;	18 days	Vila et al. (2007)
				M+G	¼ MS		
Mondia whitei Hook.f.	Leaf	Indirect	No HO	ECP+I	MS + 20 μM 2,4-D μM + 1 μM TDZ;	8 weeks	Baskaran et al. (2015)
				M+G	MS + 0.5 μM mTR + 0.25 μM IAA		
Murraya koenigii L.	Embryonic axis	Indirect	Parenchyma cells immediately below the epidermal layer cell of epidermal origin	ECP	MS + 8.88 μM BA + 2.675 μM NAA;	4 weeks	Paul et al. (2011)
	Cotyledon	Direct and secondary direct SE		I+M	MS + 4.54 or 9.08 μM TDZ;	6 weeks	
				G	1/2 MS + 2.32 μM KN + 1.07 μM NAA	4 weeks	
Ochna integerrima Lour.	Leaf and shoot explants	Direct SE and adventitious shoot	No HO	I	MS + 10.0–15.0 μM TDZ;	45 days	Ma et al. (2011)
				M	0.5 μM NAA or 5.0 μM BA	45 days	

Paulownia elongata S.Y.Hu	Leaf	Direct	No HO	I	MS + 10 mg l^{-1} TDZ;	7 days	Ipekci and Gozukirmizi (2003)
				M+G	MS	4 weeks	
	Leaf and internode	Indirect	No HO	ECP+I	MS + 4 mg/l TDZ + 0.1 mg/l Kn;	3 weeks	Ipekci and Gozukirmizi (2004)
				M	MS + 0.1 mg/l TDZ + 1 mg/l Kn + 2 mM glutamine;	4 weeks	
				G	MS	2 weeks	
Santalum album L.	Zygotic embryo	Direct and	No HO	I+M	MS + 4.5 μM TDZ;	8–10 weeks	Rai and McComb (2002)
		Secondary SE		G	MS + 1.4 μM GA_3	6 weeks	
Semecarpus anacardium L.	Cotyledon	Direct and indirect	Not found	I	WPM 9.08 μM TDZ;	4 weeks	Panda and Hazra (2012)
				M+G	WPM	4–20 weeks	
Theobroma cacao L.	Staminode	Indirect	No HO	ECP+I	DKW 9 μM 2,4-D + 22.7 nM TDZ;	2 weeks	Li et al. (1998)
					WPM + 9 μM 2,4-D + 1.4 μM kinetin;	2 weeks	
				M+G	DKW		

HO histological observations, *ECP* embryogenic callus proliferation, *I* induction of SE, *M* maturation of SE, *G* germination of SE, *MS* Murashige and Skoog medium, *WPM* woody plant medium, *DKW* Driver-Kuniyuki walnut medium

three stages instead of four or even five when using traditional PGRs. Thus, the use of TDZ optimizes the induction and maturation processes as well as reduces the production period of somatic embryos since TDZ substitutes for the auxin and cytokinin requirements of SE. Moreover, the induction of SE can be achieved with a short TDZ exposure (Visser et al. 1992).

To trigger reprogramming the fate of competent cell was noted to require low TDZ levels (from 0.5 up to 10.0 μM) which can simultaneously induce both the somatic embryos and adventitious bud formation in some species (Fiola et al. 1990; Jones et al. 2007). Furthermore, TDZ alone in range of 5–50 μM can be effective enough for embryogenic callus proliferation and SE induction in woody plants (Rai and McComb 2002; Ma et al. 2011; Panda and Hazra 2012). The duration of these stages is about 4–6 weeks, but current improvement of the technologies is toward reducing both the TDZ exposition time and its concentration in a nutrient medium. For instance, the induction and maturation of somatic embryos in the culture of immature zygotic embryos of *Melia azedarach* occurred during 6 weeks on the same nutrient medium supplemented with 13.62 μM TDZ (Vila et al. 2003). As a result, the most of embryos developed well-defined hypocotyl and cotyledons, while others showed fused or fasciated hypocotyl structures. However, it was later shown the conversion rate was increased by shorting TDZ treatment to 18 days and reducing its concentration to 4.54 μM at the stage of SE induction (Vila et al. 2007). Nevertheless, many embryos exhibited fused cotyledonary leaves. Ipekci and Gozukirmizi (2003) used successfully the 7-day exposure to 10 mg l-1 TDZ for inducing SE in the leaf explants of *Paulownia elongata*. Subsequent removing of TDZ from the medium resulted in the maturation, normal growth, and development of the embryos into plantlets on hormone-free MS.

As in the case of shoot organogenesis, TDZ can stimulate various morphological developmental anomalies in somatic embryos. Moreover, the formation of TDZ-induced somatic embryos is often asynchronous (Baskaran et al. 2015). The secondary SE often occurred when somatic embryos are formed on de novo structures (Rai and McComb 2002; Paul et al. 2011). Abnormal embryos can be with fused margin or cotyledon (D'onofrio and Morini 2005; Vila et al. 2007), fused polycotyledons, or a single cotyledon (Rai and McComb 2002; Paul et al. 2011). As a result, only a few somatic embryos were found to be of typical embryonal bipolarity and regular morphology. At the same time some of embryos were of normal morphologically but failed to convert into plantlets (D'onofrio and Morini 2005; Panda and Hazra 2012).

Unfortunately, most studies on SE of woody plants are not accompanied by morpho-histological analysis, and therefore, it is difficult to understand exactly the nature of some adverse TDZ effects. Moreover, there is a possibility of incorrect identification of the morphogenesis pathway. Dobrowolska et al. (2017) conducted a histological analysis of the structures regenerating on hypocotyls of mature zygotic embryos of *Eucalyptus globulus*. As a result, the analysis of section series revealed organogenic pathways of regeneration in this system. At the same time, Pinto et al. (2002) have early reported about effective embryogenic regeneration from zygotic embryos of *E. globulus* under similar conditions, but the study has been carried out without histological analysis. When leaf explants of *Rhododendron catawbiense* "Grandiflorum" are cultivated on a medium with 5 μM TDZ, the development of embryo-like structures was observed; however, histological analysis showed the absence of bipolarity and the presence of a common vascular system of regenerants and explants (Zaytseva et al. 2016). The embryo-like structures at the early stages of development looked like somatic embryos (Fig. 3.4a–d). Moreover, the formation of structures similar to secondary somatic embryos was marked. These results indicated that regeneration occurred through the direct organogenesis rather than SE. Woo and Wetzstein (2008) previously showed that histological evaluations of *Elliottia racemosa* cultures failed to identify true somatic embryo development. Although, some globular- and heart-shaped structures were observed suggesting the proembryo development, no bipolar structures with defined root meristems were detected as would be anticipated in later embryogenic stages of development (Woo and Wetzstein 2008). Thus, to confirm the process of SE, a histological analysis should be performed, and protocols of TDZ-induced propagation must be carefully evaluated by microscopic examination of adequate numbers of serial sections. Therefore, Bassuner et al. (2007) suggested to reevaluate many published reports on SE for clarifying the morphogenetic pathway. Moreover, the importance of morpho-histological studies of SE has been increasing because of the development of methods for plant genetic transformation. It is histological studies that can elucidate the question, as for what cells can become embryogenic in order to purposefully introduce the necessary genetic constructs into these cells for obtaining and multiplying transformed genotypes.

3.5 Concluding Remarks

Nowadays the forest tree genetic transformation and clonal propagation are a highly strategic research area (Ravikumar 2001). The genetic engineering or mutagenesis techniques would not be successfully achieved, if the basic knowledge underlying the morphogenesis processes were not well understood. The studies of localization of target cells, where initiation of de novo organogenesis or SE occurs, are of even more importance as the necessity of genetic transformation of woody plants has been increasingly growing in recent years.

The apparent progress has been made in vitro propagation of recalcitrant woody species after establishing the strong cytokinin-like activity of TDZ followed by revealing tremendous ability of TDZ to stimulate various morphogenic responses in tissue culture. Plant tissue culture techniques coupled with advanced microscopic technologies is a modern and powerful tool to insight into the dynamics of plant growth and development (Moyo et al. 2015). Despite the benefits of TDZ application for induction of in vitro woody plant regeneration, revealing the mode of action of TDZ is still a challenge. To clarify the precise mechanism of TDZ effects triggering different scenarios during in vitro woody plant regeneration, along with physiological-biochemical and genetic researches, the expansion of microscopic approaches is required. Besides histological methods, the histochemical and immunocytochemical techniques, demonstrating spatial and temporal changes in cells during morphogenesis processes and improving resolving power of microscopy, are in extreme demand. This knowledge should be critical both for establishment of efficient protocols for large-scale propagation of woody plants and basic developmental biology.

Acknowledgments The reported study was funded by the Russian Foundation for Basic Research (RFBR) according to the research project № 17-04-00782.

References

Ahmad N, Anis M (2007) Rapid clonal multiplication of a woody tree, *Vitex negundo* L. through axillary shoots proliferation. Agrofor Syst 71:195–200. https://doi.org/10.1007/s10457-007-9078-1

Ahmed MR, Anis M (2012) Role of TDZ in the quick regeneration of multiple shoots from nodal explant of *Vitex trifolia* L. – an important medicinal plant. Appl Biochem Biotechnol 168:957–966. https://doi.org/10.1007/s12010-012-9799-0

Ahuja MR (1993) Micropropagation a la carte. In: Ahuja MR (ed) Micropropagation of woody plants, forestry series, vol 41. Kluwer Academic, Dordrecht, pp 3–9

Ainsley PJ, Collins GG, Sedgeley M (2000) Adventitious shoot regeneration from leaf explants of almond (*Prunus dulcis* mill.) In Vitro Cell Dev Biol-Plant 36:470–474. https://doi.org/10.1007/s11627-000-0084-5

Atta R, Laurens L, Boucheron-Dubuisson E et al (2009) Pluripotency of *Arabidopsis* xylem pericycle underlies shoot regeneration from root and hypocotyl explants grown in vitro. Plant J 57:626–644. https://doi.org/10.1111/j.1365-313X.2008.03715.x

Azad MAK, Yokota S, Ohkubo T et al (2005) In vitro regeneration of the medicinal woody plant *Phellodendron amurense* Rupr. through excised leaves. Plant Cell Tissue Organ Cult 80:43–50. https://doi.org/10.1007/s11240-004-8809-5

Bairu MW, Kane ME (2011) Physiological and developmental problems encountered by in vitro cultured plants. Plant Growth Regul 63:101–103. https://doi.org/10.1007/s10725-011-9565-2

Baránek M, Čechová J, Raddová J et al (2015) Dynamics and reversibility of the DNA methylation landscape of grapevine plants (*Vitis vinifera*) stressed by *in vitro* cultivation and thermotherapy. PLoS One 10(5):e0126638. https://doi.org/10.1371/journal.pone.0126638

Baskaran P, Kumari A, Van Staden J (2015) Embryogenesis and synthetic seed production in *Mondia whitei*. Plant Cell Tissue Organ Cult 121:205–214. https://doi.org/10.1007/s11240-014-0695-x

Bassuner BM, Lam R, Lukowitz W, Yeung EC (2007) Auxin and root initiation in somatic embryos of *Arabidopsis*. Plant Cell Rep 26:1–11

Bhagwat B, David Lane W (2004) In vitro shoot regeneration from leaves of sweet cherry (*Prunus avium*) 'Lapins' and 'Sweetheart'. Plant Cell Tissue Organ Cult 78:173–181. https://doi.org/10.1023/B:TICU.0000022552.12449.71

Blando F, Onlu S, Colella G et al (2013) Plant regeneration from immature seeds of *Eugenia myrtifolia* Sims. In Vitro Cell Dev Biol-Plant 49:388–395. https://doi.org/10.1007/s11627-013-9502-3

Briggs BA, MCculloch SM, Edick LA (1988) Micropropagation of azalea using thidiazuron. Acta Hortic 226:205–208

Brijwal L, Pandey A, Tamta S (2015) In vitro propagation of the endangered species *Berberis aristat*a DC. via leaf-derived callus. In Vitro Cell Dev Biol-Plant 51:637–647. https://doi.org/10.1007/s11627-015-9716-7

Butenko RG (1999) Biology of higher plant cells in vitro and biotechnologies on their bases. FBC-Press, Moscow

Casanova E, Valdes AE, Fernandez B et al (2004) Levels and immunolocalization of endogenous cytokinins in thidiazuron-induced shoot organogenesis in carnation. J Plant Physiol 161(1):95–104

Cheepala S, Sharma N, Sahi S (2004) Rapid in vitro regeneration of *Sesbania drummondii*. Biol Plant 48(1):13–18. https://doi.org/10.1023/B:BIOP.0000024269.72171.42

Chhabra G, Chaudhary D, Varma M et al (2008) TDZ-induced direct shoot organogenesis and somatic embryogenesis on cotyledonary node explants of lentil (*Lens culinaris* Medik.) Physiol Mol Biol Plants 14(4):347–353. https://doi.org/10.1007/s12298-008-0033-z

Christianson ML, Warnick DA (1985) Temporal requirement for phytohormone balance in the control of organogenesis in vitro. Dev Biol 112:494–497. https://doi.org/10.1016/0012-1606(85)90423-3

Clark SE, Running MP, Meyerowitz EM (1993) CLAVATA1, a regulator of meristem and flower development in Arabidopsis. Development 119:397–418

Corredoira E, Ballester A, Vieitez AM (2008) Thidiazuron-induced high-frequency plant regeneration from leaf explants of *Paulownia tomentosa* mature trees. Plant Cell Tissue Organ Cult 95:197–208. https://doi.org/10.1007/s11240-008-9433-6

Cuenca B, Ballester A, Vieitez AM (2000) In vitro adventitious bud regeneration from internode segments of beech. Plant Cell Tissue Organ Cult 60:213–220

D'Onofrio C, Morini S (2005) Development of adventitious shoots from in vitro grown *Cydonia oblonga* leaves as influenced by different cytokinins and treatment duration. Biol Plant 49(1):17–21. https://doi.org/10.1007/s10535-005-7021-8

Dai W, Su Y, Castillo C et al (2011) Plant regeneration from in vitro leaf tissues of *Viburnum dentatum* L. Plant Cell Tissue Organ Cult 104:257–262. https://doi.org/10.1007/s11240-010-9829-y

De Oliveira C, Degenhardt-Goldbach J, de França Bettencourt GM et al (2017) Micropropagation of *Eucalyptus grandis* 3 E. urophylla AEC 224 clone. J For Res 28(1):29–39. https://doi.org/10.1007/s11676-016-0282-6

Debnath SC (2003) Improved shoot organogenesis from hypocotyl segments of lingonberry (*Vaccinium vitis-idaea* L.) In Vitro Cell Dev Biol-Plant 39:490–495. https://doi.org/10.1079/IVP2003458

Debnath SC (2009) A two-step procedure for adventitious shoot regeneration on excised leaves of lowbush blueberry. In Vitro Cell Dev Biol-Plant 45:122–128. https://doi.org/10.1007/s11627-008-9186-2

Dhavala A, Rathore TS (2010) Micropropagation of *Embelia ribes* Burm f. through proliferation of adult plant axillary shoots. In Vitro Cell Dev Biol-Plant 46:180–191. https://doi.org/10.1007/s11627-010-9285-8

Distabanjong KI, Geneve RL (1997) Multiple shoot formation from cotyledonary node segments of eastern redbud. Plant Cell Tissue Organ Cult 47:247–254

Dobrowolska I, Andrade GM, Clapham D et al (2017) Histological analysis reveals the formation of shoots rather than embryos in regenerating cultures of *Eucalyptus globulus*. Plant Cell Tissue Organ Cult 128:319–326. https://doi.org/10.1007/s11240-016-1111-5

Dolendro Singh N, Sahoo L, Sarin NB et al (2003) The effect of TDZ on organogenesis and somatic embryogenesis pigeonpea (*Cajanus cajan* L. Millsp.) Plant Sci 164:341–347. https://doi.org/10.1016/S0168-9452(02)00418-1

Durkovic J, Misalova A (2008) Micropropagation of temperate noble hardwoods: an overview. Funct Plant Sci Biotechnol 2:1–19

Elhiti M, Stasolla C, Wang A (2013) Molecular regulation of plant somatic embryogenesis. In Vitro Cell Dev Biol-Plant 49:631–642. https://doi.org/10.1007/s11627-013-9547-3

Faisal M, Ahmad N, Anis M (2005) Shoot multiplication in *Rauvolfia tetraphylla* L. using thidiazuron. Plant Cell Tissue Organ Cult 80:187–190. https://doi.org/10.1007/s11240-004-0567-x

Fehér A (2015) Somatic embryogenesis – stress-induced remodeling of plant cell fate. Biochim Biophys Acta-Gene Regul Mech 1849:385–402. https://doi.org/10.1016/j.bbagrm.2014.07.005

Feyissa T, Welander M, Negash L (2005) In vitro regeneration of *Hagenia abyssinica* (Bruce) J.F. Gmel. (*Rosaceae*) from leaf explants. Plant Cell Rep 24:392–400. https://doi.org/10.1007/s00299-005-0949-5

Fiola JA, Hassan MA, Swartz HJ et al (1990) Effect of thidiazuron, light fluence rates and kanamycin on in vitro shoot organogenesis from excised *Rubus* cotyledons and leaves. Plant Cell Tissue Organ Cult 20:223–228

Gahan PB, George EF (2008) Adventitious regeneration. In: George EF, Hall MA, De Klerk GJ (eds) Plant propagation by tissue culture, 3rd edn. Springer, Dordrecht, pp 355–401

Gaspar T, Kevers C, Bisbis B et al (2000) Loss of plant organogenic totipotency in the course of in vitro neoplastic progression. In Vitro Cell Dev-Plant 36:171–181

Geneve RL (2005) Comparative adventitious shoot induction in Kentucky coffeetree root and petiole explants treated with thidiazuron and benzylaminopurine. In Vitro Cell Dev Biol-Plant 41:489–493. https://doi.org/10.1079/IVP2005669

Genkov T, Iordanka I (1995) Effect of cytokinin-active phenylurea derivatives on shoot multiplication perioxidase and superoxide dismutase activities of in vitro cultured carnation. Bulg J Plant Physiol 21:73–83

Germana MA, Lambardi M (2016) In vitro embryogenesis in higher plants. Springer Science+Business Media, New York

Giri CC, Shyamkumar B, Anjaneyulu C (2004) Progress in tissue culture, genetic transformation and applications of biotechnology to trees: an overview. Trees 18:115–135. https://doi.org/10.1007/s00468-003-0287-6

Graner EM, Oberschelp GPJ, Brondani GE et al (2013) TDZ pulsing evaluation on the in vitro morphogenesis of peach palm. Physiol Mol Biol Plants 19(2):283–288. https://doi.org/10.1007/s12298-012-0160-4

Gu XF, Zhang JR (2005) An efficient adventitious shoot regeneration system for Zhanhua winter jujube (*Zizyphus jujuba* Mill.) using leaf explants. Plant Cell Rep 23:775–779. https://doi.org/10.1007/s00299-005-0920-5

Guo B, He W, Zhao Y et al (2017) Changes in endogenous hormones and H2O2 burst during shoot organogenesis in TDZ-treated *Saussurea involucrate* explants. Plant Cell Tissue Organ Cult 128:1–8. https://doi.org/10.1007/s11240-016-1069-3

Hazarika BN (2006) Morpho-physiological disorders in in vitro culture of plants. Sci Hortic 108:105–120

Hicks G (1994) Shoot induction and organogenesis in vitro: a developmental perspective. In Vitro Cell Dev Bio-Plant 30(1):10–15

Hosseini-Nasr M, Rashid A (2003) Thidiazuron-induced high-frequency shoot regeneration from root region of *Robinia pseudoacacia* L. seedlings. Biol Plant 47(4):593–596

Huetteman CA, Preece JE (1993) Thidiazuron – a potent cytokinin for woody plant-tissue culture. Plant Cell Tissue Org Cult 33(2):105–119

Husain MK, Anis M, Shahzad A (2007) In vitro propagation of Indian Kino (*Pterocarpus marsupium* Roxb.) using thidiazuron. In Vitro Cell Dev Biol-Plant 43:59–64. https://doi.org/10.1007/s11627-006-9011-8

Ipekci Z, Gozukirmizi N (2003) Direct somatic embryogenesis and synthetic seed production from *Paulownia elongate*. Plant Cell Rep 22:16–24. https://doi.org/10.1007/s00299-003-0650-5

Ipekci Z, Gozukirmizi N (2004) Indirect somatic embryogenesis and plant regeneration from leaf and internode explants of *Paulownia elongate*. Plant Cell Tissue Organ Cult 79:341–345

Isah T (2015) Adjustments to in vitro culture conditions and associated anomalies in plants. Acta Biol Cracov Ser Bot 57(2):9–28. https://doi.org/10.1515/abcsb-2015-0026

Isah T (2016) Induction of somatic embryogenesis in woody plants. Acta Physiol Plant 38(5):1–22. https://doi.org/10.1007/s11738-016-2134-6

Jiménez VM (2001) Regulation of in vitro somatic embryogenesis with emphasis on the role of endogenous hormones. Rev Bras Fisiol Veg 13(2):196–223

Jones MPA, Yi Z, Murch SJ et al (2007) Thidiazuron-induced regeneration of *Echinacea purpurea* L.: micropropagation in solid and liquid culture systems. Plant Cell Rep 26:13–19

José MCS, Cernadas MJ, Corredoira E (2014) Histology of the regeneration of Paulownia tomentosa (*Paulowniaceae*) by organogenesis. Rev Biol Trop 62(2):809–818

Kadota M, Niimi Y (2003) Effects of cytokinin types and their concentrations on shoot proliferation and hyperhydricity in in vitro pear cultivar shoots. Plant Cell Tissue Organ Cult 72:261–265. https://doi.org/10.1023/A:1022378511659

Khan MI, Anis M (2012) Modulation of in vitro morphogenesis in nodal segments of *Salix tetrasperma* Roxb. through the use of TDZ, different media types and culture regimes. Agrofor Syst 86:95–103. https://doi.org/10.1007/s10457-012-9512-x

Kim MK, Sommer HE, Bongarten BC et al (1997a) High-frequency induction of adventitious shoots from hypocotyl segments of *Liquidambar styraciflua* L. by thidiazuron. Plant Cell Rep 16:536–540

Kim M-S, Schumann CM, Klopfenstein NB (1997b) Effects of thidiazuron and benzyladenine on axillary shoot proliferation of three green ash (*Fraxinus pennsylvanica* Marsh.) clones. Plant Cell Tissue Organ Cult 48:45–52. https://doi.org/10.1023/A:1005856720650

Kim K-M, Kim MY, Yun PY et al (2007) Production of multiple shoots and plant regeneration from leaf segments of fig tree (*Ficus carica* L.) J Plant Biol 50(4):440–446. https://doi.org/10.1007/BF03030680

Kucharska D, Orlikowska T (2009) Enhancement of in vitro organogenetic capacity of rose by preculture of donor shoots on the medium with thidiazuron. Acta Physiol Plant 31:495–500. https://doi.org/10.1007/s11738-008-0258-z

Li Z, Traore A, Maximova S et al (1998) Somatic embryogenesis and plant regeneration from floral explants of cacao (*Theobroma cacao* L.) using thidiazuron. In Vitro Cell Dev Biol-Plant 34:293–299

Li B-Q, Feng C-H, Hu L-Y et al (2014) Shoot regeneration and cryopreservation of shoot tips of apple (*Malus*) by encapsulation–dehydration. In Vitro Cell Dev Biol-Plant 50:357–368. https://doi.org/10.1007/s11627-014-9616-2

Liu X, Pijut PM (2008) Plant regeneration from in vitro leaves of mature black cherry (*Prunus serotina*). Plant Cell Tissue Organ Cult 94:113–123. https://doi.org/10.1007/s11240-008-9393-x

Lo KH, Giles KL, Sawhney VK (1997) Histological changes associated with acquisition of competence for shoot regeneration in leaf discs of *Saintpaulia ionantha* x confuse hybrid (African violet) cultured in vitro. Plant Cell Rep 16:421–425

Lu CY (1993) The use of thidiazuron in tissue culture. In Vitro Cell Dev Biol-Plant 29:92–96

Ma G, Lu J, Da Silva JAT et al (2011) Shoot organogenesis and somatic embryogenesis from leaf and shoot explants of *Ochna integerrima* (Lour). Plant Cell Tissue Organ Cult 104:157–162

Marriott P, Sarasan V (2010) Novel micropropagation and weaning methods for the integrated conservation of a critically endangered tree species, *Medusagyne oppositifolia*. In Vitro Cell Dev Biol-Plant 46:516–523. https://doi.org/10.1007/s11627-010-9321-8

Mehta UJ, Sahasrabudhe N, Hazra S (2005) Thidiazuron-induced morphogenesis in tamarind seedlings. In Vitro Cell Dev Biol-Plant 41:240–243. https://doi.org/10.1079/IVP2004611

Meng L, Zhang S, Lemaux P (2005) Developing a molecular understanding of in vitro and in planta shoot organogenes. In: Trigiano RN, Gray DJ (eds) Plant development and biotechnology. CRC Press, Boca Raton, pp 39–54

Mithila J, Hall JC, Victor JMR, Saxena PK (2003) Thidiazuron induces shoot organogenesis at low concentrations and somatic embryogenesis at high concentrations on leaf and petiole explants of African violet (*Saintpaulia ionantha* Wendl.) Plant Cell Rep 21(5):408–414

Mok DWS, Turner JE, Mujer CV (1987) Biological and biochemical effects of cytokinin-active phenylurea derivatives in tissue culture systems. HortSci 22(6):1194–1197

Moyo M, Aremu AO, Van Staden J (2015) Insights into the multifaceted application of microscopic techniques in plant tissue culture systems. Planta 242:773–790. https://doi.org/10.1007/s00425-015-2359-4

Mulwa RM, Bhalla PL (2006) In vitro plant regeneration from immature cotyledon explants of macadamia (*Macadamia tetraphylla* L. Johnson). Plant Cell Rep 25:1281–1286. https://doi.org/10.1007/s00299-006-0182-x

Murch SJ, Saxena PK (2001) Molecular fate of thidiazuron and its effects on auxin transport in hypocotyls tissues of *Pelargonium* x*hortorum* Bailey. Plant Growth Regul 35(3):269–275

Murch SJ, KrishnaRaj S, Saxena PK (1997) Thidiazuron-induced morphogenesis of regal geraniums (*Pelargonium domesticum*): a potential stress response. Physiol Plant 101:183–191

Murthy BNS, Murch SJ, Saxena PK (1998) Thidiazuron: a potent regulator of in vitro plant morphogenesis. In Vitro Cell Dev Biol-Plant 34(4):267–275. https://doi.org/10.1007/BF02822732

Noël N, Leplé J-C, Pilate G (2002) Optimization of in vitro micropropagation and regeneration for *Populus × interamericana* and *Populus × euramericana* hybrids (*P. deltoides, P. trichocarpa, and P. nigra*). Plant Cell Rep 20:1150–1155. https://doi.org/10.1007/s00299-002-0465-9

Novikova TI, Poluboyarova TV (2013) Thidiazuron-induced shoot organogenesis of *Disanthus cercidiofolius* maxim. (*Hamamelidaceae*). In: Abstracts of the X international conference on plant cell biology in vitro and biotechnology, Kazan Institute of Biochemistry and Biophysics, Kazan, 14–18 October 2013

Pal A, Negi VS, Borthakur D (2012) Efficient in vitro regeneration of *Leucaena leucocephala* using immature zygotic embryos as explants. Agrofor Syst 84:131–140. https://doi.org/10.1007/s10457-011-9438-8

Panda BM, Hazra S (2012) In vitro morphogenic response in cotyledon explants of *Semecarpus anacardium* L. Plant Biotechnol Rep 6:141–148. https://doi.org/10.1007/s11816-011-0207-y

Parveen S, Shahzad A (2010) TDZ–induced high frequency shoot regeneration in *Cassia sophera* Linn. via cotyledonary node explants. Physiol Mol Biol Plants 16(2):201–205. https://doi.org/10.1007/s12298-010-0022-x

Paul S, Dam A, Bhattacharyya A et al (2011) An efficient regeneration system via direct and indirect somatic embryogenesis for the medicinal tree *Murraya koenigii*. Plant Cell Tissue Organ Cult 105:271–283. https://doi.org/10.1007/s11240-010-9864-8

Pavingerova D (2009) The influence of thidiazuron on shoot regeneration from leaf explants of fifteen cultivars of *Rhododendron*. Biol Plant 54:797–799

Phillips GC (2004) In vitro morphogenesis in plants – recent advances. In Vitro Cell Dev Biol-Plant 40(4):342–345. https://doi.org/10.1079/IVP2004555

Pinto G, Santos C, Neves L, Araújo C (2002) Somatic embryogenesis and plant regeneration in *Eucalyptus globulus* Labill. Plant Cell Rep 21:208–213

Raghu AV, Geetha SP, Gerald M et al (2006) Ravindran PN direct shoot organogenesis from leaf explants of *Embelia ribes* Burm. f.: a vulnerable medicinal plant. J For Res 11:57–60. https://doi.org/10.1007/s10310-005-0188-1

Rai VR (2002) Rapid clonal propagation of *Nothapodytes foetida* (Wight) sleumer – a threatened medicinal tree. In Vitro Cell Dev Biol-Plant 38:347–351. https://doi.org/10.1079/IVP2002300

Rani DN, Nair GM (2006) Effects of plant growth regulators on high frequency shoot multiplication and callus regeneration of an important Indian medicinal plant, nirgundi (*Vitex negundo* L.) In Vitro Cell Dev Biol-Plant 42:69–73. https://doi.org/10.1079/IVP2005727

Ranyaphiaa RA, Maoaa AA, Borthakurb SK (2011) Direct organogenesis from leaf and internode explants of in vitro raised wintergreen plant (*Gaultheria fragrantissima*). ScienceAsia 37:186–194. https://doi.org/10.2306/scienceasia1513-1874.2011.37.186

Rastogi S, Rizvi SMH, Singh RP et al (2008) In vitro regeneration of *Leucaena leucocephala* by organogenesis and somatic embryogenesis. Biol Plant 52:743–748. https://doi.org/10.1007/s10535-008-0144-y

Ravikumar G (2001) Forest biotechnology: development and prospects. Plant Cell Biotech Mol Biol 1:13–28

Ravishankar Rai V, McComb J (2002) Direct somatic embryogenesis from mature embryos of sandalwood. Plant Cell Tissue Organ Cult 69:65–70. https://doi.org/10.1023/A:1015037920529

Ružić DV, Vujović TI (2008) The effects of cytokinin types and their concentration on in vitro multiplication of sweet cherry cv. Lapins (*Prunus avium* L.) Hortic Sci 3:12–21

Salaj J, Petrovská B, Obert B et al (2005) Histological study of embryo-like structures initiated from hypocotyl segments of flax (*Linum usitatissimum* L.) Plant Cell Rep 24(10):590–595

Sandal I, Bhattacharya A, Ahuja PS (2001) An efficient liquid culture system for tea shoot proliferation. Plant Cell Tissue Organ Cult 65:75–80. https://doi.org/10.1023/A:1010662306067

Shaik NM, Arha M, Nookaraju A et al (2009) Improved method of in vitro regeneration in *Leucaena leucocephala* – a leguminous pulpwood tree species. Physiol Mol Biol Plants 15(4):311–318. https://doi.org/10.1007/s12298-009-0035-5

Sivanesan I, Song JY, Hwang SJ et al (2011) Micropropagation of *Cotoneaster wilsonii* Nakai—a rare endemic ornamental plant. Plant Cell Tissue Organ Cult 105:55–63. https://doi.org/10.1007/s11240-010-9841-2

Siwach P, Gill AR (2011) Enhanced shoot multiplication in *Ficus religiosa* L. in the presence of adenine sulphate, glutamine and phloroglucinol. Physiol Mol Biol Plants 17(3):271–280. https://doi.org/10.1007/s12298-011-0074-6

Skoog F, Miller CO (1957) Chemical regulation of growth and organ formation in plant tissue cultured in vitro. Symp Soc Exp Biol 11:118–131

Sriskandarajah S, Lundquist P (2009) High frequency shoot organogenesis and somatic embryogenesis in juvenile and adult tissues of seabuckthorn (*Hippophae rhamnoides* L.) Plant Cell Tissue Organ Cult 99:259–268. https://doi.org/10.1007/s11240-009-9597-8

Stevens ME, Pijut PM (2012) Hypocotyl derived in vitro regeneration of pumpkin ash (*Fraxinus profunda*). Plant Cell Tissue Organ Cult 108:129–135. https://doi.org/10.1007/s11240-011-0021-9

Sugiyama M (1999) Organogenesis in vitro. Curr Opin Plant Biol 2:61–64. https://doi.org/10.1016/S1369-5266(99)80012-0

Sujatha PK, Hazra S (2007) Micropropagation of mature *Pongamia pinnata*. In Vitro Cell Dev Biol-Plant 43:608–613. https://doi.org/10.1007/s11627-007-9049-2

Sujatha K, Panda BM, Hazra S (2008) De novo organogenesis and plant regeneration in *Pongamia pinnata*, oil producing tree legume. Trees 22:711–716. https://doi.org/10.1007/s00468-008-0230-y

Thomas JC, Katterman FR (1986) Cytokinin activity induced by thidiazuron. Plant Physiol 81:681–683

Traore A, Maximova SN, Guiltinan MJ (2003) Micropropagation of *Theobroma cacao* L. using somatic embryo-derived plants. In Vitro Cell Dev Biol-Plant 39:332–337

Tzfira T, Zuker A, Altman A (1998) Forest tree biotechnology, genetic transformation and its application to future forests. Trends Biotechnol 16:439–446

Varshney A, Anis M (2014) Trees: propagation and conservation: biotechnological approaches for propagation of a multipurpose tree, *Balanites aegyptiaca*. Springer, New Delhi

Vasil IK (2008) A history of plant biotechnology: from the cell theory of Schleiden and Schwann to biotech crops. Plant Cell Rep 27:1423–1440. https://doi.org/10.1007/s00299-008-0571-4

Vengadesan G, Pijut PM (2009) In vitro propagation of northern red oak (*Quercus rubra* L.) In Vitro Cell Dev Biol-Plant 45:474–482. https://doi.org/10.1007/s11627-008-9182-6

Vengadesan G, Ganapathi A, Anand RP et al (2003) In vitro propagation of *Acacia sinuata* (Lour.) Merr. from nodal segments of a 10-year-old tree. In Vitro Cell Dev Biol-Plant 39:409–414. https://doi.org/10.1079/IVP2003421

Verdeil JL, Alemanno L, Niemenak N et al (2007) Pluripotent versus totipotent plant stem cells: dependence versus autonomy? Trends Plant Sci 12(6):245–252. https://doi.org/10.1016/j.tplants.2007.04.002

Vieitez AM, San José MC (1996) Adventitious shoot regeneration from *Fagus sylvatica* leaf explants in vitro. In Vitro Cell Dev Biol-Plant 32(3):140–147. https://doi.org/10.1007/BF02822757

Vila S, Gonzalez A, Rey H et al (2003) Somatic embryogenesis and plant regeneration from immature zygotic embryos of *Melia azedarach* (*Meliaceae*). In Vitro Cell Dev Biol-Plant 39:283–287. https://doi.org/10.1079/IVP2002377

Vila SK, Rey HY, Mroginski LA (2007) Factors affecting somatic embryogenesis induction and conversion in "Paradise Tree" (*Melia azedar*ach L.) J Plant Growth Regul 26:268–277. https://doi.org/10.1007/s00344-007-9007-6

Vinocur B, Carmi T, Altman A et al (2000) Enhanced bud regeneration in aspen (*Populus tremula* L.) roots cultured in liquid media. Plant Cell Rep 19:1146–1154

Visser C, Qureshi JA, Gill R et al (1992) Morpho-regulatory role of thidiazuron: substitution of auxin and cytokinin requirement for the induction of somatic embryogenesis in geranium hypocotyls culture. Plant Physiol 99:1704–1707

von Arnold S, Sabala I, Bozhkov P et al (2002) Developmental pathways of somatic embryogenesis. Plant Cell Tissue Organ Cult 69:233–249

Wang SY, Steffens GL, Faust M (1986) Breaking bud dormancy in apple with a plant bioregulator, thidiazuron. Phytochemistry 25:311–317

Wang H, Liu H, Wang W et al (2008) Effects of thidiazuron, basal medium and light quality on adventitious shoot regeneration from in vitro cultured stem of *Populus alba×P. berolinensis*. J For Res 19(3):257–259. https://doi.org/10.1007/s11676-008-0042-3

Wei F, Zhao F, Tian B (2017) In vitro regeneration of *Populus tomentosa* from petioles. J For Res 28:465–471. https://doi.org/10.1007/s11676-016-0319-x

Wilhelm E (1999) Micropropagation of juvenile sycamore maple via adventitious shoot formation by use of thidiazuron. Plant Cell Tissue Organ Cult 57:57–60

Willemsen V, Scheres B (2004) Mechanisms of pattern formation in plant embryogenesis. Annu Rev Genet 38:587–614

Woo SM, Wetzstein HY (2008) Morphological and histological evaluations of in vitro regeneration in *Elliottia racemosa* leaf explants induced on media with thidiazuron. J Am Soc Hortic 133(2):167–172

Xie D, Hong Y (2001) In vitro regeneration of *Acacia mangium* via organogenesis. Plant Cell Tissue Organ Cult 66:167–173

Xu L, Liu GF, Bao MZ (2007) Adventitious shoot regeneration from in vitro leaves of formosan sweetgum (*Liquidambar formosana* L.) HortSci 42(3):721–723

Yancheva SD, Golubowicz S, Fisher E et al (2003) Auxin type and timing of application determine the activation of the developmental program during in vitro organogenesis in apple. Plant Sci 165:299–309

Zaytseva YG, Novikova TI (2015) Conservation and propagation of *Rhododendron schlippenbachii* using biotechnological methods. Rastitel'nyj Mir Aziatskoj Rossii (Plant Life of Asian Russia) 4:79–85

Zaytseva YG, Poluboyarova TV, Novikova TI (2016) Effects of thidiazuron on in vitro morphogenic response of *Rhododendron sichotense* Pojark. and *Rhododendron catawbiense* cv. Grandiflorum leaf explants. In Vitro Cell Dev Biol-Plant 52:56–63. https://doi.org/10.1007/s11627-015-9737-2

Zhang C, Fu S, Tang G et al (2013) Factors influencing direct shoot regeneration from mature leaves of *Jatropha curcas*, an important biofuel plant. In Vitro Cell Dev Biol-Plant 49:529–540. https://doi.org/10.1007/s11627-013-9530-z

Zhou H, Li M, Zhao X et al (2010) Plant regeneration from in vitro leaves of the peach rootstock 'Nemaguard' (*Prunus persica* 3 P. davidiana). Plant Cell Tissue Organ Cult 101:79–87. https://doi.org/10.1007/s11240-010-9666-z

Zhuravlev YN, Omelko AM (2008) Plant morphogenesis in vitro. Russ J Plant Physiol 55(5):579–596. https://doi.org/10.1134/S1021443708050014

Check for updates

Thidiazuron in Micropropagation of Aroid Plants

4

Jianjun Chen and Xiangying Wei

Abstract

Thidiazuron (TDZ) or phenyl-N′-(1,2,3-thiadiazol-5-yl) urea is a synthetic phenyl urea derivative and possesses strong cytokinin-like activity exceeding that of most other commonly used adenine-type cytokinins in regulating plant morphogenesis. In this article, we devote our attention to the use of TDZ in micropropagation of plants in the family Araceae, commonly known as aroids. This family has 3750 recognized species across 114 genera. A large number of genera are important ornamental plants, particularly in the foliage plant industry. Some genera are produced for edible roots or used as medicinal plants, and a few others are aquatic plants. Aroids are traditionally propagated through cutting, division, rhizomes, or tubers. Vegetative propagation not only carries plant pathogens but also significantly slows the speed in the introduction of new cultivars. Our research over the years has focused on the development of methods for micropropagating aroid plants. TDZ has been shown to be an important plant growth regulator for efficient micropropagation of aroid plants via in vitro shoot culture and plant regeneration through the route of shoot organogenesis, somatic embryogenesis, and protocorm-like bodies (PLBs). Mechanisms underlying TDZ-mediated plant regeneration are still largely unknown, but the established regeneration systems derived from our work on aroids present valuable models for molecular analysis of TDZ-mediated plant morphogenesis.

J. Chen (✉)
Environmental Horticulture Department and Mid-Florida Research and Education Center, University of Florida, Apopka, FL, USA
e-mail: jjchen@ufl.edu

X. Wei
Environmental Horticulture Department and Mid-Florida Research and Education Center, University of Florida, Apopka, FL, USA

Fujian University Key Laboratory of Plant-Microbe Interaction, College of Life Science, Fujian Agriculture and Forestry University, Fuzhou, China

N. Ahmad, M. Faisal (eds.), *Thidiazuron: From Urea Derivative to Plant Growth Regulator*, https://doi.org/10.1007/978-981-10-8004-3_4

Keywords
Araceae · Protocorm-like bodies · Shoot culture · Shoot organogenesis · Somatic embryogenesis · Thidiazuron (TDZ)

4.1 Introduction

Micropropagation is a form of tissue culture that rapidly propagates plants through in vitro shoot culture or regenerates plants through shoot organogenesis, somatic embryogenesis, and protocorm-like bodies (PLBs). In vitro shoot culture refers to in vitro propagation of plants through repeated enhanced formation of axillary shoots from shoot tips or lateral buds cultured on medium supplemented with plant growth regulators, primarily cytokinin (George et al. 2008). There are generally four distinct stages involved in shoot culture: culture initiation, shoot multiplication, in vitro rooting, and acclimatization (Murashige 1974; Rout et al. 2006).

Micropropagation through plant regeneration rests on the cell theory (Schleiden 1838; Schwann 1839) and started with Gottlieb Haberlandt's speculation of the totipotency of plant cells near the turn of the twentieth century (Haberlandt 1902). Totipotency is a single cell that has the genetic program to grow into an entire plant (Hartmann et al. 2002). Shoot organogenesis is the regeneration of plants directly from plant tissues or indirectly from callus derived from cells, tissues, and organs called explants cultured on artificial medium supplemented with plant growth regulators. In indirect shoot organogenesis, the cultured cells go through dedifferentiation, induction, and differentiation phases to produce plant shoots. Regeneration through somatic embryogenesis is a process through which undifferentiated cells are induced through the actions of cytokinins and auxins in the culture media to become embryogenically determined. When induced cells produce embryos without a callus phase, this is referred to as direct somatic embryogenesis. If cells produce callus preceding the formation of embryos, this is referred to as indirect embryogenesis. PLBs resemble protocorms structurally induced from explants or calluses which are composed of many meristematic centers that are able to differentiate into shoots and roots (Jones and Tisserat 1990; Cui et al. 2008).

Plant growth regulators, mainly cytokinins and auxins, play critical roles in both shoot culture and organogenesis. Skoog and Miller (1957) were the first to demonstrate the role of kinetin (6-furfuryladenine) in organogenesis. When the ratio of kinetin to auxin was higher, only shoots developed. Whereas when the ratio was lower, only roots were formed. Two groups of chemicals are known to have cytokinin activities: the N6-substituted adenine derivatives and certain substituted urea compounds. Thidiazuron (TDZ) or phenyl-N′-(1,2,3-thiadiazol-5-yl) urea, a substituted urea compound, was synthesized by German Schering Corporation for defoliation of cotton (*Gossypium hirsutum*) (Arndt et al. 1976). TDZ is known to have cytokinin activity (Mok et al. 1982) and to promote the growth of cytokinin-dependent callus cultures of *Phaseolus lunatus* (Capelle et al. 1983). The cytokinin activity of TDZ was reported to be similar to the most active cytokinins of the adenine type (Huetteman and Preece 1993; Murthy et al. 1998). As a result, TDZ emerged as an effective growth regulator in cell and tissue cultures in a wide array of plant species (Li et al. 2000; Hosseini-Nasr and Rashid 2002; Yancheva et al. 2003; Matand and Prakash 2007; Guo et al. 2011). This article is intended to review TDZ applications in micropropagation of aroid plants and to document the importance of this plant growth regulator in improving propagation and production of aroid plants.

4.2 Aroid Plants

The family Araceae, commonly known as aroids, encompasses 114 genera and more than 3750 species that are mostly herbaceous either as terrestrial, aquatic, or epiphytic (Mayo et al. 1997; Bown 2000). Most are indigenous to the tropics of America, Southeast Asia, the Malay Archipelago (Malaysia, Indonesia, the Philippines, Papua New Guinea, Singapore, and Brunei), and tropical Africa (Mayo et al. 1997). Some species, such as *Amorphophallus paeoniifolius*, *Colocasia esculenta*, *Cyrtosperma merkusii*, and *Xanthosoma sagittifolium*, are cultivated as sources of carbohydrate foods (Chen et al. 2003; Bown 2000) (Table 4.1). Some including *Arisaema heterophyllum*, *Pinellia ternata*, and *Typhoniumi trilobatum* are important medicinal plants (Mayo et al. 1997; Bown 2000; Chen et al. 2007). A large number of them are ornamental foliage plants, such as *Aglaonema*, *Anthurium*, *Dieffenbachia*, *Epipremnum*, *Philodendron*, *Spathiphyllum*, and *Syngonium* (Mayo et al. 1997; Bown 2000; Henny and Chen 2003; Chen et al. 2005). This group of plants is prized for their beautiful leaf forms, textures, colors, and variegation patterns as well as colorful spathes present in some genera (Chen et al. 2003; Bown 2000). Ornamental aroids are a major component of the foliage plant industry and account for about one-third of total ornamental foliage plant sales in the United States (Henny et al. 2004). Ornamental aroids have been widely used as living specimens for interior decoration or interiorscaping because of their ability to maintain an aesthetically pleasing appearance under interior low light conditions. Interior decoration with ornamental aroids brings beauty and comfort to our surroundings and also reminds us of nature (Chen et al. 2005). In addition, ornamental aroids grown in interior environments can act as natural humidifiers and have been shown to reduce indoor air pollutants (Wolverton et al. 1984; Oyabu et al. 2003).

Aroids were traditionally propagated through vegetative means, such as cuttings or divisions (Chen and Stamps 2006). Vegetatively propagated plants are often associated with the spread of diseases such as dasheen mosaic virus that can be difficult to eradicate by chemical or physical treatment (Hartman 1974) and bacterial wilt (*Ralstonia solanacearum*) carried over through the cuttings of pothos (*Epipremnum aureum*) (Norman and Yuen 1998). Hartman (1974) was the first to report the use of micropropagation for producing *Caladium bicolor*, *Xanthosoma sagittifolium*, and *Colocasia esculenta* that were free of dasheen mosaic virus. Micropropagation decreases greenhouse space needed for stock plant production and provides growers with liners (tissue-cultured plantlets grown in cell plug trays) on a year-round basis (Chen and Henny 2008). Adoption of in vitro propagation has reduced the wait for

TDZ at low concentrations (4 μM or lower) induced more axillary shoots of *Aglaonema* 'White Tip' than BA at concentrations lower than 10 μM (Chen and Yeh 2007). The genus *Acorus* is a perennial hydrophyte used as a medicinal and aromatic plant. In vitro shoot culture of two species showed that 17.8 μM BA induced 5.4 axillary shoots per explants, whereas 4.5 μM TDZ induced 11.0 shoots for *A. calamus*. The same concentrations of BA and TDZ produced 2.7 and 3.9 shoots, respectively, for *A. gramineus* (Lee and Han 2011). *Colocasia esculenta*, commonly known as taro, is an important edible crop throughout the Pacific Islands (Chand et al. 1999). Meristems of six cultivars were cultured on a modified MS medium supplemented with TDZ. In experiments with the cultivar Niue, explants cultured on the modified MS medium with 2.6 μM TDZ grew more vigorously than on the medium including BA. Subculture of explants on medium containing 4.3 μM TDZ gave a 15–25-fold increase in production of plantlets per 4-week culture period compared to a fourfold increase with BA (Chand et al. 1999). TDZ also significantly increased the shoot proliferation rate in *Alocasia amazonica* (Jo et al. 2008).

An important characteristic in shoot culture of aroids is that TDZ concentration should be lower than other commonly used cytokinins such as BA and 2iP. High TDZ concentration could either reduce multiplication rates or cause phytotoxicity. For example, TDZ at high concentrations were shown to inhibit shoot proliferation in *Spathiphyllum cannifolium* (Dewir et al. 2006). Higher concentrations of TDZ (4 or 20 μM) inhibited shoot elongation of *Aglaonema* (Chen and Yeh 2007). This phenomenon could be attributed to the following reasons: (1) TDZ is a more effective growth regulator than the commonly used other cytokinins (Preece et al. 1991). It has been shown that exposure of plant tissue to TDZ for a relatively short period is sufficient to stimulate regeneration (Visser et al. 1992; Hutchinson and Saxena 1996). Thus, a low concentration should be used for axillary shoot induction. (2) TDZ is rather stable in culture medium (Murthy et al. 1998). Radiolabeled TDZ in tissue culture remains structurally intact for up to 48 h, suggesting that the intact structure is important for its activity (Mok et al. 1982; Mok and Mok 1985). Furthermore, its activity could be the induction of cascade reactions in plant cells (Kou et al. 2016; Guo et al. 2017) as TDZ action could be retained even after transfer to fresh basal medium (Capelle et al. 1983).

4.4 TDZ in Shoot Organogenesis

TDZ has been used as a cytokinin for regeneration of plants from 11 aroid genera (Table 4.3). Explants used for the regeneration include anther, corm, leaves, petioles, spathe, and stems. Adventitious shoots are predominantly produced through callus phase, i.e., indirect shoot organogenesis. There is only one report of direct shoot organogenesis, in which adventitious shoots of a *Dieffenbachia* cultivar were induced directly from petiole explants cultured on MS medium. About 15.4% of petioles cultured with TDZ at 4.5 μM with 5.4 μM NAA produced buds compared to 10.2% of petioles cultured with 4.4 μM BA with 4.5 μM 2, 4-D (2, 4-dichlorophenoxyacetic acid) (Orlikowska et al. 1995). This study, however, did not provide any details about the claimed direct shoot organogenesis.

Table 4.3 Aroid plants regenerated through indirect organogenesis with TDZ as cytokinin in the culture medium

Scientific name	Explant	Protocol	References
Aglaonema 'Lady Valentine'	Node	MS + 9.08 μM TDZ + 2.69 μM NAA	Fang et al. (2013)
Aglaonema	Inflorescence	MS + 10 μM TDZ + 5–10 μM Dicamba	Yeh et al. (2007)
Anthurium andraeanum	Leaf	Modified MS + 1.82 μM TDZ	Gu et al. (2012)
Anthurium andraeanum	Callus	Modified MS + 0.05 μM TDZ	Kumari et al. (2011)
Anthurium spp.	Leaf	Modified MS + 2.27 μM TDZ + 2.69 μM NAA	Orlikowska and Zawadzka (2010)
Anthurium andraeanum	Node	Modified MS + 0.92 μM TDZ	Bhattacharya et al. (2015)
Anthurium andraeanum	Anther	WRM + 4.54 μM TDZ + 3.33 μM BA	Winarto et al. (2010)
Anthurium andraeanum	Half-anthers	WT-1 + 2.27 μM TDZ + 0.05 μM NAA + 4.44 μM BA	Winarto et al. (2011b)
Anthurium andraeanum	Half-anthers	NWT-3 + 6.81 μM TDZ + 0.11 μM NAA + 3.33 μM BA	Winarto and da Silva (2012)
Anthurium andraeanum	Callus	NWT + 6.81 μM TDZ + 1.13 μM 2,4-D + 0.11 μM NAA + 3.33 μM BA	Winarto et al. (2011a)
Anthurium andraeanum	Shoot	MS + 4.54 μM TDZ or 44.39 μM BA	Han and Goo (2003)
Colocasia esculenta	Corm	½ MS + 4.54 μM TDZ	Deo et al. (2009)
Colocasia esculenta	Shoot	First step: MS + 2 μM TDZ + 10 μM 2, 4-D	Verma and Cho (2007)
	Callus	Second step: MS + 5 μM TDZ	
Colocasia esculenta	Shoot tip	½ MS + 4.54 μM TDZ	Du et al. (2006)
Colocasia esculenta	Corm	First step: ½ MS + 12.67 μM 2, 4-D	Deo et al. (2010)
		Second step: ½ MS + 4.54 μM TDZ	
	Callus	MS + 4.54 μM TDZ + 2.26 μM 2,4-D	
	Callus	MS + 4.54 μM TDZ + 2.26 μM 2,4-D	
	Suspension cells	MS + 0.45 μM TDZ + 0.23 μM 2, 4-D	
Colocasia esculenta	Embryogenic callus	MS + 0.45 μM TDZ + 0.23 μM 2.4 D	Fitriani et al. (2016)
Dieffenbachia spp.	Petiole	MS + 4.54 μM TDZ + 5.37 μM NAA + 4.44 μM BAP + 4.52 μM 2, 4 D	Orlikowska et al. (1995)

(continued)

Table 4.3 (continued)

Scientific name	Explant	Protocol	References
Dieffenbachia cv. Camouflage	Leaf	Ms + 5 μM TDZ + 1 μM 2, 4-D	Shen et al. (2007a)
Dieffenbachia	Leaf	Ms + 5 μM TDZ + 1 μM 2, 4-D	Shen et al. (2007b)
Dieffenbachia spp.	Leaf	Modified MS + 5 μM TDZ + 1 μM 2, 4-D	Shen et al. (2008)
Epipremnum aureum	Leaf	MS + 5 μM TDZ + 0.5 μM NAA	Qu et al. (2002)
	Petiole	MS + 10 μM TDZ + 0.5 μM NAA	
Lemna gibba	Callus	B5 + 4.54 μM TDZ + 0.1% sucrose	Li et al. (2004)
Philodendron spp.	Stem	MS + 2.27 μM TDZ	Chen et al. (2012)
Spathiphyllum wallisii	Petiole	MS + 4.54 μM TDZ + 0.90 μM 2,4 D	Lakshmanan et al. (2011)
		MS + 9.08 μM TDZ + 4.52 μM 2,4 D	
Spathiphyllum	Petiole	MS + 4.54 μM TDZ	Orlikowska et al. (1995)
	Pith	MS + 2.27 μM TDZ	
Spirodela oligorrhiza	Callus	WP + 4.54 μM TDZ	Li et al. (2004)
Spirodela punctata	Callus	WP + 2.27 μM TDZ + 4.52 μM 2,4-D + 26.85 μM NAA	Li et al. (2004)
Syngonium podophyllum	Petiole	MS + 9.08 μM TDZ + 2.26 μM 2,4-D	Cui et al. (2008)
Xanthosoma sagittifolium	Shoot tip, petiole	B5 + MS + 0.045 μM TDZ + 13.5 μM Dicamba	Nyochembeng and Garton (1998)

Indirect shoot organogenesis has been shown to be an effective way of producing plantlets. Nyochembeng and Garton (1998) reported that addition of TDZ in a culture medium supplemented with dicamba (3,6-dichloro-2-methoxybenzoic acid) enhanced callus production from petioles of *X. sagittifolium*, but subsequent adventitious shoot regeneration occurred when the callus was cultured with dicamba alone, 2,4-D with kinetin, or dicamba with kinetin. The first detailed report on TDZ-mediated indirect shoot organogenesis came from a study from Qu et al. (2002) where adventitious shoots were regenerated from leaf and petiole explants of pothos (*E. aureum* 'Jade'). Among TDZ, 2iP, and zeatin [6-(4-Hydroxy-3-methylbut-2-enylamino) purine] tested, TDZ was the best cytokinin for pothos regeneration. Regeneration frequencies were 18% and 50% for leaf and petiole explants, respectively, on medium containing TDZ and NAA (1-naphthalene acetic acid) after only 30 days of culture, and responding explants regardless of leaf or petiole explants produced approximately 30 adventitious shoots. Regeneration on medium containing either 2iP or zeatin with NAA produced a maximum of four shoots during a 50-day culture period. The methods of TDZ-mediated shoot organogenesis and plant regeneration were then modified and used for regeneration of *Aglaonema*, *Alocasia*, *Anthurium*, *Colocasia*, *Dieffenbachia*, *Lemna*, *Philodendron*, *Spathiphyllum*, *Spirodela*, and *Syngonium* (Table 4.3).

TDZ-mediated indirect shoot organogenesis in aroid plants exhibits the following characteristics: (1) Calluses induced are usually nodular-like (Shen et al. 2007a, 2008), which are different from friable calluses. (2) TDZ is more effective in induction of callus formation than other cytokinins. In a study of *Dieffenbachia*, BA, N-(2-chloro-4- pyridyl)-N-phenylurea (CPPU), kinetin, dicamba, 4-amino-3,5,6-trichloro-2pyridinecarboxylic acid (picloram), and TDZ in combination with either 2,4-D or NAA were used for callus induction. TDZ with 2,4-D induced up to 96% of leaf explants to produce callus, while other combinations failed to produce calluses (Shen et al. 2007a). (3) TDZ alone can induce callus formation. Gu et al. (2012) reported TDZ alone induced callus formation in *Anthurium*. This could be attributed to the fact that TDZ could elicit both auxin and cytokinin responses as documented by Murthy et al. (1995), and also TDZ is a highly stable cytokinin and is resistant to degradation by cytokinin oxidase (Mok et al. 1987). (4) Medium with TDZ usually shortens the time for callus induction and adventitious regeneration such as the aforementioned pothos (Qu et al. 2002). (5) Genotypes vary in response to TDZ induction. TDZ in combination with 2,4-D was found to be an effective combination for *Dieffenbachia* shoot organogenesis; callus formation frequency and shoot numbers per callus were 96% and 6.7 for cultivar Camouflage, but these frequencies were 62% and 4.4 for 'Camille,' 66% and 0 for 'Octopus,' and 52% and 3.5 for 'Star Bright' (Shen et al. 2008). (6) Somaclonal variation could occur in regenerated populations. In general, more somaclonal variants are observed in plants regenerated via callus phase (Larkin and Scowcroft 1981; Chen et al. 2003). Three somaclonal variants were identified from regenerated populations of *Dieffenbachia* 'Camouflage' and one from *Dieffenbachia* 'Star Bright' (Shen et al. 2007b).

4.5 TDZ in Somatic Embryogenesis

Micropropagation through somatic embryogenesis has advantages over both shoot culture and organogenesis because a large number of plantlets can be produced and it can potentially scale up propagation for bioreactors and produce synthetic seeds (Rani and Raina 2000). Somatic embryos are also desirable materials for genetic transformation and cryopreservation.

Anthurium andraeanum is the first aroid species that was regenerated through somatic embryogenesis (Kuehnle et al. 1992). Leaf explants were cultured on half-strength MS medium supplemented with kinetin and 2,4-D, and embryo conversion occurred and plantlets were produced. Subsequently, plants were also regenerated from other *Anthurium* via somatic embryogenesis when culture medium was supplemented with kinetin and 2,4-D (Matsumoto et al. 1996). The use of 2,4-D alone or in combination with other growth regulators was a standard practice for inducing somatic embryos from the 1960s to the 1990s (Raghavan 2004). This is attributed to the work of Halperin and Wetherell (1964) who demonstrated that a callus induced from any vegetative part of carrot (*Daucus carota*) cultured on a medium containing a high concentration of 2,4-D could form somatic embryos upon transfer to the medium with a reduced level of the auxin (Raghavan 2004).

With the recognition that TDZ in combination with 2, 4-D was able to induce somatic embryogenesis in the 1990s, such as white ash (*Fraxinus americana* L.) (Bates et al. 1992) and watermelon (*Citrullus lanatus*) (Compton and Gray 1993), TDZ was introduced for inducing somatic embryogenesis in aroids. Somatic embryos were induced from *S. wallisii* Regel 'Speedy' when anther filaments were cultured on a modified basal medium supplemented with TDZ with 2,4-D (Eeckhaut et al. 2004). This protocol (culture media supplemented with TDZ and 2,4-D) has been successfully used for inducing somatic embryogenesis of *C. esculenta* (Deo et al. 2009; Verma and Cho 2007), *Dieffenbachia* (Shen and Lee 2009), and *Spathiphyllum* 'Supreme' (Zhao et al. 2012a) (Table 4.4).

The protocol making 2,4-D a necessary requirement for somatic embryo induction was changed in aroid plants when Werbrouck et al. (2000) reported that somatic embryos could be induced from anther filaments of *Spathiphyllum* cultured on a modified basal medium supplemented with TDZ and NAA. This TDZ and NAA combination was further refined by Zhang et al. (2005) in regeneration of *E. aureum* 'Golden Pothos.' Somatic embryos were directly induced from leaf, petiole, and stem explants cultured on MS or MK [MS ingredients in combination with Kao medium (Kao 1977) vitamins] medium supplemented with TDZ and NAA (Fig. 4.1). The frequencies of explants with embryos and explants with embryo conversion were as high as 61%, 89%, and 86% for leaf, petiole, and stem explants. The success of the protocols (culture media supplemented with TDZ with NAA) developed by Zhang et al. (2005) in 'Golden Pothos' has been modified and used for somatic embryogenesis and plant regeneration of *Syngonium podophyllum* (Zhang et al. 2006), *Zantedeschia* (Duquenne et al. 2006), and *Epipremnum aureum* 'Marble Queen' (Zhao et al. 2012b) (Table 4.4).

TDZ alone can induce somatic embryogenesis in some aroid plants such as *Colocasia* (Verma and Cho 2007). TDZ often induces direct somatic embryogenesis in aroids including *Epipremnum* (Zhang et al. 2005; Zhao et al. 2012b), *Spathiphyllum* (Eeckhaut et al. 2004; Zhao et al. 2012a), and *Syngonium* (Zhang et al. 2006). Plant species and explant types affect TDZ-mediated somatic embryogenesis. The genotype had no obvious effect on somatic embryogenesis of *C. esculenta* (Deo et al. 2009). However, the most effective combination for inducing somatic embryogenesis from leaf explants of *E. aureum* 'Marble Queen' was 4.54 μM TDZ with 1.07 μM NAA, whereas for petiole explants, it was 9.08 μM TDZ with 1.07 μM NAA (Zhao et al. 2012b). Furthermore, six times more plantlets were regenerated from petiole explants than those of leaf explants in 'Marble Queen' (Zhao et al. 2012b). Somatic embryogenesis often requires TDZ at relatively higher concentrations than those used for organogenesis. For example, somatic embryogenesis of *Epipremnum* required TDZ concentrations at 4.5 μM or higher (Zhang et al. 2005; Zhao et al. 2012b) and above 9.0 μM for *S. podophyllum* 'Variegatum' (Zhang et al. 2006).

Table 4.4 Aroid plants regenerated through somatic embryogenesis with TDZ as cytokinin in the culture medium

Scientific name	Explant	Protocol	References
Colocasia esculenta	Callus	MS + 5 μM TDZ	Verma and Cho (2007)
Colocasia esculenta	Suspension cells	MS + 0.45 μM TDZ + 0.23 μM 2,4-D	Deo et al. (2010)
Dieffenbachia 'Tiki'	Male inflorescence	½ MS + 2.27 or 4.54 μM TDZ + 18.09 μM 2,4-D	Shen and Lee (2009)
Epipremnum aureum	Leaf	MS + 4.54 μM TDZ + 1.07 μM NAA	Zhao et al. (2012b)
	Petiole	MS + 9.08 μM TDZ + 1.07 μM NAA	
Epipremnum aureum	Leaf, petiole	MS + 9.10 μM TDZ + 1.10 μM NAA	Zhao et al. (2013)
Epipremnum aureum	Petiole	MS + 11.35 μM TDZ + 2.69 μM NAA	Zhang et al. (2005)
	Stem	MS + 11.35 μM TDZ + 2.69 μM NAA	
Epipremnum aureum	Petiole	MS + 11.35 μM TDZ + 2.69 μM NAA	Wang et al. (2007)
Spathiphyllum wallisii	Anther filament	BMS + 2.5 μM TDZ + 10 μM 2,4-D	Werbrouck et al. (2000)
Syngonium podophyllum	Petiole	MS + 11.35 μM TDZ + 2.69 μM NAA	Zhang et al. (2006)
Spathiphyllum 'Supreme'	Leaf	MS + 9.08 μM TDZ + 2.26 μM 2,4-D	Zhao et al. (2012a)
	Petiole	MS + 4.54 μM TDZ + 2.26 μM 2,4-D	
Spathiphyllum wallisii	Ovules	BM + 4 μM TDZ + 15 μM IMA	Eeckhaut et al. (2001)
Syngonium podophyllum	Petiole	MS + 11.35 μM TDZ + 2.69 μM NAA	Wang et al. (2007)
Zantedeschia hybrids	Anthers	MS + 0.22 μM TDZ + 10.74 μM NAA	Duquenne et al. (2006)

Somatic embryogenesis, particularly direct somatic embryogenesis, has a low frequency of chimeras and a low probability of somaclonal variation. DNA flow cytometry analysis of randomly selected plantlets of 'Marble Queen' regenerated via direct somatic embryogenesis showed a single peak, indicating there were no mixoploids among the regenerated plantlets (Zhao et al. 2012a, b). Histological analysis of somatic embryos derived from ornamental aroids was also reported (Matsumoto et al. 1996; Hamidah et al. 1997; Zhao et al. 2012a). Longitudinal sections of a fully mature *Anthurium* somatic embryo showed clear bipolarity, with both shoot and root poles, as well as a continuous procambium and an epidermis (Matsumoto et al. 1996). Observation by Matsumoto et al. (1996) showed that somatic embryos of *A. andraeanum* originate within the mesophyll via direct embryogenesis.

Fig. 4.1 Regeneration of *Epipremnum aureum* 'Golden Pothos' from leaf explants through direct somatic embryogenesis. Somatic embryos directly appeared from cut end of a leaf explants cultured on Murashige Skoog (MS) medium supplemented with 11.35 μM TDZ with 2.69 μM NAA (**a**). Embryos developed or produced secondary embryos and appeared in clusters (**b**). Embryos were well developed structures and easy to separate; an assortment of embryos including globular, scutellum, and torpedo stages are presented (**c**). Somatic embryos were able to convert to shoots (**d**) and produced roots (**e**), which looked like seedlings (**f**). Plantlets or seedlings were transplanted singly into cell-plug trays and grown healthily in a shaded greenhouse (**g**)

4.6 TDZ-Induced Regeneration Through PLBs

TDZ has been shown to induce PLBs in some aroid plants (Table 4.5). PLBs are composed of many meristematic centers that are able to differentiate into shoots and roots (Da Silva et al. 2000). Cui et al. (2008) documented that PLBs were formed from nodal explants of *S. podophyllum* 'White Butterfly' cultured on MS medium

Table 4.5 Aroid plants regenerated through protocorm-like bodies with TDZ as cytokinin in the culture medium

Scientific name	Explant	Protocol	References
Anthurium andraeanum cv.	Shoot tip-ends	MS + 5.0 μM TDZ	Gantait et al. (2012)
Syngonium podophyllum	Node	MS + 9.08 μM TDZ + 1.07 μM NAA	Cui et al. (2008)

supplemented with TDZ and 2,4-D. Adventitious shoots were formed from PLBs and roots formed thereafter. A popular opinion about PLBs in orchid propagation is that they are somatic embryos. However, PLBs are distinguished from somatic embryos by the lack of a single embryonic axis (Norstog 1979). A recent molecular analysis of PLBs in *Phalaenopsis aphrodite* indicated that PLBs do not follow the embryogenesis pattern (Fang et al. 2016). Instead, the authors proposed that *SHOOT MERISTEMLESS*, a class I KNOTTED-LIKE HOMEOBOX gene, is likely to play a role in PLB regeneration. Thus, PLBs differ from somatic embryos. An advantage for propagation through PLBs is that a large number of plantlets (shoots with roots) can be regenerated thus enhancing propagation efficiency.

4.7 TDZ Action in Micropropagation

The effectiveness of TDZ in plant micropropagation has been attributed to its high level of activity and stability in culture media. However, the mode of action of TDZ-mediated micropropagation is still unclear. TDZ was considered as a cytokinin for its induction of natural cytokinin-like responses. Increases in endogenous auxin, ethylene, and abscisic acid (ABA) in peanut were found to be related to TDZ treatment (Murthy et al. 1995; Murch and Saxena 1997). Some evidence suggests that the action mechanism of TDZ could be closely associated with the biosynthesis and transportation of indole-3-acetic acid (IAA) (Chhabra et al. 2008). Guo et al. (2017) proposed that a combination of increased GA_3, zeatin, and H_2O_2 concentration is the basis for enhanced shoot morphogenesis in response to TDZ treatment. In a study of TDZ-mediated regeneration of rose (*Rosa canina* L.), TDZ administration affected the level of endogenous auxins and cytokinins, converted the cell fate of rhizoid tips, and triggered PLB formation and plantlet regeneration (Kou et al. 2016). Nevertheless, molecular mechanisms concerning TDZ-mediated morphogenesis are largely unknown. However, the established shoot culture and regeneration methods through shoot organogenesis, somatic embryogenesis, and PLBs in aroid plants could be valuable systems for further dissecting the molecular basis underlying shoot culture and each of the regeneration pathways.

4.8 Conclusion

Aroids are economically and environmentally high value crops. Commercial production of this group of crops was traditionally limited due to the lack of healthy starting materials. It was the application of micropropagation techniques that lead to increased commercial availability and production of healthy and disease-free propagules year-round. Aroid plants such as *Amorphophallus rivieri* were among the first reports of plants successfully micropropagated (Morel and Wetmore 1951). Since the discovery of TDZ as an effective plant growth regulator, TDZ has been used for in vitro shoot culture and for regeneration of aroid plants through shoot organogenesis, somatic embryogenesis, and PLBs. Millions of plantlets from *Alocasia, Aglaonema, Anthurium, Dieffenbachia, Homalomena, Philodendron, Spathiphyllum*, and *Syngonium* have been produced. Through our continued research on TDZ-based aroid micropropagation, more aroid plants will be in vitro cultured, and more protocols will be developed. With the advance of omics technologies, in combination with the developed protocols, the molecular basis for TDZ-mediated regeneration will be uncovered in the near future.

References

Arndt FR, Rusch R, Stillfried HV, Hanisch B, Martin WC (1976) A new cotton defoliant. Plant Physiol 57:S-99

Bates S, Preece JE, Navarrette NE, Van Sambeek JW, Gaffney GR (1992) Thidiazuron stimulates shoot organogenesis and somatic embryogenesis in white ash *(Fraxinus americana* L.) Plant Cell Tissue Organ Cult 31:21–30

Bhattacharya C, Dam A, Karmakar J, Bandyopadhyay TK (2015) Efficient organogenesis from the induced meristemoid of *Anthurium andraeanum* Linden cv. Tinora. Plant Sci Today 2:82–86

Bown D (2000) Aroids: plants of the arum family, 2nd edn. Timber Press, Portland

Capelle SC, Mok DW, Kirchner SC, Mok MC (1983) Effects of thidiazuron on cytokinin autonomy and the metabolism of N6-(Δ2-isopentenyl) [8-14C] adenosine in callus tissues of *Phaseolus lunatus* L. Plant Physiol 73:796–802

Chand H, Pearson M, Lovell PH (1999) Rapid vegetative multiplication in *Colocasia esculenta* (L) Schott (taro). Plant Cell Tissue Organ Cult 55:223–226

Chang H, Charkabarty D, Hahn E, Paek K (2003) Micropropagation of calla lily (*Zantedeschia albomaculata*) via in vitro shoot tip proliferation. In Vitro Cell Dev Biol Plant 39:29–134

Chen J, Henny RJ (2008) Role of micropropagation in the development of ornamental foliage plant industry. In: Da Silva JAT (ed) Floriculture, ornamental and plant biotechnology, vol V. Global Science Books, London, pp 206–218

Chen J, Stamps RH (2006) Cutting propagation of foliage plants. In: Dole JM, Gibson JL (eds) Cutting propagation: a guide to propagating and producing floriculture crops. Ball Publishing, Batavia, pp 203–228

Chen J, Henny RJ, Chao TC (2003) Somaclonal variation as a source for cultivar development of ornamental aroids. In: Pandalai SG (ed) Recent research development in plant science, vol 1. Research Signpost, Kerala, pp 31–43

Chen J, McConnell DB, Norman DJ, Henny RJ (2005) The foliage plant industry. Hortic Rev 31:47–112

Chen J, Henny RJ, Liao F (2007) Aroids are important medicinal plants. Acta Hortic 756:347–353

Chen F, Wang C, Fang J (2012) Micropropagation of self-heading philodendron via direct shoot regeneration. Sci Hortic 141:23–29
Chen W, Yeh D (2007) Elimination of in vitro contamination, shoot multiplication, and ex vitro rooting of *Aglaonema*. Hortscience 42:629–632
Chhabra G, Chaudhary D, Varma M, Sainger M, Jaiwal PK (2008) TDZ-induced direct shoot organogenesis and somatic embryogenesis on cotyledonary node explants of lentil (*Lens culinaris* Medik.) Physiol Mol Biol Plants 14:347–353
Compton ME, Gray D (1993) Somatic embryogenesis and plant regeneration from immature cotyledons of watermelon. Plant Cell Rep 12:61–65
Cui J, Liu J, Deng M, Chen J, Henny RJ (2008) Plant regeneration through protocorm-like bodies induced from nodal explants of *Syngonium podophyllum* 'White Butterfly'. Hortscience 43:2129–2133
Deo PC, Harding RM, Taylor M, Tyagi AP, Becker DK (2009) Somatic embryogenesis, organogenesis and plant regeneration in taro (*Colocasia esculenta* var. esculenta). Plant Cell Tissue Organ Cult 99:61–71
Deo PC, Taylor M, Harding RM, Tyagi AP, Becker DK (2010) Initiation of embryogenic cell suspensions of taro (*Colocasia esculenta* var. esculenta) and plant regeneration. Plant Cell Tissue Organ Cult 100:283–291
Dewir Y, Chakrabarty D, Hahn E, Paek K (2006) A simple method for mass propagation of *Spathiphyllum cannifolium* using an airlift bioreactor. In Vitro Cell Dev Biol Plant 42:291–297
Du H, Tang D, Huang D (2006) 'Fragrant taro'[*Colocasia esculenta* (L.) Schott var. antiquorum] micropropagation using thidiazuron and benzylaminopurine. J Hortic Sci Biotechnol 81:379–384
Duquenne B, Eeckhaut T, Werbrouck S, Van Huylenbroeck J (2006) In vitro somatic embryogenesis and plant regeneration in *Zantedeschia* hybrids. Plant Cell Tissue Organ Cult 87:329–331
Eeckhaut T, Werbrouck S, Dendauw J, Van Bockstaele E, Debergh P (2001) Induction of homozygous *Spathiphyllum wallisii* genotypes through gynogenesis. Plant Cell Tissue Organ Cult 67:181–189
Eeckhaut TG, Werbrouck SP, Leus LW, Van Bockstaele EJ, Debergh PC (2004) Chemically induced polyploidization in *Spathiphyllum wallisii* Regel through somatic embryogenesis. Plant Cell Tissue Organ Cult 78:241–246
Fang JY, Hsu YR, Chen FC (2013) Development of an efficient micropropagation procedure for *Aglaonema* 'Lady Valentine'through adventitious shoot induction and proliferation. Plant Biotech 30:423–431
Fang SC, Chen JC, Wei MJ (2016) Protocorms and protocorm-like bodies are molecularly distinct from zygotic embryonic tissues. Plant Physiol 171:2682–2700
Fitriani H, Aryaningrum PD, Hartati N (2016) Proliferation of embryogenic callus of Satoimo taro (*Colocasia esculenta* var. Antiquorum) in culture media with various level of sucrose and gelling agent. Nusantara Biosci 8:316–320
Gantait S, Sinniah UR, Mandal N, Das PK (2012) Direct induction of protocorm-like bodies from shoot tips, plantlet formation, and clonal fidelity analysis in *Anthurium andreanum* cv. CanCan. Plant Growth Regul 67:257–270
George, EF, Hall MA, De Kerk G, (2008) Plant propagation by tissue culture. Vol 1. The background, 3rd edn. Springer, Dordrecht
Gu A, Liu W, Ma C, Cui J, Henny RJ, Chen J (2012) Regeneration of *Anthurium andraeanum* from leaf explants and evaluation of microcutting rooting and growth under different light qualities. Hortscience 47:88–92
Guo B, Abbasi BH, Zeb A, Xu L, Wei Y (2011) Thidiazuron: a multi-dimensional plant growth regulator. Afr J Biotechnol 10:8984–9000
Guo B, He W, Zhao WY, Fu Y, Guo J, Wei Y (2017) Changes in endorgneous hormones and H_2O_2 bust during shoot organogeneisis in TDZ-treated *Saussurea involucrate* explants. Plant Cell Tissue Organ Cult 128:1–8
Haberlandt G (1902) Culturversuche mit isolierten Pflanzenzellen. Sitzungsberichte Akademie der Wissenschaften in Wien. Mathematisch-Naturwissenschaftliche Klasse Abteilung 111:69–92

Sama AE, Shahba MA, Hughes HG, Abbas MS (2015) Comparative growth analysis and acclimatization of tissue culture derived cocoyam (*Xanthosoma sagittifolium* L. Schott) plantlets. Am J Exp Agri 5:94–108

Schleiden MJ (1838) Beitrage zur Phytogenesis. Mullers Archives. Anatomie Physiologie 1838:137–176

Schwann Y (1839) Mikroscopishe Utersuchungen uber die Ubereinstimmung in der Struktur und dem Wanchstum der Thiere und Pflanzen, vol 176. Oswalds, Berlin, p 1910

Seetohul S, Puchooa D, Ranghoo-Sanmukhiya V (2007) Genetic improvement of taro (*Colocasia esculenta* var Esculenta) through in-vitro mutagenesis. Univ Mauritius Res J 13:79–89

Seetohul S, Maunkee V, Gungadurdoss M (2009) Improvement of taro (*Colocasia esculenta* var Esculenta) through in vitro mutagenesis. Univ Mauritius Res J A 13:79–89

Shen R, Lee N (2009) Cytokinins stimulate somatic embryogenesis and plant regeneration from male inflorescence of *Dieffenbachia* 'Tiki'. J Agri Assoc Taiwan 10:380–388

Shen X, Chen J, Kane ME (2007a) Indirect shoot organogenesis from leaves of *Dieffenbachia* cv. Camouflage. Plant Cell Tissue Organ Cult 89:83–90

Shen X, Chen J, Kane ME, Henny RJ (2007b) Assessment of somaclonal variation in *Dieffenbachia* plants regenerated through indirect shoot organogenesis. Plant Cell Tissue Organ Cult 91:21–27

Shen X, Kane ME, Chen J, Philips GC (2008) Effects of genotype, explant source, and plant growth regulators on indirect shoot organogenesis in *Dieffenbachia* cultivars. In Vitro Cell Dev Biol Plant 44:282–288

da Silva A, Moraes-Fernandes M, Ferreira A (2000) Ontogenetic events in androgenesis of Brazilian barley genotypes. Rev Bras Biol 60:315–319

Skoog F, Miller CO (1957) Chemical regulation of growth and organ formation in plant tissues cultured in vitro. Symp Soc Exp Biol 54:118–130

Verma VM, Cho JJ (2007) Plantlet development through somatic embryogenesis and organogenesis in plant cell cultures of *Colocasia esculenta* (L.) Schott. AsPac J Mol Biol Biotechnol 18:167–170

Visser C, Qureshi JA, Gill R, Saxena PK (1992) Morphoregulatory role of thidiazuron substitution of auxin and cytokinin requirement for the induction of somatic embryogenesis in geranium hypocotyl cultures. Plant Physiol 99:1704–1707

Wang X, Li Y, Nie Q, Li J, Chen J, Henny R (2007) In vitro culture of *Epipremnum aureum*, *Syngonium podophyllum*, and *Lonicera macranthodes*, three important medicinal plants. Acta Hortic 756:155–161

Werbrouck SPO, Eeckhaut TGR, Debergh PC (2000) Induction and conversion of somatic embryogenesis on the anther filament of *Spathiphyllum* Schott. Acta Hortic 520:263–270

Winarto B, da Silva JAT (2012) Influence of isolation technique of half-anthers and of initiation culture medium on callus induction and regeneration in *Anthurium andreanum*. Plant Cell Tissue Organ Cult 110:401–411

Winarto B, Mattjik NA, Da Silva JAT, Purwito A, Marwoto B (2010) Ploidy screening of anthurium (*Anthurium andreanum* Linden ex André) regenerants derived from anther culture. Sci Hortic 127:86–90

Winarto B, Rachmawati F, Da Silva JAT (2011a) New basal media for half-anther culture of *Anthurium andreanum* Linden ex André cv. Tropical. Plant Growth Regul 65:513–529

Winarto B, Rachmawati F, Pramanik D, Da Silva JAT (2011b) Morphological and cytological diversity of regenerants derived from half-anther cultures of *Anthurium*. Plant Cell Tissue Organ Cult 105:363–374

Wolverton B, McDonald RC, Watkins E (1984) Foliage plants for removing indoor air pollutants from energy-efficient homes. Econ Bot 38:224–228

Yancheva SD, Golubowicz S, Fisher E, Lev-Yadun S, Flaishman MA (2003) Auxin type and timing of application determine the activation of the developmental program during in vitro organogenesis in apple. Plant Sci 165:299–309

Yeh D-M, Yang W, Chang F, Chung M, Chen W, Huang H (2007) Breeding and micropropagation of *Aglaonema*. Acta Hortic 755:93–98

Zhang Q, Chen J, Henny R (2005) Direct somatic embryogenesis and plant regeneration from leaf, petiole, and stem explants of Golden Pothos. Plant Cell Rep 23:587–595
Zhang Q, Chen J, Henny RJ (2006) Regeneration of *Syngonium podophyllum* 'Variegatum'through direct somatic embryogenesis. Plant Cell Tissue Organ Cult 84:181–188
Zhao J, Cui J, Liu J, Liao F, Henny RJ, Chen J (2012a) Direct somatic embryogenesis from leaf and petiole explants of *Spathiphyllum* 'Supreme'and analysis of regenerants using flow cytometry. Plant Cell Tissue Organ Cult 110:239–249
Zhao J, Zhang Q, Xie J, Hung C-Y, Cui J, Henny RJ, Chen J (2012b) Plant regeneration via direct somatic embryogenesis from leaf and petiole explants of *Epipremnum aureum* 'Marble Queen'and characterization of selected variants. Acta Physiol Plant 34:1461–1469
Zhao J, Li ZT, Cui J, Henny RJ, Gray DJ, Xie J, Chen J (2013) Efficient somatic embryogenesis and agrobacterium-mediated transformation of pothos (*Epipremnum aureum*) 'Jade'. Plant Cell Tissue Organ Cult 114:237–247

Use of TDZ for Micropropagation of Some Mediterranean Crop Species

5

Mouaad Amine Mazri, Ilham Belkoura, and Reda Meziani

Abstract

Plant tissue culture is now a widely used technology for many applications. Plant growth regulators (PGRs) play an important role in this technology. Auxins and cytokinins are by far the most commonly used PGR classes in plant tissue culture. N-Phenyl-N′-1,2,3-thiadiazol-5-ylurea, also known as thidiazuron or TDZ, is a synthetic PGR and a phenylurea derivative with a strong cytokinin-like activity. TDZ has been successfully used for the micropropagation of several plant species and sometimes was reported to be more effective than adenine-based cytokinins or to fulfil both the auxin and the cytokinin requirement for in vitro growth and differentiation. In this chapter, we will focus on the use of TDZ for the micropropagation of six Mediterranean crop species. Thus, we will report some findings from past and recent studies in which TDZ was employed for in vitro culture and regeneration of *Olea europaea* L., *Citrus* spp., *Ceratonia siliqua* L., *Punica granatum* L., *Ficus carica* L., and *Prunus dulcis* Mill. We will also highlight the specific action of TDZ depending on its concentration as well as the species, the genotype, the explant, and the concentration of the associated PGRs.

M. A. Mazri (✉)
Institut National de la Recherche Agronomique, CRRA-Marrakech, UR Agro-Biotechnologie, Laboratoire de Biotechnologie Végétale, BP 533, Marrakech, Morocco

I. Belkoura
Ecole Nationale d'Agriculture, Département des Sciences de Base, Laboratoire de Culture In Vitro, BP S/40, Meknes, Morocco

R. Meziani
Institut National de la Recherche Agronomique, CRRA-Errachidia, UR Systèmes Oasiens, Laboratoire National de Culture des Tissus du Palmier Dattier,
BP 2, Avenue Moulay Ali Cherif, Errachidia, Morocco

N. Ahmad, M. Faisal (eds.), *Thidiazuron: From Urea Derivative to Plant Growth Regulator*, https://doi.org/10.1007/978-981-10-8004-3_5

Keywords
Almond · Carob · Citrus · Fig · Olive · Pomegranate · TDZ

5.1 Introduction

Thidiazuron (N-phenyl-N′-1,2,3-thiadiazol-5-ylurea; TDZ), is a phenylurea derivative with a strong cytokinin-like activity. TDZ was first used as a cotton defoliant (Arndt et al. 1976). In 1982, TDZ was reported to have a cytokinin-like activity (Mok et al. 1982). Since then, TDZ became widely used in plant tissue culture.

TDZ is a light yellow-crystalline compound highly soluble in ethanol, sparingly soluble in water, and soluble to different degrees in various organic solvents (Murthy et al. 1998). It has a molecular weight of about 220.25 and does not contain the purine ring common to adenine-type cytokinins. Instead, it contains two functional groups in its molecule: phenyl and thidiazol (Lu 1993; Murthy et al. 1998). TDZ was categorized as cytokinin because of induction of cytokinin-like activity in explants (Guo et al. 2011). In some cases, TDZ is more effective than adenine-based cytokinins, mainly in the induction of adventitious shoots (Van Staden et al. 2008). It was also reported that TDZ might play both auxin and cytokinin roles in in vitro multiplication, differentiation, and morphogenesis (Murthy et al. 1998).

It has been reported that TDZ stimulates the production and accumulation of endogenous cytokinins as well as the accumulation of phenols, catalase, and peroxidase (Guo et al. 2011; Murthy et al. 1998). In some cases, TDZ was reported to induce browning of cultured tissues. For example, in date palm (Mazri 2015). Tissue browning in date palm was positively correlated with the high peroxidase activity in explants (Meziani et al. 2016). TDZ was also reported to cause alterations in enzyme kinetics, which is responsible of morphological changes in TDZ-induced tissues (Guo et al. 2011). TDZ was used alone or in combination with other plant growth regulators (PGRs) to induce a wide range of morphogenic responses in cultured explants. For example, TDZ was used for somatic embryo induction, adventitious bud formation, and axillary bud multiplication. TDZ was also used to enhance shoot regeneration in many recalcitrant plant species (Pelah et al. 2002; Schween and Schwenkel 2002; Liu et al. 2003; Mithila et al. 2003).

The present chapter reports the findings of several studies examining the effects of TDZ on in vitro morphogenesis of some Mediterranean plant species of high agro-economic importance, namely, *Olea europaea* L., *Citrus* spp., *Ceratonia siliqua* L., *Punica granatum* L., *Ficus carica* L., and *Prunus dulcis* Mill.

5.2 TDZ and In Vitro Morphogenesis

TDZ is involved in several aspects of tissue culture: TDZ has been widely used for axillary shoot bud development, and in many plant species, TDZ showed better results in terms of axillary bud formation from stem nodes and nodal cutting explants than adenine-based cytokinins. However, many authors reported that TDZ does not stimulate shoot elongation (e.g., Rai 2002). It seems also necessary to use an auxin-supplemented medium for shoot rooting. Large-scale propagation systems using TDZ for axillary shoot bud formation and multiplication were reported in many plant species of high economic importance.

TDZ has been mainly used for adventitious shoot bud formation and proliferation. Various concentrations of TDZ were reported depending on the genotype and the explant. Generally, TDZ is very effective in terms of adventitious bud formation, even when used at low concentrations. However, its use was not recommended for shoot bud elongation (Rai 2002). Sometimes, TDZ was reported to induce adventitious roots in cultured explants (Capelo et al. 2010).

TDZ was also efficient in inducing embryogenic callus and direct somatic embryogenesis (Chen and Chang 2006; Ipekci and Gozukirmizi 2003) and was successfully used for somatic embryo maturation and germination (Ceasar and Ignacimuthu 2010). In plant breeding and genetic improvement programs, TDZ was used in protoplast culture. In fact, due to its high organogenic potential, TDZ was used as a potent cytokinin for shoot regeneration from protoplasts of various species (Böhmer et al. 1995; Tegeder et al. 1995). TDZ was also successfully used to produce haploid lines (Diao et al. 2009; Li et al. 2013).

5.3 Use of TDZ for Micropropagation of Some Mediterranean Crop Species

5.3.1 Olive

Olive (*Olea europaea* L.) is one of the most ancient cultivated fruit trees in the Mediterranean region, where it has a great socioeconomic importance. It is also one of the most important agricultural products of the Mediterranean countries and one of the most extensively cultivated fruit crops in the world (Conde et al. 2008; Hagidimitriou et al. 2005; Yaman et al. 2000). Olive is mainly propagated by leafy cuttings. However, during the last decades, many reports were published using in vitro techniques for rapid propagation of true-to-type plantlets, production of disease-free plants, cryopreservation of elite germplasm, and for genetic improvement (Rugini et al. 2011).

TDZ has been evaluated for various genotypes in various regeneration systems (Table 5.1). Rinaldi and Lambardi (1998) used TDZ at the concentration of 10 μM to evaluate the in vitro germinability of nine olive cultivars and found that TDZ significantly improves seeds germinability of cvs. Trillo, Grossolana, and Grappolo and positively affects the germination rate by shortening the germination time in seven olive cultivars: Moraiolo, Canino, Leccio, Morchiaio, Trillo, Gremignolo, and Grappolo. Therefore, these authors suggested to use TDZ to accelerate the breakage of dormancy and to shorten the germination time. Perri et al. (1994a) attempted to isolate and culture olive protoplasts from cvs. Dolce Agogia and Canino. The protoplasts were isolated from leaf- and petiole-derived calli and

Table 5.1 Effect of TDZ on in vitro morphogenesis of olive

Species	Genotype	Explant	TDZ concentration	Other growth regulators	Morphogenic response	Reference
Olea europaea L.	Trillo, Grossolana, and Grappolo	Seeds	10 μM	–	Seeds germinability	Rinaldi and Lambardi (1998)
	Moraiolo, Canino, Leccio, Morchiaio, Trillo, Gremignolo, and Grappolo	Seeds	10 μM	–	Shortening the germination time	Rinaldi and Lambardi (1998)
	Dolce Agogia and Canino	Protoplasts	5 mg/L	0.01 mg/L NAA	Cell division	Perri et al. (1994a)
	Cassanese	Anthers	5 mg/L	0.1 mg/L NAA	Callogenesis	Perri et al. (1994b)
	Conservolea	Anthers	5 mg/L	0.5 mg/L NAA	Callogenesis	Ramezani and Shekafandeh (2009)
	Conservolea	Anthers	7.5 mg/L	–	Callogenesis	Ramezani and Shekafandeh (2009)
	Moraiolo	Two-nodal apical explants	0.1–1 mg/L	–	Axillary bud formation	Zuccherelli and Zuccherelli (2002)
	Koroneiki	Single node explants	0.1–0.4 mg/L	–	Shoot and node proliferation	Roussos and Pontikis (2002)
	Arbequina, Picual, and Empeltre	Sprouted nodes derived from unimodal explants	1 μM	1 μM BAP	Shoot sprouting	García-Férriz et al. (2002)
	Moraiolo	Petioles collected from in vitro-growing shoots	5–20 μM	0–2.5 μM NAA	Shoot regeneration	Mencuccini et al. (1991)
	Moraiolo, Dolce Agogia, and Halkidikis	Petioles collected from in vitro-growing shoots	5–40 μM	0–2.5 μM NAA	Adventitious shoot formation	Mencuccini and Rugini (1993)
	Canino and Moraiolo	Petioles collected from in vitro-growing shoots	30 μM	0.54 μM NAA	Callogenesis and adventitious shoot formation	Rugini and Caricato (1995)
	Olea europaea ssp. *europaea* var. *sylvestris*	Petioles	31.78 μM	0.53 μM NAA	Organogenic callus formation	Capelo et al. (2010)
	Roghani, Zard, and Dezfooli	Leaf segments from single nodes, shoot tips, and axillary buds	1.1–3.3 mg/L	–	Meristematic protuberances and shoot regeneration	Bahrami et al. (2010)
	Dahbia	Petiole and leaf segments derived from in vitro-grown shoots	30 μM	0.5 μM NAA	Somatic embryogenesis and organogenesis	Mazri et al. (2013)
	Stop Vert and Ac18	Shoot apex and leaf primordia explants	30 μM	0.5 μM NAA	Embryogenic callus formation	Narváez et al. (2016)

cultured on MS (Murashige and Skoog 1962) medium containing 5 mg/L TDZ as well as 0.01 mg/L 1-naphthaleneacetic acid (NAA), 0.75% sucrose, and 9% mannitol. According to these authors, the cells derived from protoplasts of petiole callus were able to divide but failed to proliferate. TDZ has been also used in attempts to produce haploid plants from in vitro cultures of olive anthers. Perri et al. (1994b) used TDZ at the concentration of 5 mg/L in combination with 0.1 mg/L NAA and reported callus proliferation in 12% of cv. Cassanese explants. More recently, Ramezani and Shekafandeh (2009) evaluated the effects of various TDZ concentrations on anther culture of olive cv. Conservolea. It was found that the concentrations of 5 mg/L TDZ (in combination with 0.5 mg/L NAA) and 7.5 mg/L TDZ induced calli (18–20%). In all cases, none of these authors reported the regeneration of haploid plants.

The effect of TDZ on olive propagation through the microcutting technique was also evaluated. Zuccherelli and Zuccherelli (2002) used TDZ for large-scale propagation of 50 olive cultivars, with a focus on cv. Moraiolo. These authors used two-nodal apical explants and tried to stimulate axillary bud emission on various culture media. The use of TDZ at 0.1–1 mg/L resulted in the production of numerous shoots and a low callogenesis rate. The rooting and acclimatization phases of this cultivar were successfully achieved. In olive cv. Koroneiki, single node explants were cultured on media supplemented with various cytokinin types and concentrations (Roussos and Pontikis 2002). The use of TDZ at 0.1–0.2 mg/L induced the maximum shoot number (approximately 1.5 shoots per original explant), while the concentrations ranging from 0.2 to 0.4 mg/L induced significantly higher number of nodes per cm of shoot length (3.2 nodes per cm of shoot length) than the other cytokinins. TDZ has been also used for the micropropagation of cvs. Arbequina, Picual, and Empeltre (García-Férriz et al. 2002). Sprouted nodes derived from unimodal explants were cultured on media supplemented with various combinations of 6-benzylaminopurine (BAP) and TDZ. The highest shoot production was obtained on half-strength Rugini olive medium (OM; Rugini 1984) supplemented with 1 μM BAP and 1 μM TDZ, with 6.19, 6.07, and 5.85 shoots sprouted per explant in cvs. Arbequina, Picual, and Empeltre, respectively. The developed shoots were successfully elongated, rooted then acclimatized.

In many other works, TDZ was used for adventitious shoot induction from olive explants. Mencuccini et al. (1991) reported shoot regeneration from leaf petioles, collected from in vitro-growing shoots of cv. Moraiolo, on media supplemented with various concentrations of TDZ. The use of TDZ alone resulted in 20% and 10% shoot regeneration on media supplemented with 5 and 20 μM TDZ, respectively. The combination of 2.5 μM NAA and 10 μM TDZ exhibited a shoot regeneration rate of 10%. On the other hand, the use of leaf discs did not give any regeneration. Mencuccini and Rugini (1993) evaluated the morphogenic capacity of petioles, leaf discs, and midribs of several olive cultivars, collected from potted greenhouse plants and field-grown and in vitro shoots. These authors reported the formation of adventitious shoots only in petioles from in vitro-grown shoots of cultivars Moraiolo, Dolce Agogia, and Halkidikis. The highest shoot regeneration rate (18.7%) was obtained in cv. Moraiolo on MS medium supplemented with 5 μM TDZ. The regenerated shoots were successfully elongated, rooted, and hardened.

The shoots showed the same behavior as their mother plant and the same morphology. Rugini and Caricato (1995) used TDZ to induce adventitious shoots from petioles of in vitro-grown shoots of cvs. Canino and Moraiolo. The purpose of the study was to induce somatic embryogenesis from adventitious shoots developed in vitro. These authors reported that the use of TDZ at the concentration of 30 μM, in combination with 0.54 μM NAA resulted in the production of abundant calli as well as single shoots or group of 2–3 shoots. These shoots produced several short nodes with expanded leaves. In a more recent study, Capelo et al. (2010) used TDZ to induce somatic embryogenesis in wild olive tree *O. europaea* ssp. *europaea* var. *sylvestris*. Petiole and leaf explants were cultured on media supplemented with different PGRs combinations. The use of OM medium supplemented with 31.78 μM TDZ and 0.53 μM NAA formed organogenic calli on petioles, which developed roots with a frequency of 36.8%. Bahrami et al. (2010) evaluated the effects of various concentrations of TDZ (1.1, 2.2, and 3.3 mg/L) on direct shoot regeneration from segments of in vitro leaves and apical and axillary buds of olive cvs. Roghani, Zard, and Dezfooli. These authors reported the formation of meristematic protuberances on leaf segments, while shoot regeneration was observed on leaves growing on apical and axillary buds of the three olive cultivars. The highest shoot regeneration frequency was observed on OM medium supplemented with 1.1 mg/L TDZ. On the other hand, no shoot regeneration was observed on leaf disc explants regardless of TDZ concentration. In the Moroccan olive cultivar Dahbia, TDZ was successfully used to induce somatic embryogenesis and organogenesis (Mazri et al. 2013). Petiole and leaf segments derived from in vitro-grown shoots were cultured in various media in order to induce somatic embryogenesis. When leaf segments were used as explants, a 4-day induction period in MS/2 medium supplemented with 30 μM TDZ and 0.5 μM NAA followed by 8 weeks in a PGR-free MS/2 medium and then by a transfer to olive cyclic embryogenesis medium (ECO) supplemented with 0.25 μM indole-3-butyric acid (IBA), 0.44 μM 6-benzylaminopurine (BA), and 0.5 μM N^6-[2-isopentenyl]adenine (2iP) resulted in somatic embryo formation (Fig. 5.1a). The use of petioles resulted in callus formation and shoot bud differentiation (Fig. 5.1b). This same protocol was used by Narváez et al. (2016), who reported embryogenic callus production in two wild olive trees, Stop Vert (from shoot apex explants) and Ac18 (from leaf primordia explants).

5.3.2 Citrus Tree

Citrus is the major fruit tree species grown in the Mediterranean region (González-Mas et al. 2009). It is also one of the most important fruit crops in the world since its cultivated in more than 100 countries in Asia, America, Africa, Europe, and Oceania (Carimi and De Pasquale 2003; Sato 2015). It is a genus of high economic importance due to the high nutritional and commercial values of its products (Carimi and De Pasquale 2003). The high importance of citrus industry suggests the use of in vitro techniques to rapidly propagate new and promising genotypes (Carimi and De Pasquale 2003).

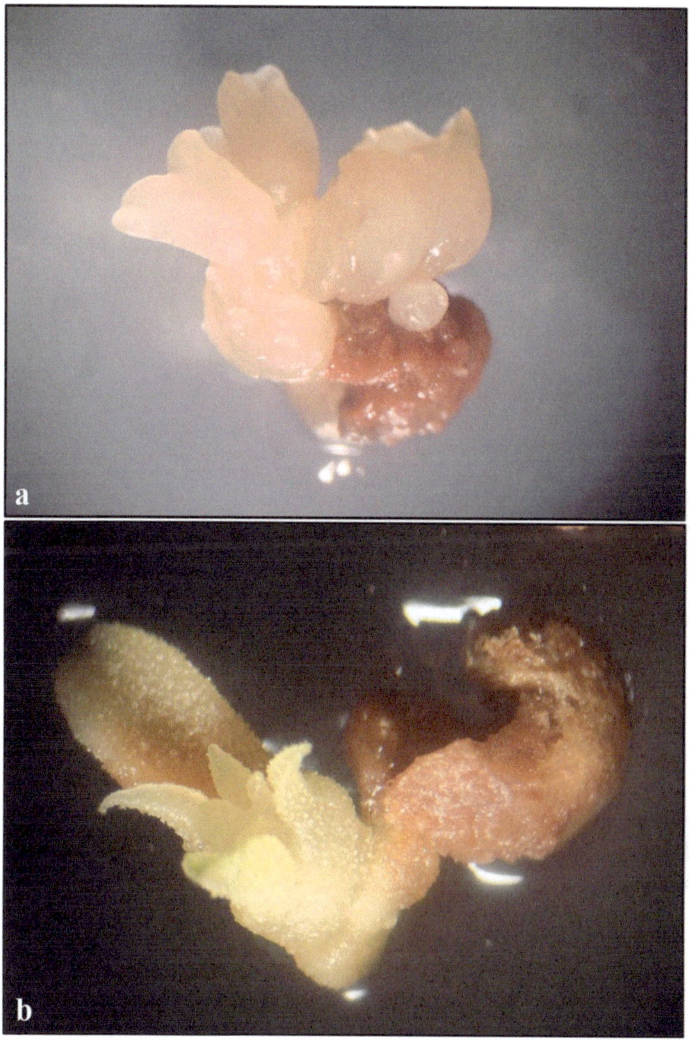

Fig. 5.1 Morphogenic responses to TDZ in olive (*Olea europaea* L. cv. Dahbia). (**a**) TDZ-induced somatic embryos on leaf segments. (**b**) TDZ-induced adventitious buds on petiole explants (Source: Mazri et al. 2013)

Propagation and regeneration systems have been established for many genotypes of *Citrus* using TDZ (Table 5.2). In order to produce haploid lines, Germanà and Chiancone (2003) evaluated the effects of TDZ at the concentration of 0.1 mg/L in the induction medium on anthers of *Citrus clementina* cvs. Nules and SRA 63. These authors reported that the use of TDZ significantly increases the percentage of anthers producing calli and the percentage of those producing embryoids and/or embryogenic calli. The haploid embryoids germinated, and the resulting plantlets

Table 5.2 Effect of TDZ on in vitro morphogenesis of citrus

Species	Genotype	Explant	TDZ concentration	Other growth regulators	Morphogenic response	Reference
Citrus clementina	Nules and SRA 63	Anthers	0.1 mg/L	0.02 mg/L 2,4-D	Callogenesis and embryoid production	Germanà and Chiancone (2003)
				0.02 mg/L NAA		
				1 mg/L kinetin		
				0.5 mg/L BA		
				0.5 mg/L zeatin		
				0.5 mg/L GA_3		
Citrus clementina	Nules	Anthers	2 µM	0.1 µM 2,4-D	Embryogenic callus production and direct somatic embryo formation	Chiancone et al. (2006)
				0.1 µM NAA		
				5 µM kinetin		
				4 µMBA		
				2 µM zeatin		
Poncirus trifoliata (L.) Raf. x *C. paradisi* McFaden	–	Node segments from branches of adult trees	0.1 µM	1.11 µM BA 1.16 µM kinetin	Callogenesis	Sen and Dhawan (2010)
Poncirus trifoliata L. Raf	–	Transverse thin cell layer explants	1 µM	0–10 µM bap	Organogenesis	Van Le et al. (1999)
Citrus macrophylla	–	Transverse sections recovered from seedlings	1 mg/L	–	Organogenesis	Germanà et al. (2008)
Citrus indica Tanaka	–	Leaves of in vitro-germinated seeds	0.01 mg/L	0.1 mg/L NAA	Organogenic callus	Laskar et al. 2009
Citrus aurantium L.	–	Epicotyls	0.05–0.2 mg/L	–	Callogenesis and organogenesis	Roussos et al. (2011)
Citrus sinensis (L.) Osb. x *Poncirus trifoliata* (L.) Raf	–	Epicotyls	1 mg/L	0.1 mg/L NAA	Organogenesis	Germanà et al. (2011)

were vigorous. Chiancone et al. (2006) investigated the influence of polyamines on androgenesis induction in anther culture of *Citrus clementina* cv. Nules. The anthers were cultured on N6 medium (Chu 1978), supplemented with Nitsch and Nitsch vitamins (Nitsch and Nitsch 1969) as well as various additives and PGRs, including TDZ at 2 μM. These authors reported the production of highly embryogenic calli that differentiated into a clump of embryos. In addition, direct embryo formation was observed. The embryos showed normal dicot developmental and were successfully converted into vigorous plantlets, while secondary embryogenesis was also reported.

In order to develop an optimal culture medium for axillary shoot proliferation from single node explants of five *Citrus* genotypes (Rangpur lime, C-35 citrange, Troyer citrange, Swingle citrumelo, and Alemow macrophylla), Sen and Dhawan (2009) evaluated the effects of various media components and PGR combinations. According to these authors, the use of TDZ at the concentration of 100 nM in a medium containing BA and kinetin causes hyperhydricity in shoots. In a different work, Sen and Dhawan (2010) attempted to develop a micropropagation protocol employing node segments from branches of adult trees of Swingle citrumelo [*Poncirus trifoliata* (L.) Raf. x *C. paradisi* McFaden]. Here again, TDZ at 0.1 μM was added to BA (1.11 μM) and kinetin (1.16 μM). The authors reported that TDZ resulted in the growth of a compact and undifferentiated mass of callus. However, no shoot regeneration was observed.

Van Le et al. (1999) used a combination of BAP and TDZ for high frequency in vitro bud regeneration in trifoliate orange (*Poncirus trifoliata* L. Raf). Transverse thin cell layer (tTCL) explants excised from the stem internodes of 1-year-old young plants were cultured on MS medium supplemented with various concentrations of BAP (1–50 μM) and TDZ (0.1–10 μM). The use of BAP at 25 μM or TDZ at 1 μM showed the best results with 87% and 72% of responsive tTCLs and 24 and 15 buds per tTCL, respectively. The combination of 10 μM BAP and 1 μM TDZ resulted in 90% organogenesis and an average of 37 buds per tTCL. Shoots were successfully elongated, rooted, and acclimatized, with no phenotypic variation.

Germanà et al. (2008) investigated the morphogenic response in vitro of *Citrus macrophylla* in response to various factors, including PGRs. In this study, 2-mm-long transverse sections recovered from seedlings were used as explants. The explants were cultured on MS medium supplemented with BAP or TDZ at 1 mg/L for shoot regeneration. The use of TDZ resulted in a bud regeneration frequency of 52.5% and an average number of buds per explant of 1.5.

In *Citrus indica* Tanaka, the effects of various PGR types and concentrations on shoot organogenesis were evaluated. Seeds were germinated in vitro and the leaves were used to induce calli. Various concentrations of TDZ (0.005–0.1 mg/L), alone or in combination with NAA, were used to induce callogenesis. The incorporation of 0.01 mg/L TDZ and 0.1 mg/L NAA in MS medium resulted in high callus formation rates, up to 93.33%. The produced calli were regenerative and gave rise to shoots (Laskar et al. 2009).

In sour orange (*Citrus aurantium* L.), Roussos et al. (2011) evaluated the effects of various cytokinins at three different concentrations on direct shoot organogenesis. Epicotyl explants were cultured on MT (Murashige and Tucker 1969) medium for 45 days. The addition of TDZ (0.05, 0.1 or 0.2 mg/L) to the culture medium resulted in 10–55% callogenesis, depending on the concentration used. However, TDZ resulted in the lowest organogenesis percentage in comparison with the other cytokinins. Germanà et al. (2011) evaluated the effects of two cytokinins, BA and TDZ, on Carrizo citrange [*Citrus sinensis* (L.) Osb. x *Poncirius trifoliate* (L.) Raf] organogenesis. Epicotyls (2-mm-long sections) excised from in vitro-derived seedlings were cultured on MS medium supplemented with 0.1 mg/L NAA and 1 mg/L TDZ or BA. The use of TDZ resulted in the production of 0.8 adventitious buds per segment and 0.1 adventitious shoots per segment, with an average length of 5 mm. Here again, TDZ showed a lower regeneration efficiency than BA.

In some cases, addition of TDZ to culture medium seems to be extremely unfavorable toward shoot formation. For example, in pummelo (*Citrus grandis* L. Osbeck), Paudyal and Haq (2000) used shoot tips derived from seedlings as explants for in vitro shoot proliferation. The explants were cultured on MS medium supplemented with various concentrations of TDZ (0.05, 0.2, 0.5, 2.3, or 4.5 μM) or BA (0.9, 1.3, 1.8, 2.2, 4.4, 8.9, or 13.3 μM), singly or in combination with NAA (0.5 or 1.1 μM). In this study, the use of TDZ resulted in tissue browning and high shoot mortality. In fact, only 30% of explants survived when TDZ was used at 0.05 μM, but their growth and proliferation were very poor. These authors concluded that TDZ is not suitable for in vitro propagation of pummelo. Similar findings were reported by Singh and Rajam (2010). These authors attempted to develop an efficient multiple shoot regeneration and rooting system in sweet orange (*Citrus sinensis* L Osbeck, var. Nagpur). Epicotyls derived from in vitro-grown seedlings were used as explants and were cultured on MT medium supplemented with various carbon sources and PGR combinations. It was reported that shoot emergence and development was accomplished under all hormonal treatments except TDZ. In fact, regardless of TDZ concentration, its addition to the culture medium resulted in profuse callusing, followed by necrosis and death of all the explants.

5.3.3 Carob

Carob (*Ceratonia siliqua* L.) belongs to the family Fabaceae. It is native to the Mediterranean basin and has spread into areas of Mediterranean-like climate of America, Africa, Asia, and Australia (Müller et al. 2010; Pérez-García 2009). Carob is an economically and ecologically important species due to its multipurpose uses: it is used for human and animal consumption, for food production, in xerogardening, to produce charcoal and bioethanol, in pharmaceutical industry, as a natural food additive, as a natural antioxidant, as an ornamental plant, and for the forestation of areas threatened by soil erosion and desertification (Cavallaro et al. 2016; Gunes et al. 2013; Pérez-García 2009). The carob tree can be male, female, or hermaphrodite. The female genotypes are the most important commercially (Patarra 2009).

The commercial propagation of carob using the conventional techniques is difficult to achieve. The use of in vitro techniques would be of great interest. Up to now, only few studies were devoted to carob micropropagation.

Few reports on the use of TDZ in in vitro culture of carob have been reported (Table 5.3). TDZ has been evaluated for haploid plant production in carob tree (Custòdio et al. 2005). Anthers of male carob trees were cultured on MS medium supplemented with 2,4-dichlorophenoxyacetic acid (2,4-D) and various types and concentrations of cytokinins, including TDZ. It was found that TDZ is the most effective cytokinin for callus induction, especially when used at 4 mg/L in combination with 0.5 mg/L 2,4-D. This combination resulted in a callogenesis rate of 100% and a haploid cell production rate of 17.2%. However, plant regeneration was not achieved.

TDZ has also been suggested for in vitro shoot multiplication of female carob trees. Mohamed et al. (2009) used cotyledon node explants for multiple shoot production and reported that TDZ at the concentration of 0.1 mg/L resulted in a shoot formation rate of 45% and the production of 1.11 shoots per cotyledon node, while 1 mg/L TDZ exhibited a shoot formation rate of 42.11% and the production of 2.12 shoots per cotyledon node. Zaen El Deen et al. (2014) found that the combination of 0.4 mg/L NAA and 0.25–2 mg/L TDZ resulted in 1–1.67 shoots per explant with lengths of 0.2–1.1 cm and 1–1.2 shoots per explant with lengths of 0.56–0.94 cm, in shoot tips and stem node segments collected from 2-month-old seedlings, respectively.

Saad and Elnour (2010) tried to induce callus from hypocotyledons and cotyledonary leaves excised from 2–4-day-old seedlings of carob. The incorporation of TDZ in the culture medium at 0.1–1 mg/L induced up to 100% callogenesis. The produced calli were morphologically different depending on TDZ concentration and explant type, and small green leaves could be regenerated from these calli.

5.3.4 Pomegranate

Pomegranate (*Punica granatum* L.) is a high-value fruit crop species that belongs to the family Punicaceae. It is native to the Himalayas, from northern India to Iran, but has been cultivated in the Mediterranean region since ancient times (Bhandary et al. 2012; Saad et al. 2012). Pomegranate has a wide therapeutic potential against various diseases including cancers, cardiovascular diseases, and diabetes, among others (Bhandary et al. 2012).

Only few studies have been published using TDZ for the micropropagation of *P. granatum* L. (Table 5.4). Al-Wasel (1999) cultured nodal segments of pomegranate cv. Al-Belehi on MS medium containing NAA in combination with various concentrations of TDZ (0.3–0.6 mg/L). Few shoots (0.12–0.87; 0.24–2 mm length) were produced in media supplemented with TDZ alone, and calli formation was observed (up to 6.07 mm diameter). The combination of NAA and TDZ showed 0.14–0.5 shoot formation, with a shoot length varying from 0.21 to 0.83 mm and a callus diameter ranging from 2.5 to 3.69 mm. For shoot proliferation, the combination of

Table 5.3 Effect of TDZ on in vitro morphogenesis of carob

Species	Genotype	Explant	TDZ concentration	Other growth regulators	Morphogenic response	Reference
Ceratonia siliqua L.	–	Anthers	4 mg/L	0.5 mg/L 2,4-D	Callogenesis and a haploid cell production	Custòdio et al. (2005)
	–	Cotyledon nodes	0.1–1 mg/L	–	Shoot formation	Mohamed et al. (2009)
	–	Shoot tips and stem node segments collected from seedlings	0.25–2 mg/L	0.4 mg/L NAA	Shoot formation	Zaen El Deen et al. (2014)
	–	Hypocotyledons and cotyledonary leaves excised from seedlings	0.1–1 mg/L	–	Callogenesis and leaf regeneration	Saad and Elnour (2010)

Table 5.4 Effect of TDZ on in vitro morphogenesis of pomegranate

Species	Genotype	Explant	TDZ concentration	Other growth regulators	Morphogenic response	Reference
Punica granatum L.	Al-Belehi	Nodal segments	0.3–0.6 mg/L	0–1 mg/L NAA	Shoot formation and callogenesis	Al-Wasel (1999)
	Al-Belehi	Nodal segments	0.5–4.5 mg/L	0–0.1 mg/L NAA	Shoot proliferation	Al-Wasel (1999)
	Ganesh	Nodal segments	0.05–0.5 mg/L	–	Bud break and shoot proliferation	Naik et al. (1999)
	Rabbab	Nodal segments	0.001–0.3 mg/L	–	Shoot proliferation	Golozan and Shekafandeh (2010)
	–	Callus derived from cotyledons and hypocotyls	3–9 μM	–	Organogenic callus and shoot formation	Kanwar et al. (2010)
	Ganesh	Petal segments and filaments derived from unopened flower buds	0.5–2 mg/L	0–0.5 mg/L IAA	Embryogenic callus formation	Khalilsaraie et al. (2015)
	Ganesh	Node segments	0.5–2 mg/L	–	Embryogenic callus formation and secondary embryogenesis	Khalilsaraee and Meti (2016)
	Ganesh	Node segments	0.1–0.2 mg/L	0–0.2 mg/LIAA	Somatic embryo germination	Khalilsaraee and Meti (2016)

0.1 mg/L NAA and 0.5 or 1.5 mg/L TDZ showed the highest number of shoots (2.5), with a shoot length of 9 and 7 mm, respectively. Naik et al. (1999) compared the effects of various cytokinins at different concentrations on bud break and shoot proliferation of cv. Ganesh using nodal segments excised from a mature tree. The use of TDZ at 0.05, 0.1, and 0.5 mg/L showed 63%, 52%, and 37% bud break, respectively, an average of one shoot produced per explant, and shoot lengths of 3.6, 3.2, and 2.1 cm, respectively. Golozan and Shekafandeh (2010) evaluated the effects of various types and concentrations of cytokinins on shoot proliferation from nodal segments (2.0–2.5 cm length, two lateral buds) of cv. Rabbab. The use of TDZ at 0.001–0.3 mg/L exhibited an average of 0.9 shoots per explant with an average length of 0.9 cm.

Kanwar et al. (2010) tried to induce shoot organogenesis from cotyledonary explants (excised from in vitro germinated seedlings) and zygotic embryos of wild pomegranate using TDZ. Cotyledon- and hypocotyl-derived calli were subcultured on media containing various PGR combinations. The use of TDZ at 3–9 μM resulted in the production of 2.25–4.05% organogenic calli and 0.82–1.22 shoots per callus clump. According to these authors, TDZ was not the most effective PGR for organogenic callus production and plant regeneration.

TDZ has been also used to induce somatic embryogenesis in pomegranate. Khalilsaraie et al. (2015) cultured petal segments and filaments derived from unopened flower buds of cv. Ganesh on media supplemented with various PGRs, including TDZ. According to these authors, the callus induction rate reached 100% in petal segments cultured on MS medium supplemented with 0.5 mg/L TDZ, while it was only 41.2% on the same medium but from filaments. These authors reported the production of embryogenic callus on media supplemented with indole-3-acetic acid (IAA) in combination with TDZ, and that TDZ was responsible for the growth of greenish, highly nodulated masses of embryogenic calli. Node segments collected from mature plants of the same cultivar were also used to induce somatic embryogenesis (Khalilsaraee and Meti 2016). A callus induction rate of 78% was observed when the culture medium was supplemented with 1 mg/L TDZ. The produced calli were compact, yellowish green, and nodulated with globular somatic embryos and secondary embryogenesis activity. TDZ has also been used in the germination medium, alone or in combination with IAA. When TDZ was used alone (0.1–0.2 mg/L), the percentage of somatic embryo germination ranged from 5% to 8%, with 0.4–0.8 germinating embryos per callus. When TDZ was combined with IAA, better results were reported: the germination frequency ranged from 15% to 27%, and the mean number of germinated embryos per callus ranged from 1.8 to 3.2.

5.3.5 Fig

Fig (*Ficus carica* L.) is a fruit tree native to Persia, Asia Minor, and Syria and has been distributed throughout the Mediterranean area (Flaishman et al. 2008). It is one of the oldest cultivated fruit trees in the world, and it is an important crop for dry and fresh consumption due to the high nutritional value of its fruits (Dueñas et al. 2008; Mars 2003).

In fig, TDZ has been mainly used for adventitious bud induction and shoot proliferation (Table 5.5). In cv. Masui Dauphine(Yakushiji et al. 2003), leaf segments derived from in vitro shoots were cultured on MS medium supplemented with various concentrations of 2,4-D (0–5 mg/L), phloroglucinol (0–1 mM), and TDZ (0–5 mg/L). In this study, callogenesis was reported after 2 weeks of culture, while organogenesis (adventitious bud formation) was observed after 4 weeks of culture. The highest percentage of adventitious shoot bud formation was 22.5% (1.8 shoot/explant) and was reported when the medium was supplemented with 1 mg/L 2,4-D, 1 mg/L TDZ, and 0.5 mM phloroglucinol, whereas the highest number of shoot buds produced per explant was 2.5 and was observed in the medium containing 0.5 mg/L 2,4-D, 0.5 mg/L TDZ, and 0.5 mM phloroglucinol (9.8% organogenesis). The use of 2,4-D alone stimulated root formation. The excised shoots were rooted successfully, and the regenerated plantlets were established in soil after acclimatization. Yancheva et al. (2005) attempted to develop an in vitro regeneration system through organogenesis that allows *Agrobacterium*-mediated transformation in two fig cultivars, Brown Turkey and Smyrna. These authors used leaves isolated from in vitro shoot cultures and evaluated the effects of various PGR types and concentrations as well as sucrose concentration and leaf surface position. They reported an organogenesis frequency of up to 100% with more than five shoots per regenerating explant on MS medium supplemented with 2 mg/L TDZ and 2 mg/L IBA. The use of TDZ alone resulted in explant expansion and compact calli formation, with very low shoot regeneration. Kim et al. (2007) reported adventitious shoot regeneration in seven fig cultivars. Leaf segments collected from apical buds of in vitro-grown plants were cultured on MS medium supplemented with various combinations of auxins and cytokinins, including TDZ. A relatively high frequency of callus induction was obtained with 2 mg/L 2,4-D and 0.1–0.5 mg/L TDZ, while the combination of IBA and TDZ showed high callogenesis rates: at 0.5 mg/L IBA and 0.5 mg/L TDZ, the callogenesis rate was 78.6%, while it was 82.1% when 1 mg/L TDZ was used. The respective frequencies of shoot regeneration were 78.6% and 67.9%, with 3.9 and 3.1 shoots per explant, respectively. In addition, these authors wounded the surfaces of leaf explants then placed them on media supplemented with combinations of IBA and TDZ. This resulted in high shoot proliferation: when 2.0 mg/L IBA was combined with TDZ (0.5–2 mg/L), high frequencies of shoot regeneration were observed, ranging from 28.6% to 92.9%, with an average number of shoots per explant up to 10.8. The regenerated shoots were successfully rooted and acclimatized. Dhage et al. (2012) used leaf and petiole explants obtained from in vitro-established shoots of four fig genotypes to induce organogenesis. The use of TDZ at 2 mg/L in combination with 4 mg/L 2iP resulted in 100% callogenesis in petioles and 85.8% callogenesis in leaves of Brown Turkey. This was followed by the emergence of shoots with a frequency of 21.4%. In a different study, calli derived from tender leaves of cv. Poona Fig were used for shoot regeneration (Dhage et al. 2015). The calli were cultured on MS medium supplemented with various PGRs combinations. The use of 7 mg/L TDZ in combination with 0.25 mg/L NAA gave the highest percentage of shoot induction (82.7%) with an average number of shoots per explant of 2.8 and an average shoot length of 20.2 mm. Higher concentrations of TDZ or its combination with 2iP did not increase the shoot induction frequency.

Table 5.6 Effect of TDZ on in vitro morphogenesis of almond

Species	Genotype	Explant	TDZ concentration	Other growth regulators	Morphogenic response	Reference
Prunus dulcis Mill.	Boa Casta	Leaves excised from in vitro cultures	9.08 µM	2.46 µM IBA	Adventitious shoot induction	Miguel et al. (1996)
	Boa Casta	Leaves of micropropagated shoots	1.5 mg/L	0.5 mg/L IAA, 0.01 mg/L 2,4-D	Adventitious shoot induction	Costa et al. (2007)
	Ne Plus Ultra, Nonpareil	Leaves of micropropagated shoots	6.8–22.7 µM	9.8 µM IBA	Adventitious shoot induction	Ainsley et al. (2000)
	Ne Plus Ultra, Nonpareil Carmel, Parkinson	Immature seed cotyledons	1–20 µM	–	Adventitious shoot induction	Ainsley et al. (2001)
	–	Dormant axillary buds	0.01–1 mg/L	–	Adventitious shoot induction	Choudhary et al. (2015)

supplemented with various PGR combinations. In cv. Ne Plus Ultra, the use of TDZ alone at the concentration of 20 μM resulted in the highest regeneration rate (93.3%). In cv. Carmel, up to 100% regeneration rate was observed when TDZ was used at 10 μM. In cv. Nonpareil, the concentrations of 1 and 20 μM TDZ resulted in a regeneration rate of 80%, whereas in cv. Parkinson a regeneration rate of 86.7% was obtained on media supplemented with 10 and 20 μM TDZ. With the highest level (20 μM) of TDZ, shoot elongation was inhibited and the incidence of shoot hyperhydricity was increased. The number of shoots per cotyledon ranged from 10.7 to 15.1. Choudhary et al. (2015) used dormant axillary buds of various almond cultivars as explants. The explants were cultured on media supplemented with various concentrations of PGRs. The use of TDZ at a concentration ranging from 0.01 to 1 mg/L resulted in a shoot induction rate ranging from 10% to 40%.

5.4 Conclusions

The presented chapter leads to the conclusion that TDZ might play various roles in in vitro morphogenesis and plant regeneration. In fact, since the late 1980s, the published reports in this research area have confirmed the significant role of TDZ in plant tissue culture. In many cases, TDZ was used alone as a PGR supplement in the culture medium. Indeed, there are several reports that suggest that TDZ may not only act as a cytokinin but that it may play both auxin and cytokinin roles. The review of the literature presented in this chapter indicates that TDZ has been used mainly for adventitious bud formation and shoot bud proliferation. Nevertheless, its effect varies considerably depending on the species and even on the genotype. It differs also with the concentration used and the associated PGR. On the other hand, reports that highlight the negative effects of TDZ are also available. For instance, in some specific cases, the use of TDZ resulted in tissue browning which leads to death of the explants. As regards the six Mediterranean plant species concerned by this review, it may be concluded that TDZ was successfully used for various purposes: adventitious organogenesis, somatic embryogenesis, protoplast culture, haploid plant production, and axillary shoot bud formation and development. This highlights again the various roles and applications of TDZ in in vitro morphogenesis and plant regeneration.

References

Ainsley PJ, Collins GG, Sedgley M (2000) Adventitious shoot regeneration from leaf explants of almond (*Prunus dulcis* mill.) In Vitro Cell Dev Biol Plant 36:470–474

Ainsley PJ, Hammerschlag FA, Bertozzi T, Collins GG, Sedley M (2001) Regeneration of almond from immature seed cotyledons. Plant Cell Tissue Organ Cult 67:221–226

Al-Wasel ASA (1999) In vitro clonal propagation of "Al-Belehi" pomegranate (*Punica granatum* L.) J King Saud Univ 11:3–14

Arndt FJ, Rusch R, Stilfried HV (1976) SN 49537, a new cotton defoliant. Plant Physiol 57:99

Bahrami MK, Azar AM, Dadpour MR (2010) Influence of thidiazuron in direct shoot regeneration from segments of in vitro leaves, and axillary and apical buds of olive (*Olea europaea*). Acta Hortic 884:383–389

Bhandary SK, Kumari SN, Bhat VS, Sharmila KP, Bekal MP (2012) Preliminary phytochemical screening of various extracts of *Punica granatum* peel, whole fruit and seeds. J Health Sci 2:34–38

Böhmer P, Meyer B, Jacobsen HJ (1995) Thidiazuron-induced high frequency of shoot induction and plant regeneration in protoplast derived pea callus. Plant Cell Rep 15:26–29

Bolling BW, Dolnikowski G, Blumberg JB, Chen CYO (2010) Polyphenol content and antioxidant activity of California almonds depend on cultivar and harvest year. Food Chem 122:819–825

Capelo AM, Silva S, Brito G, Santos C (2010) Somatic embryogenesis induction in leaves and petioles of a mature wild olive. Plant Cell Tissue Organ Cult 103:237–242

Carimi F, De Pasquale F (2003) Micropropagation of *Citrus*. In: Jain SM, Ishii K (eds) Micropropagation of woody trees and fruits. Kluwer, Deventer, pp 589–619

Cavallaro V, Barbera AC, Maucieri C, Gimma G, Scalisi C, Patanè C (2016) Evaluation of variability to drought and saline stress through the germination of different ecotypes of carob (*Ceratonia siliqua* L.) using a hydrotime model. Ecol Eng 95:557–566

Ceasar SA, Ignacimuthu S (2010) Effects of cytokinins, carbohydrates and amino acids on induction and maturation of somatic embryos in kodo millet (*Paspalum scorbiculatum* Linn.) Plant Cell Tissue Organ Cult 102:153–162

Chen JT, Chang WC (2006) Direct somatic embryogenesis and plant regeneration from leaf explants of *Phalaenopsis amabilis*. Biol Plant 50:169–173

Chiancone B, Tassoni A, Bagni N, Germanà MA (2006) Effect of polyamines on in vitro anther culture of *Citrus clementina* Hort. ex Tan. Plant Cell Tissue Organ Cult 87:145–153

Choudhary R, Chaudhury R, Malik SK, Sharma KC (2015) An efficient regeneration and rapid micropropagation protocol for almond using dormant axillary buds as explants. Indian J Exp Biol 53:462–467

Chu C (1978) The N6 medium and its applications to anther culture of cereal crops. In: Proceedings of symposium on plant tissue culture. Science Press, Peking pp 43–50

Conde C, Delrot S, Geros H (2008) Physiological, biochemical and molecular changes occurring during olive development and ripening. J Plant Physiol 165:1545–1562

Costa M, Miguel C, Oliveira MM (2007) Improved conditions for *Agrobacterium*-mediated transformation of almond. Acta Hortic 738:575–581

Custòdio L, Carneiro MF, Romano A (2005) Microsporogenesis and anther culture in carob tree (*Ceratonia siliqua* L.) Sci Hortic 104:65–77

Dhage SS, Pawar BD, Chimote VP, Jadhav AS, Kale AA (2012) In vitro callus induction and plantlet regeneration in fig (*Ficus carica* L.) J Cell Tissue Res 12:1–6

Dhage SS, Chimote VP, Pawar BD, Kale AA, Pawar SV, Jadhav AS (2015) Development of an efficient in vitro regeneration protocol in fig (*Ficus carica* L.) J Appl Hortic 17:160–164

Diao WP, Jia YY, Song H, Zhang XQ, Lou QF, Chen JF (2009) Efficient embryo induction in cucumber ovary culture and homozygous identification of the regenetants using SSR markers. Sci Hortic 119:246–251

Dueñas M, Perez-Alonso JJ, Santos-Buelga C, Escrinano-Bailon T (2008) Anthocyanin composition in fig (*Ficus carica* L.) J Food Compos Anal 21:107–115

Ferrandez-Villena M, Ferrandez-Garcia CE, Andreu-Rodriguez J, Ferrandez-Garcia MT, Garcia-Ortuño T (2013) Effect of pressing conditions on physical and mechanical properties of binderless boards made from almond hulls (*Prunus dulcis* (Mill.) DA Webb). In: Proceedings of the 41 international symposium on agricultural engineering, 19–22 February, Opatija, pp 393–397

Flaishman M, Rodov V, Stover E (2008) The fig: botany, horticulture and breeding. Hortic Rev 34:113–196

García-Férriz L, Ghorbel RR, Ybarra M, Marì A, Belaj A, Trujillo I (2002) Micropropagation from adult olive trees. Acta Hortic 586:879–882

Germanà MA, Chiancone B (2003) Improvement of *Citrus clementina* Hort. ex Tan. microspore-derived embryoid induction and regeneration. Plant Cell Rep 22:181–187

Germanà MA, Macaluso L, Patricolo G, Chiancone B (2008) Morphogenic response in vitro of epicotyl segments of *Citrus macrophylla*. Plant Biosyst 142:661–664

Germanà MA, Micheli M, Chiancone B, Macaluso L, Standardi A (2011) Organogenesis and encapsulation of in vitro-derived propagules of Carrizo citrange [*Citrus sinesis* (L.) Osb. x *Poncirius trifoliata* (L.) Raf.] Plant Cell Tissue Organ Cult 106:299–307

Golozan AB, Shekafandeh A (2010) Effects of plant growth regulators on pomegranate (*Punica granatum* L. cv. Rabbab) shoot proliferation and rooting. Adv Hortic Sci 24:207–211

González-Mas MC, Llosa MJ, Quijano A, Forner-Giner MA (2009) Rootstock effects on leaf photosynthesis in 'Navelina' trees grown in calcareous soil. Hort Sci 44:280–283

Gunes E, Gubbuk H, Ayala-Silva T, Gozlekci S, Ercisli S (2013) Effects of various treatments on seed germination and growth of carob (*Ceratonia siliqua* L.) Pak J Bot 45:1173–1177

Guo B, Abbasi BH, Zeb A, Xu LL, Wei YH (2011) Thidiazuron: a multidimensional plant growth regulator. Afr J Biotechnol 10:8984–9000

Hagidimitriou M, Katsiotis A, Menexes G, Pontikis C, Loukas M (2005) Genetic diversity of major Greek olive cultivars using molecular (AFLPs and RAPDs) marker and morphological traits. J Am Soc Hortic Sci 130:211–217

Ipekci Z, Gozukirmizi N (2003) Direct somatic embryogenesis and synthetic seed production from *Paulownia elongata*. Plant Cell Rep 22:16–24

Kanwar K, Joseph J, Deepika R (2010) Comparison of in vitro regeneration pathways in *Punica granatum* L. Plant Cell Tissue Organ Cult 100:199–207

Khalilsaraee MF, Meti NT (2016) Embryogenic cell lines and germination of shoot from somatic embryos of *Punica granatum* 'Ganesh'. Acta Hortic 1131:11–15

Khalilsaraie MF, Meti NT, Karibasappa GS (2015) Maintenance of somatic embryos and cell lines from floral parts of *Punica granatum* L. 'Ganesh'. Acta Hortic 1083:455–460

Kim KM, Kim MY, Yun PY, Chandrasekhar T, Lee HY, Song PS (2007) Production of multiple shoots and plant regeneration from leaf segments of fig tree (*Ficus carica* L.) J Plant Biol 50:440–446

Laskar MA, Hynniewta M, Rao CS (2009) In vitro propagation of *Citrus indica* Tanaka-an endangered progenitor species. Indian J Biotechnol 8:311–316

Li JW, Si SW, Cheng JY, Li JX, Liu JQ (2013) Thidiazuron and silver nitrate enhanced gynogenesis of unfertilized ovule cultures of *Cucumis sativus*. Biol Plant 57:164–168

Liu CZ, Murch SJ, Demerdash MEL, Saxena PK (2003) Regeneration of the Egyptian medicinal plant *Artemisia judaica* L. Plant Cell Rep 21:525–530

Lu C-Y (1993) The use of thidiazuron in tissue culture. In Vitro Cell Dev Biol Plant 29P:92–96

Maestri D, Martínez M, Bodoira R, Rossi Y, Oviedo A, Pierantozzi P, Torres M (2015) Variability in almond oil chemical traits from traditional cultivars and native genetic resources from Argentina. Food Chem 170:55–61

Mars M (2003) Fig (*Ficus carica* L.) genetic resources and breeding. Acta Hortic 605:19–27

Mazri MA (2015) Role of cytokinins and physical state of the culture medium to improve in vitro shoot multiplication, rooting and acclimatization of date palm (*Phoenix dactylifera* L.) cv. Boufeggous. J Plant Biochem Biotechnol 24:268–275

Mazri MA, Belkoura I, Pliego-Alfaro F, Belkoura M (2013) Somatic embryogenesis from leaf and petiole explants of the Moroccan olive cultivar Dahbia. Sci Hortic 159:88–95

Mencuccini M, Rugini E (1993) In vitro shoot regeneration from olive cultivar tissues. Plant Cell Tissue Organ Cult 32:283–288

Mencuccini M, Corona C, Mariotti M (1991) Plant regeneration and first attempt of in vitro genetic improvement of olive (cv Moraiolo). Acta Hortic 300:261–264

Meziani R, Jaiti F, Mazri MA, Hassani A, Ben Salem S, Anjarne M, Ait Chitt M, Alem C (2016) Organogenesis of *Phoenix dactylifera* L. cv. Mejhoul: influences of natural and synthetic compounds on tissue browning, and analysis of protein concentrations and peroxidase activity in explants. Sci Hortic 204:145–152

Miguel CM, Druart P, Oliveira MM (1996) Shoot regeneration from adventitious buds induced on juvenile and adult almond (*Prunus dulcis* mill.) explants. In Vitro Cell Dev Biol Plant 32:148–153

Mithila J, Hall JC, Victor JMR, Saxena PK (2003) Thidiazuron induces shoot organogenesis at low concentration and somatic embryogenesis at high concentration on leaf and petiole explants of African violet (*Saintpaulia ionantha* WEndl.) Plant Cell Rep 21:408–414

Mohamed AI, Elnour EG, Mahfouz SA (2009) Production of multiple shoots from carob tree (*Ceratonia siliqua*) using tissue culture technique. Acta Hortic 812:211–215

Mok MC, Mok DWS, Armstrong DJ, Shudo K, Isogai Y, Okamoto T (1982) Cytokinin activity of N-phenyl-N′-(1,2,3- thiadiazol-5-yl)-urea (thidiazuron). Phytochemistry 21:1509–1511

Müller GC, Xue RD, Schlein Y (2010) Seed pods of the carob tree *Ceratonia siliqua* are a favored sugar source for the mosquito *Aedes albopictus* in coastal Israel. Acta Trop 116:235–239

Murashige T, Skoog FA (1962) A revised medium for rapid growth and bioassays with tobacco tissue cultures. Physiol Plant 15:473–497

Murashige T, Tucker DPH (1969) Growth factor requirements of citrus tissue culture. In: Chapman HD (ed) Proceedings of the 1st international citrus symposium, Riverside, Mar 1968, vol 3, pp 1155–1161

Murthy BNS, Murch SJ, Saxena PK (1998) Thidiazuron: a potent regulator of in vitro plant morphogenesis. In Vitro Cell Dev Biol Plant 34:267–275

Naik SK, Pattnaik S, Chand PK (1999) In vitro propagation of pomegranate (*Punica granatum* L. cv. Ganesh) through axillary shoot proliferation from nodal segments of mature tree. Sci Hortic 79:175–183

Narváez I, Mercado JA, Jiménez-Díaz R, Pliego-Alfaro F (2016) Somatic embryogenesis in explants of adult wild olive trees. URL : http://riuma.uma.es/xmlui/handle/10630/11064

Nitsch JP, Nitsch C (1969) Haploid plants from pollen grains. Science 163:85–87

Patarra JDD (2009) Evaluation of the in vitro biological activities of extracts from carob tree and Mediterranean oaks. Master Dissertation in Biological Engineering. Universidade do Algarve, Portugal

Paudyal KP, Haq N (2000) In vitro propagation of pummelo (*Citrus grandis* L. Osbeck). In Vitro Cell Dev Biol Plant 36:511–516

Pelah D, Kaushik RA, Mizrahi Y, Sitrit Y (2002) Organogenesis in the vine cactus *Selenicereus megalanthus* using thidiazuron. Plant Cell Tissue Organ Cult 71:81–84

Pérez-García F (2009) Germination characteristics and intra-population variation in carob (*Ceratonia siliqua* L.) seeds. Spanish J Agri Res 7:398–406

Perri E, Parlati MV, Rugini E (1994a) Isolation and culture of olive (*Olea europaea* L.) cultivar protoplasts. Acta Hortic 356:51–53

Perri E, Parlati MV, Mule R, Fodale AS (1994b) Attempts to generate haploid plants from in vitro cultures of *Olea europaea* L. anthers. Acta Hortic 356:47–50

Rai RV (2002) Rapid clonal propagation of *Nothapodytes foetida* (Wight) Sleumer- a threatened medicinal tree. In Vitro Cell Dev Biol Plant 38:347–351

Ramezani S, Shekafandeh A (2009) Callus induction from anther explant of olive (*Olea europaea* L.) influenced by plant growth regulators. Adv Env Biol 3:21–24

Rinaldi LMR, Lambardi M (1998) In vitro germinability and ethylene biosynthesis in cytokinin-treated olive seeds (*Olea europaea* L.) Adv Hortic Sci 12:59–62

Roussos PA, Pontikis CA (2002) In vitro propagation of olive (*Olea europaea* L.) cv. Koroneiki. Plant Growth Regul 37:295–304

Roussos PA, Dimitriou G, Voloudakis AE (2011) N-(2-chloro-4-pyridyl)-N-phenylurea (4-CPPU) enhances in vitro direct shoot organogenesis of *Citrus aurantium* L. epicotyl segments compared to other commonly used cytokinins. Span J Agric Res 9:504–509

Rugini E (1984) In vitro propagation of some olive (*Olea europaeasativa* L.) cultivars with different root-ability, and medium development using analytical data from developing shoots and embryos. Sci Hortic 24:123–134

Rugini E, Caricato G (1995) Somatic embryogenesis and plant recovery from mature tissues of olive cultivars (*Olea europaea* L.) “Canino” and “Moraiolo”. Plant Cell Rep 14:257–260

Rugini E, De Pace C, Gutierrez-Pesce P, Muleo R (2011) Olea. In: Chittaranjan K (ed) Wild crop relatives: genomic and breeding resources, 1st edn. Springer, Heidelberg, pp 79–117

Saad AIM, Elnour GE (2010) Induction of callus from hypocotyledons and cotyledonary leaves of *Ceratonia siliqua*. Acta Hortic 865:293–296
Saad H, Charrier-El Bouhtoury F, Pizzi A, Rode K, Charrier B, Ayed N (2012) Characterization of pomegranate peels tannin extractives. Ind Crop Prod 40:239–246
Sato K (2015) Influence of drought and high temperature on citrus. In: Kanayama Y, Kochetov A (eds) Abiotic stress biology in horticultural plants. Springer, Tokyo, pp 77–86
Schween G, Schwenkel HG (2002) In vitro regeneration in *Primula* Sp. via organogenesis. Plant Cell Rep 20:1006–1010
Sen S, Dhawan V (2009) Genotypic differences in shoot multiplication among five citrus rootstocks in vitro. Acta Hortic 839:51–56
Sen S, Dhawan V (2010) Development of a highly efficient micropropagation method for the *Citrus* rootstock 'Swingle' citrumelo [*Poncirus trifoliata* (L.) Raf. x *C. paradisi* McFaden]. Int J Fruit Sci 10:65–78
Singh S, Rajam MV (2010) Highly efficient and rapid plant regeneration in *Citrus sinensis*. J Plant Biochem Biotechnol 19:195–202
Soliman HI, Gabr M, Abdallah N (2010) Efficient transformation and regeneration of fig (*Ficus carica* L.) via somatic embryogenesis. GM Crops 1:47–58
Tegeder M, Gebhardt D, Schieder O, Pickardt T (1995) Thidiazuron-induced plant regeneration from protoplast of Vicia faba cv. Mythos. Plant Cell Rep 15:164–169
Van Le B, Ha NT, Hong LTA, Van KTT (1999) High frequency shoot regeneration from trifoliate orange (*Poncirus trifoliata* L. Raf.) using the thin cell later method. CR Acad Sci Paris Life Sci 322:1105–1111
Van Staden J, Zazimalova E, George EF (2008) Plant growth regulators II: cytokinins, their analogues and antagonists. In: George EF, Hall MA, De Klerk GJ (eds) Plant propagation by tissue culture, vol I the background, 3rd edn. Springer, Dordrecht, pp 205–226
Yakushiji H, Mase N, Sato Y (2003) Adventitious bud formation and plantlet regeneration from leaves of fig (*Ficus carica* L.) J Hortic Sci Biotechnol 78:874–878
Yaman S, Sahan M, Haykiri-Acma H, Sesen K, Kucukbayrak S (2000) Production of fuel briquettes from olive refuse and paper mill waste. Fuel Process Technol 68:23–31
Yancheva SD, Golubowicz S, Yablowicz Z, Perl A, Flaishman MA (2005) Efficient *Agrobacterium*-mediated transformation and recovery of transgenic fig (*Ficus carica* L.) plants. Plant Sci 168:1433–1441
Zaen El Deen EM, El-Sayed OM, El-Sayed AEI, Hegazi GAE (2014) Studies on carob (*Ceratonia siliqua* L.) propagation. J Agri Vet Sci 7:31–40
Zuccherelli G, Zuccherelli S (2002) In vitro propagation of 50 olive cultivars. Acta Hortic 586:931–934

Thidiazuron in Micropropagation of Small Fruits

6

Samir C. Debnath

Abstract

Strawberry, raspberry, grape, blueberry, and cranberry are major small fruit crops cultivated widely across the world. They are highly appreciated and have long been enjoyed enormous popularity among consumers. Their superior nutritive components play a significant dietary role in maintaining human health that has led to a dramatic increase of their global production. There has been an immense progress in small fruit micropropagation using semisolid gelled and liquid media containing different plant growth regulators (PGRs). Thidiazuron [1-phenyl-3-(1,2,3-thiadiazol-5-yl)urea (TDZ)] is a PGR and with its cytokinin- and auxin-like effects, has significant role in in vitro propagation of small fruit crops. Bioreactor micropropagation containing liquid media with TDZ has resulted in significant progresses not only in reducing micropropagation cost but also in speeding up the process significantly for these crop species. However, the optimal plant production depends upon a number of factors including genotype, media types, types and concentration of PGR, and culture environment. The chapter deals with the progress in-depth of various aspects of small fruit micropropagation in semisolid and liquid media containing TDZ and use of TDZ in a bioreactor micropropagation for commercial production. Somaclonal variation can be a major concern in small fruit micropropagation using TDZ. Although strategies have been developed to reduce these variations, DNA-based molecular markers are promising tools to monitor clonal fidelity of TDZ-induced micropropagated small fruit plants. The chapter also describes the use of molecular markers for the assessment of genetic fidelity, stability, and true-to-typeness in small fruit tissue culture plants.

S. C. Debnath (✉)
St. John's Research and Development Centre, Agriculture and Agri-Food Canada, St. John's, Newfoundland and Labrador, Canada
e-mail: samir.debnath@agr.gc.ca

N. Ahmad, M. Faisal (eds.), *Thidiazuron: From Urea Derivative to Plant Growth Regulator*, https://doi.org/10.1007/978-981-10-8004-3_6

Keywords
Bioreactors · Genetic fidelity · Micropropagation · Molecular markers · Plant growth regulators · Small fruits

6.1 Introduction

Small fruits, also known as berry crops, are small- to moderate-sized fruits produced on perennial herbs, vines, or shrubs. Brambles (blackberry, raspberry and their hybrids), *Ribes* (currant and gooseberry), strawberries, table and wine grapes (*Vitis* spp.), and *Vaccinium* species (blueberry, cranberry, lingonberry, and others) are among the important small fruit crops worldwide (Debnath 2003a, 2016a). Native American peoples relied heavily on certain small fruits as a staple in their diet and passed on their knowledge of the fruit to the first European colonists. Many Native Americans and First Peoples of Canada combined dried meat with dried small fruits to add flavor (Trehane 2004). The production of blueberries, cranberries, raspberries, and strawberries is a profitable agricultural enterprise that began in the early nineteenth century. Regionally important minor small fruit crops include *Aronia* (*Aronia melanocarpa* [Michx.] Elliott, Rosaceae), arctic raspberry (*Rubus arcticus* L., *R. stellatus* Sm. and their hybrids; Rosaceae), cloudberry (*R. chamaemorus* L., Rosaceae), mora (*R. glaucus* Benth., Rosaceae), Juneberry/saskatoon (*Amelanchier* sp., Rosaceae), alpine strawberry (*Fragaria vesca* L., Rosaceae), edible honeysuckle (*Lonicera caerulea* L., Caprifoliaceae), elderberry (*Sambucus Canadensis* L., Caprifoliaceae), hardy kiwi (*Actinidia arguta* [Siebold & Zucc.] Planch.ex Miq., Actinidiaceae), sea buckthorn (*Hippophae rhamnoides* L., Elaeagnaceae), schisandra (*Schisandra chinensis* [Turcz.] Baill., Schisandraceae), bilberry (*Vaccinium myrtillus* L., Ericaceae), and muscadine grape (*Vitis rotundifolia* Mich., Vitaceae). Chokecherry (*Prunus virginiana* L.), highbush cranberry (*Viburnum trilobum* Marshall), serviceberry [*Amelanchier alnifolia* (Nutt.) Nutt.], and silver buffalo berry [*Shepherdia argentea* (Pursh) Nutt.] are some of the other small fruit crops that are consumed in the traditional diets of North American tribal communities (Galletta and Himelrick 1990; Finn 1999).

Diets high in small fruits have a positive impact on human health, performance, and disease. They are flavorful providing unique contributions to dietary choices of consumers. Small fruits can satisfy diverse consumer choices and tastes with their different levels of sweetness and acidity, and with a variety of flavors and textures. They are consumed in fresh, dried, juice, and processed product forms. Small fruits are a major human dietary source of phytochemicals including flavonoids and other phenolic compounds, cyanogenic glucosides, phytoestrogens (Mazur et al. 2000), and phenols that are potentially health-promoting and are believed to fight against diseases (Macheix et al. 1991). Consumption of small fruits is likely to decrease the risk of cardiovascular diseases, certain forms of cancer, hypertension, type II diabetes, and other age-related and degenerative diseases (Ames et al. 1993; Rissanen et al. 2003). Fruit and leaf extracts from some small fruit species inhibit some cancers or have strong antioxidant activities as were evident from in vitro and in vivo studies with animal models (Yau et al., 2002). Ellagic acid of small fruits (Häkkinen and Törrönen 2000; Harris et al. 2001; Cordenunsi et al. 2002) affects cell proliferation and apoptosis, suggesting a potential anticancer role. Flavonoid-rich blueberries and cranberries can limit the development and severity of certain cancers and vascular diseases including ischemic stroke, atherosclerosis, and neuro-degenerative diseases of aging (Neto 2007). Lingonberry leaves and fruits are rich in antioxidant properties (Vyas et al. 2015) and can be used to treat stomach disorders, rheumatic diseases, and bladder and kidney infections and to lower cholesterol levels (Novelli 2003). Cranberries produce proanthocyanidins (condensed tannins) that help to prevent urinary tract infections through reduced adhesion of uropathogenic *Escherichia coli* (Howell et al. 2005).

Thidiazuron (TDZ, N-phenyl-N′-1,2,3-thiadiazol-5-ylurea), first used as a cotton defoliant (Arndt et al. 1976), has been shown to exhibit strong cytokinin-like activity similar to adenine derivatives (Mok et al. 1982, 1987; Thomas and Katterman 1986). Although TDZ was categorized as a cytokinin with natural cytokinin-type response (Murthy et al. 1998), it has been found to possess both cytokinin- and auxin-like activities in in vitro culture of various plant species (Mok et al. 1982; Visser et al. 1992).While at higher concentrations, TDZ stimulates callus formation, shoot regeneration, and somatic embryo development, it induces axillary proliferation at low concentration (Huetteman and Preece 1993) although structurally TDZ is different from both auxins and purine-based cytokinins (Murthy et al. 1998).

6.2 Propagation In Vitro

Cultures in vitro (Fig. 6.1) contribute significantly to the small fruit crop development programs. In vitro propagation or micropropagation that includes plant formation from existing meristems and somatic cells has been utilized for propagation and as a part of the genetic manipulation in many small fruit crops. Although micropropagation has been successful in some small fruit crops, there are many species where in vitro methods need to be established for elite selections and to develop

Fig. 6.1 In vitro culture of blueberry on an agar-gelrite gelled medium (left) and in a bioreactor containing liquid medium (right)

genotype-independent routine procedures for increasing the propagation rates and to reduce the probability of somaclonal variation (Larkin and Scowcroft 1981).

Being genetically heterozygous, small fruit crops do not reproduce individuals from seed that are similar to the seed parent (Galletta and Himelrick 1990). Most of small fruit crop species are generally propagated vegetatively to maintain the desired genetic characteristics and to achieve rapidly a fruit-bearing condition. Although conventional vegetative propagation is successful in small fruit crops, the process is very time consuming. In vitro propagation is being used in various small fruit crops for year-round mass propagation of specific genotype and maintenance of pathogen-free (indexed) germplasm and used as the initial step in a nuclear stock crop production system. Shoot regeneration in vitro could accelerate cultivar development programs when used in combination with classical breeding. Successful application of plant tissue culture for shoot regeneration is crucial (Cao and Hammerschlag 2000), but the system can be used for genetic transformation followed by production of transgenic plants and to induce somaclonal variants. Complete plant formation using tissue culture techniques can be achieved either through shoot proliferation from pre-existing buds, through adventitious shoot regeneration, or through the formation of somatic embryos with a shoot meristem and a root (Steward et al. 1970).

Haberlandt (1902) explored plant cell culture in the early nineteenth century to study the concept of totipotency and to explore morphogenesis. He was successful to get survivability of in vitro-grown tissue. While Hannig (1904) was the first to observe plant cell division under in vitro condition, regeneration on callus tissue was first reported by Simon (1908). However, commercial micropropagation started with the work of Boxus (1974) and Anderson (1975) in strawberry and rhododendron, respectively. Since then, micropropagation with small fruit crops has been reviewed in literature by various authors (Debnath 2003a, 2006a, 2007a, 2011a, 2013, 2014a; Graham 2005; McCown and Zeldin 2005; Rowland and Hammerschlag 2005; Skirvin et al. 2005; Debnath et al. 2012).

6.3 Thidiazuron-Induced Micropropagation on Semisolid Gelled Media

6.3.1 Axillary Shoot Proliferation

Shoot tips or nodal segments can be surface sterilized and cultured on an agar or agar-gelrite solidified gelled medium containing TDZ for axillary bud production (Debnath 2005a). Plant propagation through axillary shoot proliferation is the most reliable method to produce true-to-type plants as they normally retain the genetic composition of the mother plant. A higher cytokinin concentration alone or with low levels of auxins is generally used to induce axillary budding. Cytokinins are used in culture media to overcome apical dominance and to enhance lateral bud formation from the leaf axis. More extra shoots are produced through further axillary bud growth during subculturing (Debnath 2003a). Different basal media supplemented with cytokinins such as TDZ, zeatin, 6-benzyladenine (BA), zeatin riboside, or N6-[2 isopentenyl] adenine (2iP) and possibly some auxin can be used for small fruit micropropagation (Debnath 2006a). For axillary shoot proliferation of lingonberries (Debnath 2005a), nodal explants can be cultured on Debnath and McRae's (2001a) shoot proliferation medium containing low concentration of TDZ. A concentration of 0.1–1.0 μM TDZ was found effective for shoot proliferation of lingonberries (Debnath 2005a). Shoot proliferation of strawberry was found effective with 4 μM TDZ in a semisolid culture medium (Debnath 2005b).

Explant orientation on a TDZ-containing culture medium affects shoot proliferation. Lingonberry explants when placed horizontally on the culture medium responded by callus formation around the cut ends from day 6 to day 8 of culture, while vertical placement induced callus development at the basal end of the explants only. Changing the orientation of explants from vertically upright to horizontal improved the number of shoots per explant but reduced the number of leaves per shoot and shoot height (Debnath 2005a).

Lingonberry explants on cytokinin (TDZ)-free medium produced one unbranched shoot each, suggesting the presence of apical dominance (Debnath 2005a). Apical dominance is a major problem in micropropagation of some plant species (George and Sherrington 1984). Axillary branching in nodal explants occurs only when a cytokinin, e.g., TDZ, is applied exogenously in the culture media (Debnath 2005a). TDZ has an apical dominance release that accelerates shoot proliferation (Huetteman and Preece 1993).

The genotype often profoundly affects explant shoot proliferation performance in a medium containing TDZ (Debnath 2005a). Preece et al. (1991) observed differences in axillary shoot proliferation among woody plant species when cultured on a medium containing TDZ. Lingonberry genotypes belonging to two different subspecies differed in their shoot proliferation potential (Debnath 2005a). This might be due to the fact that the cells within the same plant can have dissimilar endogenous quantities of plant growth regulators (PGRs) and additional difference in receptor affinity or cellular sensitivity to PGRs (Minocha 1987). It is, therefore, expected that in vitro response will vary from genotype to genotype.

Although TDZ promotes callus development and at low concentration promotes shoot proliferation, it inhibits shoot elongation in lowbush blueberry (Kaldmäe et al. 2006) and lingonberry (Debnath 2005a). Since TDZ possesses very high cytokinin activity, it is possible that its inhibitory effect on shoot proliferation is consistent with its high cytokinin activity as shown in cranberry (Marcotrigiano et al. 1996). The inhibition of shoot elongation can take place due to the increase of endogenous cytokinins that hinders the action of cytokinin oxidase (Hare et al. 1994).

6.3.2 Adventitious Shoot Regeneration

Regeneration of adventitious shoots in vitro can be used not only in mass multiplication of difficult-to-propagate crop plant species but also in crop improvement to produce genetically engineered plants and somaclonal variants. In vitro shoot regeneration can be either through the development of unipolar organs (shoots or roots), known as organogenesis, or of somatic embryos with a root and a shoot meristem (somatic embryogenesis) (Ammirato 1985). Plant regeneration from excised explants through organogenesis includes (i) development of adventitious bud from explants, (ii) elongation of buds to form rootable shoots, and (iii) rooting of the shoots to form plantlets (Qu et al. 2000). Factors like genotype; culture medium; type, concentration, and combination of growth regulators; physical environment; and explant development stage are important for shoot regeneration.

TDZ-induced shoot regeneration in vitro on a semisolid gelled medium has been reported in many small fruit crops including lingonberry (Debnath 2003b, 2005c), strawberry (Debnath 2005b, 2006b; Haddadi et al. 2013), ohelo and bilberry (Shibli and Smith 1996), blackberry (Vujović et al. 2010), and blueberry (Debnath 2009a). While TDZ alone was sufficient to regenerate shoots from strawberry sepal, leaves, and calyx (Debnath 2005b, 2006b), 2,4-dichlorophenoxyacetic acid (2,4-D) (Passey et al. 2003) or 1H-indole-3-butanoic acid (IBA) (Yonghua et al. 2005; Murti et al. 2012) in combination with TDZ was effective for shoot regeneration from strawberry leaves. Marcotrigiano et al. (1996) used TDZ in combination with α-naphthaleneacetic acid (NAA) for shoot regeneration from cranberry leaves but was not very successful as the shoot elongation was limited. Qu et al. (2000) developed a highly efficient shoot regeneration system from cranberry leaves on a basal medium consisting of Anderson's rhododendron salts (Anderson 1975) and Murashige and Skoog's (MS) organics (Murashige and Skoog 1962) with 10.0 μM TDZ and 5.0 μM N6-(g-g-dimethylallylamino) purine (2ip) in five cranberry cultivars. TDZ was found more effective than 6-benzylaminopurine (BAP) for inducing adventitious shoot regeneration from blackberry leaves (Vujović et al. 2010).

Debnath (2009a) developed a two-step procedure for adventitious shoot regeneration on an agar-gelrite gelled semisolid nutrient medium containing TDZ. Wild lowbush blueberry leaf segments were cultured on modified cranberry medium of Debnath and McRae (2001a) that contained three-quarter macro-salts and microsalts of Debnath and McRae's (2001b) shoot proliferation medium D. The cultures were incubated in the dark at 20 ± 2 °C for 14 days and then exposed to light and maintained at 20 ± 2 °C with a 16-h photoperiod (PPF density at culture level was 30 μmolm^{-2} s^{-1}). The TDZ concentration affected the frequency and growth of calli, buds, and shoots on leaf explants. A range of 2.3–4.5 μM of TDZ concentration on a semisolid gelled medium was found the most suitable range for shoot regeneration of wild lowbush blueberry clones (Debnath 2009a). In strawberry, TDZ at 2–4 μM induced adventitious meristem, bud, and shoot regeneration, but the formation of buds and shoots was completely stopped in a semisolid gelled medium with 8 μM TDZ (Debnath 2005b, 2006b).

Shoot regeneration on a TDZ-containing medium is influenced by a number of factors including genotype, TDZ concentration, and the polarity and orientation of the explants on the culture medium. The concentration of TDZ affects callus size and regeneration percentage, shoot number, and the vigor of regenerated shoots. In lowbush blueberry, the leaf explants produced less shoots but more callus on a nutrient medium with 4.5 μM than those treated with 2.3 μM of TDZ. Shoot vigor declined with the increase of TDZ concentration on the culture medium (Debnath

2009a). Vujović et al. (2010) reported the highest shoot regeneration rate from blackberry leaves on a medium containing 4.5 μM TDZ. Swartz et al. (1990) obtained shoot regeneration from *Rubus* leaves on a MS medium containing 10 μM thidiazuron.

Polarity of shoot regeneration can vary from genotype to genotype and can be upturned by PGR treatments (George 1993). In lowbush blueberries, TDZ was found to induce shoot formation on the whole leaf surface. However, more regeneration was observed on basal and medial segments of leaves than on apical segments (Debnath 2009a). This could be due to the fact that the distal portion of the leaf has less meristematic cells than those at the proximal portions. The effect of polarity on regeneration on a TDZ-containing medium was evident by more callus growth and higher number of buds and shoots formed from the apical than in the central and basal segments of lingonberry hypocotyl segments from seedlings (Debnath 2003b). Regenerative capacity increased substantially from the base toward the tip of the hypocotyl (Debnath 2003b). In strawberry, bud and shoot regeneration occurred on both sides of the sepals on a TDZ-containing medium (Debnath 2005b) as were observed in lingonberry (Debnath 2005c) and cranberry leaf cultures on a semisolid medium with TDZ (Marcotrigiano et al. 1996). However, shoot regeneration was on adaxial side of cranberry leaves on a medium with TDZ (Qu et al. 2000). Regeneration of lingonberry shoots from leaves was better when the adaxial side was in contact with the TDZ-containing medium (Debnath 2005c). Shoot regeneration was best when young expanding basal leaf segments of lowbush blueberry were placed with the adaxial side in contact with the culture medium supplemented with 2.3–4.5 μM TDZ and kept for 14 days in darkness (Debnath 2009a). TDZ induces shoot regeneration in various small fruit crops (Debnath 2003b, 2005b, c, 2007a, 2009a).

TDZ concentration required for the regeneration of adventitious shoots depends on genotype. A high concentration of TDZ (37.8–40.5 μM) in combination with 2.5–0.5 μM IBA was effective for strawberry shoot regeneration by Murti et al. (2012). However, excessive PGR concentration in culture media may cause somaclonal variation in micropropagated plants (Larkin and Scowcroft 1981).

6.3.3 Somatic Embryogenesis

Induction of somatic embryogenesis in blueberries has been reported recently by Ghosh et al. (2017) where callus formed from leaf segments after 4 weeks of culture on a semisolid gelled medium containing TDZ. Highest percentage (98%) of callus formation was observed in a hybrid blueberry obtained through crossing between highbush blueberry cvs. Chippewa and Patriot, at 4.5 μM of TDZ. Reports on plant regeneration via somatic embryogenesis are not available in *Vaccinium* species on gelled media, but it has been observed in the diploid (*Fragaria vesca* subspecies *vesca* "Hawaii 4") (Zhang et al. 2014) and octoploid strawberries (Donnoli et al. 2001; Biswas et al. 2007; Husaini and Abdin 2007; Husaini et al. 2008; Kordestani and Karami 2008). Strawberry shoot regeneration from leaf culture was noticed via somatic embryogenesis or direct shoot regeneration based on the concentration of TDZ (Husaini and Abdin 2007). Strawberry leaf discs were cultured on a nutrient medium containing 4.0 mg l^{-1} TDZ and maintained at 10 ± 1 °C under darkness for 1 week followed by 3 weeks under 16-h photoperiod to get somatic embryos (Husaini et al. 2008). Initiation of strawberry somatic embryos was successful with dark (Donnoli et al. 2001; Husaini et al. 2011) and cold treatments (Husaini et al. 2011) of the culture. Nakajima and Matsuda (2003) reported somatic embryogenesis from filaments of eight grape cultivars using a combination of 1 μM 2,4-dichlorophenoxyacetic acid (2,4-D) and 1 μM TDZ or 10 μM 2,4-D and 10 μM TDZ. TDZ has been used to induce somatic embryo formation from filaments in grapevines (Nakajima and Matsuta, 2003; Oláh et al. 2003). Bouamama et al. (2007) used 11.35 μM of thidiazuron and 9 μM of 2,4-D for the induction as well as the development of somatic embryos in several grapevine cultivars, using anther culture.

6.3.4 Rooting and Acclimatization

Thidiazuron-induced small fruit microshoots can be rooted either under in vitro or ex vitro conditions (Qu et al. 2000; Debnath 2005a, b, 2009a). For rooting on a gelled medium, microshoots are excised and cultured onto an auxin-free medium (Qu et al. 2000). Ex vitro rooting of micropropagated shoots can be done in shredded sphagnum moss (Qu et al. 2000) or in a peat-perlite medium (Debnath 2003b, 2005a, c, 2009a). *Vaccinium* species can be rooted under ex vitro condition, while rooting in vitro is very common for strawberries and *Rubus* species (Debnath 2005b, 2006b, 2007b, 2010). Rootings ex vitro are rapid and less expensive, but in vitro rooting reduces disease contamination and environmental stress during rooting period (Pedroso et al. 1992).

Ex vitro rooting of *Vaccinium* microshoots can be done by treating the excised shoots by 39.4 mM IBA powder and planting them in a peat-perlite medium (Debnath 2009a). In vitro-derived strawberry shoots can be planted in a potting medium and maintained in a humidity chamber with a vaporizer at a temperature of 20 ± 2 °C, humidity 95%, PPF = 55 μmol m^{-2} s^{-1}, and 16-h photoperiod. Acclimatization of the plantlets can be done by gradually dropping the humidity over 2–3 weeks. Hardened-off plants can be transferred in a greenhouse and grown at 20 ± 2 °C, humidity 85%, maximum PPF = 90 μmol m^{-2} s^{-1}, and a 16-h photoperiod (Debnath 2005c).

Debnath (2006b) observed that TDZ, in a semisolid culture medium, strongly inhibited root formation of adventitious strawberry microshoot. Media with TDZ promoted more callus formation but suppressed shoot elongation and rooting of shoots. TDZ-induced strawberry shoots when proliferated in a medium containing 1 or 2 μM zeatin rooted well (Debnath 2006b). Lower cytokinin concentration may be required to form roots as the formation of roots is generally inhibited when the cytokinin is adequately concentrated for the initiation of shoot proliferation (Gaspar and Coumans 1987). Endogenous cytokinins play a role in the formation of adventitious root (Bollmark et al. 1988). It is probable that TDZ is more efficient than zeatin to enhance endogenous cytokinin production which, in turn, might have prevented rooting (Bollmark et al. 1988). Induction of rooting for strawberry microshoots is possible without exogenous auxin. This might be due to the effect of exogenous auxins that can inhibit root growth (Scott 1972).

6.4 Bioreactor Micropropagation in Liquid Culture with Thidiazuron

Haberlandt (1902) was the first to use a liquid medium to culture isolated cells from bracts of *Lamium purpureum* in Knop's solution supplemented with sucrose (Preil 2005) although the cells did not divide. Later, Kohlenbach (1959) observed that differentiated mesophyll cells of *Macleaya* develop into cell clusters and calli forming organs and somatic embryos. Use of a liquid culture medium for in vitro propagation offers much more uniform culturing conditions than a semisolid medium (Debnath 2011a). Use of a gelled medium for propagating plants is difficult to automate and costly for large-scale production. Automated bioreactors that use liquid media are important for large-scale production of small fruit crops.

Bioreactors are designed for intensive culture and control over microenvironmental conditions (aeration, agitation, dissolved oxygen, etc.) (Paek et al. 2005). Liquid culture in a bioreactor system can cut down cost and labor in terms of agar, medium volume, and subculture periods (Sandal et al. 2001). Micropropagation can be done in bioreactors in agitated and non-agitated vessels (Ziv 2005). However, under submersed condition, a bioreactor system can limit the gas exchange to the culture that may lead to suffocation, hyperhydricity, and abnormal plantlet formation (Detrez et al. 1994) with chlorophyll-deficient glossy hyperhydrous leaves, cell hyperhydricity, hypolignification, reduced deposition of epicuticular waxes, and changes in enzyme activity and protein synthesis (Ziv 1991a, b). Temporary immersion bioreactors (TIBs) and use of growth retardants in culture media can be used to overcome some of these problems (Ziv et al. 2003). Cultures are alternately exposed to air and dipped into a liquid medium in a TIB system. Some of the other alternative procedures include putting a liquid medium on top of an established culture on agar and mist bioreactors and use of supports over stationary liquid media such as cellulose blocks, rafts, sponges, or paper bridges (Etienne and Berthouly 2002).

Bioreactor micropropagation in small fruit crop has not been used with many species but reviewed in literature (Debnath 2011a; Debnath et al. 2016). Some results that used TDZ-containing liquid media in *Fragaria*, *Rubus*, and *Vaccinium* species are presented below.

6.5 *Fragaria* (Strawberry)

The strawberry is enjoyed by millions of people across the world (Hancock et al. 1991) and is used fresh or in processed forms including jams, jellies, and frozen whole berries or sweetened juice extracts or flavorings. It is one of the most popular small fruit crops more extensively distributed than any other fruit crops (Childers 1980). The cultivated strawberry (*Fragaria* × *ananassa* Duch.) is a hybrid between *F. virginiana* Duch. and *F. chiloensis* (L.) Duch. Strawberries are a low-growing, dicotyledonous, perennial herb. They are grown in most arable regions of the world. Strawberries are a high source of vitamin C and fiber (Galletta and Bringhurst 1990).

Although strawberries can be propagated vegetatively by runner cuttings, a limited number of propagules are produced though this process and the product are not free of fungal diseases (Dijkstra 1993). Virus-free plants can be produced through micropropagation, and they can be saved under refrigeration (Mullin and Schlegel 1976). This makes it a reliable technique for germplasm storage.

Strawberry liquid culture with cell suspensions was started by Keßler et al. (1997) in bioreactors with different stirrer types. Adventitious shoot regeneration was successful in strawberry cultivars using a TIB bioreactor (RITA®) in a liquid MS medium supplemented with 9 μM TDZ and 2.5 μM IBA although regeneration frequency was not as good as on semisolid medium (Hanhineva et al. 2005). A combination of semisolid gelled medium and a liquid medium has been used by Debnath (2008a) where shoots were regenerated from leaf, sepal, or petiole explants of strawberries on a semisolid culture medium containing 2–4 μM TDZ (Debnath 2005b, 2006b), followed by culturing in the same liquid medium in a TIB bioreactor system with a 15-min immersion of explants every 4 h. Shoots can be proliferated and rooted in the bioreactor system with the same medium with 0.5–1 μM zeatin. In vitro-derived rooted shoots can be transferred planted on ProMix BX (Premier Horticulture Limited, Riviere-du-Loup, QC) potting medium and acclimatized following Debnath (2008a).

6.5.1 *Rubus* Species

The members of the genus *Rubus* (Tourn.) L. are called brambles that include raspberries, blackberries, and dewberries. They are distributed in both hemispheres except desert regions (Daubeny 1996). Ideobatus (raspberries) is the most important domesticated subgenera of genus *Rubus* containing around 200 species (Debnath 2011a, 2016a). The cloudberry (*R. chamaemorus* L., family Rosaceae) is a less known small fruit crop in *Rubus* species. This boreal circumpolar fruit species is a perennial, rhizomatous, and dioecious herb common to bogs. Cloudberries are rich in vitamin C and tannins and are used in traditional medicine to treat scurvy and diarrhea (Thiem 2003).

Bioreactor micropropagation using a liquid medium in *Rubus* species was first described by Debnath (2007b). Three cloudberry wild clones were cultured in an airlift bioreactor containing liquid medium with 0.45–2.3 lμ TDZ. A concentration of 1.1–2.3 μM TDZ was needed for shoot proliferation. Liquid culture system enhanced the micropropagation efficiency of cloudberry clones; shoot number was almost double those on semisolid gelled medium for two cloudberry wild clones. A concentration of 1.1 μM TDZ was found to produce five to seven 4-cm-high shoots per nodal explant in a bioreactor containing a liquid medium in wild cloudberries (Debnath 2007b). However, TDZ induces hyperhydricity in liquid culture; 20–30% of the cloudberry shoots were hyperhydric after 8–10 weeks of culture in liquid medium (Debnath 2007b). Generally, hyperhydricity takes place in liquid media because of high water potential of leaves (Paek and Han 1989).This happens when a culture medium is rich in cytokinin (Gaspar 1991). Liquid culture-derived hyperhydric microshoots cannot root properly and give rise to malformed plants with poor survivability. In the reversible process, it was found that the quality of the cloudberry shoots could be improved by transferring them onto a gelled medium with 8.9 μM BAP and 5.8 μM gibberellic acid (GA_3) (Debnath 2007b).

Bioreactor micropropagation in a TDZ-containing liquid medium was also reported in red raspberry (*R. idaeus* L.) (Debnath 2010, 2014b). Shoot regeneration can be achieved from raspberry leaves in a liquid medium-containing bioreactor system combined with a semisolid gelled medium with 2.3–9.0 μM TDZ (Debnath 2014b). The polarity and orientation of red raspberry leaves and TDZ concentration played a significant role for callus and bud formation and for bud and shoot number per regenerating explants. Although regeneration was observed on the whole surface of the leaf, it increased markedly from the tip toward the base of the leaf. Regeneration was more in explants from basal segments (proximal ends) than the apical segments (distal regions) (Debnath 2010). As was in semisolid gelled media with small fruit crops, TDZ also inhibits shoot elongation in liquid media. In red raspberry, shoot inhibition can be improved by culturing in a BA-added medium. BA-induced elongated shoots can be rooted in the same liquid medium that contains no plant growth regulator (Debnath 2010, 2014b).

6.5.2 *Vaccinium* Species

The genus *Vaccinium* L., with about 400–500 species, is native to all continents except Antarctica and Australia (Vander Kloet 1988; Vander Kloet and Dickinson 2009). Genetically they are dicot and heterozygous angiosperms with small- to medium-sized fleshy edible fruits on woody perennial shrubs or vines. Although *Vaccinium* species includes blueberry, cranberry, lingonberry, bilberry, huckleberry, and whortleberry, the first three are commercially cultivated *Vaccinium* fruit crops.

Not many reports are available where bioreactor micropropagation has been used in *Vaccinium* species in a TDZ-supplemented liquid medium. A bioreactor system containing a liquid medium with TDZ combined with a semisolid gelled medium was used by Debnath (2011b) to propagate wild lowbush blueberries. Leaf segments were cultured on a semisolid gelled medium with 2.3 μM TDZ for 4 weeks followed by in liquid medium containing 1.2–2.3 μM TDZ for another 4 weeks. Leaf polarity and TDZ concentration had significant influence in callus formation and shoot regeneration. Regeneration percentage was highest in the basal leaf segment followed by medial and apical leaf segments (Debnath 2011b) that collaborates the previous findings on a semisolid gelled medium (Debnath 2009a). Shoot regeneration took place on both sides of the leaves but was better when leaf segments were cultured with their adaxial surface in contact with the culture medium (Debnath 2011b).

6.6 Characteristics of Micropropagated Plants

Enhanced vegetative growth including increased branching and/or rhizome production is often observed in micropropagated small fruit crop plants (Debnath et al. 2012; Fig. 6.2). The effect of propagation methods on the morphological and biochemical properties of *Vaccinium* species was reported by various authors (Gustavsson and Stanys 2000; Debnath 2005d, 2006c, 2007c, 2008b; Foley and Debnath 2007; Debnath et al. 2012; Vyas et al. 2013; Goyali et al. 2015). In strawberry, Debnath (2009b) compared TDZ-induced regenerated strawberry tissue culture (TC) shoots that were elongated by treating with zeatin with those propagated by conventional runner cutting (RC) plants. TC plants produced more vegetative growth with more berries than those of RC plants. Berries produced by TC plants had also more anthocyanin contents and antioxidant activities than did RC plants (Debnath 2009b). This might be because the in vitro hormonal treatment (TDZ, zeatin) could have effects to increase crown, runner, leave, and berry number per plant (Debnath 2009b). However, increased vegetative growth and berry yield of TC plants over RC plants are genotype dependent, and all genotypes did not produce enhanced growth and berry yield in raspberries (Debnath 2014b). TC plants had higher berry yield and more and longer canes and more berries than root cutting plants in cultivar "Festival" but not in "Latham" indicating genotype-dependent juvenile branching characteristics of "Festival" TC plants but not in "Latham" TC plants (Debnath 2014b). Similar results with micropropagated strawberries were also reported by Dalman and Malata (1997) for overwintering. Increased resistance to frost damage was observed in micropropagated strawberries than the runner plants (Rancillac and Nourrisseau (1989).

Fig. 6.2 Greenhouse-grown root cutting (left) and tissue culture (right) raspberry plants

6.7 Clonal Fidelity and Molecular Analysis in Micropropagules

True-to-type propagules and clonal fidelity are prerequisites for commercial micropropagation. The use of in vitro propagation has concerns about genetic changes resulting from the process (Dale et al. 2008). Although production of true-to-type micropropagules is the main objective for mass propagation or conservation of a specific genotype, in vitro culture is also a tool to create new variation. In vitro culture-derived variation or somaclonal variation (Larkin and Scowcroft 1981) can broaden the genetic variation in small fruit crop plants resulting in a range of genetically stable variations useful in crop improvement (Jain 2001). Somaclonal variation can be genetic (heritable) and epigenetic (nonheritable). Somaclones were found to be regenerated from leaf culture (Popescu et al. 1997), from somatic embryogenesis (Donnoli et al. 2001), and from leaf and petiole cultures irradiated with gamma rays (Kaushal et al. 2004). Debnath (2017) reported somaclonal variations in strawberries for fruit yield under field condition. Two TC plants had higher berry yield than those of other tissue culture plants and the runner cutting mother plant. However it was not identified whether these variations were genetic or epigenetic (Debnath 2017). Somaclonal variation can be due to changes in the structure and number of chromosomes, sister chromatid exchanges, transposable element activation, DNA methylation pattern alteration and activation of hypervariable DNA regions, and point mutations including deletion, addition, or substitution of nucleotides and rearrangements in the nuclear and cytoplasmic genomes (Kaeppler et al. 1998). Factors like genotype, ploidy level, degree of departure from organized meristematic growth, explant type, donor plant age, types and concentrations of growth regulators used, auxin-cytokinin balance, duration of culture period, and number of subcultures are the possible causes for the origin of somaclonal variation (Henry et al. 1998).

There are many ways to monitor variation in micropropagated plants including evaluation at morphological, biochemical, physiological, and genetic levels. Vujović et al. (2010) used cytological, flow cytometry, and isozyme analyses to monitor somaclonal variation in blackberry regenerants. Chromosome counting in root tip meristems and flow cytometry analysis indicated identical ploidy level in all TC plants although the peroxidase patterns showed differences between some in vivo and micropropagated plants (Vujović et al. 2010).

DNA markers are independent of environmental influences (Weising et al. 1995) and can be a powerful tool for assessing clonal fidelity in micropropagated small fruit crops. Markers that are available for genetic analysis of tissue culture-raised plants include random amplified polymorphic DNA (RAPD), amplified fragment length polymorphism (AFLP), inter simple sequence repeat (ISSR), restriction fragment length polymorphism (RFLP), arbitrary primed polymerase chain reaction (AP-PCR), sequence characterized amplified region (SCAR), DNA amplification fingerprinting (DAF), simple sequence repeat (SSR), short tandem repeat (STR), sequence-tagged sites (STSs), expressed sequence tag-polymerase chain reaction (EST-PCR), and cleaved amplified polymorphic sequences (CAPS) derived from

EST-PCR markers (Debnath 2011a). While ISSR markers have been used to confirm trueness-to-type of bioreactor-derived micropropagated strawberries (Debnath 2009b), EST-PCR markers showed similar monomorphic amplification profiles in lowbush blueberry micropropagules (Debnath 2011b). SSR markers have been used for monitoring clonal fidelity in raspberry micropropagules (Debnath 2014b). Somaclonal variation is likely to be associated with regeneration of plants through unorganized callus formation (Piola et al. 1999). However, axillary buds can also produce variant plants (Soneji et al. 2002).

6.8 Conclusions

Small fruit crops are being propagated increasingly using tissue culture methods to multiply massive amounts of disease-free, genetically uniform plants. Axillary shoot proliferation is a very simple and reliable method to produce true-to-type micropropagules, and it is more preferred over adventitious shoot regeneration and somatic embryogenesis in small fruit crops. The latter two, however, are also powerful tools for rapid propagation of small fruit crops, provided clonal fidelity of the micropropagated plants is maintained. TDZ possesses both cytokinin- and auxin-like effects in in vitro culture (Mok et al. 1982; Visser et al. 1992) and can provide significant role in small fruit micropropagation. Compared to other cytokinins, TDZ at lower concentration induces proliferation of axillary shoots, while at higher concentration it promotes both axillary and adventitious shoot formation in small fruit crops (Debnath 2005a, b, c, 2008a). In vitro organogenesis was found to produce genetically uniform and true-to-type micropropagules in strawberry (Debnath 2009b), blueberry (Debnath 2011b), and raspberry (Debnath 2014b). Although bioreactor micropropagation is more cost-effective and ideal for automation, hyperhydricity including morphological and physiological disorders is common in liquid culture-derived micropropagules (Debnath 2011a). Optimization of TDZ concentration and culture conditions is needed for TDZ-induced bioreactor micropropagation in various small fruit crops.

Clonal fidelity is a major concern in small fruit micropropagation and can be monitored by DNA-based markers. Occurrence of variation during in vitro culture depends on factors like explant donor genotype, explant type, explant polarity and orientation on a culture medium, presence of chimeral tissue, media type, types and concentrations of plant growth regulators, culture duration, and cultural environment (temperature and light) (Debnath 2011a; Graham 2005). Micropropagated small fruit crops exhibit enhanced vegetative growth and can be used for rapid establishment and early fruit production. In vitro and molecular techniques are powerful tools, and combined with classical breeding, they can be used in small fruit improvement program (Debnath 2011a, 2016b; Fig. 6.3).

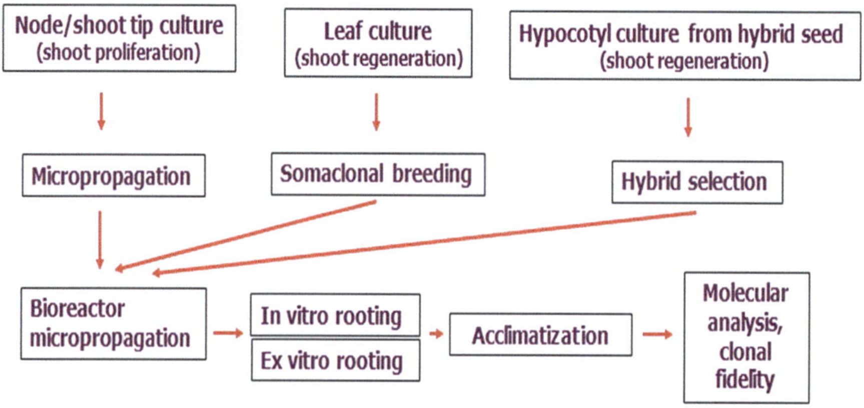

Fig. 6.3 Small fruit improvement program using in vitro and molecular techniques

References

Ames BN, Shigena MK, Hegen TM (1993) Oxidants, antioxidants and the degenerative diseases of aging. Proc Nat Acad Sci USA 90:7915–7922

Ammirato PV (1985) Patterns of development in culture. In: Henke RR, Hughes KW, Constantin MJ, Hollaender A (eds) Tissue culture in forestry and agriculture. Plenum Press, New York, pp 9–29

Anderson WC (1975) Propagation of rhododendrons by tissue culture. Part I: development of a culture medium for multiplication of shoots. Comb Proc Int Plant Prop Soc 25:129–135

Arndt FR, Rusch R, Stillfried HV, Hanisch B, Martin WC (1976) SN 49537, a new defoliant. Plant Physiol 57:s-99. Abstr

Biswas MK, Islam R, Hossain M (2007) Somatic embryogenesis in strawberry (*Fragaria* sp.) through callus culture. Plant Cell Tissue Organ Cult 90:49–54

Bollmark M, Kubat B, Eliasson L (1988) Variation in endogenous cytokinin content during adventitious root formation in pea cuttings. J Plant Physiol 132:262–265

Bouamama B, Ben Salem-Fnayou A, Ben Jouira H, Ghorbel A, Mliki A (2007) Influence of the flower stage and culture medium on the induction of somatic embryogenesis from anther culture in Tunisian grapevine cultivars. J Int Sci Vigne Vin 41(4):185–192

Boxus P (1974) The production of strawberry plants by in vitro micropropagation. J Hort Sci 49:209–210

Cao X, Hammerschlag FA (2000) Improved shoot organogenesis from leaf explants of highbush blueberry. Hortscience 35:945–947

Childers NF (1980) Foreward. In: Childers NF (ed) The strawberry: cultivars to marketing. Hort Publ, Gainesville, p ix

Cordenunsi BR, do Nascimento JRO, Genovese MI, Lajolo FM (2002) Influence of cultivar on quality parameters and chemical composition of strawberry fruits grown in Brazil. J Agr Food Chem 50:2581–2586

Dale A, Hughes BR, Donnelly D (2008) The role of micropropagation in producing specific pathogen-tested plants. Hortscience 43:74–77

Dalman P, Malata V (1997) The effect of cultivation practices on the overwintering and yield of strawberry. Acta Hort 439:881–886

Daubeny HA (1996) Brambles. In: Janick J, Moore JN (eds) Fruit breeding, Vine and small fruit crops, vol Vol II. Wiley, New York, pp 109–190

Debnath SC (2003a) Micropropagation of small fruits. In: Jain SM, Ishii K (eds) Micropropagation of woody trees and fruits. Kluwer Academic Publishers, Dordrecht, pp 465–506

Debnath SC (2003b) Improved shoot organogenesis from hypocotyl segments of lingonberry (*Vaccinium vitis-idaea* L.) In Vitro Cell Dev Biol – Plant 39:490–495

Debnath SC (2005a) Micropropagation of lingonberry: influence of genotype, explant orientation, and overcoming TDZ-induced inhibition of shoot elongation using zeatin. Hortscience 40:185–188

Debnath SC (2005b) Strawberry sepal: another explant for thidiazuron-induced adventitious shoot regeneration. In Vitro Cell Dev Biol Plant 41:671–676

Debnath SC (2005c) A two-step procedure for adventitious shoot regeneration from in vitro-derived lingonberry leaves: shoot induction with TDZ and shoot elongation using zeatin. Hortscience 40:189–192

Debnath SC (2005d) Morphological development of lingonberry as affected by in vitro and ex vitro propagation methods and source propagule. Hortscience 40:760–763

Debnath SC (2006a) Propagation of *Vaccinium* in vitro: a review. Int J Fruit Sci 6:47–71

Debnath SC (2006b) Zeatin overcomes thidiazuron-induced inhibition of shoot elongation and promotes rooting in strawberry culture in vitro. J Hort Sci Biotechnol 81:349–354

Debnath SC (2006c) Influence of propagation method and indole-3-butyric acid on growth and development of in vitro- and ex vitro-derived lingonberry plants. Can J Plant Sci 86:235–243

Debnath SC (2007a) Strategies to propagate *Vaccinium* fruit nuclear stocks for Canadian industry. Can J Plant Sci 87:911–922

Debnath SC (2007b) A two-step procedure for in vitro multiplication of cloudberry (*Rubus chamaemorus* L.) shoots using bioreactor. Plant Cell Tissue Organ Cult 88:185–191

Debnath SC (2007c) Influence of indole-3-butyric acid and propagation method on growth and development of in vitro- and ex vitro-derived lowbush blueberry plants. Plant Growth Regul 51:245–253

Debnath SC (2008a) Developing a scale-up system for the in vitro multiplication of thidiazuron-induced strawberry shoots using a bioreactor. Can J Plant Sci 88:737–746

Debnath SC (2008b) Zeatin-induced one-step in vitro cloning affects the vegetative growth of cranberry (*Vaccinium macrocarpon* Ait.) micropropagules over stem cuttings. Plant Cell Tissue Organ Cult 93:231–240

Debnath SC (2009a) A two-step procedure for adventitious shoot regeneration on excised leaves of lowbush blueberry. In Vitro Cell Develop Biol – Plant 45:122–128

Debnath SC (2009b) Characteristics of strawberry plants propagated by in vitro bioreactor culture and ex vitro propagation method. Eng Life Sci 9:239–246

Debnath SC (2010) A scaled-up system for in vitro multiplication of thidiazuron-induced red raspberry shoots using a bioreactor. J Hort Sci Biotechnol 85:94–100

Debnath SC (2011a) Bioreactors and molecular analysis in berry crop micropropagation – a review. Can J Plant Sci 91:147–157

Debnath SC (2011b) Adventitious shoot regeneration in a bioreactor system and EST-PCR based clonal fidelity in lowbush blueberry (*Vaccinium angustifolium* Ait.) Sci Hort 128:124–130

Debnath SC (2013) Propagation strategies and genetic fidelity in strawberries. Int J Fruit Sci 13:3–18

Debnath SC (2014a) Strawberry micropropagation and somaclonal variation. In: Malone N (ed) Strawberries: cultivation, antioxidant properties and health benefits. Nova Science Publishers, Hauppauge, New York, pp 93–108

Debnath SC (2014b) Bioreactor-induced adventitious shoot regeneration affects genotype-dependent morphology but maintains clonal fidelity in red raspberry. In Vitro Cell Dev Biol – Plant 50:777–788

Debnath SC (2016a) Genetic diversity and erosion in berries. In: Ahuja MR, Jain SM (eds) Genetic diversity and erosion in plants. Springer Int Publ, Switzerland, pp 75–129

Debnath SC (2016b) Corrigendum: bioreactors and molecular analysis in berry crop micropropagation – a review. Can J Plant Sci 96:382–383

Debnath SC (2017) Molecular approaches for monitoring clonal fidelity and epigenetic variation in in vitro-derived strawberry plants. Acta Hort 1156:83–87

Debnath SC, McRae KB (2001) An efficient in vitro shoot propagation of cranberry (*Vaccinium macrocarpon* Ait.) by axillary bud proliferation. In Vitro Cell Dev Biol Plant 37:243–249

Debnath SC, McRae KB (2001b) In vitro culture of lingonberry (*Vaccinium vitis-idaea* L.): the influence of cytokinins and media types on propagation. Small Fruits Rev 1:3–19

Debnath SC, Vyas P, Goyali JC, Igamberdiev AU (2012) Morphological and molecular analyses in micropropagated berry plants acclimatized under ex vitro condition. Can J Plant Sci 92:1065–1073

Debnath SC, McKenzie D, Bishop G, Percival D (2016) Strategic approaches to propagate berry crop nuclear stock using a bioreactor. Acta Hort 1113:47–52

Detrez C, Ndiaye S, Dreyfus B (1994) In vitro regeneration of the tropical multipurpose leguminous tree *Sesbania grandiflora* from cotyledon explants. Plant Cell Rep 14(2–3):87–93

Dijkstra J (1993) Research on strawberries focuses on healthy plant material. Expensive cultural method requires excellent material. Fmitteelt – Den-Haag 83:14–15

Donnoli R, Sunseri F, Martelli G, Greco I (2001) Somatic embryogenesis, plant regeneration and genetic transformation in *Fragaria* spp. Acta Hort 560:235–240

Etienne H, Berthouly M (2002) Temporary immersion systems in plant micropropagation. Plant Cell Tissue Organ Cult 69:215–231

Finn C (1999) Temperate berry crops. In: Janick J (ed) Perspectives on new crops and new uses. ASHS Press, Alexandria, pp 324–334

Foley SL, Debnath SC (2007) Influence of in vitro and ex vitro propagation on anthocyanin content and antioxidant activity of lingonberries. J Hort Sci Biotechnol 82:114–118

Galletta GJ, Bringhurst RS (1990) Strawberry management. In: Galletta GJ, Himelrick DG (eds) Small fruit crop management. Prentice Hall, Englewood Cliffs, pp 83–156

Galletta GJ, Himelrick DG (1990) The small fruit crop. In: Galletta GJ, Himelrick DG (eds) Small fruit crop management. Prentice Hall, Englewood Cliffs, pp 1–13

Gaspar T (1991) Vitrification in micropropagation. In: Bajaj YPS (ed) Biotechnology in agriculture and forestry, high-tech and micropropagation I, vol vol 17. Springer Verlag, Berlin, pp 116–126

Gaspar TH, Coumans M (1987) Root formation. In: Bonga JM, Durzan DJ (eds) Cell and tissue culture in forestry, vol. 2. Specific principles and methods: growth and developments. Martinus Nijhoff Publ, Dordrecht, pp 202–217

George EF (1993) Plant propagation by tissue culture. Part 1. The technology. Exegetics Ltd, Edington. 574 pp

George EF, Sherrington PD (1984) Plant propagation by tissue culture. Exegetics Ltd, Reading

Ghosh A, Igamberdiev AU, Debnath SC (2017) Detection of DNA methylation pattern in thidiazuron-induced blueberry callus using methylationsensitive amplification polymorphism. Biol Plant 61:511–519

Goyali GC, Igamberdiev AU, Debnath SC (2015) Propagation methods affect fruit morphology and antioxidant properties but maintain clonal fidelity in lowbush blueberry. Hortscience 50:888–896

Graham J (2005) *Fragaria* strawberry. In: Litz R (ed) Biotechnology of fruit and nut crops. Biotechnology in agriculture series no. 29. CAB International, Wallingford, pp 456–474

Gustavsson BA, Stanys V (2000) Field performance of 'Sanna' lingonberry derived by micropropagation vs. stem cuttings. Hortscience 35:742–744

Haberlandt G (1902) Kulturversuche mit isolierten Pflanzenzellen. Sitzungsber Math Naturwiss Kl Kais Akad Wiss Wien 111:69–92

Haddadi F, Aziz MA, Kamaladini H, Ravanfar SA (2013) Thidiazuron- and zeatin-induced high-frequency shoot regeneration from leaf and shoottip explants of strawberry. HortTechnology 23:276–281

Häkkinen SH, Törrönen AR (2000) Content of flavonols and selected phenolic acids in strawberries and *Vaccinium* species: influence of cultivars, cultivation site and technique. Food Res Int 33:517–524

Hancock JF, Maas JL, Shanks CH, Breen PJ, Luby JJ (1991) Strawberries (*Fragaria*). Acta Hort 290:491–548

Hanhineva K, Kokko H, Kärenlampi S (2005) Shoot regeneration from leaf explants of five strawberry (*Fragaria* × *ananassa*) cultivars in temporary immersion bioreactor system. In Vitro Cell Dev Biol – Plant 41:826–831

Hannig E (1904) Zur physiologie pflanzlicher embryonen. I. Ueber die cultur von cruciferenembryonen ausserhalb des embryosacks. Bot Ztg 62:45–80

Hare PD, Staden J, Van Staden J (1994) Inhibitory effect of TDZ on the activity of cytokinin oxidase isolated from soybean callus. Plant Cell Physiol 35:1121–1125

Harris GK, Gupta A, Nines RG, Kresty LA, Habib SG, Frankel WL, LaPerle K, Gallaher DD, Schwartz SJ, Stoner GD (2001) Effects of lyophilized black raspberries on azoxymethane-induced colon cancer and 8-hydroxy-2#-deoxyguanosine levels in the Fischer 344 rat. Nutr Cancer 40:125–133

Henry Y, Nato A, DeBuyser J (1998) Genetic fidelity of plants regenerated from somatic embryos in cereals. In: Jain SM, Brar DS, Ahloowalia BS (eds) Somaclonal variation and induced mutations in crop improvement. Kluwer Acad Publ, Dordrecht, pp 65–80

Howell AB, Reed JD, Krueger CG, Winterbottom R, Cunningham DG, Leahy M (2005) A-type cranberry proanthocyanidins and uropathogenic bacterial anti-adhesion activity. Phytochemistry 66:2281–2291

Huetteman CA, Preece JE (1993) Thidiazuron: a potent cytokinin for woody plant tissue culture. Plant Cell Tissue Organ Cult 33:105–119

Husaini AM, Abdin MZ (2007) Interactive effect of light, temperature and TDZ on the regeneration potential of leaf discs of *Fragaria × ananassa* Duch. In Vitro Cell Dev Biol Plant 43:576–584

Husaini AM, Aquil S, Bhat M, Qadri T, Kamaluddin MZ, Abdin MZ (2008) A high-efficiency direct somatic embryogenesis system for strawberry (*Fragaria × ananassa* Duch.) cultivar Chandler. J Crop Sci Biotech 11:107–110

Husaini AM, Mercado JA, da Silva JAT, Schaart JG (2011) Review of factors affecting organogenesis, somatic embryogenesis and agrobacterium tumefaciens-mediated transformation of strawberry. Gen Genom Genomics (Spec Issue I) 5:1–11

Jain SM (2001) Tissue culture-derived variation in crop improvement. Euphytica 118:153–166

Kaeppler SM, Phillips RL, Olhoft P (1998) Molecular basis of heritable tissue culture induced variation in plants. In: Jain SM, Brar DS, Ahloowalia BS (eds) Somaclonal variation and induced mutations in crop improvement. Current plant science and biotechnology in agriculture, vol 32. Kluwer Acad Publ, Dordrecht, pp 465–484

Kaldmäe H, Starast M, Karp K, Paal T (2006) Effect of donor plant physiological condition on in vitro establishment of *Vaccinium angustifolium* shoot explants. Acta Hort 715:433–438

Kaushal K, Nath AK, Kaundal P, Sharma DR (2004) Studies on somaclonal variation in strawberry (*Fragaria × ananassa* Duch.) cultivars. Acta Hortic 662:269–275

Keßler M, ten Hoopen HJG, Heijnen JJ, Fumsaki S (1997) O_2 uptake rate measurements as a novel tool to study shear effects on suspended strawberry cells. Biotechnol Tech 11:507–510

Kohlenbach HW (1959) Streckungs- und Teilungswachstum isolerter Mesophyllzellen von *Macleaya cordata*. Naturwissenschaft 46:116–117

Kordestani GK, Karami O (2008) Picloram-induced somatic embryogenesis in leaves of strawberry (*Fragaria Ananassa* L.) Acta Biol Cracov Ser Bot 50:69–72

Larkin PJ, Scowcroft WR (1981) Somaclonal variation – a novel source of variability from cell culture for plant improvement. Theor Appl Genet 60:197–214

Macheix JJ, Sapis JC, Fleuriet A (1991) Phenolic compounds and polyphenoloxidase in relation to browning in grapes and wines. Crit Rev Food Sci Nutr 30:441–486

Marcotrigiano M, McGlew SP, Hackett G, Chawla B (1996) Shoot regeneration from tissue-cultured leaves of the American cranberry (*Vaccinium macrocarpon*). Plant Cell Tissue Organ Cult 44:195–199

Mazur WM, Uehara M, Wahala K, Adlercreutz H (2000) Phyto-oestrogen content of berries, and plasma concentrations and urinary excretion of enterolactone after a single strawberry-meal in human subjects. Br J Nutr 83:381–387

McCown BH, Zeldin EL (2005) *Vaccinium* spp. cranberry. In: Litz RE (ed) Biotechnology of fruit and nut crops. CABI Publ, Wallingford, pp 247–261

Minocha SC (1987) Plant growth regulators and morphogenesis in cell and tissue culture of forest trees. In: Bonga JM, Durzan DJ (eds) Cell and tissue culture in forestry, vol. I. Martinus Nijhoff Publ, Dordrecht, pp 50–66

Mok MC, Mok DWS, Armstrong DJ, Shudo K, Isogal Y, Okamoto T (1982) Cytokinin activity of N-phenyl-N'-1,2,3-thiadiazol-5-urea (thidiazuron). Phytochemistry 21:1509–1511

Mok MC, Mok DWS, Turner JE, Mujer CV (1987) Biological and biochemical effects of cytokinin active phenylurea derivatives in tissue culture systems. Hortscience 22:1194–1196

Mullin RH, Schlegel DE (1976) Cold storage maintenance of strawberry meristem plantlets. Hortscience 11:100–101

Murashige T, Skoog F (1962) A revised medium for rapid growth and bioassays with tobacco tissue cultures. Physiol Plant 15:473–479

Murthy BNS, Murch SJ, Saxena PK (1998) Thidiazuron: a potent regulator of in vitro plant morphogenesis. In Vitro Cell Dev Biol Plant 34:267–275

Murti RH, Debnath SC, Yeoung YR (2012) Effect of high concentration of thidiazuron (TDZ) combined with 1H-indole-3-butanoic acid (IBA) on Albion strawberry (*Fragaria* × *ananassa*) cultivar plantlets induction. Afr J Biotechnol 11:14696–14702

Nakajima I, Matsuda N (2003) Somatic embryogenesis from filaments of *Vitis vinifera* L, *Vitis labruscana* Bailey. Vitis 42:53–54

Neto CC (2007) Review – cranberry and blueberry: evidence for protective effects against cancer and vascular diseases. Mol Nutr Food Res 51:652–664

Novelli S (2003) Developments in berry production and use. Bi-weekly Bull Agric Agri-Food Can 16(21):5–6

Oláh R, Szegedi E, Ruthner S, Korbuly J (2003) Thidiazuron-induced regeneration and genetic transformation of grapevine rootstocks varieties. Vitis 42:133–136

Paek KY, Han BH (1989) Physiological, biochemial and morphological characteristics of vitrified shoot regenerated in vitro. J Kor Soc Plant Tiss Cult 18:151–162

Paek KY, Chakrabarty D, Hahn EJ (2005) Application of bioreactor systems for large scale production of horticultural and medicinal plants. Plant Cell Tissue Organ Cult 81:287–300

Passey AJ, Barrett KJ, James DJ (2003) Adventitious shoot regeneration from seven commercial strawberry cultivars (*Fragaria* × *ananassa* Duch.) using a range of explant types. Plant Cell Rep 21:397–401

Pedroso MC, Oliveira MM, Pais MSS (1992) Micropropagation and simultaneous rooting of *Actinidia deliciosa* var. deliciosa 'Hayward'. Hortscience 27:443–445

Piola F, Rohr R, Heizmann P (1999) Rapid detection of genetic variation within and among in vitro propagated cedar (*Cedrus libani* Loudon) clones. Plant Sci 141:159–163

Popescu AN, Isac VS, Coman MS, Radulescu MS (1997) Somaclonal variation in plants regenerated by organogenesis from callus culture of strawberry (*Fragaria Ananassa*). Acta Hort 439:89–96

Preece JE, Huetteman CA, Ashby WC, Roth PL (1991) Micro- and cutting propagation of silver maple. II. Genotype and provenance affect performance. J Am Soc Hort Sci 116:149–155

Preil W (2005) General introduction: a personal reflection on the use of liquid media for in vitro culture. In: Hvoslef-Eide AK, Preil W (eds) Liquid culture systems for in vitro plant propagation. Springer, Dordrecht, pp 1–18

Qu L, Polashock J, Vorsa N (2000) A high efficient in vitro cranberry regeneration system using leaf explants. Hortscience 35:948–952

Rancillac M, Nourrisseau JG (1989) Micropropagation and strawberry plant quality. Acta Hort 265:343–348

Rissanen T, Voutilainen S, Virtanen J, Venho B, Vanharante M, Mursu J, Salonen J (2003) Low intake of fruits, berries and vegetables is associated with excess mortality in men: the Kuopio Ischaemic heart disease risk factor (KIHD) study. J Nutr 133:199–204

Rowland LJ, Hammerschlag FA (2005) *Vaccinium* spp. blueberry. In: Litz RE (ed) Biotechnology of fruit and nut crops. CABI Publ, Wallingford, pp 222–246

Sandal I, Bhattacharya A, Ahuja PS (2001) An efficient liquid culture system for tea shoot proliferation. Plant Cell Tissue Organ Cult 65:75–80

Scott TK (1972) Auxins and roots. Annu Rev Plant Physiol 23:235258

Shibli RA, Smith MAL (1996) Direct shoot regeneration from *Vaccinium pahalae* (ohelo) and V. myrtillus (bilberry) leaf explants. Hortscience 31:1225–1228

Simon S (1908) Experimentelle untersuchungen über die differenzierungsvorgange im callusgewebe von holzgewachsen. Jahrb Wiss Bot 45:351–478
Skirvin RM, Motoike S, Coyner M, Norton MA (2005) *Rubus* spp. cane fruit. In: Litz RE (ed) Biotechnology of fruit and nut crops. CABI Publ, Wallingford, pp 566–582
Soneji JR, Rao PS, Mhatre M (2002) Somaclonal variation in micropropagated dormant axillary buds of pineapple (*Ananas comosus* L., Merr.) J Hort Sci Biotechnol 77:28–32
Steward FC, Ammirato PV, Mapes MD (1970) Growth and development of totipotent cells: some problems, procedures and prospectives. Ann Bot 34:761–787
Swartz HJ, Bors R, Mohamed F, Naess SK (1990) The effect of in vitro pretreatments on subsequent shoot organogenesis from excised *Rubus* and *Malus* leaves. Plant Cell Tissue Organ Cult 21:179–184
Thiem B (2003) *Rubus chamaemorus* L. – a boreal plant rich in biologically active metabolites: a review. Biol Lett 40:3–13
Thomas JC, Katterman FR (1986) Cytokinin activity induced by thidiazuron. Plant Physiol 18(1):681–683
Trehane J (2004) Blueberries, cranberries and other Vacciniums. Timber Press, Portland
Vander Kloet SP (1988) The genus *Vaccinium* in North America. Agr Can Publ, 1828, Canada
Vander Kloet SP, Dickinson TA (2009) A subgeneric classification of the genus *Vaccinium* and the metamorphosis of *V.* section *Bracteata* Nakai: more terrestrial and less epiphytic in habit, more continental and less insular in distribution. J Plant Res 122:253–268
Visser C, Qureshi JA, Gill R, Saxena PK (1992) Morphoregulatory role of thidiazuron: substitution of auxin and cytokinin requirement for the induction of somatic embryogenesis in geranium hypocotyl cultures. Plant Physiol 99:1704–1707
Vujović T, RužićĐ CR, Momirović GŠ (2010) Adventitious regeneration in blackberry (*Rubus fruticosus* L.) and assessment of genetic stability in regenerants. Plant Growth Regul 61(61):265–275
Vyas P, Debnath SC, Igamberdiev AU (2013) Metabolism of glutathione and ascorbate in lingonberry cultivars during in vitro and ex vitro propagation. Biol Plant 57:603–612
Vyas P, Curran NH, Igamberdiev AU, Debnath SC (2015) Antioxidant properties of lingonberry (*Vaccinium vitis-idaea* L.) leaves within a set of wild clones and cultivars. Can J Plant Sci 95:663–669
Weising K, Nybom H, Wolff K, Meyer W (1995) DNA fingerprinting in plants and fungi. CRC Press, Boca Raton
Yau MH, Che CT, Liang SM, Kong YC, FongWP (2002) An aqueous extract of *Rubus chingii* fruits protects primary rat hepatocytes against tert-butyl hydroperoxide induced oxidative stress. Life Sci 72:329–338
Yonghua Q, Shanglong Z, Asghar S, Lingxiao Z, Qiaoping Q, Kunsong C, Changjie X (2005) Regeneration mechanism of Toyonokastrawberry under different color plastic films. Plant Sci 168:1425–1431
Zhang Q, Folta KM, Thomas M, Davis TM (2014) Somatic embryogenesis, tetraploidy, and variant leaf morphology in transgenic diploid strawberry (*Fragaria vesca* subspecies *vesca* 'Hawaii 4'). BMC Plant Biol 14:23. https://doi.org/10.1186/1471-2229-14-23
Ziv M (1991a) Quality of micropropagated plants – vitrification. In Vitro Cell Dev Biol Plant 27:64–69
Ziv M (1991b) Vitrification: morphological and physiological disorders of in vitro plants. In: Debergh PC, Zimmerman RH (eds) Micropropagation. Kluwer AcadPubl, Dordrecht, pp 45–69
Ziv M (2005) Simple bioreactors for mass propagation of plants. Plant Cell Tissue Organ Cult 81:277–285
Ziv M, Chen J, Vishnevetsky J (2003) Propagation of plants in bioreactors: prospects and limitations. Acta Hort 616:85–93

TDZ in Cereal Gametic Embryogenesis

7

Patricio Esteves and François J. Belzile

Abstract

Gametic embryogenesis is defined as the process that allows immature pollen grains – microspores – to parthenogenetically become embryos. The process can successfully be induced at a high frequency under in vitro culture conditions in a wide number of crop species. Microspores are haploid cells carrying half the somatic number of chromosomes, but if either spontaneously or artificially their chromosomal complement is doubled during the embryogenic pathway, the resulting embryos will become diploid and perfectly homozygous. The products of gametic embryogenesis are therefore called doubled-haploid plants, which are coveted materials for research and for plant breeding. Yet, to be efficiently used in a plant breeding program, doubled haploids need to be produced at a high frequency and in a reproducible manner. The efficiency and reproducibility of DH production are tied to the control of key factors intervening in the process. As is the case in many in vitro procedures, growth regulators play an important role in stimulating and guiding the process of orderly cell divisions leading to the regeneration of a complete plant from a single immature microspore. In this chapter, we review some of the key factors in this process with emphasis placed on the important role played by growth regulators, among which thidiazuron (TDZ). To illustrate the utility of TDZ in cereal gametic embryogenesis, we describe a highly efficient protocol for producing doubled haploids (either via anther culture or isolated microspore culture) that relies on an innovative combination of growth regulators: thidiazuron and dicamba. In our hands, this protocol proved successful for producing high numbers of barley, wheat, and rice doubled haploids.

P. Esteves (✉)
Département de phytologie, Pavillon Paul-Comtois, Université Laval, Québec, Canada

F. J. Belzile
Département de phytologie, Pavillon Charles-Eugène Marchand, Université Laval, Québec, Canada
e-mail: francois.belzile@fsaa.ulaval.ca

N. Ahmad, M. Faisal (eds.), *Thidiazuron: From Urea Derivative to Plant Growth Regulator*, https://doi.org/10.1007/978-981-10-8004-3_7

Keywords
Gametic embryogenesis · Cereals · In vitro culture · Doubled haploids · Thidiazuron · Dicamba

7.1 Overview

It was around 50 years ago that the first reports of doubled-haploid production in cereals started to be communicated in barley (Clapham 1973) and in rice (Guha-Mukherjee 1973). However, efficiency was then so low that these procedures could not make a meaningful contribution to breeding programs, and cereals were even considered recalcitrant to anther androgenesis (Foroughi-Wehr et al. 1976). Indeed, much higher success rates were being achieved in anther culture of species such as *Datura innoxia* and *Nicotiana tabacum*, where dozens of haploid plants could be easily obtained from just a few in vitro cultured buds (Guha and Maheshwari 1964; Nitsch and Nitsch 1969). In contrast, in barley, efficiencies of about 0.003–2 green plants per processed spike were being reported (Clapham 1973; Foroughi-Wehr et al. 1982; Foroughi-Wehr et al. 1984). In contrast, several years later very high rates of success were reported in several cereals such as wheat (Zheng et al. 2001) and barley (Kasha et al. 2001). Using isolated microspore culture in our lab, on a wide number of diverse barley F1 genotypes, we measured a mean efficiency of 111.5 green plants per processed spike (Esteves et al. 2014). This dramatic increase in efficiency of about 10,000% along these years can essentially be explained by two factors. One is the replacement of anther culture by isolated microspore culture, and the other is the successive improvements made to the protocol – including, among other parameters, changes to the type and concentration of growth regulators in culture media. Together, these lead to a very high frequency of microspore embryogenesis and green plant regeneration in cereals.

7.2 Introduction

7.2.1 Plant Hormones and In Vitro Culture of Plant Cells, Tissues, and Organs

Plant in vitro culture is a discipline which encompasses a very broad set of techniques and methods, all having in common the use of synthetic culture media onto which cells, organs, tissues, or pieces of plants, called explants, are cultured for specific purposes. The discipline started to develop after the Austrian plant physiologist Gottlieb Haberlandt (1902) hypothesized that plant cells are "totipotent," meaning that they have the capacity to regenerate a complete individual under the proper circumstances and provided with adequate conditions in culture. Also, he stated that in order to regulate the proper division, growth, and differentiation of cultured cells, "growth factors" had to be provided. Amazing as it may seem, this time, Haberlandt did not have the experimental means to test his hypothesis, but he was nonetheless beginning to establish the fundamentals of what was going to become the field of plant in vitro culture as a discipline (Murashige 1979). The reprogramming of young pollen grains (microspores) to produce embryos, also referred to as gametic embryogenesis or androgenesis, clearly confirmed Haberlandt's hypothesis about the totipotency of plant cells (Soriano et al. 2013). The process, in addition to its applied uses in breeding, provides an experimental system that allows a close insight into the causes and factors involved in the determination of fate and differentiation in plant morphogenesis. As mentioned above, from a practical point of view, DHs are exceptional resources either for research in plant genetics and genomics and to be exploited in crop breeding programs as well.

7.2.2 The Emergence of Doubled-Haploid Production

In higher plants, gametes are haploid cells where the normal, somatic number of chromosomes has been reduced by half. Embryos can be induced to form parthenogenetically from these haploid cells by a process called gametic embryogenesis (Olmedilla 2010). When male gametes are involved, the process is called "androgenesis," while "gynogenesis" describes the process where the female gametes are used (Germanà 2011). If a spontaneous doubling of the haploid number of chromosomes during the very early beginning of the embryogenesis process occurs, or if it is induced by treatment with chemical compounds such as colchicine, the resulting plants will carry perfectly identical copies of each of its chromosomes. It is this kind of product that is called a doubled haploid (DH). Because of their origin, DHs will be 100% homozygous, which means that genetic fixation can be obtained in the immediate progeny of an F1 plant. It is worth noting that between eight and ten cycles of selfing would be necessary to obtain a similar level of homozygosity (Dunwell 2010; Germanà 2011). A large population of DHs can be considered as a representative sample of the products of genetic recombination of their parental genes. Thus, DH production techniques allow the shortening of the length of a breeding cycle (from one cross to the next) by several years (Forster et al. 2007).

7.3 Obtaining Doubled Haploids

7.3.1 Current Methods for DH Production

Several methods for obtaining DHs have been developed in different species. For instance, interspecific crosses can lead to DH progeny following elimination of the set of chromosomes from the "exotic" parent (Kasha and Kao 1970), a method which is often used in barley and wheat (Devaux 2003). In maize, DHs are obtained by crosses made within the same species using as the male parent a genotype inducing haploidy (Prigge and Melchinger 2012). Also, pollinating with irradiated pollen, as reported in watermelon by Sari and Abak (1994)), is another method used to obtain haploids. Alternatively, in vitro culture of ovaries – gynogenesis – has also been reported as a method for producing DHs, as in barley by Castillo and Cistué (1993), in *Beta vulgaris* by Bossoutrot and Hosemans (1985), in *Helianthus annuus* by Cai and Zhou (1984), and in *Allium cepa* by Campion and Alloni (1990). However, to the best of our knowledge, in none of these crops, the gynogenetic method for producing DHs reached the efficiency required to be of use in a breeding program. Compared to these alternatives, in vitro androgenesis – including both the methods of anther culture (AC) and of isolated microspore culture (IMC) – is often recognized as the most efficient and thus the preferred approach for producing DHs in crop species (Devaux and Kasha 2009).

7.3.2 Anther Culture and Isolated Microspore Culture: The Techniques

Anther culture was developed starting from the work of Guha and Maheshwari (1964, 1966) in obtaining haploid embryos of *Datura innoxia* Mill. when they cultured anthers in a defined culture medium. Clapham (1973) was the first to report the production of haploid plants of barley using AC. Next, by the late 1970s and throughout the 1980s, a huge amount of research was directed toward the improvement in the efficiency of androgenesis in cereals. From the very beginning of this period, the vast majority of this work focused on rice and on barley, and, a few years later, some varieties of these crops were considered as "model responsive genotypes" for both AC and microspore embryogenesis (Forster et al. 2007; Begheyn et al. 2016). Until today, there has been an impressive gain in efficiency in obtaining the androgenetic response (Devaux and Kasha 2009), and nowadays several efficient protocols are available (Jacquard et al. 2003; Cistué et al. 2003; Szarejko 2003).

Essentially, AC consists of first culturing the donor plants under optimal conditions of mineral nutrition, watering, lighting, temperature, and humidity. Next, at the proper stage of development of the microspores, stalks are harvested and spikes "pretreated" (Hoekstra et al. 1997) with a stress treatment, so that the microspores are stimulated to switch from the gametophytic pathway to the sporophytic pathway (Touraev et al. 2001; Devaux and Kasha 2009) (Fig. 7.1). Next, anthers are extracted from the florets and put in culture onto a defined induction medium, where embryos – often called "embryo-like structures" (ELSs) – might start to form inside the anthers containing responsive microspores. The ELSs are next recovered and transferred onto a regeneration medium where they will germinate and regenerate the DH plants (Fig. 7.2).

The IMC method to obtain DHs was developed more recently. Among the first reports, the work of Hunter (1988) in barley is one of the most exhaustive. As happened with AC, the efficiency of IMC was greatly improved in the following years. At present, this latter method is considered to significantly surpass AC and the *Bulbosum* method (interspecific cross) in its potential to produce DHs (Davies

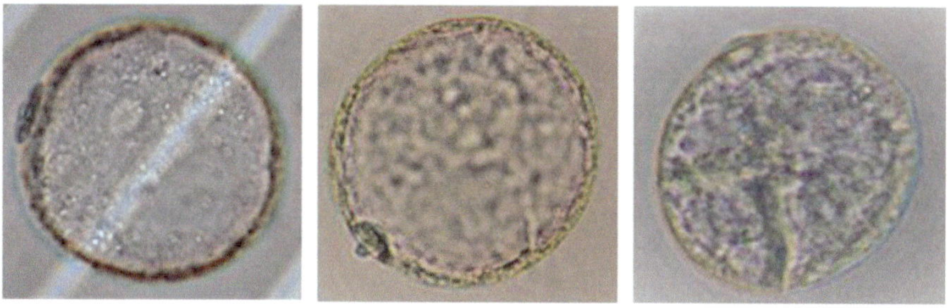

Fig. 7.1 Freshly isolated, non-pretreated microspore (left). After the pretreatment, embryogenic, competent microspores often display the "starlike" phenotype (center) (Maraschin et al. 2005) that characterizes them. After a few days in culture, divisions inside the embryogenic microspores become evident (right)

Fig. 7.2 A comparison of IMC (upper row: in rice to the left and barley at the center and to the right) and AC (row below: in wheat to the left and rice at the center and to the right). In IMC a massive production of ELSs is obtained, and both the control of the process and the manipulation of its products are much more facilitated. Both methods, in the three species mentioned here, were performed using induction medium containing TDZ and DIC

and Morton 1998; Li and Devaux 2005). In our own work (Esteves et al. 2014), we found IMC to be at least five times more efficient than AC in six-row spring barley. Indeed, IMC allows to plate large populations of selected, synchronized, viable, and embryogenic microspores (Li and Devaux 2005) through a series of steps of the protocol, including (1) blending the spikes, (2) filtering the produced extract through a 100 μm sieve, (3) rinsing several times the crude isolate of microspores, and (4) performing a discontinuous gradient centrifugation to greatly enrich for competent and viable microspores (Wenzel et al. 1975; Bedinger and Edgerton 1991) (Fig. 7.3). A detailed description of our IMC protocol is presented in Esteves et al. (2014).

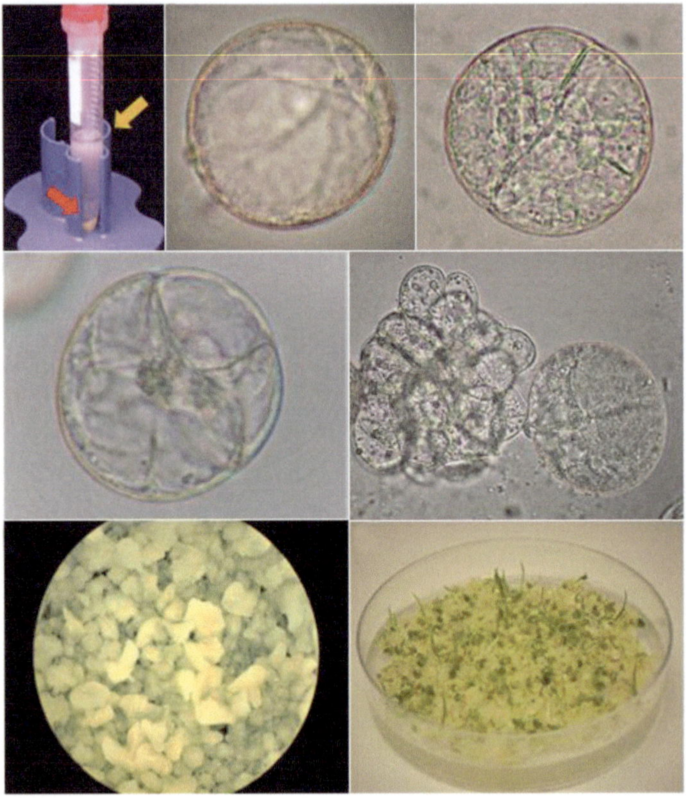

Fig. 7.3 Isolated microspore culture in barley performed with TDZ and DIC. Upper row, left: the product of the gradient centrifugation – a selection of viable, embryogenic microspores (orange upper arrow) and the debris (red arrow below). In the following panels (left to right), the development of microspores in culture is evident as they start dividing until, after 12–15 days, masses of ELSs are produced (bottom row, left). When the latter are transferred to a regeneration medium, these regenerate into green DH plants (bottom row, right)

Other protocols with slight differences are reported by Kasha et al. (2001) and by Maluszynski et al. (2003).

After harvesting the stalks and pretreating the spikes (microspores), only a small fraction of the microspores will engage in the sporophytic pathway. Among the differences existing between AC and IMC, one essential feature is that the latter, through the density gradient centrifugation, allows the selection of the fraction of viable microspores that are competent to engage in embryogenesis. At the same time, IMC allows a better follow-up and control of the cultures, because microspore embryogenesis is much better synchronized. For its part, AC requires less lab equipment and is more labor-intensive.

7.4 Gametic Embryogenesis: Impact of the Genotype

It is widely reported in the literature that success in obtaining the embryogenic response of microspores is strongly influenced by the genotype (Torp and Andersen 2009; Cistué et al. 1999; Marchand et al. 2008; Devaux and Kasha 2009; Makowska et al. 2015). Indeed, it has been proposed that microspore embryogenesis and plant regeneration might be under the control of separate classes of genes (Henry et al. 1994; Szakacs et al. 1988), as might be factors determining the regeneration of green or albino plants which very often are regenerated (Li and Devaux 2005; Weyen 2009). These are heritable traits (Foroughi-Wehr et al. 1984), highly influenced by the growth conditions of the donor plants and by the composition of the culture media (Kasha 1989; Knudsen et al. 1989; Simmonds 1989). Thus, for instance, Davies (2003) reported up to 1000 green plants regenerated per processed spike for a group of barley (winter-type) genotypes, while for spring-type barleys, the rate was only 10–20 green plants per spike. However, Atanassov et al. (1995) and Olmedilla (2010) identified factors of the protocol that, when optimized, lead to significant improvements in the production of DHs over genotypes displaying an otherwise poor response. The same has been observed in our lab when performing IMC on six-row spring barley, the type recognized as the most recalcitrant to in vitro androgenesis. We obtained a significant improvement in DH production when we optimized the pretreatment (30% increase in production) and the hormone regime used in the culture media (250% increase). Among the key changes in this regard were the use of thidiazuron (N-phenyl-N′-1,2,3-thiadiazol-5-ylurea, TDZ) and dicamba (3,6-dichloro-O-anisic acid, DIC) as growth regulators in the induction medium. In both cases the gain was due both to a higher frequency of embryogenesis and a reduction in albinism as well. Also, by these means the response to androgenesis increased very significantly in almost all the genotypes processed, such that the proportion of F1s that yielded a useful number of DH lines exceeded 90% in our work. This is in stark contrast with the situation prior to these improvements, where only a subset (typically less than a third) of F1s responded sufficiently well in androgenesis to produce useful sets of progeny (Esteves et al. 2014). Moreover, we have recently obtained evidence suggesting that this very same protocol is also efficient in producing DHs of diverse genotypes of other species, such as wheat and rice (Esteves, unpublished).

7.5 In Vitro Controlling Factors in DH Production

7.5.1 Composition of the Basal Culture Medium

Optimizing culture media composition, both at the levels of the induction of microspore embryogenesis and of the regeneration phase, is an important aspect in androgenesis of cereals as it provides nutrition and also determines the fate of microspores (Santra et al. 2012; Mishra and Rao 2016). For instance, even though green plant regeneration is admittedly under strong genetic control, adjustments to environmental conditions to which cells or explants in culture are exposed can change the genetic response (Zhou et al. 1991).

The composition of the basal culture medium for in vitro androgenesis typically consists of slight or more conspicuous modifications of the highly generalized MS medium (Murashige and Skoog 1962). Some of the commonly used basal media for the induction phase in AC or IMC in cereal species are the IMI medium for barley (Li and Devaux 2001); the N6 medium for rice, maize, and wheat (Chu et al. 1975); and the W14 medium for oats (Ouyang et al. 1989). Although the diverse species and genotypes usually show different basal medium requirements to induce plant regeneration from microspores, there are some major common improvements reported in the literature involving (1) the type of sugar (Cai et al. 1992; Navarro-Alvarez et al. 1994; Lentini et al. 1995), (2) the nitrogen source (Jähne and Lörz 1995; Raina and Zapata 1997; Lu et al. 2016), (3) the organic additives (Powell 1990; Hoekstra et al. 1992), and (4) the type and concentration of growth regulators (plant hormones) (Schulze 2007; Germanà 2011).

7.5.2 Growth Regulators and In Vitro Androgenesis

7.5.2.1 Hormones and In Vitro Gametic Embryogenesis

Plant growth regulators and hormones are known as key signaling molecules controlling plant growth and development, as well as initiating signal transduction pathways in response to environmental stimuli (Kohli et al. 2013). The plant hormones, mainly auxins and cytokinins, are known to control the dedifferentiation process in plant in vitro cultures, and the rate of success can be enhanced by manipulating the plant growth regulators (Mishra and Rao 2016). Among the auxins, 2,4-dichlorophenoxyacetic acid (2,4-D) and naphthalene-acetic acid (NAA) are the most commonly used growth regulators for the induction of callus from rice anthers (Trejo-Tapia et al. 2002). However, neither 2,4-D nor NAA can support regeneration, and the use of cytokinins like kinetin (KIN) and 6-benzyl-aminopurine (BAP) is required (Mandal and Gupta 1995). In barley AC, weaker auxins at low concentration have been commonly used, but in some cases the auxin has been omitted, leaving the cytokinin BAP as the sole growth substance (Hunter 1988; Kihara et al. 1994). In addition, Huang (1984) found that another type of hormone, abscisic acid (ABA), may promote barley green plant regeneration. Hoekstra et al. (1997) stated that the right hormone in the culture medium of microspores is necessary for the induction of plant production, and in barley they achieved this either with the application of a cytokinin (BAP) or 2,4-D in the culture medium. Alternatively, Kasha et al. (2001) using IMC in barley, and Ziauddin et al. (1992) using AC and IMC in barley and wheat, included the naturally occurring auxin phenylacetic acid (PAA) together with BAP in the induction medium and the cytokinin only in the regeneration medium, and they both improved significantly the production of ELSs and green DH plants. In our lab we tried to reproduce this PAA-based protocol, without success. It has been proposed that the action of PAA is dependent on plant growth and its associated endogenous levels of auxin (Kasha et al. 2001), and thus more experimentation might be necessary to optimize it use.

While there are some reports suggesting that growth regulators are not essential for microspore embryogenesis in most plant species (Olmedilla 2010), it has also been reported that recalcitrant and low-responding wheat and barley genotypes became responsive following a pretreatment with chemicals, including growth regulators (Liu et al. 2002), or by changing the hormone composition of the culture medium (Esteves et al. 2014). Aligned with these results, Shariatpanahi et al. (2006) obtained a significantly higher frequency of embryogenesis and higher green plant regeneration from wheat microspores when growth regulators were included in the induction medium, compared to when they were not. Either alone or in combination, various types of auxins and cytokinins have been used in media designed for in vitro androgenesis (Cistué et al. 1999; Devaux and Pickering 2005; Datta 2005), and the inclusion of these plant growth regulators has been shown to modify embryogenesis and plant regeneration frequencies (Tyankova and Zagorska 2008).

At the beginning of the development of IMC in barley, Hunter (1988) used an induction medium containing only 1.0 mg/l of the auxin IAA (indoleacetic acid) as growth regulator, and he compared it to five others containing 1.0 mg/l of the cytokinins [BAP, KIN, zeatin (ZEA), zeatin riboside (ZEAr), or 6-γ-γ-(dimethylallylamino)-purine (2iP)]. He concluded that there existed differences in androgenesis that were attributable to this factor, and it was BAP that provided the best result. Then, to a medium containing 1.0 mg/l of BAP, he compared four other media containing one of our auxins, either IAA, indole butyric acid (IBA), NAA, or 2,4-D, and concluded that IAA provided the best response. Yet, as he subsequently found a negative correlation between the concentration of IAA (between 0 and 1.0 mg/l in the medium) and the regeneration of green plants, he concluded that an optimal induction medium would only contain the cytokinin BAP at a rate of 1.0 mg/l. He named this developed induction medium as FHG, which was to become one of the most commonly used in androgenesis in barley all over the world (Devaux and Pickering 2005).

As described above, the effects of plant growth regulators in in vitro androgenesis have been considered controversial (Devaux and Pickering 2005). However, based on our own experience, hormone composition of induction and regeneration media in in vitro androgenesis is a key success factor in obtaining high frequencies of embryogenesis and the regeneration of green DH plants. Indeed, when we included TDZ as a growth regulator in the induction medium, together with the auxin DIC, a massive embryogenic response was obtained, and at the same time, both the total number of green regenerated plants and the ratio of green to albino plants increased dramatically compared to the standard BAP-based induction medium (Esteves et al. 2014).

7.5.2.2 The Role of Cytokinins in Gametic Embryogenesis

If the impact of growth regulators in inducing gametic embryogenesis has been a matter of discussion in the literature, to a certain degree, it is much less so in the case of somatic embryogenesis. Somatic embryogenesis is the process by which diploid, somatic plant cells differentiate into somatic embryos. Any differentiated plant cell that retains its nucleus has the ability to revert to the embryogenic condition and regenerate an entire plant (Reynolds 1997). Somatic embryo formation usually requires first a treatment of diploid cells with plant hormones, mostly auxin but also cytokinins, under specific culture conditions, and later auxin withdrawal to allow embryogenesis to continue (Toonen et al. 1994; Cistué and Kasha 2005). Indeed, in the majority of species studied where exogenous plant growth regulators were necessary for inducting somatic embryogenesis and embryoid formation, both auxins and cytokinins were shown to be key factors in determining the response (Jimenez 2005).

Plant regeneration, regardless of ploidy level, occurs by either somatic embryogenesis or organogenesis (Rybczynski et al. 1991). It has been proposed that both types of embryogenesis, somatic and gametic embryogenesis, are regulated by the same basic cellular mechanisms, namely, cell expansion and asymmetric cell division (De Jong et al. 1993). Also, signal molecules, such as hormones and growth regulators, have been shown to play a role during both types of embryogenesis (Dodeman et al. 1997). Considering that both TDZ and DIC are reported to be efficient growth regulators capable of inducing direct and indirect somatic embryogenesis, we decided to experiment with them in gametic embryogenesis in barley via IMC. The experiments were to be carried out in two steps: The first was to evaluate replacing the classically used cytokinin BAP (at 1.0 mg/l) by TDZ at a range of concentrations (0.1, 0.3, and 1.0 mg/l). The second step focused on the impact of replacing BAP by both TDZ (included at the concentration which had performed best in the first step) and the auxin DIC at a range of concentrations (0.1, 0.3, and 1.0 mg/l). Four contrasting genotypes were chosen for this experiment: one winter, two-row type (Igri); one spring, two-row type (Gobernadora); and two spring, six-row types (ACCA and Léger). Our results showed that when BAP was replaced by TDZ alone, we did not observe statistically significant differences for the parameters tested (total number of green plants, total number of albino plants, and green/albino plants ratio; Table 7.1). BAP and TDZ have very similar molecular weights (225.25 and 220.25 g/mol, respectively), so this taught us that even at 1/10 of its molar concentration, TDZ was as potent as BAP in inducing embryogenesis and ELS formation. Next, contrasting results were observed during the second step of the experiment: when BAP was replaced by the combination of TDZ (0.3 mg/l) and DIC (1.0 mg/l), the number of regenerated green plants increased consistently, and significantly, with increasing concentrations of DIC (Table 7.1). Averaged over the four barley cultivars, all three induction media including DIC and TDZ produced significantly more green plants than the BAP check. Indeed, relative to the control, the TDZ (0.3 mg/l) + DIC (1.0 mg/l) treatment produced greater than five times more green plants on average (176.0 vs. 34.4 green regenerated plants/10^5 microspores, respectively; $P < 0.01$), whereas the number of albino plants produced did not change significantly (197.1 vs. 162.4; $P = 0.249$). Thus, the observed increase in the total number of plants/10^5 microspores (373.1 vs. 196.8, $P < 0.001$) could be almost entirely ascribed to an increase in the production of green plants. As can also be seen in Table 7.1, the green/albino plant ratio increased dramatically

Table 7.1 Comparison of the impact of replacing BAP with three levels of TDZ (first phase of the experiment) and by 0.3 mg/l of TDZ + DIC (three levels of DIC; second phase) in induction medium of barley IMC

Growth regulator (mg/l)	Green plants	Albino plants	Total plants	Green/albino plants
First phase				
BAP 1.0	33.2^{a}	161.2^{a}	194.4^{a}	0.21^{a}
TDZ 0.1	42.1^{a}	143.1^{a}	185.2^{a}	0.29^{a}
TDZ 0.3	40.1^{a}	133.4^{a}	173.5^{a}	0.30^{a}
TDZ 1.0	37.5^{a}	101.5^{a}	138.5^{a}	0.37^{a}
Second phase				
BAP 1.0	34.4^{a}	162.4^{a}	196.8^{a}	0.21^{a}
TDZ 0.3 + DIC 0.1	66.9^{b}	173.6^{a}	240.5^{a}	0.38^{ab}
TDZ 0.3 + DIC 0.3	107.0^{c}	178.8^{a}	285.8^{b}	0.60^{bc}
TDZ 0.3 + DIC 1.0	176.0^{d}	197.1^{a}	373.1^{b}	0.89^{c}

Numbers in the table correspond to regenerated plants per 10^5 microspores plated. Different letters correspond to significant ($P < 0.05$) differences

from 0.21 for the control treatment up to 0.89 for the TDZ 0.3 mg/l+DIC 1.0 mg/l treatment ($P < 0.01$). Since the mean number of albino plants was not significantly different among all the treatments, it is clearly the increasing number of green regenerated plants, produced at increasing concentrations of the auxin, that allowed the green/albino plant ratio to increase significantly (5.1-fold).

7.5.2.3 TDZ and In Vitro Embryogenesis

In cereal somatic embryogenesis, TDZ, in combination with the auxins 2,4-D and NAA, has been shown to improve more than threefold the response of recalcitrant indica rice varieties when compared to some other cytokinins such as ZEA, BAP, KIN, and 2iP (Wenzhong et al. 1994). Improved regeneration frequencies were also described for Australian rice varieties when combining TDZ with the auxin NAA, instead of using BAP alone, for mature embryo-derived callus production (Ahzria and Bhalla 2000). Similarly, using mature embryos as explants, Parmar et al. (2012) reported that a considerable improvement in the regeneration frequency (up to 97%) and in the average number of shoots was obtained with a combination of TDZ and 2,4-D.

Although TDZ has proved useful in inducing somatic embryogenesis and even allowed improving the response of recalcitrant species in several instances, it has seldom been tested in gametic embryogenesis in cereal species, and the only existing reports in the literature proved either unsuccessful or nonconclusive. For instance, Ouédraogo et al. (1998) reported that the addition of 0.01 mg/l of TDZ in AC of *Hordeum vulgare* increased or decreased the number of embryos per 100 anthers, or it had no effect, depending on the cultivar. Also, Kiviharju et al. (2005) added TDZ as a supplement in induction media in AC of oats (0.2 and 0.5 mg/l) and found no benefits or even a reduction of the regeneration efficiency in relation to the other cytokinins such as ZEA, 2iP, KIN, and BAP. In fact, at the same concentrations as TDZ, the latter cytokinins performed much better in embryogenesis and green plant regeneration. In contrast with these previous results, in our lab, after a step-by-step adjustment of the optimal concentrations, we found that 0.3 mg/l of TDZ together with 1.0 mg/l of DIC in the induction medium dramatically increased the yield of green plants/spike, by simultaneously increasing the frequency of both gametic embryogenesis and green plant regeneration and reducing albinism as well. Further experimentation leads us to reduce the DIC concentration in the induction medium to 0.6 mg/l as, at the higher dose, regenerated plants seemed to suffer from vitrification. Using this latter formulation (TDZ at 0.3 and DIC at 0.6 mg/l) allowed us to produce hundreds of DHs of a wide array of recalcitrant barley genotypes in a reproducible fashion over several years, and this same induction medium has recently proved efficient to obtain massive embryogenesis and plant regeneration in wheat and rice (unpublished results).

7.6 Conclusion and Perspectives

It has been recently proposed that plant growth regulators, which cross talk and interact with the plant's genotype and environmental factors, play a crucial role in microspore embryogenesis, controlling microspore-derived embryo differentiation and development as well as haploid/doubled-haploid plant regeneration (Żur et al. 2015). In agreement with this, we observed a significant positive impact in gametic embryogenesis of cereal species provided by modifying the types and concentrations of auxins and cytokinins in the induction medium. To the best of our knowledge, this is the first report of a substantial benefit of the inclusion of TDZ, together with the auxin DIC, in a gametic embryogenesis protocol. Replacing the traditionally used cytokinin BAP used alone by the cytokinin TDZ + the auxin DIC allowed not only to nearly double the total number of regenerated plants but also increased by 4.1-fold the ratio of green/albino plants. Even if it is well known that cytokinins are involved in the control of chloroplast biogenesis and function (Zubo et al. 2008) and also in chlorophyll synthesis (Ricci et al. 2001), the role played by the auxin-cytokinin ratio in increasing the proportion of green regenerated plants is not clear and would deserve further research (i.e., it remains to be determined whether the same kind of improvement would be seen using other types of auxin-cytokinin combinations). Nonetheless, we observed that this same hormone formulation, including TDZ and DIC, proved useful to obtain high numbers of green DHs in a diverse array of barley genotypes and in rice and wheat as well, using either AC or IMC. Thus, we believe that the adjustment of the ratio and dosage of this highly efficient combination of auxin and cytokinin (TDZ and DIC) might be a key to overcoming the recalcitrance to DH production in other species and genotypes, such as indica races of rice and leguminous, woody, and fruit tree species. More research would be warranted to test this hypothesis, which would speed up breeding projects dealing with these species. Also, such research might shed further light on the sporophytic pathway leading to DH plants.

References

Ahzria D, Bhalla PL (2000) Plant regeneration from mature embryo derived callus of Australian rice (*Oryza sativa* L.) varieties. Aust J Agri Res 51:305–312

Atanassov A, Zagorska N, Boyadjiev P et al (1995) In vitro production of haploid plants. World J Microbiol Biotechnol Vol 11:400–408

Bedinger PA, Edgerton MD (1991) Developmental staging of maize microspores reveals a transition in developing microspore proteins. Plant Physiol 92:474–479. https://doi.org/10.1104/pp.92.2.474

Begheyn RF, Vangsgaard K, Roulund N et al (2016) Efficient doubled haploid production in perennial ryegrass (*Lolium perenne* L). In: Roldán-Ruiz I, Baert J, Reheul D (eds) Breeding in a world of scarcity. Springer, Cham

Bossoutrot D, Hosemans D (1985) Gynogenesis in *Beta vulgaris*: from in vitro culture of unpollinated ovules to the production of doubled haploid plants in soil. Plant Cell Rep 4:300–303

Cai DT, Zhou C (1984) In vitro production of haploid embryoids and plantlets from unpollinated young florets and ovules of *Helianthus annuus* L. Kexue Tongbao 29:680–682

Cai Q, Szarejko I, Polok K et al (1992) The effect of sugars and growth regulators on embryoid formation and plant regeneration from barley anther culture. Plant Breed 109:218–226. https://doi.org/10.1111/j.1439-0523.1992.tb00176.x

Campion B, Alloni C (1990) Induction of haploid plants in onion (*Allium cepa* L.) by in vitro culture of unpollinated ovules. Plant Cell Tiss Organ Cult 20:1–6

Castillo AM, Cistué L (1993) Production of gynogenic haploids of *Hordeum vulgare* L. Plant Cell Rep Jan 12(3):139–143. https://doi.org/10.1007/BF00239094

Chu CC, Wang CC, Sun CS et al (1975) Establishment of an efficient medium for anther culture of rice through comparative experiment on the nitrogen sources. Sci Sn 18:659–668

Cistué L, Kasha KJ (2005) Gametic embryogenesis in *Triticum*: a study of some critical factors in haploid (microspore) embryogenesis. In: Mujib A, Samaj J (eds) Somatic embryogenesis. Springer, Berlin, pp 321–342

Cistué L, Ramos A, Castillo AM (1999) Influence of anther pretreatment and culture medium composition on the production of barley doubled haploids from model and low responding cultivars. Plant Cell Tissue Organ Cult 55:159–166

Cistué L, Vallés MP, Echávarri B et al (2003) Barley anther culture. In: Maluszynski M, Kasha K, Foster B (eds) Doubled haploid production in crop plants, a manual. FAO/IAEA Division, Kluwer, Dordrecht, pp 29–35

Clapham D (1973) Haploid *Hordeum* plants from anther in vitro. Z Pflanzenzucht 69:142–155

Datta SK (2005) Androgenic haploids: factors controlling development and its application in crop improvement. Curr Sci 89(11):1870–1878

Davies PA (2003) Barley isolated microspore culture (IMC) protocol. In: Maluszynski M, Kasha KJ, Forster BP, Szarejko I (eds) Doubled haploid production in crop plants: a manual. Springer, Dordrecht, pp 49–52

Davies PA, Morton S (1998) A comparison of barley isolated microspore and anther culture and the influence of cell culture density. Plant Cell Rep 17:206–210

De Jong AJ, Schmidt EDL, De Vries SC (1993) Early events in higher-plant embryogenesis. Plant Mol Biol 22:367–377

Devaux P (2003) The Hordeum bulbosum (L.) method. In: Maluszynski M, Kasha KJ, Forster BP, Szarejko I (eds) Doubled haploid production in crop plants: a manual. Springer, Dordrecht, pp 15–19

Devaux P, Kasha KJ (2009) Overview of Barley doubled haploid production. In: Touraev A, Forster BP, Jain SM (eds) Advances in haploid production in higher plants. Springer Science + Business Media BV, Dordrecht

Devaux P, Pickering R (2005) Haploids in the improvement of Poaceae. In: Palmer CE, Keller WA, Kasha KJ (eds) Haploids in crop improvement II. Springer, Berlin

Dodeman VL, Ducreux G, Kreis M (1997) Zygotic embryogenesis versus somatic embryogenesis. J Exp Bot 48(313):1493–1509

Dunwell JM (2010) Haploids in flowering plants: origins and exploitation. Plant Biotechnol J 8(4):377–424

Esteves P, Clermont I, Marchand S et al (2014) Improving the efficiency of isolated microspore culture in six-row spring barley: II-exploring novel growth regulators to maximize embryogenesis and reduce albinism. Plant Cell Rep 33(6):871–879

Foroughi-Wehr B, Mix G, Gaul H, Wilson HM (1976) Plant production from cultured anthers of *Hordeum vulgare* L. Z Pflanzenzucht 77:198–204

Foroughi-Wehr B, Friedt W, Wenzel G (1982) On the genetic improvement of androgenetic haploid formation in *Hordeum vulgare* L. Theor Appl Genet 62:246–248

Foroughi-Wehr B, Friedt W, Wenzel G (1984) On the genetic improvement of androgenetic haploid formation in *Hordeum vulgare* L. Theor Appl Genet 62:233–239

Forster BP, Heberle-Bors E, Kasha KJ et al (2007) The resurgence of haploids in higher plants. Trends Plant Sci 12:368–375. https://doi.org/10.1016/j.tplants.2007.06.007

Germanà MA (2011) Gametic embryogenesis and haploid technology as valuable support to plant breeding. Plant Cell Rep 30:839–857

Guha S, Maheshwari SC (1964) In vitro production of embryos from anthers of *Datura*. Nature 204:497

Guha S, Maheshwari SC (1966) Cell division and differentiation of embryos in the pollen grains of *Datura* in vitro. Nature (London) 212:97

Guha-Mukherjee S (1973) Genotypic differences in the in vitro formation of Embryoids from Rice pollen. J Exp Bot 24(78):139–144

Haberlandt G (1902) Culturversuche mit isolierten Pflanzenzellen. Sitz-Ber Mat Nat Kl Kais Akad Wiss Wien 111:69–92

Henry Y, Vain P, De Buyser J (1994) Genetic analysis of in vitro plant tissue culture responses and regeneration capacities. Euphytica 79:45–58

Hoekstra S, van Zijderveld MH, Louwerse JD et al (1992) Anther and microspore culture of *Hordeum vulgare* L. cv. Igri. Plant Sci 86:89–96

Hoekstra S, van Bergen S, van Brouwershaven IR et al (1997) Androgenesis in *Hordeum vulgare* L.: effects of mannitol, calcium and abscisic acid on anther pretreatment. Plant Sci 126:211–218

Huang B (1984) The effects of several factors in callus induction in barley anther culture. Acta Phys Sin 10:403–405

Hunter CP (1988) Plant regeneration from microspores of barley, Hordeum vulgare L. Ph D thesis. Wye College, University London, London

Jacquard C, Wojnarowiez G, Clement C (2003) Another culture in barley. In: Maluszynski M, Kasha K, Foster B, Szarejko I (eds) Doubled haploid production in crop plants: a manual. Springer, Dordrecht, pp 21–27

Jähne A, Lörz H (1995) Cereal microspore culture. Plant Sci 109:1–12

Jimenez VM (2005) Involvement of plant hormones and plant growth regulators on in vitro somatic embryogenesis. Plant Growth Regul 47:91–110

Kasha K (1989) Production of haploids in cereals. In: Maluszynski M (ed) Current options for cereal improvement: double haploids, mutants and Heterosis. Kluwer, Dordrecht, pp 71–80

Kasha KJ, Kao KN (1970) High frequency haploid production in barley (*Hordeum vulgare* L.) Nature 225:874–876

Kasha KJ, Simion E, Oro R et al (2001) An improved in vitro technique for isolated microspore culture of barley. Euphytica 120:379–385

Kihara M, Fukuda K, Funatsuki H et al (1994) Plant regeneration through anther culture of three wild species of Hordeum (*H. murinum, H. marinum* and *H. bulbosum*). Plant Breed 112:244–247

Kiviharju E, Moisander S, Laurila J (2005) Improved green plant regeneration rates from oat anther culture and the agronomic performance of some HD lines. PCTOC Plant Cell Tiss Org Cult 81:1–9. https://doi.org/10.1007/s11240-004-1560-0

Knudsen S, Dui IK, Andersen SB (1989) Components of response in barley anther culture. Plant Breed 103:241–246

Kohli A, Sreenivasulu N, Lakshmanan P et al (2013) The phytohormone crosstalk paradigm takes center stage in understanding how plants respond to abiotic stresses. Plant Cell Rep 32:945–957. https://doi.org/10.1007/s00299-013-1461-y

Lentini Z, Reyes P, Martinez CP, Roca WM (1995) Androgenesis of highly recalcitrant rice genotypes with maltose and silver nitrate. Plant Sci 110:127–138

Li H, Devaux P (2001) Enhancement of microspore culture efficiency of recalcitrant barley genotypes. Plant Cell Rep 20:475–481

Li H, Devaux P (2005) Isolated microspore culture over performs anther culture for green plant regeneration in barley (*Hordeum vulgare* L.) Acta Physiol Plant 27:611–619

Liu W, Zheng MY, Polle EA, Konzak CF (2002) Highly efficient doubled-haploid production in wheat (*Triticum aestivum* L.) via induced microspore embryogenesis. Crop Sci 42:686–692

Lu R, Chen Z, Gao R et al (2016) Genotypes-independent optimization of nitrogen supply for isolated microspore cultures in barley. Bio Med Res Intl 2016:1801646. https://doi.org/10.1155/2016/1801646

Makowska K, Oleszczuk S, Zimny A et al (2015) Androgenic capability among genotypes of winter and spring barley. Plant Breed 134:668–674

Maluszynski M, Kasha KJ, Szarejko I (2003) Published doubled haploid protocols in plant species. In: Maluszynski M, Kasha KJ, Forster BP, Szarejko I (eds) Doubled haploid production in crop plants: a manual. Kluwer Academic Publishers, Dordrecht, pp 309–335

Mandal N, Gupta S (1995) Effect of genotype and culture medium on androgenic callus formation and green plant regeneration in indica rice. Indian J Exp Biol 33:761–765

Maraschin SF, de Priester W, Spaink HP et al (2005) Androgenic switch: an example of plant embryogenesis from the male gametophyte perspective. J Exp Bot 56:1711–1726

Marchand S, Fonquerne G, Clermont I et al (2008) Androgenic response of barley accessions and F1s with fusarium head blight resistance. Plant Cell Rep 27:443–451

Mishra R, Rao NGJ (2016) In-vitro androgenesis in rice: advantages, constraints and future prospects. Rice Sci 23(2):57–68. https://doi.org/10.1016/j.rsci.2016.02.001

Murashige T (1979) Principles of rapid propagation. In: Proceedings of symposium on propagation of higher plants through tissue culture. A Bridge between Research and Application Feb 1979. University of Tenn Knoxville TN

Murashige T, Skoog F (1962) A revised medium for rapid growth and bio assays with tobacco tissue cultures. Physiol Plant 15:473–497

Navarro-Alvarez W, Baezinger PS, Eskridge KM et al (1994) Effect of sugars in wheat anther culture media. Plant Breed 112:53–62

Nitsch JP, Nitsch C (1969) Haploid plants from pollen grains. Science 163:85–87. https://doi.org/10.1126/science.163.3862.85

Olmedilla A (2010) Microspore embryogenesis. In: Pua E-C, Davey MR (eds) Plant developmental biology – biotechnological perspectives vol 2. Springer, Berlin

Ouédraogo JT, St-Pierre C-A, Collin J, Rioux S et al (1998) Effect of amino acids, growth regulators and genotype on androgenesis in barley. Plant Cell Tissue Organ Cult 53:59–66

Ouyang J, Jia S, Zhang C et al (1989) A new synthetic medium (W14) for wheat anther culture. In: Ann Rep Inst Genet, Academia Sinica, Beijing, p 91–92

Parmar SS, Sainger M, Chaudhary D et al (2012) Plant regeneration from mature embryo of commercial Indian bread wheat (*Triticum aestivum* L.) cultivars. Physiol Mol Biol Plants 18(2):177–183. https://doi.org/10.1007/s12298-012-0101-2

Powell W (1990) Environmental and genetic aspects of pollen embryogenesis. Bajaj YPS Biotechnology in agriculture and forestry part I. Haploids in crop improvement vol 12 Springer Berlin 44–65

Prigge V, Melchinger AE (2012) Production of haploids and doubled haploids in maize. Methods Mol Biol 877:161–172

Raina SK, Zapata FJ (1997) Enhanced anther culture efficiency of indica rice (*Oryza sativa* L.) through modification of the culture media. Plant Breed 116(4):305–315

Reynolds TL (1997) Pollen embryogenesis. Plant Mol Biol 33:1–10
Ricci A, Carra A, Torelli A et al (2001) Cytokinin-like activity of N′-substituted N-phenylureas. Plant Growth Regul 34:167–172. https://doi.org/10.1023/A:1013399927783
Rybczynski JJ, Simonson RL, Baenziger PS (1991) Evidence for microspore embryogenesis in wheat anther culture. In Vitro Cell Dev Biol Plant 27:168–174
Santra M, Ankrah N, Santra DK et al (2012) An improved wheat microspore culture technique for the production of doubled haploid plants. Crop Sci 52:2314–2320. https://doi.org/10.2135/cropsci2012.03.0141
Sari N, Abak K (1994) Induction of parthenogenetic haploid embryos after pollination by irradiated pollen in watermelon. Hortic Sci 29(10):1189–1190
Schulze J (2007) Improvements in cereal tissue culture by Thidiazuron: a review. In: Fruit, vegetable and cereal science and biotechnology 1(2):64–79 Global Science Book
Shariatpanahi ME, Bal U, Heberle-Bors E et al (2006) Stresses applied for the re-programming of plant microspores towards in vitro embryogenesis. Physiol Plant 127:519–534
Simmonds J (1989) Improved androgenesis of winter cultivars of *Triticum aestivum* L. in response to low temperature treatment of donor plants. Plant Sci 65:225–231
Soriano M, Li H, Boutilier K (2013) Microspore embryogenesis: establishment of embryo identity and pattern in culture. Plant Reprod 26:181–196
Szakacs E, Kovacs G, Pauk J et al (1988) Substitution analysis of callus induction and plant regeneration from anther culture in wheat (*Triticum aestivum* L.) Plant Cell Rep 7:127–129
Szarejko I (2003) Anther culture for doubled haploid production in barley (*Hordeum vulgare* L.) In: Maluszynski M, Kasha KJ, Forster BP, Szarejko I (eds) Doubled haploid production in crop plants – a manual. Kluwer Academic Publishers, Dordrecht, pp 35–42
Toonen MAJ, Hendriks T, Schmidt EDL et al (1994) Description of somatic embryo-forming single cells in carrot suspension cultures employing video cell tracking. Planta 194:565–572
Torp AM, Andersen SB (2009) Albinism in microspore culture. In: Touraev A, Forster BP, Shri Mohan J (eds) Advances in haploid production in higher plants. Springer Science + Business Media BV, Dordrecht, pp 155–160
Touraev A, Pfosser M, Heberle-Bors E (2001) The microspore: a haploid multipurpose cell. Adv Bot Res 35:53–109
Trejo-Tapia G, Amaya UM, Morales GS et al (2002) The effects of cold-pretreatment, auxins and carbon source on anther culture of rice. Plant Cell Tissue Organ Cult 71:41–46
Tyankova ND, Zagorska NA (2008) Genetic control of in vitro response in wheat (*Triticum aestivum* L.) In Vitro Cell Dev Biol Plant 37:524–530. https://doi.org/10.1079/IVP2001213
Wenzel G, Hoffmann F, Potrykus I et al (1975) The separation of viable rye microspores from mixed populations and their development in culture. Mol Gen Genet 138:293–297
Wenzhong T, Rance I, Sivamani E et al (1994) Improvement of Plant Regeneration Frequency in vitro in Indica Rice. Chin J Genet 21(3):1–7
Weyen J (2009) Barley and wheat doubled haploids in breeding. In: Touraev A, Forster BP, Jain SM (eds) Advances in haploid production in higher plants. Springer Science+ Business Media BV, Dordrecht, pp 179–187
Zheng MY, Liu W, Weng Y et al (2001) Culture of freshly isolated wheat (*Triticum aestivum* L.) microspores treated with inducer chemicals. Plant Cell Rep 20:685–690
Zhou H, Zheng Y, Konzak CE (1991) Osmotic potential of media affecting green plant percentage in wheat anther culture. Plant Cell Rep 10:63–66
Ziauddin A, Marsolais A, Simion E et al (1992) Improved plant regeneration from wheat anther and barley microspore culture using phenylacetic acid (PAA). Plant Cell Rep 11:489–498
Zubo YO, Yamburenko MV, Selivankina SY et al (2008) Cytokinin stimulates chloroplast transcription in detached barley leaves. Plant Physiol 148(2):1082–1093. https://doi.org/10.1104/pp.108.122275
Żur I, Dubas E, Krzewska M et al (2015) Current insights into hormonal regulation of microspore embryogenesis. Front Plant Sci 6:424. https://doi.org/10.3389/fpls.2015.00424

8 Preconditioning of Nodal Explants in Thidiazuron-Supplemented Liquid Media Improves Shoot Multiplication in *Pterocarpus marsupium* (Roxb.)

Anees Ahmad, Naseem Ahmad, and Mohammad Anis

Abstract

Pterocarpus marsupium Roxb. (Fabaceae), commonly known as "Bijasal" or "Indian Kino," is a potential herbal drug-yielding tree since ancient times. In the present study, an efficient micropropagation system was developed for propagation of this valuable forest tree by pretreating the nodal explant in thidiazuron (TDZ) in half-strength Murashige and Skoog (MS) liquid medium before their inoculation onto the full-strength MS semisolid medium containing meta-Topolin (mT), a new aromatic cytokinin, at different doses either singly or in combination with auxins. Among the treatments tried, nodal explants treated with TDZ (10.0 μM) for 8 days followed by transfer to MS medium containing mT (5.0 μM) + NAA (1.0 μM) were found to be most effective combination in inducing maximum shoot number (11.16 ± 0.16) with an average shoot length (4.9 ± 0.13 cm) and maximum regeneration frequency (85%). A proliferating shoot culture was established by subculturing the original nodal explants on shoot multiplication medium supplemented with meta-Topolin after each harvest of newly formed shoots. For root induction, basal end of isolated shootlets (≥ 4.0 cm) was treated with high dose of indole-3-butyric acid (150.0 μM IBA) solutions for 1 week followed by their transfer to half-strength MS supplemented with low doses of IBA (1.5 μM) in the medium which produced an average of (6.32 ± 0.26) roots with mean root length (4.05 ± 0.27 cm) in 80% shootlets, after 4 weeks of culture transfer. The regenerated plantlets with proper root system were successfully acclimatized to field condition with 75% survival rate.

Keywords

Axillary bud · Meta-Topolin · Morphogenesis · Plant growth regulators · Pretreatment · Regeneration

---

A. Ahmad · N. Ahmad · M. Anis (✉)
Plant Biotechnology Laboratory, Department of Botany, Aligarh Muslim University, Aligarh, UP, India
e-mail: manis.bt@amu.ac.in

N. Ahmad, M. Faisal (eds.), *Thidiazuron: From Urea Derivative to Plant Growth Regulator*, https://doi.org/10.1007/978-981-10-8004-3_8

Abbreviations

BA	6-Benzyladenine
CKs	Cytokinins
IAA	Indole-3-acetic acid
IBA	Indole-3-butyric acid
MS	Murashige and Skoog medium
mT	Meta-Topolin
NAA	α-naphthalene acetic acid
NS	Nodal segment
PGRs	Plant growth regulators
TDZ	Thidiazuron

8.1 Introduction

World Health Organization (WHO) has assessed that approximately 80% of the people of developing countries depend on herbal drugs, and about 25% drugs obtained are from plants resources (Akshay et al. 2014). However, several countries including India are facing acute problems in maintaining the population of medicinal plants because of anthropogenic activities, unrestrained exploitation, pathogen attacks, plant diseases, low seed viability, or more recently climate change and uncontrolled use for making furniture goods, etc. These activities have resulted in the sharp decline in the population of many plant species. *Pterocarpus marsupium* Roxb. (Fabaceae) is a lofty, highly medicinal valued, and multipurpose forest tree of India. An aqueous extract of heartwood contains 5,7,2–4 tetrahydroxy 6–6 isoflavone-glucoside which are potent antioxidant and used to cure cardiovascular diseases, vasodilation, and inhibition of platelet aggregation (Mohire et al. 2007). Ethanolic extract of stem bark contains lupeol, tetradecanoic acid, and octadecadienoic acid which are very reputed components known to have cancer preventive and antitumor properties (Maruthupandian and Mohan 2011). One of the biggest problems with this plant is its low percentage of seed germination because of hard fruit coat and poor seed viability in natural environmental conditions. Micropropagation can offer great advantages over traditional method of plant propagation, and through this approach, a single explant can give rise to many plantlets within few months. Plant regeneration through in vitro techniques is potentially utilized in several areas like roadside tree plantation, forestry, and plant-derived pharmaceutical industries.

In the present study, an attempt has been made to improve regeneration efficiency by giving a pretreatment with thidiazuron (TDZ) followed by their transfer to a secondary medium enriched with various dose of meta-Topolin (mT). The mT, a hydroxylated 6-benzyladenine derivative, is a new phytohormone identified in different plants and extracted from poplar leaves (Strnad et al. 1997). It promotes shoot elongation and multiplication in many plant species (Kubalakova and Strnad 1992; Werbrouck et al. 1996). It has the same mode of action to that of other cytokinins such as benzyladenine, kinetin, etc. with small difference in molecular structure and has found to be effective in micropropagation system (Strnad et al. 1997). The intention of our study was to examine the role of TDZ in combination with mT on shoot induction and multiplication of axillary buds from nodal explant for developing a reliable in vitro regeneration protocol for procurement of true-to-type and healthy plantlets with maximum survival rate in natural condition.

8.2 Materials and Methods

8.2.1 Establishment of Aseptic Seedlings

Mature winged fruits of *Pterocarpus marsupium* were collected from forest area of Kharsia, District Raigarh (Chhattisgarh). The healthy seeds were excised from the fruits with the help of eastman stripper cutter and washed under running tap water for 15 min, treated with 1% Bavistin, again washed under running tap water for about 15 min, treated with a laboratory detergent (Labolene, Qualigens, India) 5% (*v/v*) for 10 min followed by 5–6 washes with sterilized distilled water. The treated seeds were imbibed in distilled water for 24 h to remove leachates and surface sterilized with 0.1% (*w/v*) $HgCl_2$ for 4 min under aseptic condition following repeated washes with sterile distilled water in order to remove traces of sterilant and placed on ½ MS (Murashige and Skoog 1962) medium for seed germination. Nodal segment (1.0–1.5 cm) was excised from 4-week-old aseptic seedlings and transferred on sterile culture medium containing various growth regulators (Fig. 8.1a, b).

8.3 Media and Culture Condition

The nutrient medium used in all the experiments consisted of MS salts and vitamins with 3% (w/v) sucrose (Qualigens, India) as a sole carbon source for plant growth in vitro and gelled with 0.75% (w/v) bacteriological grade agar (Qualigens, India) for shoot regeneration. The pH of media was adjusted to 5.8 with 1 N NaOH or HCl prior to autoclaving at 121 °C for 18 min. All culture tubes were incubated at 24 ± 2 °C under 16/8 h (light/dark) cycle with a photosynthetic photon flux density (PPFD) of 50 μmol m^{-2} s^{-1} provided by cool white fluorescent tubes (Philips, Kolkata, India) with 55 ± 5% relative humidity, regularly checked by thermohygrometer (Testo, India Pvt. Ltd.).

8.4 Experimental Design

The experiments were conducted in two sets; in the first set, 4-week-old aseptic nodal segments were treated in half-strength MS liquid medium containing various doses of TDZ (0.5, 2.5, 5.0, 10.0, or 20.0 μM; Fig. 8.1c) for different durations (4, 8, 12, or 16 days; Table 8.1). The exposure of TDZ was conduct in 50 ml liquid

Fig. 8.1 In vitro regeneration of *Pterocarpus marsupium* (Roxb.). (**a**) A 4-week-old aseptic seedling. (**b**) NS explant obtained from 4-week-old aseptic seedlings. (**c**) NS explant treatment in TDZ containing ½ MS liquid medium. (**d**) Shoots induction from NS explant on MS + mT (5.0 μM), after 4 weeks of culture. (**e**) Shoot induction from TDZ-pretreated NS on MS + mT (5.0 μM), after 4 weeks of culture. (**f**) Shoot multiplication from TDZ-pretreated NS on MS + mT (5.0 μM), after 8 weeks of culture. **g**. Pretreatment of isolated shootlet for root induction on ½ MS (liquid) + IBA (150.0 μM), employing a filter paper bridge. (**h**) An in vitro rooted shootlet on ½ MS (semisolid) + IBA (1.5 μM), after 2 weeks of culture transfer. (**i**) A 4-week-old well-rooted plantlet. (**j**) An acclimatized plantlet (2 months old) in Soilrite

media on a rotatory shaker at 120 rpm for different time periods in optimum culture condition. MS medium without TDZ supplementation served as control. In order to determine the optimal exposure concentration and duration, each culture flask was carefully observed with regard to bud breaking. Following the initial pretreatment, the explants pretreated with optimized exposure dose and duration with TDZ were transferred to a secondary medium. In the second set, TDZ-pretreated nodal explants were inoculated on full-strength MS semisolid medium supplemented with various concentrations of mT (0.5, 2.5, 5.0, 7.5, or 10.0 μM; Table 8.2) singly or in

Table 8.1 Effect of various doses of TDZ on axillary shoot bud breaking from 4-weeks old aseptic nodal segments on half-strength MS liquid medium at different durations

Hormone (μM)	Days			
TDZ	4	8	12	16
0.0	–	–	×	×
0.5	–	–	–	×
2.5	–	+	+	–
5.0	+	+	++	+
10.0	++	+++	+	–
20.0	+	–	×	×

Where, '–' no response; '+' responsive; '++' moderately responsive; '+++' best responsive; '×' explant dead

Table 8.2 Effect of various concentration of mT on multiple shoot induction from pretreated nodal segment on MS medium, after 4 weeks of incubation

Cytokinins (μM)				
TDZ- Pretreated	mT	% Response	Mean shoots/explant	Mean shoot length (cm)
–	0.00	00	0.00 ± 0.00^{f}	0.00 ± 0.00^{e}
–	0.5	10	2.24 ± 0.21^{e}	1.83 ± 0.22^{cd}
–	2.5	30	3.81 ± 0.28^{cd}	2.54 ± 0.20^{b}
–	5.0	55	5.65 ± 0.24^{b}	2.92 ± 0.04^{b}
–	7.5	25	3.37 ± 0.25^{d}	2.44 ± 0.35^{bc}
–	10.0	15	1.68 ± 0.21^{e}	1.56 ± 0.16^{d}
10.0	0.5	25	3.44 ± 0.17^{d}	1.61 ± 0.23^{d}
10.0	2.5	50	5.42 ± 0.10^{b}	2.76 ± 0.06^{b}
10.0	5.0	70	7.53 ± 0.21^{a}	3.53 ± 0.03^{a}
10.0	7.5	35	4.38 ± 0.35^{c}	2.37 ± 0.31^{bc}
10.0	10.0	20	3.70 ± 0.24^{cd}	1.49 ± 0.18^{d}

Values represented means ± SE. Means followed by the same letter within column are not significantly different ($P = 0.05$) using Duncan's multiple range test

combination with IAA or NAA at various concentrations (0.5, 1.0, or 1.5 μM; Table 8.3) for further proliferation and elongation of microshoots. Cultures were subcultured onto the same fresh media after every 3 weeks. Data on percentage of shoot regeneration, number of shoots per explant, and average shoot length were recorded after 4 weeks of subculture.

8.5 In Vitro Rooting

Healthy, well-elongated in vitro raised microshoots of about 4–5 cm in length with three to five fully expanded leaves were isolated from 8-week-old cultures, and rooting was achieved following two-step procedure previously established by Anis et al. (2005) in *P. marsupium* with few modifications. In the first step, the basal end

Table 8.3 Combined effect of various concentrations of IAA or NAA with mT (5.0 μM) for shoot proliferation from pretreated nodal segment on MS medium, after 8 weeks

Plant growth regulators (μM)					Mean shoot
TDZ- Pretreated	IAA	NAA	% Response	Mean shoots/explant	length (cm)
–	0.50		55	6.04 ± 0.21^{f}	3.63 ± 0.14^{de}
–	1.00		60	6.43 ± 0.23^{f}	3.76 ± 0.10^{de}
–	1.50		50	5.85 ± 0.21^{f}	3.55 ± 0.17^{e}
–		0.50	65	7.54 ± 0.16^{e}	3.79 ± 0.08^{de}
–		1.00	70	8.65 ± 0.14^{c}	3.92 ± 0.12^{cd}
–		1.50	60	6.24 ± 0.15^{f}	3.65 ± 0.09^{de}
10.0	0.50		70	7.95 ± 0.04^{de}	3.71 ± 0.11^{de}
10.0	1.00		80	8.80 ± 0.07^{bc}	3.84 ± 0.12^{cde}
10.0	1.50		65	7.74 ± 0.12^{e}	3.58 ± 0.10^{de}
10.0		0.50	80	9.26 ± 0.41^{b}	4.35 ± 0.04^{b}
10.0		1.00	85	11.16 ± 0.16^{a}	4.96 ± 0.13^{a}
10.0		1.50	75	8.34 ± 0.10^{cd}	4.15 ± 0.12^{bc}

Values represented means ± SE. Means followed by the same letter with in column are not significantly different ($P = 0.05$) using Duncan's multiple test

Table 8.4 Effect of higher dose (150.0 μM) of IBA pretreated microshoots followed by transfer to various low doses of IBA in half-strength MS semisolid medium for roots induction, after 4 weeks of transfer

IBA concentrations (μM)	% Response	Mean roots/microshoot	Mean root length (cm)
Control	00	0.00 ± 0.00^{e}	0.00 ± 0.00^{d}
0.0	30	2.36 ± 0.27^{d}	1.64 ± 0.19^{c}
0.5	45	3.83 ± 0.20^{c}	2.85 ± 0.38^{b}
1.0	60	4.44 ± 0.11^{b}	3.27 ± 0.14^{b}
1.5	80	6.32 ± 0.26^{a}	4.05 ± 0.27^{a}
2.0	55	3.57 ± 0.16^{c}	2.55 ± 0.25^{b}

Values represented means ± SE. Means followed by the same letter within column are not significantly different ($P = 0.05$) using Duncan's multiple test

of isolated microshoots was pretreated with high dose of IBA (50, 100, 150, or 200 μM; Fig. 8.1g) in half-strength MS liquid medium with 2% sucrose for 1 week using a filter paper bridge. Thereafter, these microshoots were transferred onto ½ MS semisolid medium supplemented with different low doses of IBA (0.0, 0.5, 1.0, 1.5, or 2.0 μM; Table 8.4) with 0.25% Phytagel and 2% sucrose. The data on percentage of rooting, mean root number, and root length were recorded, after 4 weeks of culture transfer.

8.6 Hardening and Acclimatization

Microshoots with well-developed roots were removed from the culture tube (Fig.8.1i), rinsed with running tap water to remove any adherent agar, and transferred to 10-cm-diameter thermocol cups containing sterile Soilrite™ (Keltech

Energies, Ltd., Bangalore, India). The cup was covered with transparent polythene bags as a safeguard for high humidity and irrigated with ½ MS (liquid) without organic salts and sucrose, after every 2 days. The polythene bags were opened, after a couple of week in order to harden the in vitro raised plantlets. These acclimatized plants were transferred to pots containing normal garden soil and kept at 25 ± 5 °C in incubation chamber for 64 days. Afterward, these potted plantlets were shifted to greenhouse.

8.7 Statistical Analysis

All experiments were based on a completely randomized block design (RBD) and repeated three times with ten replicates for each treatment. The data on various parameters were subjected to one-way analysis of variance (ANOVA) using SPSS version 16 (SPSS Inc., Chicago, USA). The significance of differences among means was carried out using Duncan's multiple range test at $P = 0.05$, and results were expressed as the mean ± SE of three repeated experiments.

8.8 Results

8.8.1 In Vitro Shoot Regeneration

The nodal explant did not show any response of bud break when cultured on a phytohormone-free medium and failed to induce any shoots even after 4 weeks of culture. Explants cultured on half-strength MS liquid medium with various doses of TDZ in a rotatory shaker for different days showed a discernible response on axillary shoot bud breaking. Of the five doses of TDZ tested, 10.0 μM for 8 days was found to be more effective than other treatments in respect to multiple shoot bud induction, and about 96% (data not shown) explants showed maximum response (Table 8.1). Further increase in the dose beyond the optimal level did not improve any bud breaking. When the TDZ-pretreated nodal explant was transferred to full-strength MS media supplemented with various concentrations of mT, a positive effect on shoot differentiation and elongation was recorded. Of the five treatments of mT used, 5.0 μM was found to be more effective where maximum number of shoots (7.53 ± 0.21) with mean shoot length (3.53 ± 0.03 cm) and maximum regeneration frequency (70%) were obtained, whereas in without pretreated explant, about 5.65 ± 0.24 shoots per explant with an average shoot length (2.92 ± 0.04 cm) and 55% regeneration frequency were recorded, after 4 weeks of culture transfer (Table 8.2; Fig. 8.1e, d). Basal callusing was removed during every subculturing as a precautious measure because it retarded shoot multiplication and elongation. Furthermore, we analyzed the combined effect of auxins (IAA or NAA) at various concentrations with the optimal concentration of mT (5.0 μM) for enhancing the shoot proliferation rate. Among all treatments tried, MS medium supplemented with mT (5.0 μM) + NAA (1.0 μM) showed highest number of shoots (11.16 ± 0.16) with mean shoot length (4.96 ± 0.13 cm) per nodal explant and maximum regeneration frequency (85%), after 8 weeks of culture transfer (Table 8.3; Fig. 8.1f). Average number of shoots per explant with shoot length got increased with an increased concentration up to the optimal dose, whereas gradual decrease in shoot number with shoot length per explant was observed beyond the optimal dose (Table 8.3). The results further confirm that the ratio of auxin/cytokinin and their combination play an important role in plant tissue culture system.

8.8.2 In Vitro Root Induction

In vitro rooting in the regenerated microhoots was a challenging step as none of the isolated microhoots rooted on full MS or its different reducing strengths (1/2 MS, 1/3 MS, 1/4 MS) with or without various concentrations of auxins (IAA, IBA, or NAA, data not shown) added to the medium. A two-step procedure for better induction of root was successfully used. In the first step, the basal end of isolated microshoots was treated with high dose of IBA (150.0 μM; Fig. 8.1g) for 1 week on a filter paper bridge followed by their transfer to the next stage. In the second step, the treated microhoots were transferred on a semisolid medium containing low dose of IBA (1.5 μM) where highest frequency (80%) of root formation and highest number of roots (6.32 ± 0.26) with an average shoot length (4.05 ± 0.27 cm) per microshoot were recorded, after 4 weeks of culture transfer. The roots were healthy and showed numerous secondary root hairs (Fig.8.1i), which helped in establishing plantlets in soils. Therefore, IBA (150.0 μM) pretreated microshoots followed by transfer to lower dose of IBA (1.5 μM) was found to be best as compared to other concentrations used (Table 8.4).

8.8.3 Acclimatization of In Vitro Raised Plantlets

In vitro raised plantlets with fully expanded and healthy leaves were removed from culture tube and transferred to thermocol cups containing sterile Soilrite™ for 4 weeks (Fig. 8.1j). Acclimatization and hardening procedure of in vitro regenerated plantlets was the same as described in materials and methods. The 75% of regenerated plantlets got survived during acclimatization and successfully transferred to pots containing normal garden soil. The plantlets grew well and did not show any morphological variation in growth parameters.

8.9 Discussion

A successful and efficient protocol for axillary shoot multiplication from nodal segment of *P. marsupium* has been established. The nodal segments are a good source of axillary shoot production in tissue culture system, and even a single explant can give rise to multiple copies of true-to-type plantlets within few months. The nodal explants were first given a high dose of TDZ followed by their transfer to MS medium supplemented with meta-Topolin, a new cytokinin for multiple shoot induction.

Thidiazuron (TDZ—N-phenyl- N′-1,2,3-thiadiazol-5-ylurea) is a synthetic phenylurea having potent cytokinin-like activity and widely used as a phytohormone in in vitro regeneration protocols. It is more effective as compared to other cytokinins for inducing multiple de novo shoots from different explants of several plant species such as *Paulownia* species (Chalupa 1988), many woody trees (Huetteman and Preece 1993), *Pterocarpus marsupium* (Husain et al. 2007), *Vitex negundo* (Ahmad and Anis 2007), *Embelia ribes* (Dhavala and Rathore 2010), and *Dendrocalamus strictus* (Singh et al. 2013). According to Murthy et al. (1995), TDZ modifies the endogenous level of phytohormones especially cytokinins and auxins in plant tissue culture system and affects the metabolism of cytokinins biosynthetic pathways which is responsible for the degradation of endogenous level of elevated concentrations of purine metabolites (Zhang et al. 2005). There are many reports available on TDZ-supplemented media which showed positive response on multiple shoot bud break but failed to shoot elongation in many plants including in *Rhododendron* (Preece and Imel 1991), *Adhatoda beddomei* (Sudha and Seeni 1994), *Dalbergia sissoo* (Pradhan et al. 1998), *Rauvolfia tetraphylla* (Faisal et al. 2005), *Capsicum annuum* (Ahmad et al. 2006), and *Psoralea corylifolia* (Faisal and Anis 2006). But in some cases the use of these highly active cytokinin (Ck) in plant regeneration protocol was limited (Huetteman and Preece 1993). The antagonistic effect of this Ck may be due to the presence of phenyl group which on over exposure beyond the optimal level can lead to many drawbacks such as bunching, fasciation, hyperhydricity, poor microshoot quality, and loss of rooting capacity (Singh and Syamal 2001; Lu 1993).

However, the twofold culture strategy was applied to address the problem of shoot elongation. Primarily, we have evaluated the effect of TDZ on nodal explant for axillary buds breaking. Half-strength MS liquid media supplemented with TDZ (10.0 μM) for 8 days were found to be more responsive than other doses tried. When pretreated nodal segments were transferred on a secondary medium containing mT, a positive change in enhancing shoot multiplication and elongation was observed. Some positive results of mT have been reported in several plant species (Werbrouck et al. 1996; Wojtania 2010; Aremu et al. 2012; Mala et al. 2013; Clapa et al. 2014). It has the same mode of action as BA and kinetin, but due to little difference in molecular structure, it was found to be effective on micropropagation in many plant species (Kubalakova and Strnad 1992; Werbrouck et al. 1996; Strnad et al. 1997). Among the different concentrations of mT tested, the efficient shoot elongation was recorded at 5.0 μM mT where maximum shoot numbers (7.53 ± 0.21) with an average shoot length (3.53 ± 0.03 cm) were recorded in pretreated explants, after 4 weeks of culture transfer. Similar type of culture strategy using shoot bud breaking (primary) and shoot elongation (secondary) medium was effectively applied in many plant species, such as *Malus alba* (Thomas 2003), *Acacia sinuata* (Vengadesan et al. 2002), and *Vitex negundo* (Ahmad and Anis 2007).

Furthermore, we have evaluated the combined effect of cytokinins with auxin for improving the microshoots quality or number for obtaining healthy plantlets of *P. marsupium*. TDZ-pretreated nodal explants when transferred to MS medium augmented with mT (5.0 μM) plus NAA (1.0 μM) in combination were found to be best where highest proliferation rate (11.16 ± 16 shoots) with an average shoot length (4.96 ± 0.13 cm) per explant was achieved, after 8 weeks of culture (Table 8.3; Fig. 8.1f). Our results are in agreement with earlier findings of several workers in many recalcitrant or woody species like *A. catechu* (Kaur et al. 1998), *Eucalyptus grandis* (Luis et al. 1999), *Lagerstroemia parviflora* (Tiwari et al. 2002), *Melia azedarach* (Husain and Anis 2009), *Balanites aegyptiaca* (Siddique and Anis 2009), and *Acacia ehrenbergiana* (Javed et al. 2013). This is an improved and cost-effective protocol than the earlier described by Husain et al. (2008) in *P. marsupium* where only 8.6 shoots per explant with 4.8 cm shoot length were obtained. By repeated subculturing onto the fresh medium, successful healthy shoot cultures were established.

In vitro rooting has been a difficult process in woody trees. Indole-3-butyric acid is known to have stimulatory effect on the multiple root formation due to its preferential uptake, transport, and stability as compared to other auxins and may play an important role in rooting gene activation (Ludwing-Muller 2000). Husain et al. (2008) reported the root induction in BA-regenerated microshoots where a maximum root (3.8) with average length (3.9 cm) was recorded. In several other reports, exogenous application of cytokinins like BA used in micropropagation has negative effect on rooting, resulting in poor acclimatization of many plant species (Werbrouck et al. 1995; Bairu et al. 2008; Amoo et al. 2011). These negative effects are due to the formation of biologically inactive and chemically stable metabolites such as N-glucosides or alanine conjugates (Werbrouck et al. 1995). BA-treated in vitro regenerated shootlets have the propensity of accumulation of BA metabolites as a toxic substance in a basal rooting zone, and these toxic metabolites interfere rooting and acclimatization of in vitro raised plantlets (Werbrouck et al. 1995; Valero-Aracama et al. 2010). But in our study, we have tried mT in place of other CKs for regeneration of multiple and healthy microshoots. The mT-treated regenerated shootlets gave highest positive response in respect to root number as well as root length with 75% of successful acclimatization. According to Werbrouck et al. (1996), mT produce relatively less stable and low toxic metabolites as compared to other CKs. Bairu et al. (2011) postulated that mT-derived shootlets produce more roots with respect to BA because of small differences in hydroxyl groups of meta-Topolin. Thus an improved regeneration protocol has been established using a new cytokinin, meta-Topolin.

8.10 Conclusion

The research finding emphasizes the role of TDZ and mT on in vitro regeneration of a woody legume for ensuring better transplant success due to proper continuity of conducting tissues of shoot and root in the absence of callusing. The thidiazuron-pretreated nodal explants, when transferred onto mT-supplemented medium, showed positive results on in vitro regeneration such as shoot multiplication, shoot elongation, rooting and acclimatization, and improving physiological parameters. The rapid in vitro propagation using TDZ and mT will not only help in conservation but also low-cost multiplication of this tree species. This research work might be an initial step in the breakthrough of in vitro propagation revolution which is yet to firmly establish its roots in plant biotechnology field.

Acknowledgment Anees Ahmad is thankful to the Department of Biotechnology, Govt. of India, New Delhi, for financial support as a Junior Research Fellow under the Major Research Project (Vide No. BT/PR2189/PBD/17/744/2011 Dated 19.12.2012). The award of UGC-BSR Faculty Fellowship (2017) to M. Anis is duly acknowledged.

References

Ahmad N, Anis M (2007) High frequency axillary shoot multiplication from mature tree of *Vitex negundo* L. Agrofor Syst 71:195–200

Ahmad N, Siddique I, Anis M (2006) Improved plant regeneration in *Capsicum annuum* L. from nodal segments. Biol Plant 50(4):701–704

Akshay KR, Sudharani N, Anjali KB, Deepak TM (2014) Biodiversity and strategies for conservation of rare, endangered and threatened medicinal plants. Research and Reviews: J Pharmacogn Phytochem 2:12–20

Amoo S, Finnie JF, Van Staden J (2011) The role of meta-topolins in alleviating micropropagation problems. Plant Growth Regul 63:197–206

Anis M, Husain MK, Shahzad A (2005) In vitro plantlet regeneration of *Pterocarpus marsupium* Roxb. an endangered leguminous tree. Curr Sci 88:861–863

Aremu AO, Bairu MW, Dolezal K, Finnie JF, Van Staden J (2012) Assessment of the role of meta-topolins on in vitro produced phenolics and acclimatization competence of micropropagated 'Williams' banana. Acta Physiol Plant 34:2265–2273

Bairu MW, Stirk WA, Dolezal K, Van Staden J (2008) The role of topolins in micropropagation and somaclonal variation of banana cultivars 'Williams' and 'Grand Naine' (Musa spp. AAA). Plant Cell Tissue Organ Cult 95:373–379

Bairu MW, Novak O, Dolezal K, Van Staden J (2011) Changes in endogenous cytokinins profiles in micropropagated *Harpagophytum procumbens* in relation to shoot-tip necrosis and cytokinins treatments. Plant Growth Regul 63:105–114

Chalupa V (1988) Large scale micropropagation of *Quercus robur* L. using adenine type cytokinins and thidiazuron to stimulate shoot proliferation. Biol Plant 30:414–421

Clapa D, Fira A, Simu M, Vasu LB, Buduroi D (2014) Improved *in vitro* propagation of *Paulownia elongata*, *P. fortunei* and its interspecific hybrid *P. elongata* x *P. fortunei*. Bull UASVM Hortic 71(1):6–14

Dhavala A, Rathore TS (2010) Micropropagation of *Embelia ribes* Burm f. through proliferation of adult plant axillary shoots. In Vitro Cell Dev Biol Plant 46:180–191

Faisal M, Anis M (2006) Thidiazuron induced high frequency axillary shoot multiplication in *Psoralea corylifolia*. Biol Plant 50:437–440

Faisal M, Ahmad N, Anis M (2005) Shoot multiplication in *Rauvolfia tetraphylla* L. using thidiazuron. Plant Cell Tissue Organ Cult 80:187–190

Huetteman CA, Preece JE (1993) Thidiazuron: a potent cytokinin for woody plant tissue culture. Plant Cell Tissue Organ Cult 33:105–119

Husain MK, Anis M (2009) Rapid in vitro multiplication of *Melia azedarach* L. (a multipurpose woody tree). Acta Physiol Plant 31:765–772

Husain MK, Anis M, Shahzad A (2007) In vitro propagation of Indian Kino (*Pterocarpus marsupium* Roxb.) using thidiazuron. In Vitro Cell Dev Biol Plant 43:59–64

Husain MK, Anis M, Shahzad A (2008) *In vitro* propagation of a multipurpose leguminous tree (*Pterocarpus marsupium* Roxb.) using nodal explants. Acta Physiol Plant 30:353–359

Javed SB, Anis M, Khan PR, Aref M (2013) In vitro regeneration and multiplication for mass propagation of *Acacia ehrenbergiana* Hayne: a potential reclaiment of denude arid lands. Agrofor Syst 87:621–629

Kaur K, Verma B, Kant U (1998) Plants obtained from the Khair tree (*Acacia catechu* wild) using mature nodal segment. Plant Cell Rep 17:427–429

Kubalakova M, Strnad M (1992) The effect of aromatic cytokinins (populins) on micropropagation and regeneration in vitro. Biol Plant 34:578–579

Lu CY (1993) The use of thidiazuron in tissue culture. In Vitro Cell Dev Biol Plant 29:92–96

Ludwing-Muller J (2000) Indole-3-Butyric acid in plant growth and development. Plant Growth Regul 32:219–230

Luis PBC, Adraine CMGM, Silvica BRCC, Anna Christina MB (1999) Plant regeneration from seedling explants of *Eucalyptus grandis* x *E. urophylla*. Plant Cell Tissue Organ Cult 56:17–23

Mala J, Machova P, Cvrckova H, Karady M, Novak O, Mikulık J, Dostal J, Strnad M, Dolezal K (2013) The role of cytokinins during micropropagation of wych elm. Biol Plant 57:174–178

Maruthupandian A, Mohan VR (2011) GC-MS analysis of some bioactive constituents of *Pterocarpus marsupium* Roxb. Int J Chem Tech Res 3(3):1652–1657

Mohire NC, Salunkhe VR, Bhise SB, Yadav AV (2007) Cardiotonic activity of aqueous extract of heartwood of *Pterocarpus marsupium*. Indian J Exp Biol 45:532–537

Murashige T, Skoog F (1962) A revised medium for rapid growth and bioassays with tobacco tissue cultures. Physiol Plant 15:473–497

Murthy BNS, Murch SJ, Saxena PK (1995) Thidiazuron- induced somatic embryogenesis in intact seedlings of peanut (*Arachis hypogaea*): endogenous growth regulator levels and significance of cotyledon. Physiol Plant 94:268–276

Pradhan C, Kar S, Pattnaik S, Chand PK (1998) Propagation of *Dalbergia sissoo* Roxb. through in vitro shoot proliferation from cotyledonary nodes. Plant Cell Rep 18:122–126

Preece JE, Imel MR (1991) Plant regeneration from leaf explant of *Rhododendron* P J M hybrids. Sci Hortic 48:159–170

Siddique I, Anis M (2009) Direct plant regeneration from nodal explant of *Balanites aegyptiaca* L. (Del.)- valuable medicinal tree. New For 37:53–62

Singh SK, Syamal MM (2001) A short pre-culture soak in Thidiazuron or Forchlorfenuron improves axillary shoot proliferation in rose micropropagation. Sci Hortic 91:169–177

Singh M, Jaiswal VS, Jaiswal U (2013) Thidiazuron-induced anatomical changes and direct shoot morphogenesis in *Dendrocalamus strictus* Nees. Scro Res Ann Rep 1:44–47

Strnad M, Hanus J, Vanek T, Kaminek M, Ballantine JA, Fussell B, Hanke DE (1997) Meta-Topolin, a highly active aromatic cytokinin from poplar leaves (*Populus x canadensis* Moench, cv Robusta). Phytochemistry 45:213–218

Sudha GC, Seeni S (1994) In vitro multiplication and field establishment of *Adhatoda beddomei* CB Clark, rare medicinal plant. Plant Cell Rep 13:203–207

Thomas TD (2003) Thidiazuron induced multiple shoot induction and plant regeneration from cotyledonary explants of mulberry. Biol Plant 46:529–533

Tiwari SK, Kashyap MK, Ujjaini MM, Agrawal AP (2002) In vitro propagation of *Lagerstroemia parviflora* Roxb. from adult tree. Indian J Exp Biol 40:212–215

Valero-Aracama C, Kane M, Wilson S, Philman N (2010) Substitution of benzyladenine with meta-topolin during shoot multiplication increases acclimatization of difficult- and easy– to acclimatize sea oats (*Uniola paniculata* L.) genotypes. Plant Growth Regul 60:43–49

Vengadesan G, Ganapathi A, Anand PR, Anbazhagan RV (2002) In vitro propagation of *Acacia sinuata* (Lour.) Merr. via cotyledonary node. Agrofor Syst 55:9–15

Werbrouck SPO, Van der Jeugt B, Dewitte W, Prinsen E, Van Onckelen HA, Debergh PC (1995) The metabolism of benzyladenine in *Spathiphyllum floribundum* 'Schott Petite' in relation to acclimatisation problems. Plant Cell Rep 14:662–665

Werbrouck SPO, Strnad M, Van Onckelen HA, Debergh PC (1996) Meta-topolin, an alternative to benzyladenine in tissue culture. Physiol Plant 98:291–297

Wojtania A (2010) Effect of meta-topolin on in vitro propagation of *Pelargonium* x *hortorum* and *Pelargonium* x *hederaefolium* cultivars. Acta Soc Bot Pol 79:101–106

Zhang CG, Li W, Mao YF, Zhao DL, Dong W, Guo GQ (2005) Endogenous hormonal levels in *Scutellaria baicalensis* calli induced by thidiazuron. Russ J Plant Physiol 52(3):345–351

Application of Thidiazuron in the Micropropagation of Fagaceae

9

Ma del Carmen San José, Ma Teresa Martínez, Ma José Cernadas, Raquel Montenegro, and Elena Corredoira

Abstract

The Fagaceae family consists of 7 genera and around 1000 species of trees and bushes that are mainly distributed in temperate and warm areas of the northern hemisphere, although few cross the equator in Southeast Asia. In terms of forestry, members of the Fagaceae are of most importance in forests in the temperate regions of the northern hemisphere, a dominance shared with the conifers that replace this family in cold areas and mountain tops. The genera *Quercus* (oaks and holm oaks), *Fagus* (beeches), and *Castanea* (chestnut) are commercially important sources of timber; *Castanea* and *Quercus* (holm oaks) also provide fruits that are used as human food and as animal feed. Many of these trees are also of ornamental value, mainly due to their attractive color of their leaves in autumn.

The majority of these species are difficult to propagate, particularly when the trees reach their adult stage. Biotechnology techniques, such as in vitro tissue culture, would therefore be of great use for their propagation and conservation. These techniques involve the use of growth regulators, especially cytokinins, among which is included thidiazuron (TDZ). This cytokinin has been used to stimulate the development of axillary buds and, mainly, for the induction of adventitious buds and in very few cases in somatic embryogenesis processes. This review presents a summary of the various studies in which TDZ has been used in the micropropagation of diverse species of the family Fagaceae.

Keywords

Castanea · Fagaceae · Fagus · Micropropagation · Quercus · Thidiazuron

---

M. del Carmen San José (✉) · M. Teresa Martínez · M. José Cernadas
R. Montenegro · E. Corredoira
Instituto de Investigaciones Agrobiológicas de Galicia, CSIC,
Santiago de Compostela, Spain
e-mail: sanjose@iiag.csic.es

N. Ahmad, M. Faisal (eds.), *Thidiazuron: From Urea Derivative to Plant Growth Regulator*, https://doi.org/10.1007/978-981-10-8004-3_9

9.1 Fagaceae Family

The Fagaceae are a large family of angiosperms with species that are spread throughout the northern hemisphere, from tropical areas to northern areas, with some species that cross the Equator in Southeast Asia. Beeches (*Fagus*), oaks (*Quercus*), and chestnuts (*Castanea*) are the only genera distributed in Asia, Europe, and North America, where they cover, or used to cover, large forest areas. Evergreen oaks are important members of the forests around the Gulf of Mexico, as well as in southern China and southern Japan. In Southeast Asia, the structure of the mixed mountain forest is largely determined by evergreen members of the family, particularly oaks. In total, therefore, the Fagaceae produce a colossal biomass, possibly exceeded only by the conifers. They also have a wide diversity of uses (biomass, fiber, wood and food products). As well as their economic contribution, they are important specimens in forestry ecosystems and are major drivers of terrestrial biodiversity. In the majority of countries where they are found, they are considered as major patrimonial and cultural resources (Kremer et al. 2012).

The members of this family are deciduous or evergreen trees, and rarely shrubs, with alternate simple, entire to pinnately lobed leaves, and scarious, usually deciduous stipules. The flowers are unisexual (plant monoecious) and usually arranged in catkins or small spikes that may comprise only flowers of one sex, as in oaks, or may have female flowers at the base of otherwise male inflorescences, such as in chestnuts. The perianth is bract-like, with four to seven lobes. The male flowers have as many or twice as many stamens as perianth segments, occasionally up to 40, with the filaments free, with or without a pistillode. The female flowers are in groups of one to three, each group being surrounded by a basal involucre. They appear at maturity. The fruits of the Fagaceae are animal dispersed and have a short viability. The pollen and other features, such as a strongly scented inflorescence, suggest that insect pollination is the ancestral condition in the Fagaceae, and this is retained in most members, except *Fagus* and the temperate oak species (Heywood et al. 2007).

The cupule has a wide variety of forms that surround one or more fruits, which is a unique feature of the family and the origin of which has been controversial (Fey and Endress 1983). Only with the discovery of *Trigonobalanus* in 1961, which is restricted to Sulawesi, northern Borneo, Malaysia, and northern Thailand, has it been possible to suggest firmly that the cupule is derived from a three-lobed extension of the pedicel below each flower, which has been variously fused around single flowers or groups of flowers. It is possible that the cupule provides a link with the pteridosperm ancestors of the flowering plants. The tremendous diversity of scales and spines on the cupules appears to be derived from branches spines (Nixon and Crepet 1989).

The Fagaceae provide some of the most important woods worldwide, especially the oaks (particularly the white oak of North America), beech, and chestnut varieties. Together with clearance for agriculture, this has resulted in the destruction of large areas of forests dominated by these species. *Castanopsis* and *Lithocarpus* have been little exploited, although their wood is also of a high quality. Overall, the woods obtained from the species of this family have a wide range of properties and uses, from floorboards and furniture to carbon and whisky barrels. The cork is obtained from the bark of the cork oak (*Quercus suber*), and the oak galls were an important source of tannin in Asia Minor and southeastern Europe. Many of the chestnut species, especially *Castanea sativa*, are cultivated for their edible fruits, from which purees, stuffings, and stews can be made, as well as the French delicacy *marrons glacés*. The fruits of the beech are rich in oil (46%) and are used in many regions to feed pig stock, as well as the acorns from the oaks. The acorns produced by *Q. ilex* are used for feeding the so-called Iberian pigs, the meat of which contributes to the development of a high-quality food industry (Cañellas et al. 2007). The form and the colors of many of these species in autumn, particularly the oaks, chestnuts, and beeches, make them used as ornamental in parks and gardens. Only American species of *Castanopsis* (*C. chrysophylla*) and *Lithocarpus* (*L. densiflorus*) may be cultivated in warmer regions (Heywood et al. 2007). Besides their economic importance, the Fagaceae species have a high ecological value, as illustrated by the fact that chestnut blight disease, caused by a pathogenic fungus, caused the worst ecological disaster in the history of the United States (Wheeler and Sedroff 2009).

Of the seven genera that make up the Fagaceae family (*Fagus*, *Chrysolepis*, *Castanea*, *Castanopsis*, *Lithocarpus*, *Quercus*, and *Trigonobalanus*), three of them are specially highlighted for their economic value: *Castanea*, *Fagus*, and *Quercus*. These three genera are the most widely micropropagated, and on which we are going to focus this review.

9.2 Thidiazuron

The cytokinin activity of TDZ was first reported in 1982 and, since then, has been gaining acceptance due to its efficient role in the cultivation of plant cells and tissue. TDZ has been successfully used for the induction of adventitious buds, to promote the proliferation of axillary buds, and somatic embryogenesis being especially effective in the case of recalcitrant woody species (Huetteman and Preece 1993; Lu 1993; Murthy et al. 1998; Guo et al. 2011).

There are several derivatives of urea with cytokinin capacity, with the most active compounds being N-phenyl-N′-1,2,3-thidiazol-5-ylurea or thidiazuron (TDZ), N-N′-diphenylurea (DPU), and N-(2-chloro-4-pyridyl)-N′-phenylurea (CPPU). TDZ is one of the most potent diphenylureas evaluated in plant tissue culture (Mok et al. 1982). It was initially developed by the Schering AG group as a cotton defoliant (Arndt et al. 1976) and has a high activity at concentrations as low as 10 pM. The exposure of plant tissue to TDZ for a short period of time is sufficient to stimulate regeneration (Preece et al. 1991; Visser et al. 1992; Li et al. 2000; Matand and Prakash 2007).

The structure of TDZ is nothing like the natural cytokinins, and it appears that it is directly responsible for the morphogenic responses induced by this regulator. Any change in the molecules that affects its functional groups (phenyl or thiadiazole) reduces its activity (Mok et al. 1982; Mok and Mok 1985).

Different physiological and biochemical processes are induced or increased in the cells due to the action of TDZ, but its exact mode of action is still not known (Guo et al. 2011). Some examples of the diversity of the physiological effects mediated by TDZ include efficient seed germination, expedited bud break, induction and stimulation of sprouting, cotyledonary growth and development, formation of trichomes and stomata appearance on floral parts, and berry weight of grapes. More recently, the morpho-regulatory potential of this compound has led to its application in the cultivation of plant cells, tissues, and organs, with the aim of improving regeneration protocols (Guo et al. 2011). Several studies cite the use of TDZ as a unique growth regulator to induce regeneration in different species, meeting both the auxin and cytokinin requirements (Visser et al. 1992; Murthy et al. 1996; Murthy and Saxena 1998). However, TDZ is currently considered as a cytokinin due to induction of natural cytokinin-like responses. In some cultivation systems, it has replaced the habitual cytokinins, provoking similar responses. This substitution is confirmed with the suppression of the effect of the TDZ provoked by the inhibitors of the metabolism of cytokinins with purine ring, which suggests that there is a common site of action for both types of regulators (Nagata et al. 1993; Hutchinson and Saxena 1996). Its role in morphogenesis seems be closely related to the metabolism of the endogenous growth regulators, especially cytokinins, auxins, abscisic acid, and ethylene (Yip and Yang 1986; Murthy et al. 1995, 1998).

Besides its cytokinin activity, TDZ is capable of inducing somatic embryogenesis in some species, which leads to thinking that it might have some auxin activity or it is associated with auxin metabolism (Saxena et al. 1992; Lu 1993). Murthy et al. (1995) established that TDZ can stimulate auxin synthesis. Other authors indicate the possibility that TDZ could affect the action of certain enzyme systems, such as those associated with cell walls and membranes, or the cytokinin oxidase system, in which the activity seems to be inhibited with TDZ (Wang et al. 1991; Hare et al. 1994).

Other theory is that TDZ may produce a stress situation in plant tissue, for which they respond by developing a survival mechanism in order to perpetuate the plant by means of the increase in morphogenic response (Murch et al. 1997).

TDZ has an effect on a large number of species, even in those that show little or no response to conventional cytokinins. This makes it especially effective in woody species of difficult regenerative capacity in which organogenesis is only achieved with very high concentrations of adenine-type cytokinin (Fellman et al. 1987; Preece et al. 1991; Baker and Bhatia 1993). The reason for this may be found in its greater stability, since it is more resistant to enzymatic action than the endogenous cytokinins and is degraded less during the culture medium autoclaving process (Huetteman and Preece 1993).

TDZ has been employed for the multiplication of various woody species, owing to its capacity to induce the proliferation of axillary shoots (Wojtania et al. 2011; Sedlák and Paprštein 2015; Castillo et al. 2015). In the induction of caulogenesis, it has been shown to promote the differentiation of meristemoids at concentrations much lower than other cytokinins, and the regeneration also occurs with a comparable, or even greater, efficiency (San José et al. 2014; Liu et al. 2016). Its greater organogenic capacity is also shown by producing a higher number of shoots per explant than benzyladenine (BA), kinetin (KIN), or zeatin.

TDZ also promotes callus formation, in many cases more intensely than other regulators, particularly at concentrations greater than0.02 mg/l. This callus is, in many cases, an essential step for the regeneration of adventitious buds (Huetteman and Preece 1993; Murthy et al. 1998; Traore et al. 2003). More recently, TDZ alone, or in combination with other regulators, has been used as a somatic embryogenesis stimulant in tissue culture media (Nhut et al. 2006; Jones et al. 2007; Kahia et al. 2016; Rugini and Silvestri 2016).

Prolonged exposure to TDZ can lead to problems like hyperhydricity, leaves with abnormal morphology, short and compact shoots, as well as difficulty for the elongation and rooting of the regenerated shoots (Huetteman and Preece 1993; Lu 1993; Ahmad and Anis 2007).

9.3 Biotechnological Approaches in the Propagation of Fagaceae Using TDZ

The majority of members of the Fagaceae are included in the group of long-rotation hardwoods for which no long-term improvement programs have been carried out. However, these species play a vital role in the conservation of the soil and water and make a significant contribution to the sustainability and viability of the ecosystems. The low natural regeneration using seeds, their irregular fructification, the consumption of the seeds by animals, and the loss of viability during storage exacerbates the regeneration problems suffered by most of the species of this family. Furthermore, there are great difficulties in their vegetative propagation when adult material is used.

Several reviews have described the possibilities offered by forest biotechnology as an emerging opportunity in relation to tree improvement (Vieitez and Merkle 2005; Vieitez et al. 2012; Nelson et al. 2014; Ballester et al. 2016; Corredoira et al. 2016; Monteuuis 2016). In vitro culture techniques have been applied to different woody species since they provide the appropriate tools for the rapid production of genotypes with desired traits and to capture all the genetic superiority without involving any gene segregation. These techniques may alleviate, at least in theory, the lack of seed production, as well as the difficulties in the production of offspring with desired traits, storage of the seeds, and the low rooting capacity of the cuttings, helping to rapidly increase the number of individuals in a species with reproduction problems and/or in extremely reduced populations (Vieitez et al. 2012). The plant material obtained can be of great value for research, living collections, to reduce pressure on natural populations and, where possible, for reintroduction programs (González-Benito and Martín 2011).

9.4 Genus *Castanea*

Castanea is a small genus native to the temperate zones of Asia, Europe, and Eastern United States (Camus 1929). The precise number of species is uncertain due to the use of synonyms and the lack of accurate characterization for some species of chinkapins. The most representative species and of greater economic importance within this genus are *C. crenata* Sieb. and Zucc. (Japanese chestnut), *C. dentate* (Marshall) Borkh. (American chestnut), *C. mollissima* Blume (Chinese chestnut), and *C. sativa* Mill. (European chestnut). The other species are small trees or bushes of importance only for breeding, as rootstocks, or for special uses. The chestnuts are long-living trees and can reach 500–1000 years. However, like the majority of large seeded hardwoods, they are very difficult to propagate vegetatively. These trees are also threatened by pollution, economic and social changes, and two major diseases: ink disease caused by *Phytophthora cambivora* and *P. cinnamomi* and chestnut blight caused by *Cryphonectria parasitica*, which have destroyed a large number of trees in Europe and North America (Nelson et al. 2014).

9.5 TDZ and Proliferation of Axillary Buds

Micropropagation by means of axillary bud proliferation is the preferred method for the commercial propagation of woody species, since it is usually easier to do and is also considered more appropriate in order to maintain genetic stability of the regenerated plants (Bonga and von Aderkas 1992; Monteuuis et al. 2008; Gomes and Canhoto 2009).

Although TDZ has been selected for the propagation of numerous woody species due to its enormous capacity to stimulate the proliferation of the shoots (Huetteman and Preece 1993; Guo et al. 2011; Panda et al. 2016; Singh and Agarwal 2016), however, in Fagaceae positive results with TDZ are limited, and most of the cultures of these species are maintained in media supplemented with BA or zeatin (Table 9.1).

Among the works carried out, we should mention those of Wilhem and Rodkachane (1992), who evaluated the capacity of TDZ (4–11 mg/l) with and without 0.5 mg/l BA in the proliferation of juvenile and adult material of *Castanea sativa*. The results indicate that, after induction, a subculture with a high concentration of TDZ combined with BA improved the micropropagation. However, if TDZ is maintained in the medium, it promotes massive callus growth and inhibits elongation of the shoots. Fernández-Lorenzo et al. (2001) tested the effect of 0.1 mg/l BA or 0.01 mg/l TDZ in the multiplication of cultivars of adult origin of this same species, observing that TDZ produced a significant decline in the multiplication rate and the production of vitrified deformed shoots with excessive callus. In material of juvenile origin (1 year and 2–3 months) of two chestnut cultivars, Maraval 74 (*C. sativa* x *C. crenata*) and Marigoule 15 (*C. crenata* x *C. sativa*), Soylu and Ertük (1999) tested the effect of BA and in some cases 0.1 mg/l TDZ in the proliferation of buds. They pointed out that, although this concentration of TDZ in combination with BA had been mentioned previously by Waidinger and Rodkachane (1993) as

Table 9.1 Summary of studies on micropropagation of different species of *Castanea*

Species	Treatment	Observations	References
		Axillary shoots	
C. sativa	GD + TDZ (4–11 mg/l) + BA	A subculture improves micropropagation	Wilhem and Rodkachane (1992)
	P24 + TDZ (0.01 mg/l) + BA	In combination with BA, it favors bud expansion	Waindinger and Rodkachane (1993)
	GD, WPM + TDZ 0.01 mg/l	Decreases the multiplication rate; vitrified shoots; excessive callus	Fernández Lorenzo et al. (2001)
	MS (1/2NO_3) + TDZ (0.025–0.1 mg/l)	BA more effective than TDZ	Roussos et al. (2016)
		Adventitious shoots	
	MS + 0.1 mg/l TDZ	Preconditioning for genetic transformation	Corredoira et al. (2005)
	MS + 0.2 mg/l TDZ + IBA	More effective than BA	Tafazoli et al. (2013)
		Somatic embryogenesis	
	MS, DKW + 0.1 mg/l TDZ + BA/KIN+ IBA		Sezgin and Dumanoglu (2014)
		Axillary shoots	
C. sativa x C. crenata	MS (1/2NO_3) + TDZ 0.1 mg/l	Does not stimulate growth	Soylu and Ertük (1999)
C. crenata x C. sativa			
		Adventitious shoots	
C. sativa x C. crenata	MS + 0.1–2 mg/l TDZ + NAA	Cotyledon nodes; preconditioning with BA	San José et al. (2001)
		Axillary shoots	
C. dentata	WPM + TDZ 0.1–0.5 mg/l	Bud proliferation increases (reddish color)	Yang et al. (2009)
	– + TDZ 0.01 mg/l	Positive results	Herman (1995)
		Somatic embryogenesis	
	WPM +0.1 mg/l TDZ ± 2,4-D	TDZ did not give an embryogenic response	Carraway and Merkle (1997)
		Axillary shoots	
C. crenata	BW + TDZ 1 mg/l	Less effective than zeatin	Tetsumura and Yamashita (2004)

GD Gresshoff and Doy's medium (1972), *TDZ* thidiazuron, *BA* benzyladenine, *P24*'s medium (Waidinger and Rodkachane 1993), *WPM* Lloyd and McCown's medium (1980), *MS (1/2 NO_3)* Murashige and Skoog's medium (1962) with half strength nitrates, *IBA* indole-3-butyric acid, *DKW* Driver and Kuniyuki's medium (1972), *KIN* kinetin, *NAA* naphthalene acetic acid, *2,4-D* 2,4-dichlorophenoxyacetic acid, *BW* Sato's medium (1991)

bud growth stimulator in adult chestnut material, it was not effective in these two varieties. In a recent work, Roussos et al. (2016) evaluated the efficacy of several cytokinins (BA, KIN, isopentenyladenine (2iP), forchlorfenuron (FCF), and TDZ) and growth retardants on the multiplication and rooting in in vitro cultures of the European chestnut from material of juvenile origin. BA was the most effective cytokinin, followed by KIN and 2iP. The two phenylureas used (FCF and TDZ) were not as effective as the adenine-type cytokinins.

Different culture media (Sato's medium (BW, 1991), Lloyd and McCown (WPM, 1980), and Driver and Kuniyuki (DKW 1972)) were used in the micropropagation of *C. crenata* from plantlets of 2 months (Tetsumura and Yamashita 2004). These media were supplemented with BA, zeatin, or TDZ (1 mg/l). The best results in the establishment phase were obtained with BW medium supplemented with zeatin and thus were subsequently used for shoot multiplication.

According to the study by Herman (1995), the use of TDZ (0.01 mg/l) has given positive results in American chestnut (*C. dentata*). In this same species, Yang et al. (2009) investigated the role of different growth regulators in increasing shoot proliferation and callus formation, although no details were given on the type of explant or the age of the material used. TDZ and CPPU cytokinins at low concentrations (0.1 and 0.5 mg/l) produced the best multiplication rates compared with KIN or zeatin. However, the shoots obtained from explants cultivated in medium with CPPU or TDZ medium showed a reddish coloration, while with KIN or zeatin, they had a more normal morphological appearance.

9.6 TDZ and Differentiation Adventitious Buds

Although the formation of adventitious buds is not desirable for clonal propagation, it does offer an excellent opportunity for the regeneration of genetically manipulated plants using biotechnology. The in vitro regeneration via de novo differentiation of buds may be used for the production of more resistant plants and/or more productive genotypes after the insertion of new genes into the cells and could accelerate tree breeding programs. TDZ is currently considered to be one of the most effective triggers of morphogenesis in differentiated cells of woody species, having been successfully used in several species (Pavingerova 2009; Lenz et al. 2016; Zaytseva et al. 2016), although, as with the proliferation of axillary buds, there are few works on Fagaceae (Table 9.1).

The works found in the literature only mention the European chestnut or its hybrids. Thus, San José et al. (2001) studied TDZ on the adventitious bud formation capacity in preconditioned explants of *C. sativa* x *C. crenata*. In order to do this, they germinated the embryonic axes isolated in Murashige and Skoog medium (1962) supplemented with 0.1 mg/l TDZ or 1 mg/l BA. Segments of the hypocotyl, epicotyl, and cotyledon nodes were isolated from these embryos preconditioned and cultivated in induction medium with 0.01 mg/l of naphthalene acetic acid (NAA) and different concentrations of TDZ (0.1–2 mg/l). The best results were obtained with cotyledon nodes preconditioned in BA for 12–14 days and subsequently treated with 1–2 mg/l TDZ for 4 weeks (Fig. 9.1a, b). These shoots were elongated in medium with 0.1–0.05 mg/l BA. Cotyledon nodes pre-cultivated with 0.1 mg/l of TDZ have subsequently been used in genetic transformation works with *Agrobacterium tumefaciens* in *C. sativa* (Corredoira et al. 2005). Tafazoli et al. (2013) also established a protocol for the regeneration of adventitious buds from

Fig. 9.1 Micropropagation of Fagaceae using TDZ. (**a**) Longitudinal section of axillary bud meristem of chestnut formed at the cotyledonary node region cultured in preconditioning medium with 0.1 mg/l TDZ (×66.9); (**b**) multiple shoot formation in a BA-preconditioned explant of chestnut cultured on 1 mg/l TDZ induction medium (×124.5); (**c**) adventitious shoot differentiation in *Fagus sylvatica* leaf explant treated with 0.5 mg/l IAA and 0.5 mg/l TDZ; (**d**) adventitious shoot regeneration in the cotyledon of a somatic embryo of *Quercus robur* after 1 week culture in 0.1 mg/l TDZ and 7 weeks in germination medium; (**e**) germinated embryos of *Quercus robur* following 1 week exposure to TDZ and 7 weeks in germination medium

roots, nodes, and internodes of 3-month-old plantlets of *C. sativa*. The results showed the superiority of TDZ compared to BA as a shoot – inducing cytokinin in the in vitro induction of adventitious shoots from nodal segments of chestnut. The highest regeneration rates were achieved with the concentration of 0.2 mg/l.

9.7 TDZ and Somatic Embryogenesis

Somatic embryogenesis is an alternative regeneration process that offers numerous advantages as regards the differentiation of adventitious buds, is currently considered as one of the main biotechnology techniques for the mass propagation of plants, and is potentially of enormous use in genetic improvement programs (Lelu-Walter et al. 2013). In woody species, somatic embryogenesis is the most promising propagation method, enabling the implementation of multi-varietal forestry (Park et al. 2016) and biotechnological approaches such as the large-scale propagation of selected material, genetic transformation, and cryopreservation of elite genotypes (Corredoira et al. 2006; Bonga 2016; Guan et al. 2016).

TDZ was used for the induction of somatic embryogenesis in the genus *Castanea* by Sezgin and Dumanoglu in 2014 (Table 9.1). These authors established a protocol for the induction of somatic embryogenesis and regeneration of plants from immature cotyledons in two cultivars of *C. sativa* using different combinations of growth regulators (BA, TDZ, KIN, indole-3-butyric acid (IBA), 2,4-dichlorophenoxyacetic acid (2,4-D), and NAA). For both cultivars (Osmanoglu and Sariaslama), the best results were obtained with the MS and DKW formulas and the combination of different cytokinins and auxins (BA+KIN+IBA, BA+TDZ+IBA, KIN+TDZ+IBA). According to these authors, TDZ at 0.1 mg/l has been used in binary combinations with other auxins (2,4-D, indole-3-acetic acid (IAA), or NAA) in the somatic embryogenesis of different chestnut species and hybrids. Carraway and Merkle (1997), using *C. dentata*, tested the effect of the combination of three auxins (2,4-D, IAA, and NAA) with BA or TDZ (0.1 mg/l) in the induction of somatic embryos in ovules and immature zygotic embryos, obtaining the best results with 3 mg/l of 2,4-D. NAA or TDZ gave no embryogenic response.

9.8 Genus *Fagus*

The genus *Fagus* consists of ten species of monoecious trees, natives of the Northern hemisphere temperate zones of Eurasia and Eastern North America. *Fagus sylvatica* L. (European beech) is the most important species of the genus and one of the economically most important of central Europe, which together with the oaks define the climax vegetation of this region. *F. orientalis* Lipski (oriental beech) is a native of the temperate zones of Eastern Europe, the Balkan Peninsula, the Caucasus, and Asia Minor. *F. grandiflora* Ehrh. (American beech), of great ornamental value and with a high price as timber, is a native of Eastern North America. *F. japonica* Max. and *F. sieboldi* Engl. are Japanese species similar to the European beech (Chalupa 1996). Beech trees are important from an economic, as well as an ecosystem point of view. Its fine grain wood is used for flooring, furniture, and veneers, as well as an excellent wood for burning. The ornamental varieties are also of great economic importance (Vieitez et al. 2003).

The beeches are slow-growing trees with a life-span of 150–200 years, but some examples have been reported that have reached 300 years. Some ornamental beeches are propagated by bench grafting, with conventional vegetative propagation of beech generally being difficult (Meier and Reuther 1994; Chalupa 1996; Vieitez et al. 2003). They are extremely difficult to propagate if the cuttings are of adult origin, particularly if they are taken from branches of the crown (Ahuja 1984). The beeches are mainly propagated by seeds, which can be stored at 5 °C with a humidity of 20–25% in order to maintain their viability. However, good seed harvests are only obtained every 4–6 years, and storage over long periods has its problems. Furthermore, the use of seeds in tree improvement can take 30–50 years for trees to attain maturity; a similar period of time is needed in the propagation via cuttings taken from seedlings. Hybridization is very limited; thus the genetic basis for the selection of elite trees is also restricted. New methods for the rapid clonal propagation of the beech is thus necessary due to the lack of techniques for the production of large quantities of selected plant material using vegetative propagation (Vieitez et al. 2003).This need is especially acute in the case of *F. grandiflora* affected by the so-called beech bark disease, an introduced insect-fungus disease complex incited by an initial infestation by the scale insect, *Cryptococcus fasifuga* Lind., followed by an infection with one of the *Nectria* fungi, primarily *Nectria coccinea* var. faginata Lohman et al. (Ramírez et al. 2007). Thus, attention has been focused on the development of in vitro culture techniques, viable and rapid propagation methods that will enable new individuals to be obtained from the few tolerant/resistant trees that can be identified in the zones affected.

9.9 Proliferation of Axillary Buds and TDZ

The first attempts at propagating beeches employing in vitro axillary bud proliferation were carried out by Chalupa (1979, 1985) using apices and nodal segments of plantlets of the *F. sylvatica* species (Table 9.2).The explants were cultivated in different media (MS, WPM, and DKW) supplemented with various cytokinins: N-(phenylmethyl)-9-(tetrahydro-2H-pyran-2-yl)-9H-purin-6-amine (PBA), BA,

Table 9.2 Summary of studies on micropropagation of different species of *Fagus*

Species	Treatment	Observations	References
		Axillary shoots	
F. sylvatica	WPM+ TDZ 0.001–0.005 mg/l	PBA and BA more effective than TDZ	Chalupa (1979, 1985)
		Adventitious shoots	
	WPM + 0.5 mg/l TDZ + IAA	Leaves. TDZ 3 w more effective than BA	Vieitez and San José (1996)
	WPM + 1 mg/l TDZ + IAA	Internodes. More effective than BA	Cuenca et al. (2000)
		Adventitious shoots	
F. orientalis	WPM + 0.5–1 mg/l TDZ + IAA	Leaves, originating from the callus	Cuenca and Vieitez (1999)
	WPM + 1 mg/l TDZ + IAA	Internodes. More effective than BA	Cuenca et al. (2000)
	WPM + 1 mg/l TDZ + IAA	Leaves, internodes. Glucose 3–4%	Cuenca and Vieitez (2000)

WPM Lloyd and McCown's medium (1980), *TDZ* thidiazuron, *PBA* N-(phenylmethyl)-9-(tetrahydro-2H-pyran-2-yl)-9H-purin-6-amine, *BA* benzyladenine, *IAA* indole-3-acetic acid and TDZ. The best results were obtained with PBA and BA in combination with IBA. Low concentrations of TDZ (0.001–0.005 mg/l) stimulated the growth of axillary buds, but used in higher concentrations, it promoted callus formation and inhibited shoot elongation (Chalupa 1985). When the explants came from selected adult trees, the in vitro development was more difficult, and the majority of the works have used media supplemented with BA (Vieitez et al. 2003).

9.10 Differentiation of Adventitious Buds and TDZ

Although Chalupa (1996) mentions a preliminary study in which he obtained adventitious and axillary buds in embryonic axes, the first protocol for organogenesis induction in *F. sylvatica* was developed by Vieitez et al. (1993). These authors observed the induction of adventitious buds in hypocotyls of plantlets resulting from the germination of zygotic embryos cultivated in the presence of BA. The capacity to differentiate adventitious buds was also observed in cotyledon and hypocotyl segments isolated from plantlets of 3 weeks and cultivated in media with BA and NAA. These observations led to subsequent studies for the development of adventitious buds in leaves and internodes of juvenile cultures maintained in vitro.

Vieitez and San José (1996) describe the induction of adventitious buds in foliar explants of shoots cultivated in vitro originally established from 2-month-old and 3-year-old plants of *F. sylvatica*. The explants were cultivated in WPM medium supplemented with 0.5 mg/l IAA and different concentrations of BA (1, 2, or 4 mg/l) or TDZ (0.05, 0.5, 1, or 2 mg/l) (Table 9.2). The efficiency of the regeneration depended on the type of explant (proximal or distal half of the leaf), the cytokinin added to the culture medium, and the genotype. TDZ at 0.5 mg/l was more effective than BA (Fig. 9.1c). The concentration of TDZ and the time of exposure were critical for optimizing the results. The continued exposure to TDZ produced stunted and compact bud clusters, thus more difficult to elongate than those induced by BA. The problem can be overcome by reducing exposure time to 1–3 weeks, in accordance to the two-stage procedure proposed by Huetteman and Preece (1993) (culture in TDZ medium followed by culture in a second medium without TDZ or with other growth regulators). The histology study confirmed the direct or indirect origin of the adventitious buds.

Cuenca and Vieitez (1999) also achieved the induction of adventitious buds in foliar explants of shoots cultivated in vitro established from plants of 2 months and 4 years of *F. orientalis*. These authors used WPM supplemented with 0.5 mg/l IAA and 0.5–1 mg/l TDZ. Histologically, it could be shown that, although some buds developed directly from the epidermis or sub-epidermis, the majority of them originated from cell file proliferation produced by the periclinal division of cells subjacent to the epidermis.

Cuenca et al. (2000) evaluated the formation of adventitious buds in internodes of shoots cultivated in vitro from European and Oriental beech (*F. sylvatica* and *F. orientalis*, respectively). Induction medium consisted of WPM supplemented with

0.5 mg/l IAA and 0.1, 0.5, 1, 2, and 4 mg/l TDZ or BA. As in the case of foliar explants (Vieitez and San José 1996), TDZ was much more effective than BA for the induction of adventitious buds in this type of explant, with 1 mg/l being the optimum dose for TDZ. An increase in the concentration of TDZ leads to the formation of clusters of tiny buds that subsequently do not develop. The frequency of explants forming buds and the number of adventitious buds was significantly affected by the genotype and the concentration of TDZ. These results are in agreement with those of Murthy et al. (1998), who confirm that TDZ is more effective for the induction of adventitious buds than other purine-type cytokinins. The combination of TDZ with IAA or IBA at 0.5 mg/l increases the capacity in forming buds in the internodes.

In a study carried out by Cuenca and Vieitez (2000) to see the effect of the carbon source on adventitious bud differentiation in leaves and internodes of *F. orientalis* with 0.5 mg/l IAA and 1 mg/l TDZ, these authors showed that both the percentage and number of adventitious buds per explant were significantly affected by the type and the concentration of sugar used, with 3–4% glucose giving the best results.

The elongation of the adventitious buds was promoted by storing them under a dim light at 3–4 °C for 2 months (Vieitez et al. 2003). The importance of a cold storage period for elongation of shoots induced with TDZ has been described previously by Huetteman and Preece (1993).

9.11 Genus *Quercus*

The genus *Quercus* consists of about 400 species distributed throughout the temperate regions of the world (Johnson et al. 2002). These trees can reach a height of 30–40 m and live for 800 years or more. Of the 24 species of oak and their hybrids that inhabit Europe, *Q. robur* L. and *Q. petraea* L. have a higher number of specimens and are the most important from an economic and ecological point of view. Together with these species, *Q. suber* L. and *Q. ilex* L. should also be mentioned as species characteristic of Mediterranean ecosystems. In America, *Q. rubra* L. and *Q. alba* L. are the most important and most widely represented (Vieitez et al. 2012). There are also more than 35 species of *Quercus* reported in the Himalayan region (Singh et al. 2011).

As in the majority of the Fagaceae, the regeneration of *Quercus* spp. is reported to be gradually deteriorating due to extensive harvesting, irregular fructification, unavailability of seed every year, and predation, as well as the difficulties encountered with their propagation using conventional methods, particularly when they are adult specimens (Vengadesan and Pijut 2009). There has also been the loss of a large number of specimens in the last few years, among other causes, due to that known as oak decline syndrome. This is a gradual and episodic phenomenon characterized by a loss of vigor that seems to be caused by an interaction between several biotic and abiotic factors (Vieitez et al. 2012). The need for plantations with high value genetic material has favored the application of micropropagation techniques in this genus.

9.12 Proliferation of Axillary Buds and TDZ

Benzyladenine, in different concentrations, is the most widely used cytokinin for the proliferation of axillary buds in the genus *Quercus*. However, Chalupa (1988) evaluated the effect of TDZ in cultures of juvenile (3–6 months) and adult (30–50 years) origin from *Q. robur*. TDZ significantly affected the elongation and morphology of the shoots. At very low concentrations (0.001–0.004 mg/l), it promoted proliferation, but at higher concentrations (0.01–0.02 mg/l), it stimulated the formation of a large amount of callus, and the shoots developed were short and fewer in number (Table 9.3).

In *Q. euboica* Pap., an endemic species of Greece, Kartsonas and Papafotiou (2007) studied the effect of different cytokinins (KIN, zeatin, 2iP, BA, and TDZ) in the multiplication of shoots established from material of adult and juvenile origin.

Table 9.3 Summary of studies on micropropagation of different species of *Quercus*

Species	Treatment	Observations	References
		Axillary shoots	
Q. robur	WPM o BTM + 0.001–0.004 mg/l TDZ	Promotes shoot proliferation	Chalupa (1988)
		Adventitious shoots	
	WPM o BTM + 0.01–0.02 mg/l TDZ	Also combinations of TDZ with BA or PBA	Chalupa (1988)
	MS + 0.1–1 mg/l TDZ	Cotyledons of somatic embryos	Martínez et al. (2008)
		Somatic embryogenesis	
	MS + 0.01–0.02 mg/l TDZ	Favors the conversion of the somatic embryos	Martínez et al. (2008)
		Axillary shoots	
Q. euboica	WPM + 0.02 mg/l TDZ	Inhibits bud development	Kartsonas and Papafotiou (2007)
	WPM + 0.1 mg/l TDZ	Totally ineffective	Kartsonas and Papafotiou (2009)
		Axillary shoots	
Q. serrata	WPM + 1.5 mg/l TDZ	Very short shoots and small leaves	Pandey and Tamta (2014)
		Axillary shoots	
Q. rubra	MS + 0.1 mg/l TDZ+ BA	Elongation of the shoots in medium with BA and GA_3	Vengadesan and Pijut (2009)
		Somatic embryogenesis	
Q. ilex	MS + 0.2 mg/l TDZ+ picloram	TDZ did not produce somatic embryos. Better treatments: NAA or IAA with BA	Martínez et al. (2017)

WPM Lloyd and McCown's medium (1980), *BTM* broad-leaved tree medium (Chalupa 1981), *TDZ* thidiazuron, *BA* benzyladenine, *PBA* N-(phenylmethyl)-9-tetrahydro-2H-pyran-2-yl)-9H-purin-6-amine, *MS* Murashige and Skoog's medium (1962), *GA_3* gibberellic acid, *NAA* naphthalene acetic acid, *IAA* indole-3-acetic acid

The percentage of explants that formed shoots was affected by the cytokinin type and concentration. The cultures in medium with BA and zeatin gave the best results; with TDZ (0.02 mg/l) in the medium, the percentage of shoot formation was zero. This result was corroborated in a subsequent work carried out on this same species (Kartsonas and Papafotiou 2009).

With material of juvenile origin from *Q. serrata* Thunb., a native species of east Asia, Pandey and Tamta (2014) studied the effect of different cytokinins (BA, 2iP, KIN, and TDZ) (0.5, 1, 1.5 mg/l) in the proliferation of shoots. Although TDZ (1.5 mg/l) enabled the maximum number of explants to be obtained, these were short and with swollen bases and small leaves without expanding. The medium supplemented with BA (1 mg/l) produced a relatively lower number of shoots per explant, but these reached the greatest lengths, and their leaves appeared healthy and well expanded.

The in vitro propagation of *Q. rubra* was achieved from plantlets of 8 weeks by Vengadesan and Pijut (2009) using MS medium supplemented with 1 mg/l BA and 0.1 mg/l TDZ. TDZ together with BA stimulates the development of a large number of shoots, but this combination did not allow the subsequent growth and development of the shoots. According to these authors, TDZ produced cell division, but not elongation as occurred in *Acer grandidentatum* (Bowen-O'Connor et al. 2007). The addition of BA (1 mg/l) and gibberellic acid (0.05 mg/l) to the medium was necessary in order to stimulate the elongation of the shoots.

9.13 Differentiation of Adventitious Buds and TDZ

Adventitious buds were differentiated in the callus developed on the basal part of oak shoots (*Q. robur*) cultivated in medium with TDZ (0.01–0.02 mg/l). A higher concentration of TDZ (0.05–0.1 mg/l) promoted the formation a large quantity of callus and inhibited the elongation of the buds. The combinations of PBA or BA with TDZ also promoted callus formation and the differentiation of adventitious buds (Chalupa 1988) (Table 9.3).

Martínez et al. (2008) observed the differentiation of adventitious buds in cotyledons of somatic embryos of *Q. robur* treated with0.1–1 mg/l TDZ (Fig. 9.1d).

9.14 Somatic Embryogenesis and TDZ

Martínez et al. (2017) studied the effect of different growth regulators (BA, IAA, NAA, IBA, picloram, and TDZ) in order to induce the formation of somatic embryos in leaves and apices from in vitro shoots cultures of *Q. ilex*. The best results were obtained with the apical explants and the combination of NAA or IAA (4 mg/l) with BA 0.5 mg/l. Thidiazuron was not effective in the induction of somatic embryogenesis in this species (Table 9.3). Martínez et al. (2008) determined the ability of TDZ to improve the conversion capacity of somatic embryos induced in adult material of

Q. robur, incorporating it into the germination medium (MS) for short periods of time. According to these authors, the addition of TDZ to the germination medium induces the formation of multiple buds in the epicotyl area. The exposure of somatic embryos to 0.01–0.02 mg/l TDZ for 7 days increases the frequency of somatic embryos with elongated shoots (Fig. 9.1e). TDZ at 0.02 mg/l increases the conversion percentages by up to 64% in one of the genotypes studied (Sainza).

9.15 Concluding Remarks and Future Prospects

In forestry terms, the members of the Fagaceae family are considered to be of great economic and ecological importance, occupying a wide ecological niche in the northern hemisphere. However, the vegetative propagation of the majority of the species of this family presents considerable difficulties, especially when the selected material is of adult origin. From a few decades ago, biotechnology offers new methods for the propagation, improvement, and conservation of plant material and is particularly suitable in the case of recalcitrant species.

In the in vitro culture of plant material, we are able to consider (1) the proliferation of axillary buds, considered as the simplest and efficient method for maintaining the genetic stability of the regenerated plants; (2) the differentiation of adventitious buds (although it is not desirable for clonal propagation, it does offer an excellent opportunity to capture somaclonal variations, obtain chimeral modifications, and apply selection and mutagenic pressures due to the adventitious nature of regeneration); and (3) the somatic embryogenesis, which is considered as an ideal regeneration system for genetic transformation studies and for the mass propagation of plants.

The application of micropropagation requires the use of culture media supplemented with various growth regulators, mainly auxins and cytokinins. Among these latter, TDZ has been used successfully in the micropropagation of numerous species and is considered as one of the most powerful diphenylureas evaluated in plant tissue culture. Despite this, it has not been widely used in the micropropagation of Fagaceae, being replaced by other purine-type cytokinins, such as benzyladenine or zeatin. Further studies will lead to the development and use of better and more powerful growth regulators that will make it easier for the in vitro propagation of material collect from selected adult trees or the differentiation of adventitious buds and somatic embryos that will allow carrying out works on genetic improvement and conservation of these species.

Acknowledgments To all the members who, during all these years, have been part of the Biotechnology and Forest Improvement Group, having contributed in one way or another to the success of the micropropagation of these species. These works have been partially funded with different projects from CICYT, MINECO, and Xunta de Galicia (Spain).

References

Ahmad N, Anis M (2007) Rapid clonal multiplication of a woody tree, *Vitex negundo* L., through axillary shoot proliferation. Agrofor Syst 71:195–200

Ahuja MR (1984) In vitro induction of organogenesis in juvenile and mature beech. Silv Genet 33:241–242

Arndt FR, Rusch R, Stillfried HV, Hanisch B, Martin WC (1976) SN 49537. A new defoliant. Plant Physiol 57:s-99. (abstr)

Baker BS, Bhatia SK (1993) Factors affecting adventitious shoot regeneration from leaf explants of quince (*Cydonia oblonga*). Plant Cell Tissue Organ Cult 35:273–277

Ballester A, Corredoira E, Vieitez AM (2016) Limitations of somatic embryogenesis in hardwoods trees. In: Park Y-S, Bonga JM, Moon H-K (eds) Vegetative propagation of Forest trees. NIFoS, Seoul, pp 56–74

Bonga JM (2016) Can explant choice help to resolve recalcitrance problems in in vitro propagation, a problem still acute especially for adult conifers? Trees. https://doi.org/10.1007/s00468-016-1509-z

Bonga JM, von Aderkas P (1992) In vitro culture of trees. Kluwer Academic Publishers, Dordrecht

Bowen-O'Connor CA, Hubstenberger J, Killough C, Van Leeuwen DM, St. Hilaire R (2007) In vitro propagation of *Acer grandidentatum* Nutt. In Vitro Cell Dev Biol Plant 43:40–50

Camus A (1929) Les chataigniers. Monographie des genres *Castanea* et *Castanopsis*. In: Le Chevalier P (ed) Encyclopédie Economique de Sylviculture. Paul Lechevalier, Paris

Cañellas I, Roig S, Poblaciones MJ, Gea-Izquierdo G, Olea L (2007) An approach to acorn production in Iberian dehesas. Agrofor Syst 70:3–9

Carraway DT, Merkle SA (1997) Plantlet regeneration from somatic embryos of American chestnut. Can J For Res 27:1805–1812

Castillo A, Cabrera D, Rodríguez P, Zoppolo R, Robinson T (2015) In vitro micropropagation of CG41 apple rootstock. Acta Hortic 1083:569–576

Chalupa V (1979) In vitro propagation of some broad-leaved forest trees. Commun Inst For Czech 11:150–170

Chalupa V (1981) Clonal propagation of broad-leaved forest trees in vitro. Commun Inst For Czech 12:255–271

Chalupa V (1985) In vitro propagation of *Larix*, *Picea*, *Pinus*, *Quercus*, *Fagus* and other species using adenine-type cytokinins and thidiazuron. Commun Inst For Czech 14:65–90

Chalupa V (1988) Large scale micropropagation of *Quercus robur* L. using adenine-type cytokinins and thidiazuron to stimulate shoot proliferation. Biol Plant 30:414–421

Chalupa V (1996) *Fagus sylvatica* L. (European beech). In: Bajaj YPS (ed) Biotechnology in agriculture and forestry, Trees IV, vol 35. Springer, Berlin/Heidelberg, pp 138–154

Corredoira E, San José MC, Ballester A, Vieitez AM (2005) Genetic transformation of *Castanea sativa* Mill. by *Agrobacterium tumefaciens*. Acta Hortic 693:387–393

Corredoira E, Ballester A, Vieitez FJ, Vieitez AM (2006) Somatic embryogenesis in chestnut. In: Mujib A, Samaj J (eds) Plant cell monographs, Somatic Embryogenesis, vol 2. Springer, Berlin/Heidelberg, pp 177–199

Corredoira E, Vieitez AM, San José MC, Vieitez FJ, Ballester A (2016) Advances in somatic embryogenesis and genetic transformation of European chestnut: development of transgenic resistance to ink and blight disease. In: Park Y-S, Bonga JM, Moon H-K (eds) Vegetative propagation of forest trees. NIFoS, Seoul, pp 279–301

Cuenca B, Vieitez AM (1999) Histological study of in vitro development of adventitious buds on leaf explant of Oriental beech (*Fagus orientalis* Lipski). In Vitro Cell Dev Biol Plant 35:326–332

Cuenca B, Vieitez AM (2000) Influence of carbon source on shoot multiplication and adventitious bud regeneration in in vitro beech cultures. Plant Growth Regul 32:1–12

Cuenca B, Ballester A, Vieitez AM (2000) In vitro adventitious bud regeneration from internode segments of beech. Plant Cell Tissue Organ Cult 60:213–220

Driver JA, Kuniyuki AH (1972) In vitro propagation of Paradox walnut rootstock. Hortscience 19:507–509

Fellman CD, Read PE, Hosier MA (1987) Effects of TDZ and CPPU on meristem formation and shoot proliferation. Hortscience 22:1197–1200

Fernández-Lorenzo JL, Rodríguez S, Viega M (2001) Micropropagación de dos cultivares de fruto de *Castanea sativa* Mill. In: Proc III Congreso Forestal Español. Vol II. Mejora Genética, Viveros y Repoblación Forestal, Granada (Spain), pp 742–749

Fey BS, Endress PK (1983) Development and morphological interpretation of the cupule in Fagaceae. Flora 173:451–468

Gomes F, Canhoto JM (2009) Micropropagation of strawberry tree (*Arbutus unedo* L.) from adult plants. In Vitro Cell Dev Biol Plant 45:72–82

González-Benito ME, Martín C (2011) In vitro preservation of Spanish biodiversity. In Vitro Cell Dev Biol Plant 47:46–54

Gresshoff PM, Doy CH (1972) Development and differentiation of haploid *Lycopersicon esculentum*. Planta 107:161–170

Guan Y, Li S-G, Fan X-F, Su Z-H (2016) Application of somatic embryogenesis in woody plants. Front Plant Sci 7:1–12

Guo B, Abbasi BH, Zeb A, Xu LL, Wei YH (2011) Thidiazuron: a multi-dimensional plant growth regulator. Afr J Biotech 10:8984–9000

Hare PD, Staden J, van Staden J (1994) Inhibitory effect of TDZ on the activity of cyotkinin oxidase isolated from soybean callus. Plant Cell Physiol 35:1121–1125

Herman EB (1995) Recent advances in plant tissue culture III. Agritech Consultants, Shrub Oak

Heywood VH, Brummitt RK, Culham A, Seberg O (2007) Flowering plant families of the world. Royal Botanic Gardens, Kew

Huetteman CA, Preece JE (1993) Thidiazuron: a potent cytokinin for woody plant tissue culture. Plant Cell Tissue Organ Cult 33:105–119

Hutchinson MJ, Saxena PK (1996) Role of purine metabolism in TDZ-induced somatic embryogenesis of geranium (*Pelargonium* x *hortorum*) hypocotyls cultures. Physiol Plant 98:517–522

Johnson PS, Shifley SR, Rogers R (2002) The ecology and silviculture of oaks. CABI, New York

Jones MPA, Cao J, O'Brien R, Murch SJ, Saxena PK (2007) The mode of action of thidiazuron: auxins, indoleamines, and ion channels in the regeneration of *Echinacea purpurea* L. Plant Cell Rep 26:1481–1490

Kahia J, Kirika M, Lubabali H, Mantel S (2016) High-frequency direct somatic embryogenesis and plantlet regeneration from leaves derived from in vitro-germinated seedlings of a *Coffea arabica* hybrid cultivar. Hortscience 51:1148–1152

Kartsonas E, Papafotiou M (2007) Mother plant age and seasonal influence on in vitro propagation of *Quercus euboica* Pap., an endemic, rare and endangered oak species of Greece. Plant Cell Tissue Organ Cult 90:111–116

Kartsonas E, Papafotiou M (2009) Micropropagation of *Quercus euboica* Pap., a rare endemic oak species in Greece. Acta Hortic 813:485–490

Kremer A, Abbott AG, Carlson JE, Manos PS, Plomion C, Sisco P, Staton ME, Ueno S, Vendramin GG (2012) Genomics of Fagaceae. Tree Genet Genome 8:583–610

Lelu-Walter MA, Thompson D, Harvengt L, Sánchez L, Toribio M, Pâques LE (2013) Somatic embryogenesis in forestry with a focus on Europe: state-of-the art, benefits, challenges and future direction. Trees Genet Genomes 9:883–899

Lenz RR, Magnusson VA, Dai W (2016) Plant regeneration of 'Amethyst' purple raspberry (*Rubus occidentalis* x *R. idaeus* 'Amethyst') from in vitro leaf tissues. Acta Hortic 1133:491–496

Li H, Murch SJ, Saxena PK (2000) Thidiazuron-induced de novo shoot organogenesis on seedlings, etiolated hypocotyls and stem segments of Huang-qin. Plant Cell Tissue Organ Cult 62:169–173

Liu Y, Lu J, Zhu H, Li L, Shi Y, Yin X (2016) Efficient culture protocol for plant regeneration from cotyledonary petiole explants of *Jatropha curcas* L. Biotechnol Biotechnol Equip 30:907–914

Lloyd G, McCown BH (1980) Commercially-feasible micropropagation of mountain laurel, *Kalmia latifolia*, by use of shoot tip culture. Comb Proc Int Plant Propagators' Soc 30:421–427

Lu C-Y (1993) The use of thidiazuron in tissue culture. In Vitro Cell Dev Biol Plant 29P:92–96
Martínez MT, Corredoira E, Valladares S, Jorquera L, Vieitez AM (2008) Germination and conversion of somatic embryos derived from mature *Quercus robur* trees: the effects of cold storage and thidiazuron. Plant Cell Tissue Organ Cult 95:341–351
Martínez MT, San José MC, Vieitez AM, Cernadas MJ, Ballester A, Corredoira E (2017) Propagation of mature *Quercus ilex* L. (holm oak) trees by somatic embyogenesis. Plant Cell Tissue Organ Cult 131:321–333
Matand K, Prakash CC (2007) Evaluation of peanut genotypes for in vitro plant regeneration using thidiazuron. J Biotechnol 130:202–207
Meier K, Reuther G (1994) Factors controlling micropropagation of mature *Fagus sylvatica*. Plant Cell Tissue Organ Cult 39:231–238
Mok MC, Mok DWS (1985) The metabolism of [^{14}C]-TDZ in callus cultures of *Phaseolus lunatus*. Physiol Plant 65:427–432
Mok MC, Mok DWS, Armstrong DJ, Shudo K, Isogai Y, Okamoto T (1982) Cytokinin activity of N-phenyl-N′-1,2,3-thidiazol-5-ylurea (TDZ). Phytochemistry 21:1509–1511
Monteuuis O (2016) Micropropagation and production of forest trees. In: Park Y-S, Bonga JM, Moon H-K (eds) Vegetative propagation of Forest trees. NIFoS, Seoul, pp 32–55
Monteuuis O, Doulbeau S, Verdeil JL (2008) DNA methylation in different original clonal offspring from mature *Sequoiadendron giganteum* genotype. Trees 22:779–784
Murashige T, Skoog F (1962) A revised medium for rapid growth and bioassays with tobacco tissue cultures. Physiol Plant 15:473–497
Murch SJ, Krishnaraj S, Saxena PK (1997) TDZ-induce morphogenesis of Regal Geranium (*Pelargonium domesticum*): a potential stress response. Physiol Plant 101:183–191
Murthy BNS, Saxena PK (1998) Somatic embryogenesis and plant regeneration of Neem (*Azadirachta indica* A. Juss). Plant Cell Rep 17:469–475
Murthy BNS, Murch SJ, Saxena PK (1995) TDZ-induced somatic embryogenesis in intact seedlings of peanut (*Arachis hypogaea*): endogenous growth regulator levels and significance of cotyledons. Physiol Plant 94:268–276
Murthy BNS, Singh RP, Saxena PK (1996) Induction of high frequency somatic embryogenesis in geranium (*Pelargonium* x *hortorum* Bailey cv. Ringo Rose) cotyledonary cultures. Plant Cell Rep 15:423–426
Murthy BNS, Murch SJ, Saxena PK (1998) Thidiazuron: a potent regulator of in vitro plant morphogenesis. In Vitro Cell Dev Biol Plant 34:267–275
Nagata R, Kawachi E, Hashimoto Y, Shudo K (1993) Cytokinins-specific binding protein in etiolated mung bean seedlings. Biochem Biophys Res Commun 191:543–549
Nelson CD, Powell WA, Merkle SA, Carlson JE, Hebard FV, Islam-Faridi N, Staton ME, Georgi L (2014) Biotechnology of trees: chestnut. In: Ramawat KG, Mérillon J-M, Ahuja MR (eds) Tree biotechnology. CRC Press, Boca Raton, pp 3–34
Nhut DT, Hahn NTM, Tuan PQ, Nguyet TM, Tram NTH, Chinh NC, Nguyen NH, Vinh DN (2006) Liquid culture as a positive condition to induce and enhance quality and quantity of somatic embryogenesis of *Lilium longiflorum*. Sci Hortic 110:93–97
Nixon KC, Crepet WL (1989) *Triganobalanus* (Fagaceae): taxonomy status and phylogenetic relationships. Am J Bot 76:828–841
Panda BM, Mehta UJ, Hazra S (2016) Micropropagation of *Semecarpus anacardium* L. from mature tree-derived nodal explants. Plant Biosyst 150:942–952
Pandey A, Tamta S (2014) In vitro propagation of the important tasar oak (*Quercus serrata* Thunb.) by casein hydrolysate promoted high frequency shoot proliferation. J Sustain Forest 33:590–603
Park YS, Beaulieu J, Bousquet J (2016) Multi-varietal forestry integrating genomic selection and somatic embryogenesis. In: Park Y-S, Bonga JM, Moon H-K (eds) Vegetative propagation of forest trees. National Institute for Forest Science (NIfoS), Seoul, pp 302–322
Pavingerova D (2009) The influence of thidiazuron on shoot regeneration from leaf explants of fifteen cultivars of *Rhododendron*. Biol Plant 54:797–799

Preece JE, Huetteman CA, Ashby WC, Roth PL (1991) Micro- and cutting preparation of silver maple I. Results with adult and juvenile propagules. J Am Soc Hortic Sci 116:142–148
Ramírez M, Krasowski MJ, Loo JA (2007) Vegetative propagation of American beech resistant to beech bark disease. Hort Sci 40:320–324
Roussos PA, Archimandriti A, Beldekou I (2016) Improving in vitro multiplication of juvenile European chestnut (*Castanea sativa* Mill.) explants by the use of growth retardants. Sci Hortic 198:254–256
Rugini E, Silvestri C (2016) Somatic embryogenesis in olive (*Olea europaea* L. subsp *europaea* var. *sativa* and var. *sylvestris*). Methods Mol Biol 1359:341–349
San José MC, Ballester A, Vieitez AM (2001) Effect of thidiazuron on multiple shoot induction and plant regeneration from cotyledonary nodes of chestnut. J Hortic Sci Biotechnol 76:588–595
San José MC, Cernadas MJ, Corredoira E (2014) Histology of the regeneration of *Paulownia tomentosa* (Paulowniaceae) by organogenesis. Rev Biol Trop 62:809–818
Sato T (1991) Basic studies of organ and callus culture in woody plants. Bull For Prod Res Inst 360:35–119
Saxena PK, Malik KA, Gill R (1992) Induction by TDZ of somatic embryogenesis in intact seedlings of peanut. Planta 187:421–424
Sedlák J, Paprštein F (2015) In vitro multiplication of old pear cultivars. Acta Hortic 1094:163–167
Sezgin M, Dumanoglu H (2014) Somatic embryogenesis and plant regeneration from immature cotyledons of European chestnut (*Castanea sativa* Mill). In Vitro Cell Dev Biol Plant 50:58–68
Singh A, Agarwal PK (2016) Enhanced micropropagation protocol of ex vitro rooting of a commercially important crop plant *Simmondsia chinensis* (Link) Schneider. J For Sci 62:107–115
Singh G, Rai ID, Rawat GS (2011) The year 2010 was 'mast sed year' for the Kharsu oak (*Quercus semecarpifolia* Sm.) in the western Himalaya. Curr Sci 100:1275
Soylu A, Ertük Ü (1999) Researches on micropropagation of chestnut. Acta Hortic 494:247–253
Tafazoli M, Nasr SMH, Jalilvand H, Bayat D (2013) Plant regeneration through organogenesis of chestnut (*Castanea sativa* Mill.) Afr J Biotechnol 12:7063–7069
Tetsumura T, Yamashita K (2004) Micropropagation of Japanese chestnut (*Castanea crenata* Sieb. et Zucc.) seedlings. Hort Sci 39:1684–1687
Traore A, Maximova SN, Guiltinan MJ (2003) Micropropagation of *Theobroma cacao* L. using embryo-derived plants. In Vitro Cell Dev Biol Plant 39:332–337
Vengadesan G, Pijut PM (2009) In vitro propagation of northern red oak (*Quercus rubra* L.) In Vitro Cell Dev Biol Plant 45:474–482
Vieitez FJ, Merkle SZ (2005) *Castanea* spp. chestnut. In: Litz (ed) Biotechnology of fruit and nut crops. CAB International, Wallingford, pp 265–296
Vieitez AM, San José MC (1996) Adventitious shoot regeneration from *Fagus sylvatica* leaf explants in vitro. In Vitro Cell Dev Biol Plant 32:140–147
Vieitez AM, Ferro E, Ballester A (1993) Micropropagation of *Fagus sylvatica* L. In Vitro Cell Dev Biol Plant 29P:183–188
Vieitez AM, San José MC, Sánchez MC, Ballester A (2003) Micropropagation of *Fagus* spp. In: Jain SM, Ishii K (eds) Micropropagation of woody trees and fruits. Kluwer Academic Publishers, Dordrecht, pp 181–215
Vieitez AM, Corredoira E, Martínez MT, San José MC, Sánchez C, Valladares S, Vidal N, Ballester A (2012) Application of biotechnological tools to *Quercus* improvement. Eur J For Res 131:519–539
Visser C, Qureshi JA, Gill T, Saxena PK (1992) Morphoregulatory role of TDZ. Substitution of auxin and cytokinin requirement for the induction of somatic embryogenesis in geranium hypocotyl cultures. Plant Physiol 99:1704–1707
Waidinger E, Rodkachane P (1993) Investigations on micropropagation of adult chestnut. Proc Int Cong on Chestnut. Spoleto, Italy, pp 205–210
Wang SY, Jiao HJ, Faust M (1991) Changes in metabolic enzyme activities during TDZ-induced lateral bud break of apple. Hort Sci 26:171–173
Wheeler N, Sedroff R (2009) Role of genomics in the potential restoration of the American chestnut. Tree Genet Genomes 5:181–187

Wilhem E, Rodkachane P (1992) Micropropagation of juvenile and adult *Castanea sativa* by using thidiazuron. Proc Int Chestnut Conference, Morgantown, West Virginia, pp 129

Wojtania A, Gabryszewska E, Podwyszynska M (2011) The effect of growth regulators and sucrose concentration on in vitro propagation of *Camellia japonica* L. Propag Ornam Plants 11:177–183

Yang G, Zhongge L, Asante TM (2009) In vitro responses of American chestnut to plant growth regulators in culture medium. Acta Hortic 844:229–234

Yip WK, Yang SF (1986) Effect of TDZ, a cytokinin-active urea derivative, in cytokinin-dependent ethylene production systems. Plant Physiol 80:515–519

Zaytseva Y, Poluboyarova TV, Novikova TI (2016) Effects of thidiazuron on in vitro morphogenic response of *Rhododendron sichotense* Pojark. and *Rhododendron catawbiense* cv. Grandiflorum leaf explants. In Vitro Cell Dev Biol Plant 52:56–63

In Vitro Morphogenesis of Woody Plants Using Thidiazuron

10

A. Vinoth and R. Ravindhran

Abstract

Thidiazuron (TDZ) has been in use for in vitro shoot regeneration, in particular, recalcitrant woody perennials. Owing to its superiority over natural cytokinins in plant regeneration, TDZ has been the plant growth regulator (PGR) of choice for mature tissues. In majority of the tree species, TDZ has induced regeneration via axillary shoot proliferation, adventitious shoot organogenesis and somatic embryogenesis. Interestingly, TDZ has evoked different regeneration routes from the same explant at different concentrations. In addition, various other factors like pretreatment, explant type, maturity, orientation, TDZ concentration, combination with other PGRs and organic additives interact synergistically to promote shoot regeneration. Despite being potent PGR, supra-optimal level of TDZ produces shoot abnormalities like vitrification/hyperhydricity (stunted shoots) or fasciation (fused shoots). In shoot organogenesis and somatic embryogenesis, prolonged exposure to TDZ resulted in callus necrosis or reversal of shoot buds or somatic embryos to callus. Therefore, this review paper is intended to bring out the effectiveness of TDZ in woody plant tissue culture. The authors also emphasize on various interacting factors to minimize the negative consequences of TDZ treatment.

Keywords

Thidiazuron · Woody plants · Shoot regeneration · Hyperhydricity · Fasciation

A. Vinoth · R. Ravindhran (✉)
T.A.L. Samy Unit for Plant Tissue Culture and Molecular Biology, Department of Plant Biology and Biotechnology, Loyola College (Autonomous), Nungambakkam, Chennai, India

N. Ahmad, M. Faisal (eds.), *Thidiazuron: From Urea Derivative to Plant Growth Regulator*, https://doi.org/10.1007/978-981-10-8004-3_10

Abbreviations

BA	Benzyladenine
GA_3	Gibberellic acid
IAA	Indole-3-acetic acid
IBA	Indole-3-butyric acid
Kin	Kinetin
MS	Murashige and Skoog
NAA	α-naphthalene acetic acid
PGR	Plant growth regulator
SE	Somatic embryo
SSE	Secondary somatic embryo
TDZ	Thidiazuron

10.1 Introduction

Discovery and the application of plant growth regulators (PGRs) have revolutionized the field of plant tissue culture. Shoot regeneration of plants under in vitro conditions depends on the specific balance of PGRs (natural and synthetic) such as auxins and cytokinins. Cytokinins are N^6-substituted adenines that regulate cell division and cell differentiation in plants. Based on the chemical structure of the side chain, cytokinins fall into two categories: isoprenoid and aromatic. Benzyladenine (BA) or kinetin (6-furfurylaminopurine, Kin) are the most frequently used aromatic natural cytokinins (Kieber and Schaller 2014) in tissue culture systems but with limited success in woody plant regeneration (van Staden et al. 2008). On the other hand, synthetic analogues of natural cytokinins have evoked striking regeneration potential in recalcitrant woody tissues (Ďurkovič and Mišalová 2008). Thidiazuron (TDZ) is a synthetic phenyl urea (N-phenyl-N′-1,2,3-thiadiazol-5-yl-urea) with growth-promoting activity like cytokinins in plant cell cultures (Mok et al. 1982). Manufactured primarily for its use as a defoliant for cotton (Arndt et al. 1976), TDZ has induced a variety of morphogenetic responses in plant cells in vitro till date (Huetteman and Preece 1993; Zhang et al. 2016). TDZ was highly effective than other cytokinins at very minimal concentrations because of its stability to withstand degradation by cytokinin oxidases (Mok et al. 1987).

Propagation of woody plants by in vitro techniques is a challenging task, as majority of the tree species are recalcitrant. TDZ supplementation in the culture media significantly reduced the limitations encountered with the regeneration of recalcitrant tissues. Explants from mature trees that are non-regenerative using other cytokinins underwent rapid proliferation in the presence of TDZ (Huetteman and Preece 1993). Thus TDZ has played a significant role in ex situ conservation of trees by facilitating regeneration via micropropagation, shoot organogenesis and somatic embryogenesis (Murthy et al. 1998; Cuenca et al. 2000; Bunn et al. 2005). The morphogenetic route in TDZ media is greatly dependent on the factors like pretreatment, explant type, maturity, orientation, TDZ concentration, combination with other PGRs and organic additives. This paper comprehends the successful application of TDZ in woody plant tissue culture with critical reviews on various parameters to be considered under different culture conditions when the media is amended with TDZ.

10.2 Micropropagation

Woody perennials are predominantly exploited for their timber products. Forest trees find their utilization in agroforestry (Kuzovkina and Volk 2009), phytoremediation (Perttu and Kowalik 1997; Vervaeke et al. 2003) and production of biofuels (Mashkina et al. 2010). Continued over-exploitation and indiscriminate logging of trees for the welfare of humankind had resulted in the shrinking of forest resources (Newton et al. 1999). Around 9000 tree species worldwide are threatened with extinction (Oldfield et al. 1998). In vitro propagation has emerged as a boon for the ex situ germplasm conservation of economically important tree species (Vinoth and Ravindhran 2013). It circumvents the need to exploit natural resources, thus preserving the tree populations in their natural habitats. Micropropagation is a useful means of mass propagating plantlets from young or mature tree tissues with a lower risk of genetic instability, than those obtained by other regeneration pathways (Rao and Lee 1986). Establishment of plantlets via micropropagation involves three phases, namely, axillary shoot proliferation, shoot elongation and rooting of in vitro shoots. TDZ plays a promising role in the entire regeneration processes. Protocols employing TDZ for the axillary shoot proliferation of various tree species until the 1990s have been described in Huetteman and Preece (1993). Table 10.1 reports the use of TDZ in micropropagation of woody plants from the late 1990s till date.

Table 10.1 Successful application of TDZ in micropropagation of woody plants

Species	Explant	Effective TDZ concentration (μM)	References
Dendrocalamus strictus	Node	2.27	Singh et al. (2001)
Leucaena leucocephala	Immature zygotic embryo	0.26	Pal et al. (2012)
	Cotyledonary node	0.23	Shaik et al. (2009)
Pinus massoniana	Node	4.0	Wang and Yao (2017)
Salix tetrasperma	Node	2.5	Khan and Anis (2012)
Sterculia urens	Node, cotyledonary node	0.90	Devi et al. (2011)
Vitex trifolia	Node	5.0	Ahmed and Anis (2012)

10.2.1 Axillary Shoot Proliferation

In micropropagation, TDZ revives the meristematic activity of lateral buds suppressed by apical dominance and promotes multiple shoot formation (Table 10.1). TDZ concentration and exposure are pivotal in modulating axillary shoot proliferation in woody plants (Fig. 10.1). In majority of the woody species, shoot proliferation showed an increasing trend with the increase in TDZ concentration (Singh et al. 2001; Faisal et al. 2005; Ahmad and Anis 2007). But the optimum concentration varied depending on the species as indicated in Table 10.1. In *Salix tetrasperma*, commonly called as 'Indian Willow', nodal explants cultured on 2.5 μM TDZ produced 90% regeneration response with 4.53 shoots per explant (Khan and Anis 2012). In *Leucaena leucocephala*, a fast-growing tree legume, the optimum concentration of TDZ varied based on the type of explants used (immature zygotic embryo (1.5 μM) and cotyledonary node (0.23 μM)) (Shaik et al. 2009; Pal et al. 2012). Unlike other cytokinins which are active at higher concentrations, TDZ is highly effective in axillary shoot proliferation at very low concentrations ranging from 0.1 nM to 10 μM. The physiological role of TDZ in shoot proliferation is attributed to the accumulation of endogenous cytokinins, conversion of cytokinin nucleotides to potent biologically active nucleosides and enhanced translocation of auxins to axillary meristems (Capelle et al. 1983; Thomas and Katterman 1986; Murch and Saxena 2001).

TDZ promotes shoot proliferation in different explants which are otherwise unresponsive to other PGRs. For example, in bamboo, nodal explants collected from the base to the shoot apex of in vitro germinated seedlings produced multiple shoot buds in TDZ liquid media. On the contrary, in TDZ-free media, only basal node showed slight response, while nodes closer to the apex failed to regenerate (Singh et al. 2001). In *Sterculia urens*, cotyledonary node from 15-day-old seedlings and nodal explants from 1-month-old seedlings were comparatively regenerative. The difference was observed only in the mean number of regenerated shoots per se with TDZ × explant age interaction, where the juvenile tissues produced maximum number of shoots (Devi et al. 2011). It gave a clear understanding that the effect of TDZ in promoting regeneration competence is well pronounced irrespective of the physiological gradients along the stem and the maturity of explants.

Prolonged culture under in vitro conditions will lead to epigenetic variation, thereby affecting the genetic integrity of micropropagated plantlets. TDZ favours rapid shoot proliferation over other adenine-type cytokinins by reducing the culture cycle. For instance, in *S. urens*, TDZ produced 19 shoots per cotyledonary node in 2 harvests within 45 days when compared to 16 shoots in 3 harvests (63 days) using BAP (Devi et al. 2011). This study signified the high cytokinin activity of TDZ as reported by Huetteman and Preece (1993).

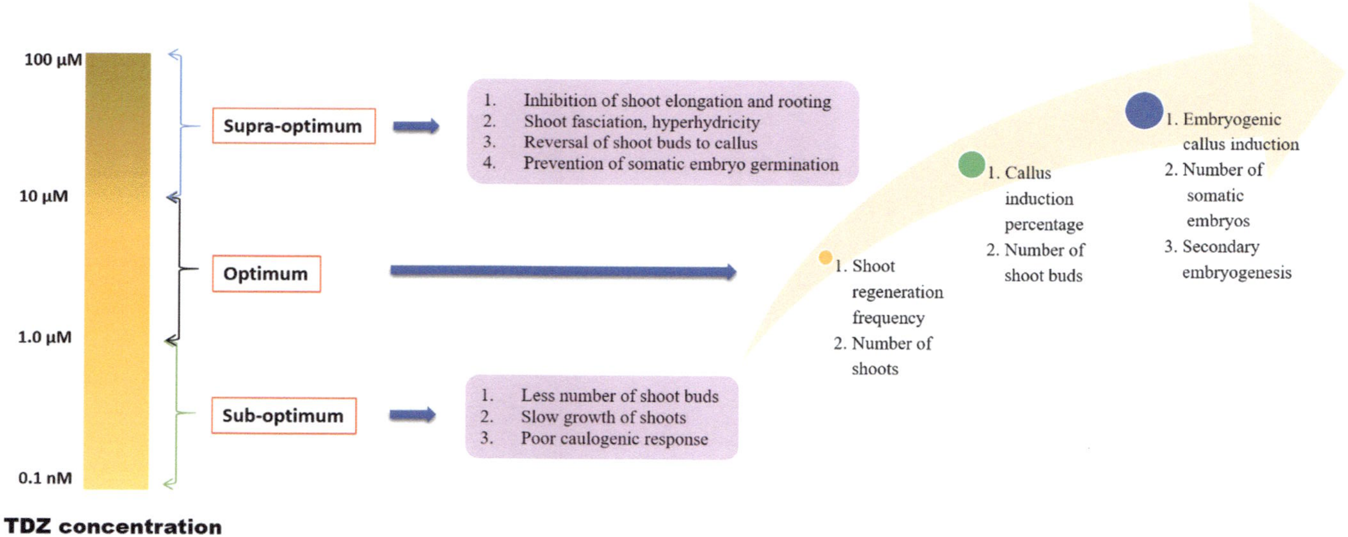

Fig. 10.1 Dose-dependent role of TDZ in shoot regeneration of woody plants. Within the optimum range, increase in TDZ had a positive correlation with different morphogenetic responses. At supra-optimal range, negative consequences were observed, while at suboptimal range, poor regeneration response is recorded. This schematic diagram does not take into account the genotype dependency

10.2.2 Shoot Elongation

Shoot elongation was found to have an inverse relation to TDZ concentration and exposure in solid as well as liquid medium (Fig. 10.1). Shoot buds initiated at concentrations greater than the optimum level (0.26 μM) of TDZ failed to elongate (Pal et al. 2012) or showed basal callusing, shoot fasciation and necrosis (Shaik et al. 2009), even after the removal of TDZ from the culture media in further subcultures. The mean number of elongated shoots was highly dependent on the dosage of TDZ in the shoot initiation medium (Singh et al. 2001; Shaik et al. 2009). Inhibition of shoot elongation by TDZ, in general, was found to be reversed upon transfer to PGR-free medium or to media supplemented with adenine-type cytokinins and auxins. In *Vitex trifolia*, a shrub by tree used by tribes and native medical practitioners, highest shoot regeneration frequency was achieved when explants exposed to TDZ for 7 days were repeatedly subcultured on MS media containing a combination of BA (1.0 μM) and NAA (0.5 μM). Prolonged exposure in TDZ media for more than 7 days resulted in fasciated or distorted shoots with occasional reversal of shoots into callus or necrotic tissues (Ahmed and Anis 2012). Similar results were observed in *S. tetrasperma,* primarily used for fuel wood and timber, when nodal explants were cultured in TDZ media for more than 4 weeks (Khan and Anis 2012). Shoot fasciation (fused shoots) is attributed to the high cytokinin activity and resistance/inhibition of cytokinin oxidases by TDZ (Huetteman and Preece 1993).

10.2.3 Rooting

Shoots regenerated from TDZ medium often failed to root in auxin-free basal medium. A negative correlation was commonly observed between TDZ concentration and mean rooting percentage (Fig. 10.1). In vitro shoots of *L. leucocephala* regenerated on lower concentrations of TDZ (0.05–0.23 μM) showed poor rooting response compared to control, while those from higher concentrations (0.45–2.27 μM) did not root at all (Shaik et al. 2009). This phenomenon was observed due to the residual effect of TDZ on the cells at the shoot base which retained their shoot organogenic capacity even after the removal of TDZ (Meyer and van Staden 1988). Rooting of such shoots required transfer to media containing different dosages of auxins like IBA, NAA or IAA. In a study by Khan and Anis (2012), in vitro regenerated shoots of willow species underwent rooting only in medium containing 0.5 μM IBA. Contrastingly, shoots induced from media supplemented with TDZ higher than 0.6 μM showed no root development in rooting medium containing NAA or IBA and Kin (Pal et al. 2012). Rooting of in vitro shoots is a vital phase which affects their subsequent acclimatization and survivability in field conditions. Though TDZ promotes rapid proliferation of axillary shoots, dosage and exposure time in the shoot induction medium are two critical parameters to be monitored to facilitate proper root development.

10.3 Adventitious Shoot Organogenesis

Adventitious shoot regeneration is a promising approach for genetic transformation in tree species. Standardized protocol for adventitious shoot bud initiation is a prerequisite to introduce desired traits into elite tree genotypes. Shoot organogenesis occurs through three vital developmental stages, namely, competence (dedifferentiation), determination (redifferentiation) and morphogenesis (Sugiyama 1999). Transition to different stages occurs by the balance of exogenously supplied PGRs. Among the PGRs, TDZ is highly influential in regenerating shoot buds from the explants of woody plants, in particular, recalcitrant genotypes (Table 10.2).

10.3.1 Callus Induction (Dedifferentiation)

Callus induction is the preliminary step to achieve shoot organogenesis from woody plants as explants from mature tissues fail to produce shoot buds de novo. TDZ is an effective PGR known for its greater ability to induce callus (Table 10.2). Various explants respond differently to TDZ concentration in the media. In a study on regeneration of *Eucalyptus grandis* × *E. urophylla*, 2.0 μM TDZ induced callus in 95–100% of the explants (hypocotyl, cotyledon, primary leaves and cotyledonary node) (Barrueto Cid et al. 1999). Despite the difference in the types of explant, callus was fast-growing, homogeneous and highly regenerative. The regeneration potential was however maintained only until 30 days of TDZ exposure, and further extension beyond this time frame resulted in callus necrosis. Shoot regeneration in beeches (*Fagus* sp.), commercially important trees for timber production, was achieved using internodal explants through callus culture (Cuenca et al. 2000). In this study, callus with higher organogenic capacity was induced from different genotypes of *F. orientalis* and *F. sylvatica*. TDZ was thus able to overcome the limitations offered by explant type and genotypes in producing organogenic callus.

10.3.2 Shoot Bud Initiation (Redifferentiation)

Induction of adventitious shoot buds occurs either from the explant (direct) or from the organogenic callus (indirect). Leaf, petiole and cotyledon were the most suitable explants for adventitious shoot organogenesis. In *Robinia pseudoacacia*, seeds and hypocotyl were used as explants, wherein seeds formed shoots via callusing from the root region, while hypocotyl exhibited de novo meristematic activity in TDZ medium (Hosseini-Nasr and Rashid 2004). Explants from juvenile tissues of in vitro grown seedlings were highly competent to shoot regeneration compared to the in vivo explants from mature trees (Mante et al. 1989; Liu and Pijut 2008; Kumar et al. 2010a, b; Aggarwal et al. 2015). Explant preparation and orientation in medium are vital factors to be considered in shoot organogenesis. Shoot initials were induced from cotyledon explants primarily in the proximal regions devoid of embryonal axis (Mante et al. 1989; Sujatha et al. 2008). Excision of proximal part from cotyledon

Table 10.2 Adventitious shoot organogenesis in woody plants using TDZ

Regeneration route	Species	Explant	Effective TDZ concentration (μM)	References
Direct	*Acacia crassicarpa*	Phyllode	2.27	Yang et al. (2006)
	Alstroemeria sp.	Leaf	10.0	Lin et al. (1997)
	Hagenia abyssinica	Leaf	<1.0	Feyissa et al. (2005)
	Nothapodytes foetida	Leaf	1.36	Thengane et al. (2001)
		Hypocotyl	2.27	
	Populus alba × *P. berolinensis*	Stem	0.45	Wang et al. (2008)
	Paulownia tomentosa	Leaf	22.7	Corredoira et al. (2008)
	Pongamia pinnata	Cotyledon	11.35	Sujatha et al. (2008)
	Populus tremula	Root	0.04	Vinocur et al. (2000)
	Prunus serotina	Leaf	4.4	Hammatt and Grant (1998)
	P. avium	Leaf	9.08	Liu and Pijut (2008)
	P. serotine		22.7	
	Ricinus communis	Hypocotyl	1.0	Ahn et al. (2007)
Indirect	*Acacia mangium*	Embryo axes, cotyledon, leaf, petiole and stem	4.55	Xie and Hong (2001)
	Eucalyptus grandis × *E. urophylla*	Hypocotyl, cotyledon, primary leaves and cotyledonary node	2.0	Barrueto Cid et al. (1999)
	Fagus sylvatica	Leaf	2.3	Vieitez and San José (1996)
	F. orientalis	Internode	4.5	Cuenca et al. (2000)
	F. sylvatica			
	Hagenia abyssinica	Leaf	>1.0	Feyissa et al. (2005)
	Jatropha curcas	Leaf	2.27	Deore and Johnson (2008)
	Pinus strobus	Zygotic embryo	6.0	Tang and Newton (2005)
	Prunus persica × *P. davidiana*	Leaf	9.08	Zhou et al. (2010)
	Robinia pseudoacacia	Seed	1.0	Hosseini-Nasr and Rashid (2004)
	Santalum album	Node	0.6	Singh et al. (2016)
	Ulmus americana	Leaf	22.5	George and Tripepi (1994)

explants failed to induce shoot buds even with the supplementation of TDZ. Culturing of cotyledon explants with abaxial surface in the medium was more effective and generated more buds than the adaxial side (Sujatha et al. 2008). Leaf explants responded contrastingly with higher shoot regeneration frequency when the adaxial side was in contact with the media (Kim et al. 2007). With petiole explants, horizontal positioning induced more shoot buds than vertical placement as more surface area was in contact with the medium (Kumar et al. 2010b). Leaf explants closer to the shoot apex formed adventitious buds not only on the petiolar end but also on the laminar end (Corredoira et al. 2008). Similarly, internodal explants displayed decrease in regeneration frequency and shoot bud number basipetally along the stem (Cuenca et al. 2000).

Culturing of explants in TDZ media under varied photoperiod conditions also affected the regeneration frequency. Pre-culturing of leaf explants under darkness for a short duration improved the regeneration percentage in *Ficus carica* (Kim et al. 2007). Genotype dependency was another notable factor that interplayed with TDZ in adventitious shoot regeneration. In trees like apple, cherry, beech and poplars, TDZ was prominently superior over other cytokinins in shoot bud induction across different species. However, the range of TDZ concentration varied greatly across different genotypes (Hammatt and Grant 1998; Cuenca et al. 2000; Magyar-Tábori et al. 2010; Aggarwal et al. 2015).

In majority of the tree species, amendment of TDZ in the shoot induction medium (liquid or solid) hastened the regeneration process (Sriskandarajah and Goodwin 1998). TDZ supplementation exhibited dose-dependent response with increase in the regeneration efficiency until a saturation point at 10 μM (Fig. 10.1). Cuenca et al. (2000) reported significant interaction between bud-forming capacity and TDZ concentration in beech, with optimal concentration being 4.5 μM. There are exceptions where the optimum TDZ concentration shifted to above 10 μM. In vitro regeneration frequency of *Paulownia tometosa* (empress tree) from leaf explants was maximum in the induction media supplemented with 22.7–27.3 μM TDZ (Corredoira et al. 2008) and that of *Ulmus americana* (American elm) at 22.5 μM TDZ (George and Tripepi 1994).

TDZ pretreatment promoted autonomous competence in cells, that is, to regenerate in the absence of PGRs. An interesting observation was recorded in *Hagenia abyssinica*, where TDZ concentrations less than 1.0 μM induced direct shoot buds while above 1.0 μM produced callus. The calli exhibited 96–100% regeneration when TDZ was removed or supplemented at lower concentration (0.1 μM) (Feyissa et al. 2005). In another study, TDZ-pretreated calli derived from cotyledons, hypocotyls or cotyledonary nodes were able to regenerate shoot buds in PGR-free medium (Barrueto Cid et al. 1999). Therefore, it is clearly evident that TDZ could act as a sole PGR that positively influenced the regeneration ability without requiring the need for combination with other auxins/cytokinins in the induction medium. Even the report discussed below describing the combination of auxins/cytokinins with TDZ for shoot bud induction indicated the prominent role of TDZ in the synergistic interaction. In a study by Deore and Johnson (2008), TDZ (2.27 μM) + BA (2.22 μM) + IBA (0.49 μM) induced maximum adventitious shoot buds (53.5%) in leaf disc explants. Elimination of TDZ from this combination drastically reduced the shoot bud induction and resulted in higher callus induction, thereby displaying its necessity for the promotion of organogenic competence.

Conclusively, in spite of TDZ being potent cytokinin to initiate shoot organogenesis in recalcitrant tissues, explant type, explant age, preprocessing of explants, physiological gradients in parent tissue, orientation in the medium, photoperiod and genotype exhibited synergistic interactions to promote regeneration competence in cells.

10.3.3 Shoot Morphogenesis and Rooting

Amendment of TDZ greater than 10 μM in the shoot induction medium suppressed the elongation of shoots, thereby reducing the number of harvestable shoots. In *Fagus* sp., TDZ greater than 8.9 μM produced bud clusters on callus, which were dense and compact mass of minute buds. At still higher concentration, shoot buds appeared as green nodular organogenic patches which failed to elongate into shoots (Cuenca et al. 2000). This was attributed to the compaction of shoot buds due to spatial constraint. The shoot clusters also required longer time to develop into individual shoots (Liu and Pijut 2008; Kumar et al. 2010b). In some cases, shoots were hyperhydric and fasciated (Pawlicki and Welander 1994; Caboni et al. 1996; Dobránszki et al. 2002).

Combination of auxins, adenine-type cytokinins or organic additives was found to be necessary to promote elongation of shoot buds that were suppressed by TDZ. Regeneration of macadamia trees (*Macadamia tetraphylla*) from cotyledon explants underwent three developmental stages including callus formation, shoot bud initiation and shoot regeneration. Supplementation of coconut milk (2%) + TDZ (15 μM) improved both callus and shoot bud induction. However, the conversion of buds into individual shoots required the elimination of TDZ and coconut milk from the shoot initiation medium and subsequent transfer to shoot development medium containing BA alone or combined with GA_3 (Mulwa and Bhalla 2006). Likewise, in Himalayan poplar, shoot buds initiated from MS medium supplemented with 0.02 μM TDZ + 79.7 mg/L adenine were transferred to elongation medium containing BAP and GA_3 (Aggarwal et al. 2015). Barrueto Cid et al. (1999) reported shoot regeneration and elongation from TDZ-derived callus only upon transfer to medium containing BAP, NAA and GA_3. Shoot elongation in *Jatropha curcas* was best achieved in medium containing 2.25 μM BA and 8.5 μM IAA (Kumar et al. 2010b). In all the above reports, the shoot buds did not elongate simultaneously, and the elongated shoots have to be excised continuously to reduce the growth suppression of young buds. As described in Sect. 10.2.3, rooting of in vitro shoots produced from TDZ media was prominent in auxin-based medium.

10.4 Somatic Embryogenesis

Somatic embryogenesis is the most preferred regenerative pathway for mass propagation of recalcitrant woody plants. As somatic embryos (SE) originate by bipolar development of somatic cells, SE-derived plantlets are genetically identical to the seed-grown plantlets. Somatic embryogenesis occurs through three stages comprising of embryogenic callus induction, SE maturation and conversion of SE into plantlets. In addition, SEs undergo cyclic embryogenesis resulting in the formation of secondary somatic embryos (SSE) originating directly from primary SEs. Auxins, in general, are potent inducers of somatic cells transition to embryogenic cells in vitro. TDZ, though being a cytokinin, mimics auxins by stimulating direct and indirect somatic embryogenesis in several woody plants (Table 10.3).

10.4.1 Embryogenic Callus Induction

Embryogenic callus induction is the primary approach to generate SEs from woody tissues which are otherwise recalcitrant to de novo somatic embryogenesis. Embryogenic calli are characterized by nodular structures called 'pro-embryogenic masses' (Jiménez and Bangerth 2001). Supplementation of TDZ had a positive influence on indirect somatic embryogenesis from staminode explants of 19 genotypes in *Theobroma cacao* (Li et al. 1998). Previously published reports on *T. cacao* expressed poor embryogenic response using other PGRs. In the study by Li et al. (1998), development of SEs required three different regeneration media: primary

Table 10.3 In vitro regeneration of woody plants through somatic embryogenesis using TDZ

Regeneration route	Species	Explant	Effective TDZ concentration (μM)	References
Direct	*Melia azedarach*	Immature zygotic embryo	13.62	Vila et al. (2003)
	Santalum yasi × *S. album*	Node	4.55	Zhang et al. (2016)
Indirect	*Eucalyptus microtheca*	Green twigs	0.45	Mamaghani et al. (2009)
	Murraya koenigii	Zygotic embryo, cotyledon	4.54	Paul et al. (2011)
	Paulownia elongata	Leaf, internode	0.45	Ipekci and Gozukirmizi (2004)
	Theobroma cacao	Staminode	0.02	Li et al. (1998)
Secondary embryogenesis	*Cinnamomum camphora*	Immature zygotic embryo	0.90	Shi et al. (2010)
	M. koenigii	Zygotic embryo, cotyledon	9.08	Paul et al. (2011)
	Prunus avium × *P. pseudocerasus*	Transgenic roots	0.45	Pesce and Rugini (2004)

callus induction, secondary callus growth and somatic embryo development. TDZ (22.7–454.5 nM) was amended only in the primary callus induction medium for a short exposure of about 2 weeks. Inclusion of TDZ in the initial stages displayed a positive correlation between embryogenic callus induction and conversion of pro-embryogenic masses into SEs compared to TDZ-free medium. The highest number of SEs (46 per responsive staminode) was obtained in medium containing minimal concentration (22.7 nM) of TDZ. This dosage of TDZ was 20–600-folds less, when compared to other studies in Table 10.3, thus providing a clear evidence of its high growth-promoting activity even at negligible concentrations. Deviation above this optimum concentration (22.7 nM) resulted in necrotic tissues while below produced poor callogenic response (Li et al. 1998). In *Paulownia elongata*, higher concentration of TDZ (18.16 μM) favoured embryogenic callus induction from leaf and internode explants of micropropagated plantlets (Ipekci and Gozukirmizi 2004). Nevertheless, the dosage of TDZ required to induce embryogenic competence varied according to the tree species.

10.4.2 Somatic Embryo Maturation

Maturation of SEs occurs through four developmental stages (globular, heart-shaped, torpedo and cotyledonary stage) commencing from small rounded structure and terminating into bipolar structure bearing shoot apical meristem embedded between a pair of cotyledons, hypocotyl and a root axis. Differentiation of SEs is characterized by accumulation of lipid-rich globular bodies (Bandyopadhyay and Hamill 2000). Usually, SEs of various medicinal plants undergo maturity when transferred to media containing cytokinins like BA and Kin, while TDZ resulted in reversion to callus (Baskaran and van Staden 2012). In woody plants, TDZ had higher embryogenic activity than that of adenine-type derivatives.

Development of globular, heart-shaped and torpedo stage SEs in *P. elongata* was achieved on medium containing a combination of 0.45 μM TDZ and 4.64 μM Kin (Ipekci and Gozukirmizi 2004). Inclusion of TDZ modulated the embryogenic pathway as direct or indirect in *Murraya koenigii*, commonly referred as curry plant (Paul et al. 2011). Culturing of cotyledon and zygotic embryo explants in TDZ medium produced direct SEs, while embryogenic calli obtained from medium containing 4.44 μM BA and 2.675 μM NAA formed mature SEs on exposure to TDZ. Concentration of TDZ above or below the optimum level (4.54 μM) could not trigger the sequential development of SEs from globular to cotyledonary stages (Paul et al. 2011). Likewise, in *Eucalyptus microtheca*, inclusion of TDZ facilitated SE formation in 81% of embryogenic calli obtained from 18.56 μM Kin + 2.68 μM NAA (Mamaghani et al. 2009). In total, 244 plantlets were regenerated from SEs with the supplementation of 0.45 μM TDZ.

10.4.3 Somatic Embryo Germination

In several woody plant species' embryogenic systems, the limiting step is the conversion of SEs into plantlets. The reduction in plant recovery from SE is due to poor embryo quality, lack of maturation and desiccation tolerance (Etienne et al. 2013). Germination of SEs thus depends on the cellular organization of meristem regions and the sizes of vacuoles in embryo cells (Nickle and Yeung 1993; Taylor and Vasil 1996). Prolonged culture and exposure of SEs to higher concentration of TDZ prevented their precocious germination. Conversion of SEs into plantlets thus required transfer of cotyledonary stage SEs either to basal medium or to medium containing other PGRs. In *Murraya koenigii*, direct SEs that originated in TDZ medium failed to regenerate into plantlets in the same medium, while germination was achieved in medium containing NAA and Kin (Paul et al. 2011). Similarly, direct SEs from zygotic embryo of *Melia azedarach* germinated into well-developed plantlets in quarter-strength MS basal medium containing activated charcoal (Vila et al. 2003).

10.4.4 Secondary Somatic Embryogenesis

Cyclic secondary somatic embryogenesis maintains the embryogenic competence of cells for longer time period. TDZ facilitates production of secondary somatic embryos (SSE) by interplaying with various factors like carbon source, photoperiod conditions, etc. (Fig. 10.1). In transgenic cherry rootstock 'Colt' (*Prunus avium* ×*P. pseudocerasus*), interaction of TDZ with maltose produced higher number of SSEs. On the contrary, combination with sucrose dedifferentiated the embryogenic masses into non-morphogenic callus. Dark incubation of embryogenic calli in TDZ medium was superior to light conditions in promoting secondary somatic embryogenesis (Pesce and Rugini 2004). In cherry and apple, higher TDZ concentration resulted in reversal of SSE to callus, thereby indicating the modifications in the endogenous auxin/cytokinin ratio (Daigny et al. 1996; Pesce and Rugini 2004). Higher concentration of TDZ (9.08 μM) induced secondary embryogenesis in *M. koenigii* but produced abnormal embryos with fused margin, fused polycotyledons and a single cotyledon (Paul et al. 2011). Lower concentration of TDZ (0.90 μM) formed opaque white SSEs in camphor tree (*Cinnamomum camphora*), an indicative character of embryo maturation (Shi et al. 2010). Recurrent embryogenesis was observed when primary SE was cultured back in TDZ medium (Vila et al. 2003; Shi et al. 2010).

10.5 TDZ-Induced Stress

TDZ, being a cytokinin analogue, has severe negative consequences on in vitro morphogenesis (Fig. 10.1). Though it has maximum cytokinin activity, prolonged exposure of explants resulted in the production of fasciated or vitrified shoots and sometimes leading to reversal of shoots into callus or necrotic tissues (Poudyal et al. 2008). On the contrary, suboptimal exposure results in the dormancy or slow growth of shoot buds. In somatic embryogenesis, prolonged subculture and higher concentration of TDZ reduced the frequency of embryogenesis and resulted in reversal of embryogenic masses into non-morphogenic callus (Pesce and Rugini 2004). This may be attributed to the changes in the levels of the endogenous auxin/cytokinin and elevated levels of ethylene induced by TDZ (Lu 1993).

TDZ pretreatment also suppressed the rooting of regenerated shoots (Khan and Anis 2012). A report indicated the negative consequence of TDZ on seed germination which resulted in rapid expansion of cotyledons, stunted shoots and impairment of root growth (Murthy and Saxena 1998). Higher concentration of TDZ produced an abnormality termed 'burr knots' in leaf explants of pear. Burr knots are brown, radicle-like structures that cannot differentiate into roots or shoots (Poudyal et al. 2008).

10.6 Conclusion and Recommendations

Plant tissue culturists have exploited TDZ to the maximum potential to induce a wide variety of morphogenetic responses in recalcitrant woody plants. Because of its greater stability and high cytokinin activity at very low concentrations, TDZ stands alone as a potent PGR influencing regeneration in recalcitrant tissues compared to its natural counterparts. However, the use of TDZ in plant cell cultures has its own limitations. TDZ produces toxic effects at higher concentration and prolonged exposure, thereby resulting in the growth of abnormal shoots. TDZ also failed to overcome the differential explant response and genotype dependency, a similar phenomenon exhibited by other PGRs. Henceforth, it is recommended to carefully consider all the interplaying factors to negate the shoot abnormalities in using TDZ. Further investigations are also needed to elucidate the molecular mechanism behind in vitro shoot regeneration by TDZ. This could provide a better understanding for experimental design to achieve regeneration via suitable regeneration pathway.

Acknowledgements The project is financially supported by University Grants Commission (UGC), New Delhi [F. 39-259/2010 (SR) dated 27 December 2010] under the scheme Major Research Project granted to Dr. R. Ravindhran. The authors are also thankful to the Loyola College management, Chennai, for providing the laboratory and infrastructure facilities. We also express our sincere gratitude to Dr. G. Ganesan and Dr. S. Selvalakshmi for their critical comments and suggestions on the manuscript.

Author Contributions AV and RR conceptualized the manuscript. AV prepared the manuscript, while RR edited and provided critical inputs to the manuscript for publication.

Conflict of Interest The authors declare that the review paper was written in the absence of any commercial or financial relationships that could be construed as a potential conflict of interest.

References

Aggarwal G, Gaur A, Srivastava DK (2015) Establishment of high frequency shoot regeneration system in Himalayan poplar (*Populus ciliata* Wall. ex Royle) from petiole explants using thidiazuron cytokinin as plant growth regulator. J For Res 26:651–656. https://doi.org/10.1007/s11676-015-0048-6

Ahmad N, Anis M (2007) Rapid plant regeneration protocol for cluster bean (*Cyamopsis tetragonoloba* L. Taub.) J Hortic Sci Biotechnol 82:585–589. https://doi.org/10.1080/14620316.2007.11512277

Ahmed MR, Anis M (2012) Role of TDZ in the quick regeneration of multiple shoots from nodal explant of *Vitex trifolia* L.-an important medicinal plant. Appl Biochem Biotechnol 168:957–966. https://doi.org/10.1007/s12010-012-9799-0

Ahn YJ, Vang L, McKeon TA, Chen GQ (2007) High-frequency plant regeneration through adventitious shoot formation in castor (*Ricinus communis* L.) In Vitro Cell Dev Biol Plant 43:9–15. https://doi.org/10.1007/s11627-006-9009-2

Arndt FR, Rusch R, Stillfried HV, Hanisch B, Martin WC (1976) SN 49537, a new cotton defoliant. Plant Physiol 57:S–99

Bandyopadhyay S, Hamill JD (2000) Ultrastructural studies of somatic embryos of *Eucalyptus nitens* and comparisons with zygotic embryos found in mature seeds. Ann Bot 86:237–244. https://doi.org/10.1006/anbo.2000.1192

Barrueto Cid LP, Machado AC, Carvalheira SB, Plant BAC (1999) Regeneration from seedling explants of *Eucalyptus grandis* × *E. urophylla*. Plant Cell Tissue Organ Cult 56:17–23. https://doi.org/10.1023/A:1006283816625

Baskaran P, van Staden J (2012) Somatic embryogenesis of *Merwilla plumbea* (Lindl.) Speta. Plant Cell Tissue Organ Cult 109:517–524. https://doi.org/10.1007/s11240-012-0118-9

Bunn E, Senaratna T, Sivasithamparam K, Dixon KW (2005) In vitro propagation of *Eucalyptus phylacis* L. Johnson and K. Hill., a critically endangered relict from western Australia. In Vitro Cell Dev Biol Plant 41:812–815. https://doi.org/10.1079/IVP2005700

Caboni E, Tonelli M, Falasca G, Damiano C (1996) Factors affecting adventitious shoot regeneration in vitro in the apple rootstock 'Jork 9'. Adv Hortic Sci 10:146–150

Capelle SC, Mok DWS, Kirchner SC, Mok MC (1983) Effects of thidiazuron on cytokinin autonomy and the metabolism of N^6-(Δ^2-isopentenyl)[8-^{14}C] adenosine in callus tissues of *Phaseolus lunatus* L. Plant Physiol 73:796–802. https://doi.org/10.1104/pp.73.3.796

Corredoira E, Ballester A, Vieitez AM (2008) Thidiazuron-induced high-frequency plant regeneration from leaf explants of *Paulownia tomentosa* mature trees. Plant Cell Tissue Organ Cult 95:197–208. https://doi.org/10.1007/s11240-008-9433-6

Cuenca B, Ballester A, Vieitez AM (2000) In vitro adventitious bud regeneration from internode segments of beech. Plant Cell Tissue Organ Cult 60:213–220. https://doi.org/10.1023/A:1006428717309

Daigny G, Paul H, Sangwan RS, Sangwan-Norreel BS (1996) Factors influencing secondary somatic embryogenesis in *Malus x domestica* Borkh. (cv 'Gloster 69'). Plant Cell Rep 16:153–157. https://doi.org/10.1007/BF01890857

Deore AC, Johnson TS (2008) High-frequency plant regeneration from leaf-disc cultures of *Jatropha curcas* L.: an important biodiesel plant. Plant Biotechnol Rep 2:7–11. https://doi.org/10.1007/s11816-008-0042-y

Devi PS, Arundathi A, Rao TR (2011) Multiple shoot induction and regeneration of whole plants from cotyledonary node and nodal explants of *Sterculia urens* Roxb., a gum yielding tree. J Plant Biochem Biotechnol 20:161–165. https://doi.org/10.1007/s13562-011-0041-x

Dobránszki J, Magyar-Tábori K, Jámbor-Benczúr E, Kiss E, Lazányi J, Bubán T (2002) Effect of conditioning apple shoots with meta-topolin on the morphogenic activity of in vitro leaves. Acta Agron Hung 50:117–126. https://doi.org/10.1556/AAgr.50.2002.2.1

Ďurkovič J, Mišalová A (2008) Micropropagation of temperate noble hardwoods: an overview. Func Plant Sci Biotechnol 2:1–19

Etienne H, Bertrand B, Georget F, Barry-Etienne D (2013) Development of coffee somatic and zygotic embryos to plants differs in the morphological, histochemical and hydration aspects. Tree Physiol 33:640–653. https://doi.org/10.1093/treephys/tpt034

Faisal M, Ahmad N, Anis MF (2005) Shoot multiplication in *Rauvolfia tetraphylla* L. using thidiazuron. Plant Cell Tissue Organ Cult 80:187–190. https://doi.org/10.1007/s11240-004-0567-x

Feyissa T, Welander M, Negash L (2005) In vitro regeneration of *Hagenia abyssinica* (Bruce) J.F. Gmel. (Rosaceae) from leaf explants. Plant Cell Rep 24:392–400. https://doi.org/10.1007/s00299-005-0949-5

George MW, Tripepi RR (1994) Cytokinins, donor plants and time in culture affect shoot regenerative capacity of American elm leaves. Plant Cell Tissue Organ Cult 39:27–36. https://doi.org/10.1007/BF00037588

Hammatt N, Grant NJ (1998) Shoot regeneration from leaves of *Prunus serotina* Ehrh. (black cherry) and *P. avium* L. (wild cherry). Plant Cell Rep 17:526–530. https://doi.org/10.1007/s002990050436

Hosseini-Nasr M, Rashid A (2004) Thidiazuron-induced high-frequency shoot regeneration from root region of *Robinia pseudoacacia* L. seedlings. Biol Plant 47:593–596. https://doi.org/10.1023/B:BIOP.0000041068.19770.95

Huetteman C, Preece J (1993) Thidiazuron: a potent cytokinin for woody plant tissue culture. Plant Cell Tissue Organ Cult 33:105–119. https://doi.org/10.1007/BF01983223

Ipekci Z, Gozukirmizi N (2004) Indirect somatic embryogenesis and plant regeneration from leaf and internode explants of *Paulownia elongata*. Plant Cell Tissue Organ Cult 79:341–345. https://doi.org/10.1007/s11240-003-4632-7

Jiménez VM, Bangerth F (2001) Endogenous hormone levels in explants and in embryogenic and non-embryogenic cultures of carrot. Physiol Plant 111:389–395. https://doi.org/10.1034/j.1399-3054.2001.1110317.x

Khan MI, Anis M (2012) Modulation of in vitro morphogenesis in nodal segments of *Salix tetrasperma* Roxb. through the use of TDZ, different media types and culture regimes. Agrofor Syst 86:95–103. https://doi.org/10.1007/s10457-012-9512-x

Kieber JJ, Schaller GE (2014) Cytokinins. Arabidopsis Book 12:e0168. https://doi.org/10.1199/tab.0168

Kim KM, Kim MY, Yun PY, Chandrasekhar T, Lee HY, Song PS (2007) Production of multiple shoots and plant regeneration from leaf segments of fig tree (*Ficus carica* L.) J Plant Biol 50:440–446. https://doi.org/10.1007/BF03030680

Kumar S, Kumaria S, Tandon P (2010a) Efficient in vitro plant regeneration protocol from leaf explant of *Jatropha curcas* L. – a promising biofuel plant. J Plant Biochem Biotechnol 19:273–275. https://doi.org/10.1007/BF03263356

Kumar N, Vijay Anand KG, Reddy MP (2010b) In vitro plant regeneration of non-toxic *Jatropha curcas* L.: direct shoot organogenesis from cotyledonary petiole explants. J Crop Sci Biotechnol 13:189–194. https://doi.org/10.1007/s12892-010-0039-2

Kuzovkina YA, Volk TA (2009) The characterization of willow (*Salix* L.) varieties for use in ecological engineering applications: co-ordination of structure, function and autecology. Ecol Eng 35:1178–1189. https://doi.org/10.1016/j.ecoleng.2009.03.010

Li Z, Traore A, Maximova S, Guiltinan MJ (1998) Somatic embryogenesis and plant regeneration from floral explants of cacao (*Theobroma cacao* L.) using thidiazuron. In Vitro Cell Dev Biol Plant 34:293–299. https://doi.org/10.1007/BF02822737

Lin HS, De Jeu MJ, Jacobsen E (1997) Direct shoot regeneration from excised leaf explants of in vitro grown seedlings of *Alstroemeria* L. Plant Cell Rep 16:770–774. https://doi.org/10.1007/s002990050317

Liu X, Pijut PM (2008) Plant regeneration from in vitro leaves of mature black cherry (*Prunus serotina*). Plant Cell Tissue Organ Cult 94:113–123. https://doi.org/10.1007/s11240-008-9393-x

Lu CY (1993) The use of thidiazuron in tissue culture. In Vitro Cell Dev Biol Plant 29:92–96. https://doi.org/10.1007/BF02632259

Magyar-Tábori K, Dobránszki J, da Silva JAT, Bulley SM, Hudák I (2010) The role of cytokinins in shoot organogenesis in apple. Plant Cell Tissue Organ Cult 101:251–267. https://doi.org/10.1007/s11240-010-9696-6

Mamaghani MS, Assareh MH, Omidi M, Matinizadeh M, Ghamari-Zare A, Shahrzad S et al (2009) The effect of thidiazuron level on in vitro regeneration type and peroxidase profile in *Eucalyptus microtheca* F. muell. Plant Growth Regul 59:199–205. https://doi.org/10.1007/s10725-009-9404-x

Mante S, Scorza R, Cordts JM (1989) Plant regeneration from cotyledons of *Prunus persica*, *Prunus domestica*, and *Prunus cerasus*. Plant Cell Tissue Organ Cult 19:1–11. https://doi.org/10.1007/BF00037771

Mashkina OS, Tabatskaya TM, Gorobets AI, Shestibratov KA (2010) Method of clonal micropropagation of different willow species and hybrids. Appl Biochem Microbiol 46:769–775. https://doi.org/10.1134/S0003683810080065

Meyer HJ, Van Staden J (1988) In vitro multiplication of *Ixia flexuosa*. Hortscience 23:1070–1071

Mok MC, Mok DWS, Armstrong DJ, Shudo K, Isogai Y, Okamoto T (1982) Cytokinin activity of N-phenyl-N′-1,2,3-thiadiazol-5-ylurea (thidiazuron). Phytochemistry 21:1509–1511. https://doi.org/10.1016/S0031-9422(82)85007-3

Mok MC, Mok DWS, Turner JE, Mujer CV (1987) Biological and biochemical effects of cytokinin-active phenylurea derivatives in tissue culture systems. Hort Sci 22:1194–1196

Mulwa RMS, Bhalla PL (2006) In vitro plant regeneration from immature cotyledon explants of macadamia (*Macadamia tetraphylla* L. Johnson). Plant Cell Rep 25:1281–1286. https://doi.org/10.1007/s00299-006-0182-x

Murch SJ, Saxena PK (2001) Molecular fate of thidiazuron and its effects on auxin transport in hypocotyls tissues of *Pelargonium hortorum* Bailey. Plant Growth Regul 35:269–275. https://doi.org/10.1023/A:1014468905953

Murthy BNS, Saxena PK (1998) Somatic embryogenesis and plant regeneration of neem (*Azadirachta indica* A. Juss.) Plant Cell Rep 17:469–475. https://doi.org/10.1007/s002990050427

Murthy BNS, Murch SJ, Saxena PK (1998) Thidiazuron: a potent regulator of in vitro plant morphogenesis. In Vitro Cell Dev Biol Plant 34:267–275. https://doi.org/10.1007/BF02822732

Newton AC, Allnutt TR, Gillies ACM, Lowe AJ, Ennos RA (1999) Molecular phylogeography, intraspecific variation and the conservation of tree species. Trends Ecol Evol 14:140–145. https://doi.org/10.1016/S0169-5347(98)01555-9

Nickle TC, Yeung E (1993) Failure to establish a functional shoot meristem may be a cause of conversion failure in somatic embryos of *Daucus carota* (Apiaceae). Am J Bot 80:1284–1291. https://doi.org/10.2307/2445712

Oldfield S, Lusty C, Mac Kinven A (1998) The world list of threatened trees. World Conservation Press, Cambridge

Pal A, Negi VS, Borthakur D (2012) Efficient in vitro regeneration of *Leucaena leucocephala* using immature zygotic embryos as explants. Agrofor Syst 84:131–140. https://doi.org/10.1007/s10457-011-9438-8

Paul S, Dam A, Bhattacharyya A, Bandyopadhyay TK (2011) An efficient regeneration system via direct and indirect somatic embryogenesis for the medicinal tree *Murraya koenigii*. Plant Cell Tissue Organ Cult 105:271–283. https://doi.org/10.1007/s11240-010-9864-8

Pawlicki N, Welander M (1994) Adventitious shoot regeneration from leaf segments of in vitro cultured shoots of the apple rootstock 'Jork 9'. J Hortic Sci 69:687–696. https://doi.org/10.1080/14620316.1994.11516501

Perttu KL, Kowalik PJ (1997) *Salix* vegetation filters for purification of waters and soils. Biomass Bioenergy 12:9–19. https://doi.org/10.1016/S0961-9534(96)00063-3

Pesce PG, Rugini E (2004) Influence of plant growth regulators, carbon sources and iron on the cyclic secondary somatic embryogenesis and plant regeneration of transgenic cherry rootstock Colt' (*Prunus avium* × *P. pseudocerasus*). Plant Cell Tissue Organ Cult 79:223–232. https://doi.org/10.1007/s11240-004-0663-y

Poudyal BK, Zhang Y, Du G (2008) Adventitious shoot regeneration from the leaves of some pear varieties (*Pyrus* spp.) grown in vitro. Front Agric China 2:82–92. https://doi.org/10.1007/s11703-008-0016-4

Rao AN, Lee SK (1986) An overview of the in vitro propagation of woody plants and plantation crops. In: Withers LA, Aldeson PG (eds) Plant tissue culture and its agricultural application. University Press, Cambridge, pp 123–138

Shaik NM, Arha M, Nookaraju A, Gupta SK, Srivastava S, Yadav AK et al (2009) Improved method of in vitro regeneration in *Leucaena leucocephala* – a leguminous pulpwood tree species. Physiol Mol Biol Plants 15:311–318. https://doi.org/10.1007/s12298-009-0035-5

Shi X, Dai X, Liu G, Zhang J, Ning G, Bao M (2010) Cyclic secondary somatic embryogenesis and efficient plant regeneration in camphor tree (*Cinnamomum camphora* L.) In Vitro Cell Dev Biol Plant 46:117–125. https://doi.org/10.1007/s11627-009-9272-0

Singh M, Jaiswal U, Jaiswal VS (2001) Thidiazuron-induced shoot multiplication and plant regeneration in bamboo (*Dendrocalamus strictus* nees). J Plant Biochem Biotechnol 10:133–137. https://doi.org/10.1007/BF03263122

Singh CK, Raj SR, Jaiswal PS, Patil VR, Punwar BS, Chavda JC et al (2016) Effect of plant growth regulators on in vitro plant regeneration of sandalwood (*Santalum album* L.) via organogenesis. Agrofor Syst 90:281–288. https://doi.org/10.1007/s10457-015-9853-3

Sriskandarajah S, Goodwin P (1998) Conditioning promotes regeneration and transformation in apple leaf explants. Plant Cell Tissue Organ Cult 53:1–11. https://doi.org/10.1023/A:1006044429449

van Staden J, Zazimalova E, George EF (2008) Plant growth regulators. II. Cytokinins, their analogues and antagonists. In: George EF, Hall MA, De Klerk GJ (eds) Plant propagation by tissue culture, 3rd edn. Springer, Dordrecht, pp 205–226

Sugiyama M (1999) Organogenesis in vitro. Curr Opin Plant Biol 2:61–64. https://doi.org/10.1016/s1369-5266(99)80012-0

Sujatha K, Panda BM, Hazra S (2008) De novo organogenesis and plant regeneration in *Pongamia pinnata*, oil producing tree legume. Trees 22:711–716. https://doi.org/10.1007/s00468-008-0230-y

Tang W, Newton RJ (2005) Plant regeneration from callus cultures derived from mature zygotic embryos in white pine (*Pinus strobus* L.) Plant Cell Rep 24:1–9. https://doi.org/10.1007/s00299-005-0914-3

Taylor MG, Vasil IK (1996) The ultrastructure of somatic embryo development in pearl millet (*Pennisetum glaucum*, Poaceae). Am J Bot 83:28–44

Thengane SR, Kulkarni DK, Shrikhande VA, Krishnamurthy KV (2001) Effect of thidiazuron on adventitious shoot regeneration from seedling explants of *Nothapodytes foetida*. In Vitro Cell Dev Biol Plant 37:206–210. https://doi.org/10.1007/s11627-001-0036-8

Thomas JC, Katterman FR (1986) Cytokinin activity induced by thidiazuron. Plant Physiol 81:681–683

Vervaeke P, Luyssaert S, Mertens J, Meers E, Tack FMG, Lust N (2003) Phytoremediation prospects of willow stands on contaminated sediment: a field trial. Environ Pollut 126:275–282

Vieitez AM, San-José MC (1996) Adventitious shoot regeneration from *Fagus sylvatica* leaf explants in vitro. In Vitro Cell Dev Biol Plant 32:140–147. https://doi.org/10.1007/BF02822757

Vila S, Conzalez A, Rey H, Mroginski L (2003) Somatic embryogenesis and plant regeneration from immature zygotic embryos of *Melia azedarach* (Meliaceae). In Vitro Cell Dev Biol Plant 39:239–243. https://doi.org/10.1079/IVP2002377

Vinocur B, Carmi T, Altman A, Ziv M (2000) Enhanced bud regeneration in aspen (*Populus tremula* L.) roots cultured in liquid media. Plant Cell Rep 19:1146–1154. https://doi.org/10.1007/s002990000243

Vinoth A, Ravindhran R (2013) In vitro propagation – a potential method for plant conservation. Int J Comput Algorithm 2:268–272

Wang Y, Yao R (2017) Plantlet regeneration of adult *Pinus massoniana* Lamb. trees using explants collected in March and thidiazuron in culture medium. J For Res:1–7. 10.1007/s11676-017-0412-9

Wang HM, Liu HM, Wang WJ, Zu YG (2008) Effects of thidiazuron, basal medium and light quality on adventitious shoot regeneration from in vitro cultured stem of *Populus alba* × *P. berolinensis*. J For Res 19:257–259. https://doi.org/10.1007/s11676-008-0042-3

Xie D, Hong Y (2001) In vitro regeneration of *Acacia mangium* via organogenesis. Plant Cell Tissue Organ Cult 66:167–173. https://doi.org/10.1023/A:1010632619342

Yang M, Xie X, He X, Zhang F (2006) Plant regeneration from phyllode explants of *Acacia crassicarpa* via organogenesis. Plant Cell Tissue Organ Cult 85:241–245. https://doi.org/10.1007/s11240-006-9082-6

Zhang X, Zhao J, Teixeira da Silva JA, Ma G (2016) In vitro plant regeneration from nodal segments of the spontaneous F1 hybrid *Santalum yasi* × *S. album* and its parents *S. album* and *S. yasi*. Trees 30:1983–1994. https://doi.org/10.1007/s00468-016-1426-1

Zhou H, Li M, Zhao X, Fan X, Guo A, Zhang X et al (2010) Plant regeneration from in vitro leaves of the peach rootstock 'Nemaguard' (*Prunus persica* × *P. davidiana*). Plant Cell Rep 101:79–87. https://doi.org/10.1007/s11240-010-9666-z

Thidiazuron Influenced Morphogenesis in Some Medicinal Plants

11

Zishan Ahmad and Anwar Shahzad

Abstract

Thidiazuron (*N*-phenyl-*N*′-1,2,3-thiadiazol-5-ylurea; TDZ) is an artificial plant growth regulator that is widely used in plant tissue culture. Due to its dynamic role in plant tissue culture, it has gained ample attention for several workers since the past decades. Wide array of TDZ-influenced physiological responses are reported in different medicinal plant species. TDZ has shown both auxin- and cytokinin-like effects, although, chemically, it is totally different from commonly used auxins and cytokinins. A number of physiological and biochemical events in cells are induced or enhanced by TDZ, but the mode of action of TDZ is yet to explore. However, varieties of underlying mechanisms have been revealed in several reports to defend the morphogenic events induced by the application of TDZ. Some reports emphasized that TDZ may modify endogenous plant growth regulators, either directly or indirectly, and produce reactions in cell/tissue, necessary for its division/regeneration. Other possibilities include modification in cell membrane, fluidity, nutrient uptake, transport and assimilation, etc. In this review, recent advancements in TDZ application in plant sciences are discussed.

Keywords

Thidiazuron · Plant growth regulators · Morphogenesis · Somatic embryogenesis

Z. Ahmad · A. Shahzad (✉)
Plant Biotechnology Section, Department of Botany, Aligarh Muslim University, Aligarh, UP, India
e-mail: ashahzad.bt@amu.ac.in

N. Ahmad, M. Faisal (eds.), *Thidiazuron: From Urea Derivative to Plant Growth Regulator*, https://doi.org/10.1007/978-981-10-8004-3_11

11.1 Introduction

Thidiazuron (*N*-phenyl-*N*′-1,2,3-thidiazol-5-yl-urea; TDZ) a multitask plant growth regulator has played a vital role to trigger differential physiological response in plant cell and tissue culture. Its peculiar feature is the capacity to act as the substitute for the both auxins and cytokinin (Casanova et al. 2004). The TDZ is a substituted phenylurea compound known to act as cotton defoliant (Arndt et al. 1976) but later was found to mimic the cytokinin-like activity (Wang et al. 1986). The response of TDZ alone in plant tissue culture has become more advanced and continued to increase over the decades. The action of TDZ directly depends upon its concentration, exposure time, and cultured explants. According to Murthy et al. (1998), the effect of TDZ is 20 times more advanced as compared to other cytokinins, and hence, the comparison of TDZ and purine-based cytokine is complicated. The supremacy of the TDZ among other phytohormone is might be due to nutrients uptake capacity of the cell with the alteration in cell membrane and enhanced purine and cytokinin metabolism in the cell (Capelle et al. 1983). The production and accumulation of phenols and enzymes like peroxidase and catalase is one of the major effects of TDZ activity in to the cell (Wang et al. 1991a). Moreover alteration in several enzyme concentrations such as ribulose diphosphate, carboxylase oxidase, and pentose enzymes is also an aftereffect of TDZ action (Mok et al. 1987). Wang et al. (1991a, b) reported that most of the TDZ-influenced enzymes are related to the cell wall, cell membrane, and its fluidity. They found that TDZ-influenced organogenesis leads a metabolic cascade which affects directly or indirectly to the other endogenous plant hormone. TDZ has been proved to be an effective plant growth regulator for shoot proliferation and adventitious shoot organogenesis in various plant species (Table 11.1).

Several factors including genotype, type of culture medium and explants, plant growth hormones, their concentration and exposure time, and environmental condition affect the adventitious shoot induction in vitro (Casanova et al. 2008; Casas et al. 2010). The action of TDZ has been found to promote both the organogenesis and somatic embryogenesis in vitro.

The concentration and duration of exposure of TDZ to the explants is well documented by several plant biotechnologists. The short time exposure of TDZ with low concentration has been effective for morphogenesis, while higher levels, on the other hand, promote callus and somatic embryo formation (Rida et al. 2001; Fengyen and Han 2002; Tulac et al. 2002). The abnormal morphogenesis, stunted growth of shoot, hyperhydricity, and fasciculation to the cell were the consequences of TDZ when the exposure was extended beyond the optimum level (Huetteman and Preece 1993; Faisal et al. 2005; Ahmad and Anis 2007). Shirani et al. (2009) also reported the deleterious effect of higher concentration of TDZ in regenerated shoots of banana and plantain (*Musa* spp.) after in vitro multiplication with TDZ and BAP from excised shoot tips. Additionally, continuous or more than optimal exposure of TDZ resulted in the inhibition of shoot elongation and formation of fasciated/distorted shoot development.

Table 11.1 TDZ-influenced morphogenesis in some medicinal plant

	Plant name	Family	Explants	Treatment	References
1	*Acacia mangium*	Mimosacea	Cotyledons	MS + 4.55 μM TDZ + 1.43 μM IAA	Xie and Hong (2001)
2	*Acacia sinuate*	Mimosacea	Nodal segment	MS + 0.6 μM TDZ + 0.1 I μM AA	Shahzad et al. (2006)
3	*Acanthophyllum sordidum*	Caryophyllaceae	Leaf segment	MS + 2.69 μM NAA + 4.54 μM TDZ + 2.46 μM IBA	Meratan et al. (2009)
4	*Aerva lanata*	Amaranthaceae	Leaf segment	MS + 2.0 mg L^{-1} TDZ	Varutharaju et al. (2014)
5	*Arachis correntina*	Leguminosae	Leaf segment	MS + 5.0 μM TDZ	Mroginski et al. (2004)
6	*Bactris gasipaes*	Arecaceae	Shoot	MS + 0.36 μM TDZ	Graner et al. (2013)
7	*Bauhinia tomentosa*	Fabaceae	Cotyledons/nodal segment	MS + 0.8 μM TDZ	Naaz et al. (2012)
8	*Beta vulgaris*	Amaranthaceae	Petiole	MS + 4.6 μM TDZ	Zhang et al. (2001)
9	*Cannabis sativa*	Cannabaceae	Nodal segments	MS + 0.5 μM TDZ	Lata et al. (2009)
10	*Cassia sophera*	Fabaceae	Cotyledonary node	MS + 2.5 μM TDZ	Parveen and Shahzad (2010)
11	*Ceropegia ensifolia*	Apocynaceae	Nodal (in vitro derived)	MS + 20 μM TDZ	Reddy et al. (2015)
12	*Cicer arietinum*	Fabaceae	Cotyledons	MS + 10.0 μM TDZ	Murthy et al. (1996)
13	*Citrus sinensis*	Apocynaceae	Epicotyl/hypocotyl	MS + 5.0 μM TDZ	Kumari et al. (2008)
14	*Chirita swinglei*	Gesneriaceae	Leaf	MS + 2.0 μM TDZ + 2.5 μM BA	Chen et al. (2016)
15	*Cotoneaster wilsonii*	Rosaceae	Nodal segment	MS + 2.1 μM TDZ + 0.4 μM NAA	Sivanesan et al. (2011)
16	*Echinacea purpurea*	Asteraceae	Petiole	MS + 2.5 μM BA + 0.5 μM TDZ	Choffe et al. (2000)
17	*Embelia ribes*	Primulaceae	In vitro derived leaf	MS + 0.27 μM TDZ	Raghu et al. (2006)
18	*Fraxinus americana*	Oleaceae	Cotyledon/hypocotyl	MS + 10.0 μM TDZ	Bates et al. (1992)
20	*Glycine max*	Fabaceae	Cotyledonary nodes	MS + 5.4 μM TDZ	Radhakrishnan et al. (2009)
21	*Hypericum perforatum*	Hypericaceae	Hypocotyl	MS + 5.0 μM TDZ	Murch et al. (2000)
22	*Jatropha curcas*	Euphorbeacea	Leaf	MS + 2.27 μM TDZ + 2.22 μM BA + 0.49 μM IBA	Deore and Johnson (2008)
23	*Kigelia pinnata*	Bignoniaceae	Nodal segment	MS + 3.0 μM TDZ + 0.5 μM NAA	Thomas and Puthur (2004a, b)

(continued)

Table 11.1 (continued)

	Plant name	Family	Explants	Treatment	References
24	*Kigelia pinnata*	Bignoniaceae	Nodal segment	MS + 3 μM TDZ + 0.5 μM NAA	Thomas et al. (2004)
25	*Metabriggsia ovalifolia*	Gesneriaceae	Leaf	MS + 2.5 μM BA + 5.0 μM TDZ	Ouyang et al. (2016)
26	*Morus alba*	Moraceae	Leaf	MS + 18.17 μM TDZ	Chitra and Padmaja (2005)
27	*Nothapodytes foetida*	Icacinaceae	Shoot/hypocotyl	L2 + TDZ 0.44 μM + BAP 2.22 μM + L- glutamine 0.03 μM	Tejavathi et al. (2011)
28	*Ochna integerrima*	Ochnaceae	Leaf/shoot	MS + 10.0 μM TDZ	Ma et al. (2011)
29	*Pterocarpus marsupium*	Fabaceae	Cotyledonary nodes	MS + 0.4 μM TDZ	Husain et al. (2007)
30	*Psoralea corylifolia*	Fabaceae	Nodal segment	MS + 2.0 μM TDZ	Faisal and Anis (2006)
31	*Rauvolfia tetraphylla*	Apocynaceae	Nodal segment	MS + 5.0 μM TDZ	Faisal et al. (2005)
32	*Ricinus communis*	Euphorbeacea	Cotyledon	MS + 2.5 mg dm^{-3} TDZ + 0.4 mg dm^{-3} NAA + 15 mg dm^{-3} glutamine	Kumari et al. (2008)
33	*Saintpaulia ionantha*	Gesneriaceae	Leaf/petiole	MS + 2.5 μM TDZ	Mithila et al. (2003)
				MS + 5.0 μM TDZ	
34	*Santalum album*	Santalaceae	Nodal	WPM + 0.6 mg/l TDZ + 1.5 mg/l 2,4-D	Singh et al. (2016)
35	*Saussurea involucrata*	Asteraceae	Leaf	MS + 0.5 μM TDZ	Guo et al. (2017)
36	*Selenicereus megalanthus*	Cactaceae	Cotyledons	MS + 200 μM TDZ	Pelah et al. (2002)
37	*Stevia rebaudiana*	Asteraceae	Nodal segments	MS + 1.0 μM TDZ	Lata et al. (2013)
38	*Tylophora indica*	Apocynaceae	Leaf callus	MS + 8.0 μM TDZ	Thomas and Philip (2005)
39	*Vitex trifolia*	Verbenaceae	Nodal explants	MS + 5.0 μM TDZ + 1.0 μM BA + 0.5 μM NAA	Ahmed and Anis (2012)
40	*Vitex negundo*	Verbenaceae	Nodal segment	MS + 1.0 μM TDZ + 1.0 μM BA + 0.5 μM NAA	Ahmad and Anis (2007)

A highest number of shoot were produced in *Artemisia judaica* when TDZ concentration was used at 1 μM for 20 days; when the exposure time was stretched, further differentiation of shoot was restricted coupled with abnormality in the shoot (Liu et al. 2003). Besides the magnificent response of TDZ in plant tissue culture, its deleterious responses were also known as the days advance. The deleterious effect of the continued presence of TDZ on the growth and multiplication has been earlier reported from time to time for several species. However the investigators have found a solution to overcome the harmful effect of TDZ by transferring the regenerated shoot to TDZ-free medium after the optimum exposure. The technique brings a balanced morphogenesis to the in vitro plant, and it is used by various workers including Huetteman and Preece (1993), Shahzad et al. (2006), Siddique and Anis (2007a, b), Faisal et al. (2008), Jahan and Anis (2009), Makara et al. (2010), and Jahan et al. (2011), Saeed and Shahzad (2015). The present chapter documents the detailed account of physiological and morphological effects of TDZ in several plant species.

11.2 Mechanism of Action of TDZ

There are several reports available dealing the physiological role of TDZ in different medicinal plant species. However, the mechanism of action of TDZ is not well documented, and only preliminary reports are available. The pioneer work of Hare and Cress (1997) for the mechanism of action of TDZ established that proline itself being as a stress marker was involve in the activity of TDZ (Fig. 11.1). The higher proline accumulation in the cell controls the NAD(P)′/NAD(P)H ratios as a consequence of plant undergoing stress which favors the oxidative pentose phosphate pathway leading to the production of precursor for auxin and cytokinin biosynthesis. In this way, the accumulations of the plant growth regulators occur as a result of the cascade of biochemical reactions initiated by TDZ. Murthy et al. (1996a) have reported high level of accumulation of proline during TDZ-induced regeneration via embryogenesis. In another study of Casanova et al. (2004), they found that the application of TDZ at a very low concentration (0.0–0.005 μM) leads to the formation of zeatin (ZT) while at higher concentration (0.5 μM) induces isopentyl adenine (IP) production in carnation petals. In the recent study of Jones et al. (2007) on the regeneration of *Echinacea purpurea*, they concluded the probable role of auxin, indolamines, and ion signaling in the morphogenesis. They found that the level of endogenous indoleamines is potentially influenced by the exposure of TDZ and enhanced level of the calcium and sodium transport in the cell was also found by the TDZ activity, and hence a positive effect was shown in regeneration.

There is another report on role of IAA published by Chhabra et al. (2008). They proposed that the involvement of the phytohormone is closely related to the biosynthesis and transportation of IAA. These reports indicate that TDZ-influenced morphogenesis is the demonstration of metabolic cross talk that includes a primary signaling, accumulation, and transport of endogenous plant signals such as auxin and cytokinin and enhanced transport of secondary messengers.

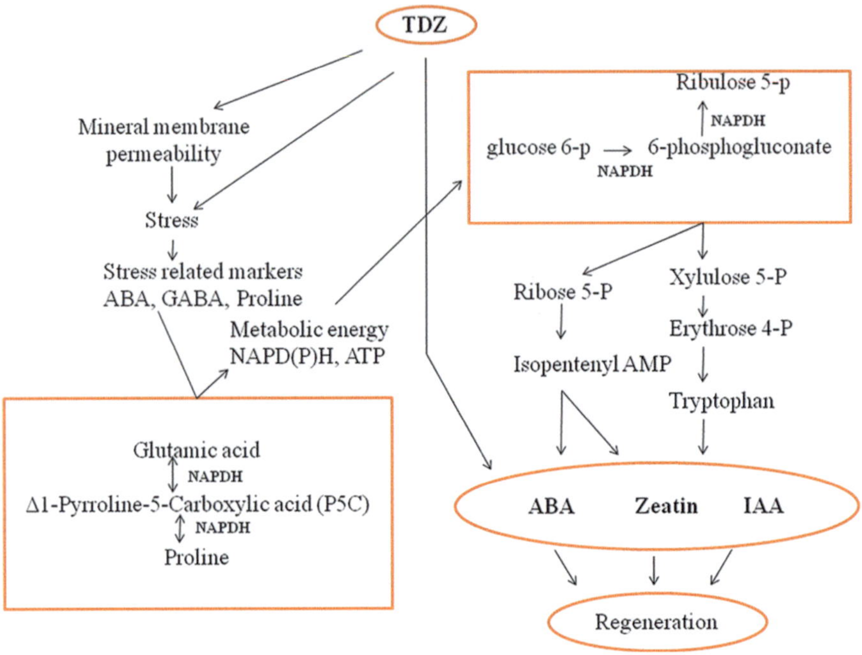

Fig. 11.1 Diagrammatic representation of proposed mechanism of action of TDZ

11.3 Effect of TDZ on Organogenesis

TDZ supposes to be less susceptible to enzymatic degradation in vivo than other naturally occurring amino purine cytokinins and has proved to be effective at lower concentrations (0.0091–3.99 μM) for the micropropagation of several plant species (Lu 1993). It has been shown to induce high bud regeneration rates than purine-based cytokinins and also has capability of fulfilling both the cytokinin and auxin requirements of regeneration responses in a number of woody plants (Jones et al. 2007). However, there was another report of Augustine and D'Souza in (1997) for the in vitro propagation of *Zanthoxylum rhetsa* using TDZ at higher concentration (2.27–145.41 μM). The use of TDZ for shoot regeneration from different explants has been widely reported at a great extent for a number of woody plant species such as *Hydrangea quercifolia* (Ledbetter and Preece 2004), *Cassia angustifolia* (Siddique and Anis 2007a, b), *Pterocarpus marsupium* (Husain et al. 2007), and *Vitex negundo* (Ahmad and Anis 2007).

Ahmed and Anis (2014) investigated the prompt response of TDZ and developed a rapid and commercially applicable regeneration protocol for *Cassia alata.* They tried various concentrations of TDZ with different duration of exposure; however, harmful effect was also shown by the in vitro culture when exposure time stretches beyond the optimum period. The highest number of shoots 17.9 ± 0.3 with shoot length of 4.6 ± 0.1 cm was achieved when the explants were exposed to TDZ (5.0 μM) for 4 weeks. To avoid the deleterious effect of TDZ, they were transferred to TDZ-free medium. Likewise in another species of *Cassia*, Parveen and Shahzad (2010) found that MS medium fortified with TDZ (2.5 μM) was optimum for the production of 6.7 ± 0.2 shoot per explants. To avoid the aftereffect of TDZ, the microshoot was consequently transferred to TDZ-free medium containing BA for proper multiplication, proliferation, and shoot elongation.

Sharma and Shahzad (2008) reported TDZ-induced organogenesis in *Abelmoschus moschatus* using cotyledonary explants. TDZ alone proved to be superior in comparison to the combination of BA and NAA. MS medium augmented with low concentration of TDZ (0.01 mg L^{-1}) was optimum for the multiple shoot induction in A. moschatus, and a maximum of 16.8 ± 1.46 shoot per explants were achieved. Faisal and Anis (2006) studied the effect of TDZ on in vitro axillary shoot proliferation from nodal explant of *Psoralea corylifolia*, an endangered medicinal plant. Proliferation of shoots was achieved on MS medium supplemented with different concentration of 0.5, 1, 2, 3, 4, and 5 μM TDZ. The maximum number (13.6 ± 1.4) of shoots per explants was obtained from nodal segments on TDZ (2 μM) after 4 weeks of culture and followed by the transfer to hormone-free MS medium wherein the shoot differentiation significantly induced to 29.7 ± 2.1 after 8 weeks. In another study on *Cassia siamea* by Parveen et al. (2010), it was found that TDZ could not be able to evoke a significant response in the terms of shoot multiplication. They applied distinct concentration of cytokinin, viz., BA, Kn, and TDZ, alone or in combination singly or in combination with auxins for regeneration from excised codeledonary nodal explants, and MS + BA (1.0 μM) found to be best for direct shoot regeneration as it induced an average of 8.20 ± 0.66 shoots per explant. The regeneration frequency further improved with synergistic response of BA with auxin. In the highest frequency for shoot regeneration (90%), the maximum number of shoots per explants (12.20 ± 0.73) was obtained on the medium which consisted of MS + BA (1.0 μM) + NAA (0.5 μM) in C. siamea.

Shahzad et al. (2006) established a protocol for the organogenesis in *Acacia sinuata* using cotyledon. All the concentration of TDZ with MS was able to generate callusing to the explants, and MS + TDZ (0.6 μM) was found to be better in the terms of maximum callus formation in *A. sinuata*. However, the callus was further transferred to the shooting medium augmented with BA (3.0 μM) for optimum shoot induction wherein 6.60 ± 0.54 shoots were produced. Cocu et al. (2004) recorded highest frequency of adventitious shoot regeneration in *Calendula officinalis* in MS medium containing TDZ (0.75 mg dm^{-3}). Likewise, Phippen and Simon (2000) reported both callus and shoot induction with TDZ (16.8 μM) alone in *Ocimum basilicum* via using leaf explants. Murthy et al. (1996) observed direct organogenesis and somatic embryogenesis in *Cicer arietinum* when cotyledonary explants were inoculated on BA- and TDZ-amended MS medium. Multiple shoots formed de novo without an intermediary callus phase at the cotyledonary notch of the seedlings within 2–3 weeks of culture initiation. TDZ was found to be more effective as compared to BA as an inductive signal of regeneration. The TDZ induced multiple shoot formation at all the concentrations tested (1.0–10.0 μM), although maximum morphogenic response was observed at 10.0 μM of TDZ.

De novo shoot organogenesis was reported in *Artemisia judaica* using TDZ (1.0 μM) by Liu et al. (2003). The role of TDZ has also been reported in several herbs and shrub like, *Bacopa monnieri* (Tiwari et al. 2001), *Artemisia judaica* (Liu et al. 2003), *Hordeum vulgare* (Ganeshan et al. 2003), *Cineraria maritime* (Banerjee et al. 2004), *Hyoscyamus niger* (Uranbey 2005), *Psoralea corylifolia* (Faisal and Anis 2006), *Rauvolfia tetraphylla* (Faisal et al. 2005), *Ricinus communis* (Kumari et al. 2008), *Hypericum perforatum* (Murch et al. 2000), *Embelia ribes* (Raghu et al. 2006), *Ochna integerrima* (Ma et al. 2011), Morus alba (Chitra and Padmaja 2005), *Bauhinia tomentosa* (Naaz et al. 2012), *Bactris gasipaes* (Graner et al. 2013), *Ceropegia ensifolia* (Reddy et al. 2015), and *Cassia sophera* (Parveen and Shahzad 2010).

11.4 Synergistic Effect of TDZ and Cytokinin

The synergistic effect of TDZ with other cytokinin found to be very useful to trigger organogenesis significantly (Chen et al. 2016). Lee and Pijut (2017) proposed an efficient regeneration system through adventitious shoot organogenesis in black ash (*Fraxinus nigra*), an endangered hardwood. In their study the MS medium augmented with BA (22.2 μM) + TDZ (31.8 μM) was found good with the production of 1.9 ± 0.65 adventitious shoots per leaf explant. Similarly Ouyang et al. (2016) reported the efficiency of combined treatment of TDZ + BA on the improvement of regenerability and somatic embryo formation from the leaf of *Metabriggsia ovalifolia*. Chen et al. (2016) reported a positive effect on shoot bud regeneration in *Chirita swinglei*. A maximum of 23.1 ± 0.20 shoot bud per explants were produced on MS + TDZ (2.0 μM) + BA (2.5 μM). The shoot bud obtained in *C. swinglei* depends upon the exposure and concentration of the TDZ. The first observation they recorded was the swallowing of leaf explants after culture for 15 days at (2.0 μM) TDZ. Some shoot buds were observed after 20 days of culture. Shoot buds were clearly visible as culture period was extended from 35 to 45 days. Callus could also be induced from leaves when α-naphthalene acetic acid (NAA) was used alone or in combination with TDZ and BA.

Parveen and Shahzad (2011) established a protocol for the in vitro propagation of the *Cassia angustifolia*. MS medium supplemented with TDZ (1.0 μM) was used for the production of organogenic calli followed by subsequent transfer to the TDZ-free medium augmented with different cytokinin, viz., BA, Kn, or TDZ for proper regeneration of shoot. They achieved a maximum of 35.63 ± 0.75 shoot per explants on MS + BA (2.5 μM) + NAA (0.6 μM) from the TDZ-induced calli. Zeng et al. (2008) reported an efficient micropropagation system for *Tigridiopalma magnifica* using leaves as explants. Up to 7.6 adventitious buds formed per leaf explant after a 40-day culture on MS + BA (2.0 mg^{-}1) + TDZ (0.1 mg^{-1}). To avoid the aftereffect of TDZ, the culture were transferred to the TDZ-free medium containing other cytokinin-like BA for enhanced proliferation rate of adventitious buds, and it reached to 5.7 on MS medium supplemented with 2.0 mg^{-1} of BA.

11.5 Effect of TDZ with Auxin/Growth Additives

The role of TDZ with different auxin and growth additives is also well documented by several workers. The auxin-like NAA, 2, 4-D, IBA, and IAA at various concentrations with optimum TDZ concentration was reported to play an important role in both direct and indirect organogenesis. In recent study of Baskaran et al. (2016) on developing a regeneration protocol for *Ledebouria ovatifolia* through direct and indirect organogenesis by using leaf explants demonstrated that the adventitious shoot was best produced on MS + TDZ (5 μM) + NAA (2 μM), while organogenic callus was obtained on MS + IAA (2.0 μM) + TDZ (5.0 μM) + glutamine (30 μM). A maximum of 26.8 ± 1.06 and 32.0 ± 1.73 shoot per explants were achieved via direct and indirect organogenesis in *L. ovatifolia*.

A micropropagation protocol was developed by Babaei et al. (2014) for *Curculigo latifolia*. They used distinct concentration of auxin with optimum TDZ concentration for direct and indirect organogenesis using shoot tip explants. MS medium augmented with TDZ (0.5 mg L^{-1}) + IBA (0.25 mg L^{-1}) was found to be best for direct regeneration in terms of percentage of explants producing shoot, shoot number, and shoot length. Prathanturarug et al. (2012) studied the in vitro propagation of *Stemona hutanguriana* via using nodal and intermodal segment as explants. MS medium augmented with TDZ alone or in combination with NAA was able to promote regeneration in the *S. hutanguriana*. A regeneration frequency of 91.67% with shoot regeneration rate of 5.46 shoots/responding explant was observed when nodal segment inoculated on MS + TDZ (18.16 μM) + NAA (0.54 μM) for 8 weeks and followed by transferred to the PGR-free medium to avoid the adverse effect of TDZ.

In another study by Ma et al. (2011) on *Metabriggsia ovalifolia*, TDZ at higher concentration (5.0 μM) was found to be better for efficient propagation and regeneration of 36.7 shoots per leaf explants; however, the regeneration efficiency was further enhanced when auxin was supplemented with optimum TDZ. Among the various auxins, NAA at 0.5 μM with optimum TDZ concentration was efficient to induce a maximum of 79.1 adventitious shoots from each leaf explants. TDZ-mediated indirect organogenesis was also achieved by Siddique et al. (2010) in *Cassia angustifolia* via using petiole explants excised from 21-day-old axenic seedlings. They used MS medium fortified with 2, 4-D (5.0 μM) and TDZ (2.5 μM) for the organogenic callus induction. TDZ at higher concentration (5.0 μM) was able to induce calli differentiation to the adventitious shoot with the highest of 8.5 ± 0.98 shoots per culture. However, the regeneration efficiency of the explants was significantly improved when combination of TDZ (5 μM) + IAA (1.5 μM) was applied and produces a maximum of 12.5 ± 1.10 shoots per culture.

Sujatha and Dinesh Kumar (2007) compared the efficacy of cytokinin with TDZ for direct organogenesis in the species of *Carthamus*. The MS medium fortified with TDZ (0.2 mg dm^{-3}) + NAA (0.2 mg dm^{-3}) was more efficient for the induction of shoot from the leaf explants of *C. tinctorius*. On the other hand Radhika et al. (2006) found that optimum TDZ (0.2 mg dm^{-3}) with high concentration of NAA (1.0 mg dm^{-3}) was proved to be better for regeneration in *C. arborescens*. Faisal and Anis (2005) has set a protocol for the in vitro propagation *Tylophora indica* using petiole as an explant. They obtained optimum callus from the explants when inoculated on to the MS + 2,4-D (10 μM) + TDZ (2.5 μM). To achieve the shoot induction, TDZ-derived callus was transferred to the shoot induction medium. TDZ alone found to be best for the shoot multiplication in *T. indica* and a highest of 56 ± 3.6 adventitious shoot were obtained from the surface of the callus when MS medium fortified with TDZ (2.5 μM) was used. In another study of Thomas and Puthur (2004a, b) on a multipurpose tree, *Kigelia pinnata*, they used nodal segment and inoculated to the MS medium augmented with 2,4-D (3 μM) for callus induction. The obtained calli were then transferred to the shooting medium fortified with TDZ (3.0 μM) + NAA (0.5 μM) for the proliferation and multiplication of the shoot where 21 ± 0.3 shoots per culture were obtained.

11.6 Effect of TDZ on Somatic Embryogenesis

Somatic embryogenetic systems are of growing interest for medicinal, ornamental, and horticultural plants (Ji et al. 2011). Dedifferentiation of cells, activation of cell division, reprogramming of cell physiology, metabolism, and gene expression patterns occurred during unique developmental pathways of somatic embryogenesis. However, morphological abnormalities such as embryo fusion and lack of suitable apical meristems or loss of bipolarity have occurred resulting in poor yields (Benelli et al. 2010). TDZ-influenced regeneration via somatic embryogenesis is well documented by several workers for different medicinal plant species. In the recent study of Baskaran and Staden (2017), they were able to get friable embryogenic callus (FEC) from the leaf explants of *Lachenalia montana* through suspension culture for the first time. Liquid MS medium (MS_L) supplemented with 2, 4-D (0.5 μM) + TDZ (1 μM) was optimum for the formation of somatic embryos of different stages (globular to cotyledonary stages, respectively). However, the enhanced concentration of 2,4-D and TDZ was needed for the germination of somatic embryos, and liquid MS medium augmented with 2,4-D (1.0 μM) + TDZ (2.0 μM) was proved to be best in terms of enhanced germination frequency.

Naaty et al. (2017) found best response for somatic embryo production in *Schizozygia coffaeoides* on the medium comprises of MS + BA (2.0 mg/l) + Kn (0.8 mg^{-1}) + NAA (0.4 mg^{-1}) + TDZ (0.5 mg^{-1}), which survived to maturity and formed shoot. Baskaran et al. (2016) achieved embryogenic callus induced on liquid MS augmented with sucrose (15 g L^{-1}) + TDZ (0.2 Mm) + picloram (0.1 μM) + glutamine (10 μM) with the highest numbers of somatic embryos, 43.2–35.6

(globular to cotyledonary stages, respectively). Baskaran and Staden (2014) were able to achieve different developmental stages of somatic embryos, globular embryos, partial pear-shaped embryos and club-shaped embryos obtained from leaf explants of *Drimia robusta* on MS + Picloram (10 μM) + TDZ (1 μM) + glutamine (20 μM). Sahai et al. (2010) developed a protocol for the in vitro propagation of an endangered medicinal climber *Tylophora indica* through leaf explants. Different types of calli produced on BA and TDZ-augmented MS basal medium were selected for shoot induction and somatic embryogenesis studies. Calli when transferred from BA (5.0 μM) + TDZ (2.5 μM) to the MS medium containing BA (5.0 μM) resulted in high-frequency shoot induction (26.8 ± 0.97 shoots/culture) along with somatic embryogenesis (10.20 ± 0.37 embryoids/culture) up to three subculture passages. Embryoids transformed into complete plantlets when transferred to growth regulator-free half-strength MS medium.

Dhandapani et al. (2008) were able to achieve plant regeneration via somatic embryogenesis in *Catharanthus roseus*. The highest regeneration percentage through somatic embryogenesis was achieved from mature zygotic embryo on MS + TDZ (7.5 μM), and further the mature embryo also regenerated efficiently via organogenesis in MS medium fortified with TDZ (2.5 μM) + BA (2.2 μM). Joshi et al. (2008) found that failure of peanut somatic embryos to convert into plantlets is attributed to the abnormal development of the plumule. TDZ was effective in the conversion of peanut somatic embryos to plantlets by triggering morphogenetic activity in the abnormal plumules of the rooted somatic embryos. Bud-like projections appeared in the embryogenic masses when these were cultured in media containing combinations of 2,4-D and TDZ. These projections developed into buds, which subsequently formed shoots and plantlets. The response varied with the concentration and exposure of TDZ. At lower concentrations, the buds appeared in a defined row in the equatorial region of the explant, and with extended incubation, more and more buds appeared in rows alongside the initial row. Induction of multiple buds in a defined row in this specific site (equatorial region) suggested the presence of potent cells around this region. At higher concentrations, these projections appeared in large numbers spread over the whole upper part of the embryogenic mass starting from the equatorial region. The ability of embryogenic mass to convert into organogenic mass and to produce large number of organogenic buds provides an excellent system for basic studies and for the genetic transformation of peanut.

Mithila et al. (2003) observed TDZ-mediated regeneration using leaf and petiole explants from in vitro grown African violet plants. The response of cultures to other growth regulators over a range of 0.5–10 μM was 50% less than that observed with TDZ. A comparative study among several cultivars of African violet indicated that "Benjamin" and "William" had the highest regeneration potential. In "Benjamin," higher frequencies of shoot organogenesis (two fold) and somatic embryogenesis (a 50% increase) were observed from in vitro and greenhouse-grown plants, respectively. At concentrations lower than 2.5 μM, TDZ induced shoot organogenesis, whereas at higher doses (5–10 μM) somatic embryos were formed.

11.7 Conclusions

Regulation of cell division and cell differentiation is necessary for the morphogenesis either in vivo or in vitro. Auxin and cytokinin are believed to be responsible for this synergistic control. The present review deals the importance of TDZ, another class of plant growth regulators, significantly different from the cytokinin. It also attempts to integrate the vast amount of knowledge generated on TDZ-induced responses in a myriad of systems. Application of TDZ results in a wide variety of responses in in vitro cultured tissues, but the biochemical and physiological basis of the modulation of morphogenic response induced by TDZ are poorly understood. However, studies encompassing a wide array of species, techniques, and physiological responses have led to several tentative models to explain the regulatory role of TDZ. A complete picture concerning the mechanism of action of TDZ is not likely to occur, and many mysteries of auxin- and cytokinin-related morphogenesis are resolved. Nevertheless, the recent advancement in biochemical and molecular characterization of auxin and cytokinin mutants and general enthusiasm in plant growth regulator research promises very exciting results in the next decade. A complete understanding of the biochemical and physiological responses of plant tissues to TDZ will broaden our understanding of morphogenesis and further help in improvement of tissue culture technology.

References

Ahmad N, Anis M (2007) Rapid clonal multiplication of a woody tree, *Vitex negundo* L. through axillary shoots proliferation. Agrofor Syst 71:195–200

Ahmed MR, Anis M (2012) Role of TDZ in the quick regeneration of multiple shoots from nodal explants of *Vitex trifolia* L. – an important medicinal plant. Appl Biochem Biotechnol 168:957–966

Ahmed MR, Anis M (2014) Changes in activity of antioxidant enzymes and photosynthetic machinery during acclimatization of micropropagated *Cassia alata* L. plantlets. In Vitro Cell Dev Biol-Plant 50:601–609

Arndt FR, Rusch R, Stillfried HV, Hanisch B, Martin WC (1976) A new cotton defoliant. Plant Physiol 57:S-99

Augustine AC, D'Souza L (1997) Micropropagation of an endangered forest tree *Zanthoxylum rhetsa* Roxb. Phytomorphology 47:319–323

Babaei N, Abdullah NAP, Saleh G, Abdullah TL (2014) An efficient in vitro plantlet regeneration from shoot tip cultures of *Curculigo latifolia*, a medicinal plant. Hindawi Publishing Corporation http://dx.doi.org/10.1155/2014/275028

Banerjee S, Tripathi J, Verma PC, Dwivedi PD, Khanuja SPS, Bagchi GD (2004) Thidiazuron-induced high-frequency shoot proliferation in *Cineraria maritima*. Linn Curr Sci 87:1287–1290

Baskaran P, Staden JV (2014) Plant regeneration via somatic embryogenesis in *Drimia robusta*. Plant Cell Tissue Organ Cult 119:281–288

Baskaran P, Staden JV (2017) Ultra structure of somatic embryo development and plant propagation for *Lachenalia montana*. South Afr J Bot 109:269–274

Baskaran P, Kumari A, Naido D, Staden JV (2016) In vitro propagation and ultrastructural studies of somatic embryogenesis of *Ledebouria ovatifolia*. In Vitro Cell Dev Biol Plant. https://doi.org/10.1007/s11627-016-9762-9

Bates S, Preece JE, Navarrete NE, Sambeek JWV, Gaffney GR (1992) Thidiazuron stimulates shoot organogenesis and somatic embryogenesis in white ash (*Fraxinus americana* L.) Plant Cell Tissue Organ Cult 31:21–29

Benelli C, Germana MA, Camino T, Beghe D, Fabbri A (2010) Morphological and anatomical observations of abnormal somatic embryos from anther cultures of *Citrus reticulate*. Biol Plant 54:224–230

Capelle SC, Mok DWS, Kirchner SC, Mok MC (1983) Effects of thidiazuron on cytokinin autonomy and the metabolism of N6-(DELTA2-isopentenyl) [8- 14C] adenosine in callus tissues of Phareolus *Zunahs* L. Plant Physiol 73:796–802

Casanova E, Valdes E, Fernandez B, Moysset L, Trillas M (2004) Levels and in situ localization of endogenous cytokinins in thiadizuron induced shoot organogenesis in carnation. J Plant Physiol 16:95–104

Casanova E, Moysset L, Trillas M (2008) Effect of agar concentration and vessel closure on the organogenesis and hyperhydricity of adventitious carnation shoots. Biol Plant 52:1–8

Casas J, Olmos E, Piqueras A (2010) In vitro propagation of carnation (*Dianthus caryophyllus* L.) In: Protocols for in vitro propagation of ornamental plants. Methods Mol Biol 589:109–116

Chen Y, Zhang Y, Cheng O, Niu M, Liang H, Yan H, Zhang X, Silva JAT, Ma G (2016) Plant regeneration via direct and callus-mediated organogenesis from leaf explants of *Chirita swinglei* (Merr.) W. T. Wang. In Vitro Cell Dev Biol-Plant 52:521–529

Chhabra G, Chaudhary D, Varma M, Sainger M, Jaiwal PK (2008) TDZ induced direct shoot organogenesis and somatic embryogenesis on cotyledoray node explants of lentil (*Lens culinaris Medik.*) Physiol Mol Biol Plants 14:347–353

Chitra DSV, Padmaja G (2005) Shoot regeneration via direct organogenesis from in vitro derived leaves of mulberry using thidiazuron and 6-benzylaminopurine. Sci Hortic 106:593–602

Choffe K, Victor JMR, Murch SJ, Saxena PK (2000) In vitro regeneration of *Echinacea Purpurea* L.: direct somatic embryogenesis and indirect shoot organogenesis in petiole culture. In Vitro Cell Dev Biol 36:30–36

Cocu S, Urabey S, Iek A, Khawar KM, Sarihan EO, Kaya MD, Parmaksiz ÖS (2004) Adventitious shoot regeneration and micropropagation in *Calendula officinalis* L. Biol Plant 48:449–451

Deore AC, Johnson TS (2008) High-frequency plant regeneration from leaf-disc cultures of *Jatropha curcas* L.: an important biodiesel plant. Plant Biotechnology Reports 2(1):7–11

Dhandapani M, Kim DH, Hong SB (2008) Efficient plant regeneration via somatic embryogenesis and organogenesis from the explants of *Catharanthus roseus*. In Vitro Cell Dev Biol Plant 44:18–25

Faisal M, Anis M (2005) An efficient in vitro method for mass propagation of *Tylophora indica*. Biol Plant 49:257–260

Faisal M, Anis M (2006) Thidiazuron induced high frequency axillary shoot multiplication in *Psoralea corylifolia*. Biol Plant 50:437–440

Faisal M, Ahmad N, Anis M (2005) Shoot multiplication in *Rauvolfia tetraphylla* L. using thidiazuron. Plant Cell Tissue Organ Cult 80:187–190

Faisal M, Shahzad A, Anis M (2008) Somatic embryogenesis and plant regeneration from nodal explants in *Psoralea corylifolia* L. IJPDB 2(2):111–113

Fengyen L, Han L (2002) Effect of exogenous hormones on micropropagation of in vitro virus free potato plantlets. Chines Potato J 16:214–216

Ganeshan S, Monica B, Bryan LH, Brian GR, Graham JS, Ravindra NC (2003) Production of multiple shoots from thidiazuron-treated mature embryos and leafbase/apical meristem of barley (*Hordeum vulgare*). Plant Cell Tissue Organ Cult 73:57–64

Graner ME, Oberschelp JPG, Brondani EG, Batagin-Piotto DK, de Almeida VC, de Almeida M (2013) TDZ pulsing evaluation on the in vitro morphogenesis of peach palm. Physiol Mol Biol Plants 19:283–288

Guo B, He W, Zhao Y, Wu Y, Fu Y, Guo J, Wei Y (2017) Changes in endogenous hormones and H_2O_2 burst during shoot organogenesis in TDZ-treated *Saussurea involucrata* explants. Plant Cell Tissue Organ Cult 128:1–8

Hare PD, Cress WA (1997) Metabolic implications of stress-induced proline accumulation in plants. Plant Growth Regul 21:79–102

Huetteman CA, Preece JE (1993) Thidiazuron: a potent cytokinin for woody plant tissue culture. Plant Cell Tissue Organ Cult 33:105–119

Husain MK, Anis M, Shahzad A (2007) In vitro propagation of Indian Kino (Pterocarpus marsupium Roxb.) using Thidiazuron. In Vitro Cel Dev Biol 43:59–64

Jahan AA, Anis M (2009) In vitro rapid multiplication and propagation of *Cardiospermum halicacabum* L. through axillary bud culture. Acta Physiol Plant 31:133–138

Jahan AA, Anis M, Aref MI (2011) Assessment of factors affecting micropropagation and ex vitro acclimatization of *Nyctanthes arbor-tristis* L. Acta Biol Hung 62:45–56

Ji A, Geng X, Zhang Y, Yang H, Wu G (2011) Advances in somatic embryogenesis research of horticultural plants. American J Plant Sci 2:727–732

Jones MPA, Yi Z, Murch SJ, Saxena PK (2007) Thidiazuron-induced regeneration of *Echinacea purpurea* L- micropropagation in solid and liquid culture systems. Plant Cell Rep 26:13–19

Joshi M, Sujatha K, Hazra S (2008) Effect of TDZ and 2,4-D on peanut somatic embryogenesis and in vitro bud development. Plant Cell Tissue Organ Cult 94:85–90

Kumari KG, Ganesan M, Jayabalan N (2008) Somatic organogenesis and plant regeneration in *Ricinus communis*. Biol Plant 52:17–25

Lata H, Chandra S, Khan I, ElSohly MA (2009) Thidiazuron-induced high-frequency direct shoot organogenesis of *Cannabis sativa* L. In Vitro Cell Dev Biol-Plant 45:12–19

Lata H, Chandra S, Wang YH, Raman V, Khan IA (2013) TDZ-induced high frequency plant regeneration through direct shoot organogenesis in *Stevia rebaudiana* Bertoni: an important medicinal plant and a natural sweetener. Amer J Plant Sci 4:117–128

Ledbetter DI, Preece JE (2004) Thidiazuron stimulates adventitious shoot production from *Hydrangea quercifolia* Bart. Leaf explants. Sci Hortic 101:121–126

Lee JH, Pijut PM (2017) Adventitious shoot regeneration from in vitro leaf explants of *Fraxinus nigra*. Plant Cell Tissue Organ Cult 130:335–343

Liu CZ, Murch SJ, Demerdash ME, Saxena PK (2003) Regeneration of Egyptian medicinal plant *Artemisia judaica* L. Plant Cell Rep 21:525–530

Lu CY (1993) The use of thidiazuron in tissue culture. In Vitro Cell Dev Biol 29:92–96

Ma G, da Silva JAT, Lü J, Zhang X, Zhao J (2011) Shoot organogenesis and plant regeneration in *Metabriggsia ovalifolia*. Plant Cell Tissue Organ Cult 105:355–361

Makara AM, Rubaihayo PR, Magambo MJS (2010) Carry-over effect of Thidiazuron on banana in vitro proliferation at different culture cycles and light incubation conditions. Afr J Biotechnol 9:3079–3085

Meratan AA, Ghaffari SM, Niknam V (2009) In vitro organogenesis and antioxidant enzymes activity in *Acanthophyllum sordidum*. Biol Plant 53:5–10

Mithila J, Hall JC, Victor JMR, Saxena PK (2003) Thidiazuron induces shoot organogenesis at low concentrations and somatic embryogenesis at high concentrations on leaf and petiole explants of African violet (*Saintpaulia ionantha* Wendl.) Plant Cell 21:408–414

Mok MC, Mok DWS, Turner JE, Mujer CV (1987) Biological and biochemical effects of cytokinin-active phenylurea derivatives in tissue culture system. Hortic Sci 22:1194–1197

Mroginski E, Rey HY, Gonzalez AM, Luis A, Mroginski (2004) Thidiazuron promotes in vitro plant regeneration of *Arachis correntina* (Leguminosae) via organogenesis. J Plant Growth Regul 23:129–134

Murch SJ, Choffe KL, Victor JMR, Slimmon TY, Raj SK, Saxena PK (2000) Thidiazuron-induced plant regeneration from hypocotyl cultures of St. John's wort (*Hypericum perforatum*'). Plant Cell Rep 19:576–581

Murthy BNS, Victor J, Singh R, Fletcher RA, Saxena PK (1996) In vitro regeneration of chickpea (*Cicer arietimrn* L.) stimulation of direct organogenesis and somatic embryogenesis by thidiaairon. Plant Growth Regul 19:233–240

Murthy BNS, Murch SJ, Saxena PK (1998) Thidiazuron: a potent regulator of in vitro plant morphogenesis. In Vitro Cell Dev Biol Plant 34:267–275

Naaty B, Mbithe CM, Nyende AB, Njenga P, Muli JK (2017) In vitro regeneration via somatic embryogenesis of *Schizozygia Coffaeoides* Baill (Mpelepele) Amer. J Plant Biol 2:66–72
Naaz R, Anis M, Aref IM (2012) Assessment of the potentiality of TDZ on multiple shoot induction in *Bauhinia tomentosa* L., a woody legume. Acta Biol Hung 63:474–482
Ouyang Y, Chen Y, Lü J, Teixeira da Silva JA, Zhang X, Ma G (2016) Somatic embryogenesis and enhanced shoot organogenesis in *Metabriggsia ovalifolia* W. T. Wang. Sci Rep 19:6–24
Parveen S, Shahzad A (2010) TDZ-induced high frequency shoot regeneration in *Cassia sophera* Linn. via cotyledonary node explants. Physiol Mol Biol Plants 16:201–206
Parveen S, Shahzad A (2011) A micropropagation protocol for *Cassia angustifolia* Vahl. from root explants. Acta Physiol Plant 33:789–796
Parveen S, Shahzad A, Saema S (2010) In vitro plant regeneration system for *Cassia siamea* Lam., a leguminous tree of economic importance. Agrofor Syst 80:109–116
Pelah D, Kaushik RA, Mizrahi Y, Sitrit Y (2002) Organogenesis in the vine cactus *Selenicereus megalanthus* using thidiazuron. Plant Cell Tissue Organ Cult 71:81–84
Phippen W, Simon JE (2000) Shoot regeneration of young leaf explants from basil (*Ocimum basilicum* L.) In Vitro Cell Dev Biol Plant 36:250–254
Prathanturarug S, Pheakkoet R, Jenjittikul T, Chuakul W, Saralamp P (2012) In vitro propagation of *Stemona hutanguriana* W. Chuakul, an endangered medicinal plant. Physiol Mol Biol Plants 18:281–286
Radhakrishnan R, Ramachandran A, Kumari BDR (2009) Rooting and shooting: dual function of Thidiazuron in vitro regeneration of soybean (*Glycine max.* L.) Acta Physiol Plant 31:1213–1217
Radhika K, Sujatha M, Nageshwar Rao T (2006) Thidiazuron stimulates adventitious shoot regeneration in different safflower explants. Biol Plant 50:174–179
Raghu AV, Geetha SP, Martin G, Ravindran PN (2006) Direct shoot organogenesis from leaf explants of Embelia ribes Burm. f.: a vulnerable medicinal plant. J Forest Res 11:57–60
Reddy CM, Bramhachari PV, Murthy KSR (2015) Optimized plant tissue culture protocol for in vitro morphogenesis of an endangered medicinal herb *Ceropegia ensifolia* Bedd. Trop Subtrop Agroecosyst 18:95–101
Rida A, Shibli AM, Suwwan A, Ajlouni M (2001) In vitro multiplication of virus free Spunta potato. Pak J Bot 33:35–41
Saeed T, Shahzad A (2015) High frequency plant regeneration in Indian Siris via cyclic somatic embryogenesis with biochemical, histological and SEM investigations. Ind Crop Prod 76:623–637
Sahai A, Shahzad A, Sharma S (2010) Histology of organogenesis and somatic embryogenesis in excised root cultures of an endangered species *Tylophora indica*. Aus J Bot 58:198–205
Shahzad A, Ahmad N, Anis M (2006) An improved method of organogenesis from cotyledon callus of *Acacia sinuata* (Lour.) Merr. using thidiazuron. J Plant Biotechnol 8:1–5
Sharma R, Shahzad A (2008) Thiadiazuron (TDZ) induced regeneration from cotyledonary node explant of Abelmoschus moschatus Medik. L. (a valuable medicinal plant). World J Agr Sci 4:449–452
Shirani S, Mahdavi F, Maziah M (2009) Morphological abnormality among regenerated shoots of banana and plantain (*Musa* spp.) after in vitro multiplication with TDZ and BAP from excised shoot-tips. African J Biotechnol 8:5755–5761
Siddique I, Anis M (2007a) Rapid micropropagation of *Ocimum basilicum* L. using shoot tip explants pre-cultured in thidiazuron supplemented liquid medium. Biol Plant 51:757–790
Siddique I, Anis M (2007b) High frequency multiple shoot regeneration and plantlet formation in *Cassia angustifolia* (Vahl.) using thidiazuron. Med Arom Plant Sci Biotechnol 2:282–284
Siddique I, Anis M, Aref IM (2010) In vitro adventitious shoot regeneration via indirect organogenesis from petiole explants of *Cassia angustifolia* Vahl. – a potential medicinal plant. Appl Biochem Biotechnol 162:2067–2074
Singh CK, Raj SR, Jaiswal PS, Patil VR, Punwar PS, Chavda JC, Subhash N (2016) Effect of plant growth regulators on in vitro plant regeneration of sandalwood (*Santalum album* L.) via organogenesis. Agrofor Syst 90:281–288

Sivanesan L, Song JY, Hwang SJ, Jeong BR (2011) Micropropagation of *Cotoneaster wilsonii* Nakai – a rare endemic ornamental plant. Plant Cell Tissue Organ Cult 105:55–63

Sujatha M, Dinesh Kumar V (2007) In vitro bud regeneration of *Carthamus tinctorius* and wild Carthamus species from leaf explants and axillary buds. Biol Plant 51:782–786

Tejavathi DH, Raveesha HR, Shobha K (2011) Organogenesis is from the cultures of *Nothapodytes foetida* (Wight) Sleumer raised on TDZ supplemented media. Indian J Biotechnol 11:205–209

Thomas TD, Philip B (2005) Thidiazuron-induced high frequency shoot organogenesis from leaf derived callus of a medicinal climber, *Tylophora indica* (Burm. f.) Merrill. In Vitro Cell Dev Biol Plant 41:124–128

Thomas TD, Puthur JT (2004a) Thidiazuron induced high frequency shoot organogenesis in callus from Kigelia pinnata L. Botl Bull Acad Sin 45:307–313

Thomas TD, Puthur JT (2004b) Thidiazuron induced high frequency shoot organogenesis in callus from *Kigelia pinnata L.* Bot Bull Acad Sin 45:307–313

Tiwari V, Tiwari KN, Singh BD (2001) Comparative studies of cytokinins on in vitro propagation of *Bacopa monniera*. Plant Cell Tissue Organ Cult 66:9–16

Tulac S, Lejak-Levanic D, Krsnik-Rasol M, Jelaska S (2002) Effect of BAP, TDZ and CPPU on multiple shoot formation in pea (*Pisum sativum* L.) in culture in vitro. Acta Biol Craco Series Bot 44:161–168

Uranbey S (2005) Thidiazuron induced adventitious shoot regeneration in *Hyoscyamus niger*. Biol Plant 49:427–430

Varutharaju K, Soundar CR, Thilip C, Aslam A, Shajahan A (2014) High efficiency direct shoot organogenesis from leaf segments of Aerva lanata (L.) Juss. Ex Schult by using Thidiazuron. Sci World J 2014:1. Article ID 652919, 6 pages

Wang SY, Steffens GL, Faust M (1986) Breaking bud dormancy in apple with a plant bioregulator, thidiazuron. Phytochemistry 25:311–317

Wang SY, Iio HJ, Faust M (1991a) Changes in metabolic enzyme activities during thidiazuron-induced lateral bud break in apple. Hortic Sci 26:171–173

Wang SY, Jiao HJ, Faust M (1991b) Changes in activities of catalase, peroxidase and polyphenol oxîdase in apple buds during bud break induced by thidiaziron. J Plant Growth Regul 10:33–39

Xie D, Hong Y (2001) In vitro regeneration of *Acacia mangium* via organogenesis. Plant Cell Tissue Organ Cult 66:167–173

Zeng SJ, Duan J, LI LN (2008) Plant regeneration from leaf explants of *Tigridiopalma magnifica* (Melastomataceae). Pak J Bot 40:1179–1184

Zhang CL, Chen DF, Elliott MC, Slater A (2001) Thidiazuron- induce organogenesis and somatic embryogenesis in sugar beet (*Beta vulgaris* L.) In Vitro Cell Dev Biol Plant 37:305–310

In Vitro Morphogenesis of Some Himalayan Flora Using TDZ: A Potential Plant Growth Regulator

12

Nisha Dhiman, Nitu Gautam, Bhuvnesh Sareen, Pooja Kumari, Sapna Rajouria, and Amita Bhattacharya

Abstract

Thidiazuron or TDZ is a substituted phenylurea derivative, N-phenyl-N′-1, 2, 3-thidiazol-5-ylurea, extensively used in the in vitro morphogenesis of different plant species, including those found in Himalayas. The plants have ranged from the advanced monocotyledonous to dicotyledonous angiospermic plants to gymnosperms and even pteridophytes. While some of these plants are endemic to the Himalayas and are medicinally important, others have been introduced into the region for their immense commercial values. Optimal morphogenic responses obtained in these plants have varied with the concentration of TDZ used and its duration of exposure. These in turn have depended upon the explant and the media used, and also the plant species targeted. While in some cases TDZ alone evoked the desired responses, a combination of TDZ with other PGRs was required in other cases. The findings of these reports have been reviewed in the present article. Attempts have also been made to collate these findings into a ready reckoner for researchers working on in vitro morphogenesis of various plants inhabiting the Himalayas.

N. Dhiman · A. Bhattacharya (✉)
Academy of Scientific and Innovative Research, New Delhi, India

Division of Biotechnology, CSIR-Institute of Himalayan Bioresource Technology, Palampur, HP, India
e-mail: amitabhatta@ihbt.res.in

N. Gautam · B. Sareen · P. Kumari · S. Rajouria
Division of Biotechnology, CSIR-Institute of Himalayan Bioresource Technology, Palampur, HP, India

N. Ahmad, M. Faisal (eds.), *Thidiazuron: From Urea Derivative to Plant Growth Regulator*, https://doi.org/10.1007/978-981-10-8004-3_12

Keywords
Thidiazuron · Continuous exposure · Somatic embryogenesis · Cytokinin-like activity · Morphogenesis

Abbreviations

2, 4-D	2, 4-Dichlorophenoxyacetic acid
B5	Gamborg medium
BAP	Benzylamino purine
DCR	Douglas-fir cotyledon revised medium
Kn	Kinetin
LS	Linsmaier and Skoog medium
MRM	Modified rhododendron medium
MS	Murashige and Skoog medium
NAA	Naphthalene acetic acid
nM	Nanomole
PGR	Plant growth regulator
PLBs	Protocorm-like bodies
pM	Picomole
TDZ	Thidiazuron
WPM	Woody plant medium
μM	Micromole

12.1 Introduction

Significant progress has been made in the field of plant tissue culture since the conceptualization of cellular totipotency by Gottlieb Haberlandt in 1902 (Haberlandt 1902). Considerable progress has been also made since the establishment of the regulatory role of 'plant growth regulators' (PGRs) in morphogenesis during the twentieth century (Laimer and Rücker 2003). Ever since then, PGRs have continued to be the most important underlying factor dictating morphogenesis, or in other words, the formation of discrete organs and/or somatic embryos in cultured cells and tissues (Roberts and Hooley 1988; Jime'nez 2005; George et al. 2008). Initially, the first natural auxin, indole-3-acetic acid and the first cytokinin, i.e. 6-furfuryl-aminopurine or kinetin, were discovered by Kögl et al. (1934) and Miller et al. (1955), respectively. These paved the way for the identification of a number of other naturally occurring PGRs. Thus, only naturally occurring PGRs were used for the induction of in vitro morphogenesis in plants (Murashige 1979; Laimer and Rücker 2003). With advances in chemical sciences, however, the use of synthetic PGRs gained increasing popularity

(Gianfagna 1987; Nickell 1994). The compounds opened up multiple opportunities for the use of plant tissue culture techniques in fields as diverse as basic botany, biochemistry, agriculture, crop improvement and biotechnology. Since then, a wide variety of synthetic PGRs have been synthesized and successfully employed for in vitro morphogenesis and plant regeneration (Murthy et al. 1998; Gaba 2005). The substituted phenylurea derivative, 'thidiazuron' (N-phenyl-N′-1, 2, 3-thidiazol-5-ylurea) or TDZ, is one such synthetic PGR that has been popularly used in cultured cells and tissues of plants for in vitro morphogenesis. This moderately soluble, light yellow-coloured crystalline compound was first used by Arndt et al. (1976) for the abscission of turgid green leaves and for easing the plucking of cotton bolls.

Structurally, TDZ has two functional groups, the phenyl and the thidiazol groups. Any replacement of these functional groups results in significant reduction in the activity of TDZ. Hence, these functional groups are considered to be absolutely essential for high cytokinin-like activity in cultured cells and tissues (Mok et al. 1982). This cytokinin-like activity of TDZ is retained even at micro, nano as well as picomole concentrations (Preece et al. 1991; Mondal et al. 1998; Sandal et al. 2001). This property of TDZ distinguishes it from all other synthetic PGRs.

According to various reports published during the last 17 years, the cytokinin-like activity of TDZ has been extensively exploited for the regeneration and micropropagation of different plant species. The PGR has been particularly effective in a number of herbaceous to woody tree species (Murthy et al. 1996, 1998). While many of these plants had remained unresponsive to other PGRs and were generally recalcitrant to in vitro morphogenesis, TDZ was extremely effective in evoking regeneration responses in such plants. Significant improvement in already established micropropagation systems by TDZ has been also reported. For example, a two fold improvement in micropropagation of tea (Mondal et al. 1998; Sandal et al. 2001), banana (Lee 2005) and rapeseed (Roh et al. 2012) but four- to five fold improvement in case of *Morus indica* (Gupta et al. 2009) was reported.

12.2 TDZ in the Micropropagation of Himalayan Plants

The Himalayas being the hot spot of biodiversity are also the treasure trove of medicinally important plants (Sharma et al. 2009). These high-value medicinal plants are the subject of indiscriminate harvesting and ruthless exploitation. Thus, their conservation programmes have often required tissue culture interventions. Consequently, various in vitro approaches were employed, and a number of high-value rare, endangered and threatened medicinal plants were effectively regenerated (Faisal and Anis 2004; Anis et al. 2009). Successful tissue culture of these endangered Himalayan plants invariably depended upon the use of different types of natural as well as synthetic PGRs. Among these, the use of TDZ was significant (Sandal et al. 2001; Faisal et al. 2005; Ahmad and Anis 2007, 2012; Patial et al. 2017). Although the in vitro regeneration response of plants generally vary from species to species (Ochatt et al. 2010), a number of herbs, shrubs and woody tree species that inhabit different altitudinal gradients of Himalayas have benefited from the use of TDZ. The endangered Himalayan plants where TDZ has been used for in vitro morphogenesis and plant regeneration is summarized in Table 12.1. The list includes gymnospermous as well as monocotyledonous and dicotyledonous plants from diverse altitudinal gradients ranging from the foothills of Himalayas to steep slopes, grassy plains and river banks covering warm subtropical to cooler temperate, alpine, subalpine and cold desert regions (Fig. 12.1).

12.2.1 TDZ in the Micropropagation of Lower Plants

TDZ has played a significant role in the in vitro morphogenesis of several fern species including ornamentals, medicinal as well as tree species. Earlier in 1998, Thakur and co-workers first reported the use of TDZ (2.27 μM) and 8.06 μM of N-(4-Pyridyl)- N′-phenylurea (4-PU) in half-strength liquid MS medium for high rates of meristem multiplication in an edible fern of temperate region, i.e. *Matteuccia struthiopteris* (L.) Todaro (Thakur et al. 1998). In another study, TDZ (2.27 μM) in combination with NAA (1.07 μM), 2, 4-D (1.13 μM) and BA (4.43 μM) were used by Winarto and da Silva (2012) for improved micropropagation of the leatherleaf fern, *Rumohra adiantiformis*, using rhizomes as explants. Later, Taha et al. (2011) further demonstrated an increased prothallus growth of a rare epiphytic fern, *Platycerium coronarium* (up to 2.77 mm), by culturing young sporophytic and gametophytic leaves for 6 weeks on medium containing 4.54 μM TDZ. Recently, 4.54 μM TDZ and 1.61 μM NAA were again shown to promote highest rates of induction and multiplication of globular green bodies from in vitro juvenile sporophytes of an endangered tree fern, *Cibotium barometz* (Yu et al. 2017). However, the use of TDZ in the conservation of endangered ferns of Himalayas is yet to find any application.

12.2.2 TDZ in the Micropropagation of Threatened Gymnosperms of Himalayas

The gymnospermous plants that inhabit the different altitudinal gradients of Himalayas constitute a striking component of the Himalayan flora. These plants are generally found at altitudes ranging from 800 to 4000 m amsl in both the eastern and western regions of the Himalayas spanning India. Many of these have high medicinal and commercial value and have become threatened. Thus, it is not surprising that some of these plants have been efficiently micropropagated using TDZ (Table 12.1). It was Mathur and Nadgauda (1999) who first demonstrated the use of 0.025 μM TDZ in combination with 2.5 μM BA in Douglas-fir cotyledon revised medium for adventitious shoot bud regeneration in *Pinus wallichiana* from zygotic embryos (Gupta and Durzan 1985). However, Datta et al. (2006) failed to achieve any in vitro morphogenic response in *Taxus wallichiana* zygotic embryos cultured on half-strength WPM medium containing vitamins of SH medium and 0.045–0.18 μM TDZ. Much later however, Bhat et al. (2014)

Table 12.1 Effect of TDZ on the in vitro regeneration of plants inhabiting the Himalayas

S. no	Plant name	Family name	Native to	Natural distribution	Distribution in Indian Himalayas	Altitude (m amsl)	Status	Explant type	Basal media used	TDZ conc. used (μM)	Other PGRs	Response	Remarks on effect of TDZ	References
1	*Abies pindrow*	Pinaceae	Temperate and boreal regions of northern hemisphere	Afghanistan, Pakistan, India and Nepal	Western region	2100–3600	Vulnerable	Nodal and internodal segments	MS	13.62	–	Indirect shoot regeneration (20.00%)	Efficient for indirect shoot regeneration	Bhat et al. (2014)
2	*Pinus wallichiana*	Pinaceae	Karakoram and Hindu Kush mountains, Himalayas	Eastern Afghanistan, Northern Pakistan, Southwest China and Northwest India	Western region	2000–3500	Vulnerable	Zygotic embryos	1/2 DCS	0.025	BA (2.5 μM)	Adventitious shoot bud induction	First report on effect of TDZ on adventitious shoot bud induction in pines	Mathur and Nadgauda (1999)
3	*Taxus wallichiana*	Taxaceae	Himalayas	China, Tibet, Nepal, Bhutan, India and Myanmar	Eastern region	900–3700	Endangered	Zygotic embryos	1/2 WPM + vitamins of SH medium	0.045–0.18	–	–	Negligible effect of TDZ on adventitious bud induction	Datta et al. (2006)
4	*Aconitum balfourii*	Ranunculaceae	Himalayas spanning the Indian sub-continent	India, Nepal, South Tibet, Bhutan and Pakistan	Western region	2800–4200	Endangered	Leaf and petiole segments	MS	2.27	NAA (5.37 μM)	Indirect shoot induction (86.67%)	Synergistic effect of TDZ and NAA promoted faster regeneration	Gondval et al. (2016)
5	*Arnebia euchroma*	Boraginaceae	Himalayas	China, Nepal, India, Pakistan and also Afghanistan and Iran	Western region	3000–4000	Critically endangered	Cotyledon and hypocotyl	LS	4.5	–	Shoot regeneration	TDZ promoted maximum shoot organogenesis, irrespective of explant type	Jiang et al. (2005)
6	*A. euchroma*	Boraginaceae	Do	Do	Do	Do	Do	Rhizome buds	MS	1.0	–	Direct shoot regeneration	30 days exposure resulted in 14-fold increase in shoot number	Malik et al. (2010)
7	*Bacopa monnieri*	Scrophulariaceae	Subtropical areas world wide	India, Nepal, Sri Lanka, China, Pakistan, etc.	Eastern region	3300–5000	Endangered	Leaves	MS	1.13	–	Indirect rhizogenesis	Promotive effect of 0.272 μM TDZ	Vijayakumar et al. (2010)

(continued)

Table 12.1 (continued)

S. no	Plant name	Family name	Native to	Natural distribution	Distribution in Indian Himalayas	Altitude (m amsl)	Status	Explant type	Basal media used	TDZ conc. used (μM)	Other PGRs	Response	Remarks on effect of TDZ	References
8	*Berberis aristata*	Berberidaceae	Himalayas in India and Nepal	Subtropical regions of Asia, Europe and America	Western and eastern regions	1500–3450	Endangered	Leaf	WPM	0.5	–	Indirect shoot regeneration	Low concentration of TDZ effective in shoot regeneration	Brijwal et al. (2015)
9	*Bergenia ciliata*	Saxifragaceae	Central Asia	Afghanistan, China, India and Bhutan	North eastern region	1200–1300	Vulnerable	Leaf and node	½ B5	9.08	NAA (10.74 μM)	Callus development	TDZ promoted best callusing	Rafi et al. (2016)
10	*Camellia sinensis*	Theaceae	China	China, India, Sri Lanka, Kenya, etc.	Foothills of western and eastern regions	800–2600	–	Nodal segments	MS	1 0 pM-100.0 nM	NAA (10 μM)	Shoot multiplication	Very low TDZ concentration increased the micropropagation by twofold	Mondal et al. (1998)
11	*C. sinensis*	Theaceae	Do	Do	Do	Do	Do	Do	Liquid MS	2.5–5.0	–	Shoot multiplication	Micropropagation increased by twofold	Sandal et al. (2001)
12	*Crocus sativus*	Iridaceae	Southern Europe, Greece, Asia minor and Persia	Iran, Algeria, France, Greece, India, Italy, Monaco, Persia, Russia, Spain and Turkey	Himalayan foothills of Kashmir in India	2000	–	Corms	MS	2.5	Picloram (2.0 μM)	Somatic embryogenesis	Synergistic effect of TDZ and picloram induced somatic embryogenesis	Devi et al. (2014)
13	*Gaultheria fragrantissima*	Ericaceae	Temperate regions of the world	India, Nepal, west China, etc.	Eastern region	1800–2500	Endangered	Leaf and internode	MRM	4.54–9.08	–	Direct shoot regeneration and multiplication	Only TDZ could induce direct shoot regeneration	Ranyaphiaa et al. (2011)
14	*Gaultheria fragrantissima*	Ericaceae	Do	Do	Do	Do	Do	Adult shoot tips	WPM	1.0	–	Indirect shoot multiplication	Callus induction took place 20 days after inoculation	Bantawa et al. (2011)
15	*Gentiana kurroo*	Gentianaceae	Himalayas	India, Nepal and Pakistan	North-western region	1800–2700	Critically endangered	Leaf, petiole and roots	MS	4.54	NAA (0.53 μM)	Direct and indirect regeneration	Synergism of TDZ with NAA induced direct organogenesis	Sharma et al. (2014)

16	*Picrorhiza kurroa*	Scrophulariaceae	Himalayas	India, Nepal, Bhutan, China, Tibet and Pakistan	Western region	3000–5300	Endangered	Nodal segments	MS	0.5	2, 4-D (2.26 μM) and IBA (2.46 μM)	Indirect somatic embryogenesis	Only callusing in TDZ with 2, 4-D	Sharma et al. (2010)
17	*P. kurroa*	Scrophulariaceae	Do	Do	Do	Do	Do	Leaves		0.5	–	42 shoots per leaf segment	15 days exposure to TDZ essential	Patial et al. (2012)
18	*P. kurroa*	Scrofulariaceae	Do	Do	Do	Do	Do	Nodal segments		0.5	–	Shoot multiplication	15 days exposure to TDZ for 1.15-fold micropropagation efficiency	Patial et al. (2017)
19	*Populus ciliata*	Salicaceae	Himalayas	India, Nepal, Sikkim, Bhutan Myanmar and Pakistan	Entire region from west to east	1700–3000	Endangered	Leaf	MS	0.10	–	Shoot regeneration (80%)	TDZ more effective than other cytokinins	Aggarwal et al. (2012)
20	*Quercus rubra*	Fagaceae	North of eastern USA and Southeastern Canada	North America and Canada	Western region	150–1800	–	Cotyledonary node	MS	0.45	BA (4.4 μM)	Shoot regeneration (64.7%)	Synergistic effect of TDZ and BA	Vengadesan and Pijut (2009)
21	*Rauvolfia serpentina*	Apocynaceae	Tropical to subtropical regions and also Himalayas	India	Entire region from west to east	1000	Endangered	Nodal segments	MS	50	–	Shoot regeneration (90%)	Faster regeneration	Alatar (2015)
22	*Rheum emodi*	Polygonaceae	Temperate and subtropical regions of Himalayas	India and Tibet	Western region	3200–5200	Critically endangered	Seeds, leaves, shoots and rhizomes	MS	10.00	BAP (5.00 μM)	Callus formation (94%)	TDZ with BAP produced maximum callus	Tabin et al. (2014)
23	*Rhododendron dalhousiae*	Ericaceae	Himalayas	India, Bhutan, Nepal and China	Eastern region	1800–2300	Threatened	Rhizome nodes	MS	4.54–22.7		Shoot multiplication	Less effective in shoot multiplication than other cytokinins	Gurung and Singh (2010)
24	*Saussurea lappa*	Asteraceae	South Asia	India, Pakistan and China	Western region	2500–3500	Endangered	Shoot tips	MS	0.45	–	Multiple shoot formation (90%)	Lower concentration more effective	Johnson et al. (1997)

(continued)

Table 12.1 (continued)

S. no	Plant name	Family name	Native to	Natural distribution	Distribution in Indian Himalayas	Altitude (m amsl)	Status	Explant type	Basal media used	TDZ conc. used (μM)	Other PGRs	Response	Remarks on effect of TDZ	References
25	*Semecarpus anacardium*	Anacardiaceae	Himalayan and sub-Himalayan regions of India	India	Entire region from west to east	1050	–	Nodal segments	WPM	2.27 and 4.45	–	Shoot differentiation and multiple shoot bud induction	Preculturing for 4 weeks on 2.27 μM TDZ best for maximum shoot multiplication	Panda et al. (2016)
26	*Swertia chirayita*	Gentianaceae	Himalayas	India and Bhutan	Entire region from west to east	1200–2100	Critically endangered	Leaf	MS	2.27	BAP (2.22 μM), 2, 4-D (4.53 μM)	Callus induction	Synergistic effect of TDZ, BAP and 2,4-D produced friable callus	Kumar et al. (2014)
27	*Trichosanthes tricuspidata*	Cucurbitaceae	Tropical to subtropical regions in South Asia	India, Bangladesh, Nepal, China, Sri Lanka, and Malaysia	Eastern region	1200–2300	Endangered	Nodal segments	MS	2.27	BAP (17.75 μM)	Shoot multiplication	TDZ more effective than BAP	Rajender et al. (2017)
28	*Tylophora indica*	Apocynaceae	India	India	Western region	1260	Endangered	Petiole	MS	2.27	2,4D (9.95 μM)	Indirect shoot regeneration (90%)	TDZ more effective than other cytokinins	Faisal et al. (2005)
29	*Valeriana wallichii*	Valerianaceae	Himalayas	India and Pakistan	Northwest region	1300–3300 m	Endangered	Seedling	MS	0.1	–	Shoot multiplication	Low concentration of TDZ produced best results	Singh et al. (2015)
30	*Bambusa nutans*	Poaceae	Himalayas	Sub-Himalayan regions of India, Nepal, Bhutan, Bangladesh, Thailand and Vietnam	Both eastern and western regions	600–1500	–	Nodal segments	MS	0.50 and 2.00	NAA (0.74 μM)	Embryoid development and germination	60 days exposure to TDZ effective for shoot development	Mehta (2012)
												Shoot development	Combination of NAA and TDZ facilitated highest germination	

31	*Cymbidium aloifolium*	Orchidaceae	Temperate to tropical regions of Asia	China, Burma and Sumatra	North-western regions	1000	Endangered	Shoot segments	MS	2.20	–	Direct shoot proliferation	Lower concentration effective for induction and differentiation of shoot buds	Nayak et al. (1997)
32	*Cymbidium giganteum*	Orchidaceae	Nepal	Subtropical regions of India, China, Japan, Malaysia, Philippines, Borneo and Northern Australia	Eastern regions	1700–2300	Endangered	Pseudostem segments	MS	0.90	–	Multiplication of protocorm-like bodies and plantlet formation	Low concentration effective for induction of quality PLBs and healthy plantlets	Roy et al. (2012)
33	*Dendrocalamus hamiltonii*	Poaceae	Nepal	China, Nepal, Bhutan, India, Pakistan and Sri Lanka	North-eastern and western regions	1000–2000	Endangered	Nodal cuttings with axillary buds	MS	2.27	Kn (5.00 μM) BA (4.44 μM)	Bud sprouting (90%)	Synergistic effect of TDZ with BAP and Kn on micropropagation and in vitro bud break	Kapruwan et al. (2014)
34	*Dendrobium chrysanthum*	Orchidaceae	Himalayas (South China to Indo-China border)	China, Tibet, Bhutan, Nepal, India, Myanmar and Thailand	Foothills of western and eastern regions	350–2200	Endangered	Nodal segments	MS	5.00	BAP (5.00 μM)	Induction of shoot buds (100%) and shoot multiplication	Synergistic effect of TDZ and BAP for enhanced shoot proliferation and elongation	Hajong et al. (2013)
35	*Dendrobium nobile*	Orchidaceae	Himalayas	South China to Indo-China border, Bangladesh, Nepal, Sikkim, Bhutan, Myanmar, Thailand, India and Vietnam	Eastern region including foothills	1500–2000	Endangered	Pseudostem segments	MS	6.80	–	Induction of protocorm-like bodies (PLBs)	Exposure for 60 days to low to moderate concentration of TDZ effective for high frequency PLB proliferation	Bhattacharyya et al. (2014)
36	*Dendrobium wangliangii*	Orchidaceae	Tropical and subtropical regions of South Asia	China, and Indo-China border	Eastern region	650–1500	Critically endangered	Green pods	MS	9.08	–	Inflorescence induction	TDZ induced 100% inflorescence	Zhao et al. (2013)

(continued)

Table 12.1 (continued)

S. no	Plant name	Family name	Native to	Natural distribution	Distribution in Indian Himalayas	Altitude (m amsl)	Status	Explant type	Basal media used	TDZ conc. used (μM)	Other PGRs	Response	Remarks on effect of TDZ	References
37	*Geodorum densiflorum*	Orchidaceae	Temperate to tropical Asia and Australia	Bangladesh, India, Sri Lanka, Myanmar, Thailand, etc.	Western region	1100	Endangered	Protocorms	MS	2.27	2, 4-D (9.06 μM)	Callus development and regeneration of plantlets	TDZ alone facilitated shoot bud regeneration but callusing in presence of 2, 4-D	Lu (2010)
38	*Herminium lanceum*	Orchidaceae	Eastern Asia	Mongolia, Korea, Taiwan and India	Eastern region	1200–2000	Critically endangered	Seeds	Mitra	4.00	–	Induction of shoot buds	TDZ better than other cytokinins	Singh and Babbar (2016)
39	*Hedychium coronarium*	Zingiberaceae	Tropical and subtropical areas worldwide	Himalayas spanning South China, Nepal, Bhutan and Myanmar	Eastern region	750	Endangered	Rhizome buds	MS	4.54	–	Multiple shoot regeneration	Lower concentration effective for faster induction of in vitro regeneration	Verma and Bansal (2014)
40	*H. muluense*	Zingiberaceae	Tropical to temperate regions of Central and Southeast Asia	South China and the Himalayas	–	–	–	Shoot apex	MS	0.60	BA (8.90 μM)	Indirect somatic embryogenesis	Stimulatory effect of 30 days of exposure to BA and TDZ	Sakhanokho et al. (2008)
41	*H. spicatum*	Zingiberaceae	Himalayas	Nepal and India	Western region	800–3000	Vulnerable	Shoots	MS	1.0	–	Shoot multiplication (83.33%)	Lower concentration of TDZ evoked maximum shoot multiplication	Giri and Tamta (2011)
42	*Hyoscyamus niger*	Solanaceae	Temperate to tropical regions	Central and Southern Europe, Western Asia and Siberia	Western region	2100–3300	Threatened	Shoot tips	MS	10	–	Friable callus formation	10 μM TDZ enhanced the friable callus formation	Quadri et al. (2011)

43	*Malaxis muscifera*	Orchidaceae	Hindu Kush Himalayan region	Afghanistan, Pakistan, India, Bhutan and Southeast Tibet	North-western region	2500–4000	Threatened	Seeds	Mitra	4.54	NAA (5.37 μM)	Seed germination and PLB formation	Lower concentration of TDZ effective for PLBs formation	Kant (2015)
44	*Polygonatum verticillatum*	Liliaceae	Temperate regions	Temperate Himalayas and Europe	Western region	2000–3000	Threatened	Stem discs	MS	1.13–13.62		Comparatively higher shoot length	TDZ promoted higher shoot length than BA with not much effect on shoot multiplication	Bisht et al. (2012)
45	*Stemona tuberosa*	Stemonaceae	Southeast Asia and New Guinea	South China, India, Thailand, Malaysia, Indonesia, Philippines, etc.	Eastern region	250–1200	Endangered	Nodal segments	MS	18.16	Kn (32.55 μM)	Enhanced shoot multiplication	Tuberous root and multiple shoot formation	Murthy et al. (2013)
46	*Viola odoratum*	Violaceae	Europe and Asia	India, Iran, Pakistan, Afghanistan, Iraq, Mediterranean regions and Caucasia	Western and eastern regions	1500–1800	Rare	Leaf, petiole and root	MS	9.08	GA_3 (1.44 and 2.88 μM)	Indirect shoot regeneration	TDZ and GA_3 effectively regulated shoot morphogenesis	Mokhtari et al. (2015)

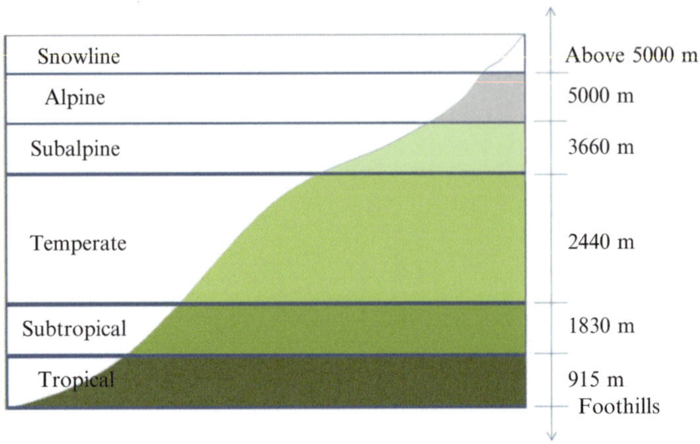

Fig. 12.1 Different altitudinal zones inhabited by different lower and higher plants reviewed in the article

used a considerably higher concentration of TDZ (i.e. 13.62 μM) for induction of indirect shoot buds in 20% nodal and internodal segments as explants from *Abies pindrow* (Table 12.1).

12.2.3 TDZ in the Micropropagation of Endangered Dicotyledonous Plants of Himalayas

TDZ has also been effectively used for an appreciable number of endangered Himalayan dicotyledonous plants belonging to various families (Fig. 12.2). The concentration of the TDZ used for these species ranged between 0.009 and 50 μM. Of these, 0.4–0.6 μM followed by 2.27 μM were the concentrations of choice for evoking maximal in vitro shoot regeneration. These concentrations of the PGR promoted either direct or indirect regeneration or both in the various explants used by different workers (Table 12.1). In all these plants, MS medium was invariably preferred for 72% species followed by WPM (Lloyd and McCown 1981) for 14%, MRM (Modified Rhododendron medium) and ½ Gamborg medium (B5) (Gamborg et al. 1968) for 4% each and LS (Linsmaier and Skoog 1965) and ½ DCS media for 3% plant species each (Fig. 12.3a).

12.2.4 TDZ in the Micropropagation of Endangered Monocotyledonous Plants of Himalayas

Among the different monocot plant species reviewed in the present article, the members of the Orchidaceae family were found to benefit maximally from the use of TDZ. This was followed by three members of Zingiberaceae, two members of Poaceae and one each of families, Solanaceae, Stemonaceae and Liliaceae (Table 12.1). Of the different media where TDZ was added, MS (Murashige and

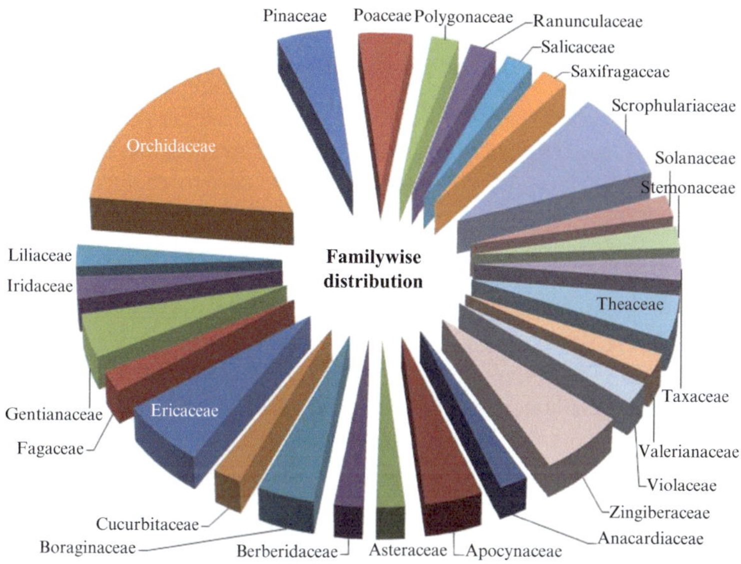

Fig. 12.2 Family-wise distribution of different plants (gymnosperms as well as angiosperms) that have been benefited by TDZ reviewed in the article

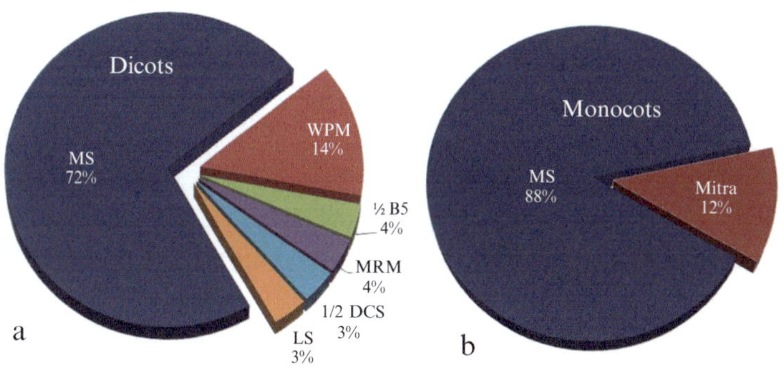

Fig. 12.3 Percent distribution of media usage for in vitro culturing of plants belonging to different families with TDZ as a PGR supplement (**a**) dicots, (**b**) monocots

Skoog 1962) was the medium of choice for 88% of plants reviewed, whereas, Mitra medium (Mitra et al. 1976) was invariably used for 12% plants belonging to the Orchidaceae family (Fig. 12.3b).

While TDZ concentration as low as 0.90 μM was effective for pseudostem segments of the orchid, *Cymbidium giganteum* (Roy et al. 2012), the concentration used for other orchids ranged between 2.2 and 6.81 μM. In the case of the family Zingiberaceae, 1.0 μM was used for the plant, *Hedychium spicatum*, whereas 0.6 and 4.54 μM were required for its other species, i.e. *H. muluense* and *H. coronarium*, respectively. *Hyoscyamus niger* (family Solanaceae) and *Stemona tuberosa* (family Stemonaceae) are the other two endangered monocots that required higher concentrations of TDZ ranging between 10 and 18.16 μM (Table 12.1).

12.3 Efficiency of TDZ Concentrations

The endogenous levels of auxins and cytokinins vary from plant to plant and tissue to tissue (Bhojwani and Dantu 2013; Sang et al. 2016). Therefore, the requirements of TDZ by a particular explant or plant also varies from either too low, moderate or too high. This explains why the concentration of TDZ used for the induction of direct as well as indirect in vitro shoot regeneration in different plant species inhabiting the Himalayas has varied with the explant used, the plant species targeted and the culture medium used. While concentrations ranging from 0.5 to 10.0 μM were used by different workers to evoke in vitro morphogenesis in different plants, concentrations as high as 13.6–18.16 μM were used for some other plants and explants. Still higher concentration like 50 μM was required for yet other plants (Table 12.1). TDZ is also known to trigger the autonomy of cytokinin biosynthesis in plant cells (Huetteman and Preece 1993; Guo et al. 2011; Patial et al. 2017). Thus, even lower concentrations of TDZ is sufficient for triggering cell division in plant tissue(s) and in turn callusing followed by indirect morphogenesis (Lu 1993; Murthy et al. 1998; Gondval et al. 2016). For example, concentrations as low as 1.0 picomole to 100 nanomole was necessary for promoting faster and improved rates (two fold) of shoot multiplication in tea in solid cultures (Mondal et al. 1998). However, shoot multiplication of the same plant in liquid cultures required 2.5–5.0 μM (Sandal et al. 2001).

12.4 Duration of TDZ Exposure and Explant Response

Depending on the endogenous levels of phytohormones within a cell or tissue, TDZ behaves like an auxin or cytokinin (Guo et al. 2011). It is also believed to convert cytokinin ribonucleotides into more potent cytokinin ribonucleosides (Capelle et al. 1983). This explains why TDZ is invariably more effective than BA, particularly at lower concentrations (Nielsen et al. 1993; Guo et al. 2011). Nevertheless, irrespective of the concentration, the continuous presence of TDZ exerts an adverse effect on the morphogenic responses of different explants and plants (Murthy et al. 1996; Mondal et al. 1998; Faisal et al. 2005; Faisal and Anis 2006). Thus, besides concentration, the length of time to which an explant is exposed to TDZ is an important factor governing the morphogenic responses in a plant (Mok and Mok 1985; Murthy et al. 1998; Faisal et al. 2005; Ahmed et al. 2006). TDZ's ability to trigger autonomy in cytokinin biosynthesis probably results in creating a supra-optimal level of cytokinins within a cell. This, thereby, exerts an adverse effect on morphogenic responses of a cell. TDZ's role as a strong cytokinin oxidase inhibitor during prolonged culture (Mok et al. 1982, 1987; Huetteman and Preece 1993; Hare and Staden 1994) is probably the other reason for the adverse effect of prolonged exposure of explants to TDZ. Necrosis or cell death due to continuous build-up of cytokinins during prolonged exposure to TDZ can thus be assumed. In order to circumvent this problem, different researchers have attempted short periods of exposure to TDZ followed by their transfer to various regeneration supporting media (Malik and Saxena 1992;

Mondal et al. 1998; Siddique and Anis 2007; Patial et al. 2017). In this regard, even 3 days of exposure to TDZ has been found sufficient for evoking a morphogenic response in explants (Hutchinson and Saxena 1996). However, the duration for which an explant needs to be exposed to TDZ varies from plant to plant (Singh and Syamal 2001; Prathanturarug et al. 2003; Tang and Newton 2005; Tao et al. 2011). Thus, Aggarwal et al. (2012) recommended only 2 weeks of exposure to 0.1 μM TDZ for in vitro morphogenesis in leaf explants of Himalayan poplar, *Populus ciliata.* On the other hand, a 30-day exposure to TDZ followed by their transfer to medium containing 5.0 μM kinetin was required for the micropropagation of *Arnebia euchroma* rhizome buds obtained from the trans-Himalayan region at Kibber (4200 m amsl) in Himachal Pradesh, India (Malik et al. 2010). In a yet another report, 60 days of exposure to 6.81 μM TDZ was required for *Dendrobium nobile* collected from eastern Himalayan region of Pakyong, Sikkim (Bhattacharyya et al. 2014). Sixty days of exposure to TDZ was also required for shoot development of *Bambusa nutans* (Mehta 2012). In each of these plants, a carry-over or residual effect of TDZ was observed. Since TDZ is a urea-based cytokinin, it is not amenable to degradation by cytokinin oxidase. Hence, it tends to remain in plant tissues to thereby impart a residual effect on explants (Makara et al. 2010). Probably, because of this the exposure of explants to TDZ beyond 8 weeks was not recommended by Lu (1993).

12.5 Explant Types and Variation in TDZ-Induced Morphogenic Responses

As in the case of other PGRs, the TDZ-dependent morphogenic responses in plant tissues have been found to vary with the explant type and the concentration used for a particular explant (Table 12.1). Depending upon the endogenous requirements of a tissue, TDZ behaves like an auxin or a cytokinin (Guo et al. 2011). Thus, TDZ can modulate the endogenous levels of both auxins and cytokinins within a plant cell (Visser et al. 1992; Murthy et al. 1998; Guo et al. 2011; Kou et al. 2016). This accounts for the differential responses of different explants of the same plant to the same concentration of TDZ. It also explains the differential requirement of TDZ concentrations by different tissues of the same plant. Depending upon the competence of the explant and the concentration that is used, a morphogenic response is generally directed towards a certain phenology (Christianson and Warnick 1983; Sharma et al. 2007; Pulianmackal et al. 2014). Thus, many a times, the same or different explants of a plant may require different concentrations of TDZ for morphogenic responses as varied as direct or indirect organogenesis or even somatic embryogenesis. For example, the same explant, i.e. the nodal segments of the endangered Himalayan plant, *Picrorhiza kurroa*, showed callus formation at 0.44 μM TDZ (Sharma et al. 2010) but 1.15-fold increase in shoot multiplication at 0.5 μM (Patial et al. 2017). Again, the leaf explants of the same plant were found to support the regeneration of about 42 shoots per explant on PGR-free medium but after a short exposure of 15 days to

0.5 μM TDZ (Patial et al. 2012). At times, even different explants of the same plant are found to show the same response despite the use of different concentrations of TDZ (Table 12.2). For example, two different explants, i.e. hypocotyl segments and rhizome buds of the critically endangered Himalayan plant, *Arnebia euchroma*, showed shoot development, albeit at 4.5–1.0 μM TDZ (Jiang et al. 2005; Malik et al. 2010). On the other hand, when the same concentration of TDZ, i.e. 0.45 μM, was used for two different plant species, i.e. *Saussurea lappa* and *Quercus rubra*, efficient shoot proliferation was recorded, irrespective of explant types (Johnson et al. 1997; Vengadesan and Pijut 2009). In a similar example, 4.54 μM TDZ supported indirect as well as direct shoot regeneration in two different Himalayan plants, i.e. *Hedychium coronarium* and *Gentiana kurroo*, albeit from different explants, i.e. rhizome buds and leaf as well as petiole explants, respectively. However, 0.53 μM NAA was additionally required for *Gentiana kurroo* (Sharma et al. 2014), whereas, no such additional PGR was necessary for *H. coronarium* (Verma and Bansal 2014).

12.6 Synergistic Effect of TDZ and Other PGRs

Despite being structurally dissimilar to auxins and cytokinins, TDZ induces morphogenic responses similar to these PGRs by mechanism(s) still unknown to researchers (Lu 1993; Guo et al. 2011). TDZ is believed to modulate the endogenous levels of auxins and cytokinins (Mok et al. 1982). Auxins or cytokinins in turn perform certain function(s) in a living cell or tissue, either individually or in synergy, to bring about various physiological and biochemical changes required for cell division and regeneration (Guo et al. 2011). A classic example to this effect is the demonstration of cytokinin-dependent cell division in soybean by Thomas and Katterman (1986). TDZ-induced callus growth as well as cytokinin activity higher than that of zeatin in *Phaseolus lunatus* cv. Kingston was also shown by Mok et al. (1982).

The synergistic effect of TDZ and other PGRs is also important and has been demonstrated in a number of endangered Himalayan plants (Table 12.1). Thus, a combination of TDZ and BAP was found to promote bud sprouting as well as high number of shoot formation in almost 90–100% explants in a variety of plants like the solid culmed bamboo, *Dendrocalamus hamiltonii* (Kapruwan et al. 2014), *Dendrobium chrysanthum*, an orchid from Sikkim Himalayas in India (Hajong et al. 2013) and *Trichosanthes tricuspidata*, an inhabitant of eastern Himalayas (Rajender et al. 2017) etc. Similarly, a combination of NAA or IBA and TDZ were required for plant regeneration in a number of endangered inhabitants of Himalayas like *Malaxis muscifera*, *Aconitum balfourii*, etc. (Kant 2015; Gondval et al. 2016). In a separate study, a combination of TDZ (2.5 μM) and the growth inhibitor, picloram (2.0 μM), was required for induction of somatic embryogenesis from leaf bases of *Crocus sativus* or saffron, a plant introduced from Iran into the Himalayan foothills of Kashmir in India (Devi et al. 2014).

Table 12.2 Explant type and its response to TDZ

Plant	Explant type	PGR (conc.)	Response type							Reference
			Callus	Somatic embryos	Shoots	Shoot buds	Root	PLBs	Inflorescence	
Abies pindrow	Nodal segments	13.6			√					Bhat et al. (2014)
Camellia sinensis										Mondal et al. (1998)
Picrorhiza kurroa		0.5	√							Sharma et al. (2010)
Picrorhiza kurroa		0.5			√					Patial et al. (2017)
Rauvolfia serpentina		50.0								Alatar (2015)
Semecarpus anacardium		2.27				√				Panda et al. (2016)
Trichosanthes tricuspidata		2.27								Rajender et al. (2017)
Bambusa nutans		0.5		√						Mehta (2012)
Dendrocalamus hamiltonii.		2.27				√				Kapruwan et al. (2014)
Dendrobium chrysanthum		5.0			√					Hajong et al. (2013)
Stemona tuberosa		18.16								Murthy et al. (2013)
Arnebia euchroma	Rhizome buds	1.0			√					Malik et al. (2010)
Rhododendron dalhousiae		4.54								Gurung and Singh (2010)
Hedychium coronarium		4.54								Verma and Bansal (2014)
Aconitum balfourii	Leaf and petiole segments	2.27			√					Gondval et al. (2016)
Gentiana kurroo		4.54								Sharma et al. (2014)
Viola odoratum		9.08								Mokhtari et al. (2015)

(continued)

Table 12.2 (continued)

Plant	Explant type	PGR (conc.)	Response type							Reference
			Callus	Somatic embryos	Shoots	Shoot buds	Root	PLBs	Inflorescence	
Pinus wallichiana	Seeds and seedling	0.025				√				Mathur and Nadgauda (1999)
Taxus wallichiana		0.045								Datta and Jha (2006)
Rheum emodi		10.0	√							Tabin et al. (2014)
Valeriana wallichii		0.1			√					Singh et al. (2015)
Herminium lanceum		4.0				√				Singh and Babbar (2016)
Malaxis muscifera		4.54						√		Kant (2015)
Quercus rubra	Cotyledonary node	0.45			√					Vengadesan and Pijut (2009)
Gaultheria fragrantissima	Adult shoot tips	1.0	√		√					Bantawa et al. 2011
Saussurea lappa		0.45								Johnson et al. (1997)
Hyoscyamus niger		10.0	√							Quadri et al. (2011)
Cymbidium aloifolium	Shoots	2.2			√	√				Nayak et al. (1997)
Hedychium muluense		0.60		√						Sakhanokho et al. (2008)
Hedychium spicatum		1.0			√					Giri and Tamta (2011)
Cymbidium giganteum	Pseudostem segments	0.90				√		√		Roy et al. (2012)
Dendrobium nobile		6.81				√				Bhattacharyya et al. (2014)

Bacopa monnieri	Leaf and leaves	1.13	√		√		√			Vijayakumar et al. (2010)
Berberis aristata		0.5								Brijwal et al. (2015)
Bergenia ciliata		9.08	√							Rafi et al. 2016
Picrorhiza kurroa		0.5			√					Patial et al. (2012)
Populus ciliata		0.10								Aggarwal et al. (2012)
Swertia chirayita		2.27	√							Kumar et al. (2014)
Geodorum densiflorum	Protocorm-derived callus	2.27	√		√					Lu (2010)
Arnebia euchroma	Dicotyledonous hypocotyls	4.5			√					Jiang et al. (2005)
Tylophora indica	Petiole-derived callus	2.27			√					Faisal et al. (2005)
Dendrobium wangliangii	Green pods	9.08							√	Zhao et al. (2013)
Crocus sativus	Corms	2.5		√						Devi et al. (2014)
Polygonatum verticillatum	Stem discs	1.13			√					Bisht et al. (2012)

12.7 Conclusion

In conclusion, it is evident from this article that TDZ has played a vital role in the efficient regeneration of several endangered plants inhabiting the Himalayas. Some of these plants (tea, saffron and some bamboo species like *Bambusa nutans*) were introduced into Himalayas from other regions of the world and are of immense commercial value. TDZ helped in improving their existing micropropagation methods for generation of quality planting materials. Still other plants for which TDZ was employed are ruthlessly exploited natives of the Himalayas and required urgent conservation.

Acknowledgements The authors thank Director, CSIR-IHBT, Palampur for providing the infrastructure. ND thanks the University Grants Commission, Govt. of India for providing Senior Research Fellowship. NG thanks the Department of Biotechnology, Govt. of India for providing her fellowship; BS thanks the Indian Council of Medical Research (ICMR), Govt. of India for providing Senior Research Fellowship. The CSIR-IHBT communication number for the present article is 4167.

References

Aggarwal G, Sharma C, Srivastava DK (2012) Thidiazuron: a potent cytokinin for efficient plant regeneration in Himalayan poplar (*Populus ciliata* Wall.) using leaf explants. Ann For Res 55(2):147

Ahmad N, Anis M (2007) Rapid clonal multiplication of a woody tree, *Vitex negundo* L. through axillary shoots proliferation. Agrofor Syst 71:195–200. https://doi.org/10.1007/s10457-007-9078-1

Ahmed MR, Anis M (2012) Role of TDZ in the quick regeneration of multiple shoots from nodal explants of Vitex *trifolia* L. – an important medicinal plant. Appl Biochem Biotechnol 168:957–966

Ahmed N, Srivastava R, Anis M (2006) Improvement in carnation shoot multiplication using thidiazuron in vitro. Propag Ornam Plants 6:109–113

Alatar AA (2015) Thidiazuron induced efficient in vitro multiplication and ex vitro conservation of *Rauvolfia serpentina* – a potent antihypertensive drug producing plant. Biotechnol Equip 29(3):489–497

Anis M, Husain MK, Faisal M, Shahzad A, Ahmad N, Siddique I, Khan H (2009) In vitro approaches for plant regeneration and conservation of some potential medicinal plants. In: Kumar A, Sopory SK (eds) Recent advances in plant biotechnology & its applications, vol 14. I.K. International Pvt. Ltd., New Delhi, pp 397–410

Arndt FR, Rusch R, Stillfried HV, Hanisch B, Martin WC (1976) A new cotton defoliant. Plant Physiol 57:S-99

Bantawa J, Da Silva JAT, Ghosh SK, Mondal TK (2011) Determination of essential oil contents and micropropagation of *Gaultheria fragrantissima*, an endangered woody aromatic plant of India. J Hortic Sci Biotechnol 86(5):479–485

Bhat SJA, Gangoo SA, Geelani SM, Qasba SS, Parray AA (2014) Callus culture and organogenesis in fir (*Abies pindrow* Royle). J Cell Tissue Res 14(3):4653–4658

Bhattacharyya P, Kumaria S, Diengdoh R, Tandon P (2014) Genetic stability and phytochemical analysis of the in vitro regenerated plants of *Dendrobium nobile* Lindl., an endangered medicinal orchid. Meta Gene 2:489–504

Bhojwani SS, Dantu PK (2013) Plant tissue culture: an introductory text. Springer, London. http://dx.doi.org/10.1007/978-81-322-1026-9

Bisht S, Bisht NS, Bhandari S (2012) In vitro micropropagation in *Polygonatum verticillatum* (L.) All. An important threatened medicinal herb of Northern India. Physiol Mol Biol Plants 18(1):89–93. https://doi.org/10.1007/s12298-011-0091-5

Brijwal L, Pandey A, Tamta S (2015) In vitro propagation of the endangered species *Berberis aristata* DC. via leaf-derived callus. In Vitro Cell Dev Biol Plant 51(6):637–647

Capelle SC, Mok DWS, Kirchner SC, Mok MC (1983) Effects of thidiazuron on cytokinin autonomy and the metabolism of N6-(Y2- isopentyl) [8-14c] adenosine in callus tissues of *Phaseolus lunatus* L. Plant Physiol 73:796–802

Christianson ML, Warnick DA (1983) Competence and determination in the process of in vitro shoot organogenesis. Dev Biol 95(2):288–293

Datta MM, Majumder A, Jha S (2006) Organogenesis and plant regeneration in *Taxus wallichiana* (Zucc.) Plant Cell Rep 25:11–18. https://doi.org/10.1007/s00299-005-0027-z

Devi K, Sharma M, Ahuja PS (2014) Direct somatic embryogenesis with high frequency plantlet regeneration and successive cormlet production in saffron (*Crocus sativus* L.) S Afr J Bot 93:207–216

Faisal M, Anis M (2004) In vitro mass propagation and conservation of some endangered medicinal plants. In: D'Souza L, Anuradha M, Nivas S, Hegde S, Rajendra K (eds) Biotechnology for a better future. SAC Publications, Mumbai, pp 82–91

Faisal M, Singh S, Anis M (2005) In vitro regeneration and plant establishment of *Tylophora indica* (Burm. f.) Merrill: petiole callus culture. In Vitro Cell Dev Biol Plant 41(4):511–515

Faisal M, Anis M (2006) Thidiazuron induced high frequency axillary shoot multiplication in *Psoralea corylifolia*. Biol Plant 50(3):437–440

Gaba VC (2005) Plant growth regulators in tissue culture and development. In: Trigiano RN, Gray DJ (eds) Plant development and biotechnology. CRC Press, Washington, DC, pp 87–99

Gamborg OL, Miller RA, Ojima O (1968) Nutrient requirements of suspension cultures of soybean root cell. Exp Cell Res 50:151–158

George EF et al (2008) Plant growth regulators I: introduction: auxins, their analogues and inhibitors. In: Plant propagation by tissue culture, 3rd edn. Springer, Netherlands, pp 175–204

Gianfagna JT (1987) Natural and synthetic growth regulators and their use in horticultural and agronomic crops. In: Davies PJ (ed) Plant hormones and their role in plant growth and development. Martinus Nijhoff Publishers, Dordrecht, pp 614–635

Giri D, Tamta S (2011) Effect of plant growth regulators (PGRs) on micropropagation of a vulnerable and high value medicinal plant *Hedychium spicatum*. Afr J Biotechnol 10(20):4040–4045

Gondval M, Chaturvedi P, Gaur AK (2016) Thidiazuron-induced high frequency establishment of callus cultures and plantlet regeneration in *Aconitum balfourii* Stapf.: an endangered medicinal herb of North-West Himalayas. Indian J Biotechnol 15:251–255

Guo B, Abbasi BH, Zeb A, Xu LL, Wei YH (2011) Thidiazuron: a multi-dimensional plant growth regulator. Afr J Biotechnol 10(45):8984–9000

Gupta PK, Durzan DJ (1985) Shoot multiplication from mature trees of Douglas fir (*Pseudotsuga menziesii*) and sugar pine (*Pinus lambertiana*). Plant Cell Rep 4:177–179

Gupta S, Singh P, Ananda RA (2009) Micropropagation of mulberry (*Morus indica* L.) using explants from mature tree: effects of plant growth regulators on shoot multiplication and rooting. Progress Hortic 41:136–144

Gurung B, Singh KK (2010) In vitro mass propagation of Sikkim Himalayan *Rhododendron* (*R. dalhousiae* Hook. f.) from nodal segment. J Am Soc Hortic Sci 12(1):42–45

Haberlandt G (1902) Kulturversuche Mit Isolierten Pflanzenzellen. Sitzungsber Akad Wiss Wien Math-Naturwiss Kl Abt 111:69–92

Hajong S, Kumaria S, Tandon P (2013) Effect of plant growth regulators on regeneration potential of axenic nodal segments of *Dendrobium chrysanthum* Wall. ex Lindl. J Agric Sci Technol 15:1425–1435

Hare PD, Staden JV (1994) Inhibitory effect of thidiazuron on the activity of cytokinin oxidase isolated from soybean callus. Plant Cell Physiol 35(8):1121–1125. https://doi.org/10.1093/oxfordjournals.pcp.a078704

Huetteman CA, Preece JE (1993) Thidiazuron: a potent cytokinin for woody plant tissue culture. Plant Cell Tissue Organ Cult 33(20):105–119

Hutchinson MJ, Saxena PK (1996) Acetylsalicylic acid enhances and synchronizes thidiazuron-induced somatic embryogenesis in *Geranium* (*Pelargonium x hortorum Bailey*) tissue cultures. Plant Cell Rep 15(7):512–515

Jiang BO, Yang YG, Guo YM, Guo ZC, Chen YZ (2005) Thidiazuron-induced in vitro shoot organogenesis of the medicinal plant *Arnebia euchroma* (Royle) Johnst. In Vitro Cell Dev Biol Plant 41(5):677–681

Jime'nez VM (2005) Involvement of plant hormones and plant growth regulators on in vitro somatic embryogenesis. Plant Growth Regul 47:91–110

Johnson TS, Narayan SB, Narayana DBA (1997) Rapid in vitro propagation of *Saussurea lappa*, an endangered medicinal plant, through multiple shoot cultures. In Vitro Cell Dev Biol Plant 33(2):128–130

Kant R (2015) Survival threats and conservation of *Malaxis muscifera* (Lindl.) Kuntze, a threatened medicinal orchid at Fagu, Himachal Pradesh. Int J App Basic Med Res 1:2395–3373

Kapruwan S, Bakshi M, Kaur M (2014) Effect of growth regulators on the in vitro multiplication of *Dendrocalamus hamiltonii*. Int J Eng Res Appl 1(4):83–86

Kögl F, Haagen-Smit AJ, Erxleben H (1934) Über ein neues auxin (Heteroauxin) aus Harn. 11. mitteilung über pflanzliche wachstumsstoffe. Hoppe-Seyler's Z Physiol Chem 228:90–103

Kou Y, Yuan C, Zhao Q, Liu G, Nie J, Ma Z, Cheng C, Silva JAT, Zhao L (2016) Thidiazuron triggers morphogenesis in *Rosa canina* L. protocorm-like bodies by changing incipient cell fate. Front Plant Sci. 04 7:557. https://doi.org/10.3389/fpls.2016.00557

Kumar A, Kaushal S, Sharma S, Sood H (2014) In vitro callus induction and plantlet regeneration of *Swertia chirayita*: a critically endangered medicinal plant. Int J Pharm Sci Rev Res 27(1):94–96

Laimer M, Rücker W (eds) (2003) Plant tissue culture 100 years since Gottlieb Haberlandt. Springer-Verlag Wien, New York, p 260

Lee SW (2005) Thidiazuron in the improvement of banana micropropagation. Acta Hortic 692:67–74

Linsmaier EM, Skoog F (1965) Organic growth factor requirements of tobacco tissue cultures. Physiol Plant 18:100–127

Lloyd G, McCown B (1981) Commercially feasible micropropagation of mountain laurel, *Kalmia latifolia*, by use of shoot tip culture. Combined Proc Int Plant Propagator's Soc 30:421–427

Lu CY (1993) The use of thidiazuron in tissue culture. In Vitro Cell Dev Biol 29:92–96

Lu MC (2010) Plant regeneration from callus culture of an endangered orchid, *Geodorum densiflorum*. Int Orchid Sym 878:175–178

Makara AM, Rubaihayo PR, Magambo MJS (2010) Carry-over effect of thidiazuron on banana in vitro proliferation at different culture cycles and light incubation conditions. Afr J Biotechnol 9(21):3079–3085

Malik KA, Saxena PK (1992) Regeneration in *Phaseolus vulgaris* L.: high frequency induction of direct shoot formation on intact seedlings by N6-benzylaminopurine and TDZ. Planta 186:384–389

Malik S, Sharma S, Sharma M, Ahuja PS (2010) Direct shoot regeneration from intact leaves of *Arnebia euchroma* (Royle) Johnston using thidiazuron. Cell Biol Int 34:537–542

Mathur G, Nadgauda R (1999) In vitro plantlet regeneration from mature zygotic embryos of *Pinus wallichiana* A.B. Jacks. Plant Cell Rep 19:74–80

Mehta R (2012) Development of efficient regeneration systems in economically important edible bamboos. Ph.D thesis. Guru Nanak Dev University, Amritsar

Miller CO, Skoog F, von Saltza MH, Strong M (1955) Kinetin, a cell division factor from deoxyribonucleic acid. J Am Chem Soc 77:1329–1334

Mitra GC, Prasad RN, Roychowdhury A (1976) Inorganic salts and differentiation of protocorms in seed callus of orchid and correlative changes in its free amino acid content. Indian J Exp Biol 14:350–351

Mok MC, Mok DWS (1985) The metabolism of [14C]-tMdiaziiroii in callus tissues of *Phaseolus lunatus*. Physiol Plant 65:427–432. https://doi.org/10.1111/j.1399-3054.1985.tb08668.x

Mok MC, Mok DWS, Armstrong DJ, Shudo K, Isogai Y, Okamoto T (1982) Cytokinin activity of N-phenyl-N′-1,2,3-thidiazol-5-yl urea (thidiazuron). Phytochemistry 21:1509–1511

Mok MC, Mok DWS, Turner JE, Mujer CV (1987) Biological and biochemical effects of cytokinin-active phenylurea derivatives in tissue culture systems. Hortic Sci 22:1194–1197

Mokhtari A, Otroshy M, Barekat T (2015) Plant regeneration through callus induction on medicinal herb *Viola odorata* – role of plant growth regulators and explants. Agric For 61(3):161–170. https://doi.org/10.17707/AgricultForest.61.3.16

Mondal TK, Bhattacharya A, Sood A, Ahuja PS (1998) Micropropagation of tea (*Camellia sinensis* (L.) O. Kuntze) using thidiazuron. Plant Growth Regul 26:57–61

Murashige T. (1979). Plant growth substances in commercial uses of tissue culture. Plant Growth Subst 426–434. https://doi.org/10.1007/978-3-642-67720-5_43

Murashige T, Skoog F (1962) A revised medium for rapid growth and bio assays with tobacco tissue cultures. Physiol Plant 15:473–497

Murthy BNS, Victor J, Singh RP, Fletcher RA, Saxena PK (1996) In vitro regeneration of chickpea (*Cicer arietinum* L.): stimulation of direct organogenesis and somatic embryogenesis by thidiazuron. Plant Growth Regul 19:233–240

Murthy BNS, Murch SJ, Saxena PK (1998) Thidiazuron: a potent regulator of in vitro plant morphogenesis. In Vitro Cell Dev Biol Plant 34:267. https://doi.org/10.1007/BF02822732

Murthy K, Reddy MC, Kondamudi R, Pullaiah T (2013) Micropropagation of *Stemona tuberosa* Lour. – an endangered and rare medicinal plant in Eastern Ghats of India. Indian J Biotechnol 12:420–424

Nayak NR, Rath SP, Patnaik S (1997) In vitro propagation of three epiphytic orchids, *Cymbidium aloifolium* (L.) Sw., *Dendrobium aphyllum* (Roxb.) Fisch. and *Dendrobium moschatum* (Buch-Ham) Sw. through thidiazuron-induced high frequency shoot proliferation. Sci Hortic 71(3–4):243–250

Nickell LG (1994) Plant growth regulators in agriculture and horticulture. In: Hedin P (ed) Bioregulators for crop protection and pest control, ACS symposium series, vol 557, pp 1–14. https://doi.org/10.1021/bk-1994-0557.ch001

Nielsen JM, Kirsten B, Hansen J (1993) Long-term effects of thidiazuron are intermediate between Benzyladenine, kinetin or isopentenyladenine in *Miscanthus sinensis*. Plant Cell Tissue Organ Cult 35:173–179

Ochatt SJ, Atif RM, Patat-Ochatt EM, Jacas L, Conreux C (2010) Competence versus recalcitrance for in vitro regeneration. Not Bot Hort Agrobot Cluj 38:102–108

Panda BM, Mehta UJ, Hazra S (2016) Micropropagation of *Semecarpus anacardium* L. from mature tree-derived nodal explants. Plant Biosyst 150(5):942–952

Patial V, Devi K, Sharma M, Bhattacharya A, Ahuja PS (2012) Propagation of *Picrorhiza kurroa* Royle ex Benth: an important medicinal plant of Western Himalaya. J Med Plant Res 6(34):4848–4860

Patial V, Sharma M, Bhattacharya A (2017) Potential of thidiazuron in improved micropropagation of *Picrorhiza kurroa* – an endangered medicinal herb of alpine Himalaya. Plant Biosyst 151(4):729–736

Prathanturarug S, Soonthornchareonnon N, Chuakul W, Phaidee Y, Saralamp P (2003) High-frequency shoot multiplication in *Curcuma longa* L. using thidiazuron. Plant Cell Rep 21:1054–1059

Preece JE, Huetteman CA, Ashaby WC, Roth PL (1991) Micro- and cutting propagation of silver maple I Results with adult and juvenile propagules. J Am Soc Hortic Sci 116:142–148

Pulianmackal AJ, Kareem AVK, Durgaprasad K, Trivedi ZB, Prasad K (2014) Competence and regulatory interactions during regeneration in plants. Front Plant Sci 5:142. https://doi.org/10.3389/fpls.2014.00142

Quadri RR, Da Silva JAT, Kamili AN, Shah AM (2011) Effect of 6-benzyladenine, kinetin and thidiazuron on in vitro shoot proliferation of *Hyoscyamus niger* L. Med Aromat Plant Sci Biotechnol 6(1):81–83

Rafi S, Kamili AZ, Ganai BA, Mir MY, Parray JA (2016) In vitro culture and biochemical attributes of *Bergenia ciliata* (Haw.) Sternb. Proc Natl Acad Sci India, Sect B Biol Sci. https://doi.org/10.1007/s40011-016-0797-9

Rajender G, Bheemanna D, Sreenu P, Prasada B, Rajender K, Reuben T, Christopher (2017) In vitro clonal propagation of vulnerable ethnomedicinal cucurbit, red ball snake gourd (*Trichosanthes tricuspidata* lour.) Int J Pharm Bio Sci 8(2):205–209

Ranyaphia RA, Maoa AA, Borthakurb SK (2011) Direct organogenesis from leaf and internode explants of in vitro raised wintergreen plant (*Gaultheria fragrantissima*). Sci Asia 37:186–194

Roberts JA, Hooley R (1988) Cellular differentiation and morphogenesis. In: Plant growth regulators. Tertiary level biology. Springer, Boston

Roh KH, Kwak BK, Kim JB, Lee KR, Kim HU, Kim SH (2012) The influence of silver thiosulfate and thidiazuron on shoot regeneration from cotyledon explants of *Brassica napus* Kyung. J Plant Biotechnol 39:133–139

Roy AR, Sajeev S, Pattanayak A, Deka BC (2012) TDZ induced micropropagation in *Cymbidium giganteum* Wall. Ex Lindl. and assessment of genetic variation in the regenerated plants. Plant Growth Regul 68(3):435–445

Sakhanokho HF, Kelly RW, Rajashekaran K (2008) First report of plant regeneration via somatic embryogenesis from shoot apex-derived callus of *Hedychium muluense*. J Crop Imp 21(2):191–200. https://doi.org/10.1080/15427520701885758

Sandal I, Bhattacharya A, Ahuja PS (2001) An efficient liquid culture system for tea shoot proliferation. Plant Cell Tissue Organ Cult 65:75–80

Sang YL, Cheng ZJ, Zhang XS (2016) Endogenous auxin biosynthesis and de novo root organogenesis. J Exp Bot 67(14):4011–4013. https://doi.org/10.1093/jxb/erw250

Sharma VK, Hansch R, Mendel RR, Schulze J (2007) Node-derived cultures with high-morphogenic competence in barley and wheat. Plant Cell Tissue Organ Cult 88:21–33. https://doi.org/10.1007/s11240-006-9172-5

Sharma E, Karma T, Chettri N, Shrestha A (2009) Biodiversity in the Himalayas – trends, perception and impacts of climate change. IMBC-Plenary session 1: climate change and its implications for mountain

Sharma S, Katoch V, Rathour R, Sharma TR (2010) In vitro propagation of endangered temperate Himalayan medicinal herb *Picrorhiza kurroa* Royle ex benth using leaf explants and nodal segments. J Appl Biochem Biotechnol 19(1):111–114

Sharma A, Kaur R, Sharma N (2014) In vitro morphogenic response of different explants of *Gentiana kurroo* Royle from Western Himalayas—an endangered medicinal plant. Physiol Mol Biol Plants 20(2):249–256

Siddique I, Anis M (2007) Rapid micropropagation of *Ocimum basilicum* using shoot tip explants pre-cultured in thidiazuron supplemented liquid medium. Biol Plant 51:787–790

Singh DK, Babbar SB (2016) In vitro propagation and chemical profiling of *Herminium lanceum* (Thunb. ex Sw.) Vuijk, a medicinally important orchid, for therapeutically important phenolic acids. Plant Biotechnol J 33(3):153–160

Singh SK, Syamal MM (2001) A short pre-culture soak in thidiazuron or forchlorfenuron improves axillary shoot proliferation in rose micropropagation. Sci Hortic 91:169–177

Singh S, Purohit VK, Prasad P, Nautiyal AR (2015) Micropropagation of *Valeriana wallichii* DC. (Indian Valerian) through nodes. Indian J Biotechnol 14:127–130

Tabin S, Gupta RC, Kamili AN (2014) In vitro micro propagation of Rheum explants supplemented with various types of growth hormones. J Agric Vet Sci 7(3):97–100

Taha RM, Haron NW, Wafa SN (2011) Morphological and tissue culture studies of *Platycerium coronarium*, a rare ornamental Fern species from Malaysia. Am Fern J 101(4):241–251

Tang W, Newton RJ (2005) Peroxidase and catalase activities are involved in direct adventitious shoot formation induced by thidiazuron in Eastern white pine (*Pinus strobus* L.) zygotic embryos. Plant Physiol Biochem 43(8):760–769

Tao J, Yu L, Kong F, Zhao D (2011) Effect of plant growth regulators on in vitro propagation of *Cymbidium faberi* Rolfe. Afr J Biotechnol 10:15639–15646

Thakur RC, Hosoi Y, Ishii K (1998) Rapid in vitro propagation of *Matteuccia struthiopteris* (L.) Todaro – an edible fern. Plant Cell Rep 18:203–208

Thomas JC, Katterman FR (1986) Cytokinin activity induced by thidiazuron. Plant Physiol 81(2):681–683

Vengadesan G, Pijut PM (2009) In vitro propagation of northern red oak (*Quercus rubra* L.) In Vitro Cell Dev Biol-Plant 45:474–482. https://doi.org/10.1007/s11627-008-9182-6

Verma M, Bansal YK (2014) Effect of a potent cytokinin thidiazuron (TDZ) on in vitro regeneration of *Hedychium coronarium* J. Koenig – a valuable medicinal plant. Int J Rec Biotech 2(1):38–44

Vijayakumar M, Vijayakumar R, Stephen R (2010) In vitro propagation of *Bacopa monnieri* L. – a multipurpose medicinal plant. Indian J Sci Technol 3(7):782–787

Visser C, Qureshi JA, Gill R, Saxena PK (1992) Morphoregulatory role of thidiazuron' substitution of auxin and cytokinin requirement for the induction of somatic embryogenesis in geranium hypocotyl cultures. Plant Physiol 99:1704–1707

Winarto B, da Silva JT (2012) Improved micropropagation protocol for leatherleaf fern (*Rumohra adiantiformis*) using rhizomes as donor explants. Sci Hortic 140:74–80

Yu R, Zhang G, Li H, Cao H, Mo X, Gui M, Zhou X, Jiang Y, Li S, Wang J (2017) In vitro propagation of the endangered tree fern *Cibotium barometz* through formation of green globular bodies. Plant Cell Tissue Organ Cult 128:369–379

Zhao D, Hu G, Chen Z, Shi Y, Zheng L, Tang A, Long C (2013) Micropropagation and in vitro flowering of *Dendrobium wangliangii*: a critically endangered medicinal orchid. J Med Plant Res 7(28):2098–2110

13 Thidiazuron-Induced Protocorm-Like Bodies in Orchid: Progress and Prospects

Suprabuddha Kundu and Saikat Gantait

Abstract

Thidiazuron (N-phenyl-N′-1,2,3-thidiazol-5-ylurea, TDZ) is a substituted phenylurea chemical compound that was initially produced for mechanical harvesting of cotton bolls. Since recent past, it has become well known as a successful plant growth regulator (PGR) in plant cell, tissue and organ culture. The unique property of TDZ mimicking both auxin and cytokinin stimulates organogenic and callogenic growth of explants. In this chapter, numerous of such potentials are discussed, considering the recently published reports on characterization of TDZ-induced orchid protocorm-like bodies (PLBs) and their subsequent conversion. The summarized results clearly exhibit that TDZ significantly influences diverse culture responses starting from induction of PLB from a wide array of explants, rate of shoot formation, shoot number per explant and the time duration needed for PLB formation compared to other PGRs. A very low concentration (<1 μM) of TDZ is much effective in comparison to many other PGRs; though TDZ may hold back shoot elongation after culturing for long duration in the same medium. The unwanted side effect of TDZ is that the fasciated shoots are formed occasionally in some orchid species after prolonged culture. Besides the minor disapproval, the high proliferative activity and positive response to organogenesis, TDZ has evolved to be the most active cytokinin-like PGR for in vitro regeneration of orchid species.

S. Kundu
Department of Agricultural Biotechnology, Faculty of Agriculture, Bidhan Chandra Krishi Viswavidyalaya, Mohanpur, Nadia, West Bengal 741252, India

S. Gantait (✉)
All India Coordinated Research Project on Groundnut, Directorate of Research Bidhan Chandra Krishi Viswavidyalaya, Kalyani, Nadia, West Bengal 741235, India

Department of Genetics and Plant Breeding, Faculty of Agriculture, Bidhan Chandra Krishi Viswavidyalaya, Mohanpur, Nadia, West Bengal 741252, India
e-mail: saikatgantait@bckv.edu.in

N. Ahmad, M. Faisal (eds.), *Thidiazuron: From Urea Derivative to Plant Growth Regulator*, https://doi.org/10.1007/978-981-10-8004-3_13

Keywords
Conversion · Orchid · PLB · TDZ

13.1 Introduction

Since the last few decades, significant attention has been given to thidiazuron (N-phenyl-N′-1,2,3-thidiazol-5-ylurea, TDZ) on account of its effective activity in plant cell and tissue culture. When TDZ was applied to different species of plants, assorted forms of physiological responses were detected. Being completely dissimilar to the chemistry of auxins and cytokinins, TDZ has many effects, very similar to both the groups of plant growth regulator (PGR). German Schering Corporation synthesized this compound for cotton defoliation (*Gossypium hirsutum*) (Arndt et al. 1976). As a cotton defoliant, it is marketed under the trade name DROPP® (NOR-AM Technical Bulletin 1987). As it induced many reactions similar to natural cytokinins, it was categorized to be a type of cytokinin initially. Later, it was shown that unlike traditional phytohormones, TDZ performed different regenerative responses in various plant species, independently. Its solubility in water is very low but it becomes highly soluble when dissolved in organic solvents such as benzene, DMSO, acetone and in ethanol. Colour is light yellow and is crystalline in nature. The chemical structure of TDZ is very much different from both auxins and adenine-type cytokinins since two functional groups, phenyl and thidiazol are present in TDZ molecule, and any substitute to these groups with other ring configurations reduces its performance. Here, the symmetrical compound, N,N′-di-thiadiazolurea, has the lowest activity similar to cytokinin that suggests the two ring structures perform complimentary roles in TDZ-induced responses (Mok et al. 1982). The physiological reactions can also be infered when stock solutions have been stored for long since TDZ is fairly a stable compound. Latest studies show that during its extended storage period, short-length polymers are formed, which are also noticed in media and solutions following autoclaving and growth incubation. It may be thus possible that TDZ-efficacy, in part, can be ascribed to its storage in the form of short-length polymers followed by successive discharge of these molecules in culture period later. During the 1980s, TDZ was used in tissue culture systems in form of potent cytokinin for regeneration of adventitious shoot, which raised an incredible curiosity of explaining the morphogenic competence of this chemical and subsequent influences on plant growth. In tissue culture system of *Geranium*, both auxin and cytokinin requirement was fulfilled using active replacement by TDZ (Visser et al. 1992). Even at a low concentration of 10 μM, TDZ shows a great level of activity when performed in vitro (Preece et al. 1991), and when plant tissue is exposed to TDZ for a lesser period, it is adequate enough to induce revival (Visser et al. 1992; Hutchinson and Saxena 1996).

13.2 What Is PLB?

For micropropagation of selected elite orchids and their genetic transformation studies, a plant regeneration system in the form of protocorm-like bodies (PLBs) is very effective (Arditti and Ernst 1993). Morel (1960) observed PLBs for the first time during *Cymbidium* orchid shoot-tip culture. As the overall structural and growth features are analogous to protocorms, hence the regenerated structures are called 'PLB' (Fig. 13.1a). They are actually composed of differentiated cells regarded as orchid embryos (Ng and Saleh 2011) that grow with distinct bipolar structures (Fig. 13.1b), the upper portion becoming shoot and the lower portion developing into root meristem (Jones 2009). Several explants like root tips, shoot tips, leaf segments and flower stalk buds can be induced directly to generate PLBs. A thorough ontogenetic study of PLBs is not yet available, though the morphological and structural developments are known. PLBs obtained from *Phalaenopsis* orchids have a single-cell origin as shown by Tokuhara and Mii (2003), and an analogous observation was also reported by Jheng et al. (2006) in case of *Oncidium*.

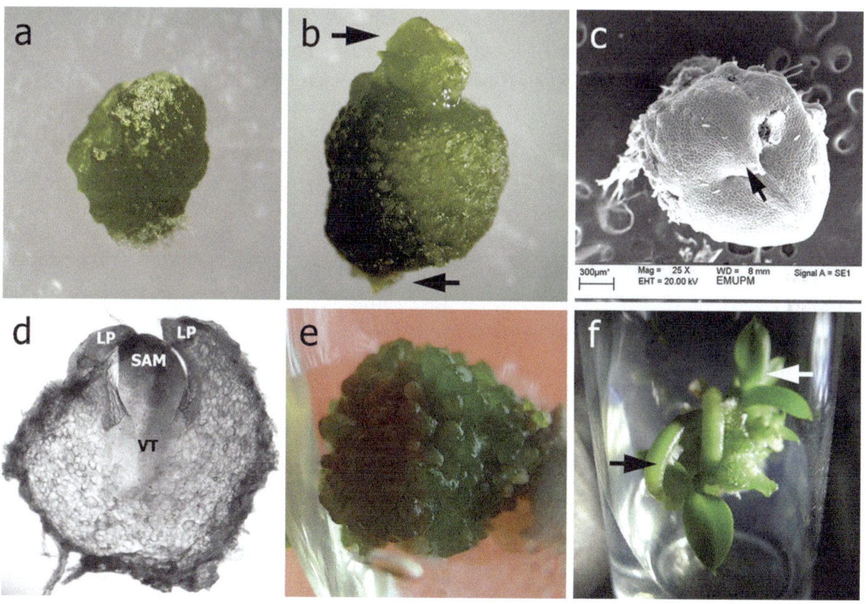

Fig. 13.1 Orchid protocorm-like bodies (PLBs): influence of TDZ on its development and conversion. (**a**) A typical PLB, (**b**) PLB showing its distinct bipolar (arrows) structure (for shoot and root formation during conversion), (**c**) a scanning electron microscopic image of PLB at an early developmental stage, characterized by globular shape, distinct protoderm in addition to leaf primordia (arrow), (**d**) a thin vertical section of PLB exhibiting a conventional structure with shoot apical meristem (SAM), leaf primordia (LP) origin and vascular tissue (VT), (**e**) induction and proliferation of PLBs through the influence of TDZ and, (**f**) conversion with shoot (white arrow) and root (black arrow) of PLBs induced by TDZ (Source: unpublished photographs of Saikat Gantait)

Explants' epidermal cell differentiates to generate PLBs (Vij et al. 1984; Chen et al. 1999) where PLBs are likely to have a single-cell origin. In some cases, subepidermal cells can also lead to the formation of PLBs, which are substantiated to have multicellular origin (Park et al. 2002, 2003). From these examples, it can be concluded that the mode of PLB induction varies significantly from species-to-species. From the published literature, it has often been advocated that PLBs are somatic embryos virtuously depending on its morphology (Ishii et al. 1998; Huan et al. 2004), or even the first step in PLB development is somatic embryogenesis (Zhao et al. 2008). As Zimmerman (1993) pointed out that somatic embryogenesis is the growth of somatic cells into differentiated plants via embryogenic phases under suitable conditions. Then, whether PLBs are somatic embryos or not and whether they have the features alike zygotic embryos, prior to developing into PLBs, became questions. To answer these, Lee et al. (2013) conceded a comprehensive ontogenetic analysis of PLBs using friable embryogenic callus of *Phalaenopsis* by histochemical and histological techniques, followed by comparison of their ontogeny to normal in situ zygotic embryo formation and successive protocorm development. Furthermore, the same group employed cell wall markers of zygotic embryo specific to JIM11 and JIM20 antibodies to investigate if analogous wall proteins are also present during the PLB developmental pathway. They elucidated that the cytoplasm of the meristemoids and globular cell clusters took a vacuolated appearance at the early stages of PLB formation as well as starch granules and protein bodies were observed. Also, JIM11 and JIM20 were positively localized in the developing PLBs resembling zygotic embryos. For more confirmation, an inhibitor to HRGPs was incorporated which hindered PLB formation. The report clearly justifies the statement that PLBs are indeed somatic embryos of orchids, as the cells of early stages of PLB show cytological attributes and cell wall markers akin to zygotic embryo. A detailed histological study has also been carried out in *Anthurium andreanum* (Gantait et al. 2012) and monopodial orchid hybrid (*Aranda* Wan Chark Kuan 'Blue' x *Vanda coerulea* Grifft. ex. Lindl.) (Gantait and Sinniah 2012). PLBs, beginning from induction to maturation stages, were demonstrated by thin free-hand sections and scanning electron microscopy. At an early developmental stage, PLBs were characterized by globular shape, distinct protoderm in addition to a constricted basal tissue (Fig. 13.1c). Histological study showed that the anterior side of the PLB has an area containing dense cytoplasm. PLBs' transverse and longitudinal sections at proliferating phase demonstrated meristematic zone with elliptical shape and surrounded by a homogeneous, parenchymatous tissue with thin-walled cells (Gantait and Sinniah 2012). This parenchymatic 'tubercle' was projected from the exterior of the upper part of the protocorm and linked via vascular tissue. In later segment of this phase, the axis turned into densely stained identifiable configuration in the inner portion of the PLB with distinct strands of meristematic cells that finally developed shoot apical meristem, and the other axis of the core produced leaf primordia. At this point, PLB exhibited a conventional structure with shoot apical meristem, leaf primordia origin and vascular tissue (Fig. 13.1d). Cells of the shoot apical meristem were smaller in dimension and compactly organized. Further growth of the tissue gave shoot apical meristem a dome shape. At the shoot apex, leaf primordia were noticeable surrounding the shoot primordia, comprising of small cells, undertaking anticlinal and periclinal divisions yet bordered by large isodiametric parenchyma cells.

13.3 Orchid PLB: Influence of TDZ on Induction

Numerous investigations in a range of orchid species documented the high effectiveness of TDZ in inducing PLBs. Comparative analysis of N^6-benzyladenine (BA), 6-furfurylaminopurine (kinetin), N^6-(Δ^2-isopentenyl) adenine (2iP) and TDZ demonstrated that TDZ performed superior to the other PGRs (Park et al. 2002; Niknejad et al. 2011). Exogenous PGRs or an alteration of the cytokinin/auxin ratio is normally required for the induction of PLBs from somatic cells. Nonetheless, TDZ alone was reported to initiate PLB formation and thus bypassed the requirement of both classes of PGRs (Chen and Chang 2002; Roy et al. 2007, 2012; Palama et al. 2010; Gantait et al. 2012; Wang and Tian 2014). These findings also justified a role of TDZ to be involved in modulation of auxin metabolism. Murch and Saxena (2001) conducted an experiment, employing radiolabelled versions of TDZ, and the results evidenced that TDZ mediates the accumulation along with translocation of indole-3-acetic acid (IAA) within the plant tissues. Table 13.1 presents a collection of research reports that were accomplished to examine the efficacy of TDZ on the induction of PLB in different orchid species. TDZ has been used in the range of 0.45–18.16 μM to stimulate PLBs (Fig. 13.1e) from flower-stalk segments (Chen and Chang 2000), seed-derived protocorms (Lin et al. 2000), root tip (Park et al. 2003; Guo et al. 2010), leaf (Kuo et al. 2005; Gantait et al. 2012), shoot tip (Roy et al. 2007; Winarto and Teixeira da Silva 2015) as well as nodal segment (Cui et al. 2008; Hong et al. 2010). Malabadi et al. (2004) noticed that the induction of PLBs was improved by increasing the concentration of TDZ from 2.27 to 11.35 μM. The maximum number of explants producing PLBs was observed with 11.35 μM TDZ-supplemented medium, which subsequently initiated the highest number of shoots from PLBs. Some researchers reported that supplementation of auxin was necessary for PLB induction (Guo et al. 2010; Chen 2012; Winarto and Teixeira da Silva 2015). Guo et al. (2010) cultured root tip segment in Murashige and Skoog (1962) (MS) medium containing 10.2 μM IAA along with 9.0 μM TDZ. Both IAA- and TDZ-enriched media promoted the induction of PLBs, but TDZ alone did not form PLB, representing a synergistic impact of the two PGRs. On the other hand, Mayer et al. (2010) found that the presence of auxin such as 2,4-dichlorophenoxy acetic acid (2,4-D) and α-naphthalene acetic acid (NAA) hindered the induction of PLBs. The highest rate of PLBs induction was attained when they were inoculated in a 1.5 μM TDZ-supplemented MS medium under dark conditions. Culture medium containing TDZ along with 2,4-D induced callus formation in different orchids, such as *Cymbidium* (Chang and Chang 1998), *Phalaenopsis* (Chen et al. 2000), *Paphiopedilum* (Lin et al. 2000; Hong et al. 2008) and *Oncidium* (Chen and Chang 2000; Jheng et al. 2006). Later, Wu et al. (2012) noticed NAA in combination with TDZ inhibited the formation of PLBs from leaf explants. Contrastingly,

Table 13.1 Influence of TDZ on induction and regeneration of orchid protocorm-like bodies

Orchid species	Explant	Basal media	Concentration of TDZ	Supplementation	Outcome	Reference
Phalaenopsis and *Doritaenopsis*	FST	XER medium	11.35 μM TDZ	–	PLB	Ernst (1994)
Phalaenopsis hybrid	Shoot tip	Vacin and Went	–	15% (*v/v*) coconut water	PLB	Chen and Piluek (1995)
Oncidium Gower Ramsey	Leaf	½MS	0.3 mg l^{-1} (1.37 μM) TDZ	–	PLB	Chen et al. (1999)
	PLB		–	0.5 mg l^{-1} NAA	Reg	
Oncidium Sweet Sugar	FST	½MS	1 mg l^{-1} (4.55 μM) TDZ	1 mg l^{-1} NAA	PLB	Chen and Chang (2000)
	PLB		–	–	Reg	
Phalaenopsis Nebula	SDP	½MS	0.45 μM TDZ	–	PLB, Reg	Chen et al. (2000)
Paphiopedilum hybrid (Paphiopedilum callosum 'Oakhi' x Paph. lawrenceanum 'Tradition')	SDP	½MS	1.0 mg l^{-1} (4.55 μM) TDZ	5.0 mg l^{-1} 2,4-D	PLB, Reg	Lin et al. (2000)
Oncidium 'Gower Ramsey'	Leaf tips	½MS	1 mg l^{-1} (4.55 μM) TDZ	–	PLB	Chen and Chang (2001)
Lilium longiflorum	Transverse young stem sections	MS	1.1 μM TDZ	–	PLB	Nhut et al. (2001)
	PLB		2.2 μM TDZ	–	Reg	
Oncidium 'Gower Ramsey'	Leaf	½MS	1 mg l^{-1} (4.55 μM) TDZ	–	PLB	Chen and Chang (2002)
Epidendrum radicans	FST	½MS	0.45 μM TDZ	–	PLB, Reg	Chen et al. (2002)
Doritaenopsis	Leaf	½MS	9.0 μM TDZ	–	PLB	Park et al. (2002)
	PLB	Modified Hyponex medium	–	30 g l^{-1} potato homogenate, 2 g l^{-1} peptone, 0.5 g l^{-1} AC	Reg	

Doritaenopsis	Root tip	MS	2.3 μM TDZ	–	PLB	Park et al. (2003)
	PLB	Modified Hyponex medium	–	30 g l^{-1} potato homogenate, 2 g l^{-1} peptone, 0.5 g l^{-1} AC	Reg	
Phalaenopsis amabilis var. formosa	SDP	½MS	13.62 μM TDZ	–	SE/PLB	Chen and Chang (2004)
	PLB		–	–	Reg	
Pleione formosana	Protocorms	½MS	0.5 mg l^{-1} (2.28 μM) TDZ	5 mg l^{-1} 2,4-D	PLB	Lu (2004)
Vanda coerulea	Shoot tip sections	Vacin and Went	11.35 μM TDZ	–	PLB; Reg	Malabadi et al. (2004)
	Leaf	½MS	18.16 μM TDZ	–	PLB	Chung et al. (2005)
	PLB		–	–	Reg	
Phalaenopsis ‘Little Steve’	Leaf	½MS	4.54 μM TDZ	–	PLB	Kuo et al. (2005)
	PLB		–	–	Reg	
Aerides crispum	Protocorm; leaf	½MS	0.5 μM TDZ	–	PLB	Sheelavanthmath et al. (2005)
	PLB		–	–	Reg	
Phalaenopsis amabilis var. formosa	Leaf	½MS	3.0 mg dm^{-3} (13.65 μM) TDZ	–	SE/PLB	Chen and Chang (2006)
	PLB		–	–	Reg	
Dendrobium cv. Chiengmai Pink	Leaf	MS	1.0 mg dm^{-3} (4.55 μM) TDZ	–	SE/PLB; Reg	Chung et al. (2007)
Dendrobium chrysotoxum	Shoot tip	MS	1 μM TDZ	–	PLB	Roy et al. (2007)
	PLB		–	–	Reg	
Syngonium podophyllum ‘white butterfly’	Nodal segment	MS	9.08 μM TDZ	2.26 μM 2,4-D	PLB; Reg	Cui et al. (2008)
Phalaenopsis amabilis and Phalaenopsis ‘Nebula’	Leaf	½MS	3.0 mg dm^{-3} (13.65 μM) TDZ		SE/PLB	Gow et al. (2009)

(continued)

Table 13.1 (continued)

Orchid species	Explant	Basal media	Concentration of TDZ	Supplementation	Outcome	Reference
Phalaenopsis amabilis and Phalaenopsis 'Nebula'	Leaf	½MS	3.0 mg l^{-1} (13.65 μM) TDZ	–	SE/PLB	Gow et al. (2010)
	PLB		–	0.5 mg l^{-1} BA	Reg	
Cyrtopodium paranaense	Root tip	MS	9.0 μM TDZ	5.1 μM IAA	PLB	Guo et al. (2010)
	PLB		–	–	Reg	
Zygopetalum mackayi	Nodal segments	½MS	4.54 μM TDZ	–	PLB	Hong et al. (2010)
	PLB		–	–	Reg	
Oncidium flexuosum	Leaf		1.5 μM TDZ	–	PLB	Mayer et al. (2010)
Vanilla planifolia	Seeds	MS	0.5 mg l^{-1} (2.28 μM) TDZ	–	PLB	Palama et al. (2010)
	PLB		–	0.5 mg l^{-1} NAA	Reg	
Lycaste hybrid	Intact buds	½MS	1.0 mg l^{-1} (4.55 μM) TDZ	0.5 mg l^{-1} BA	PLB	Huang and Chung (2011)
	PLB		–	–	Reg	
Phalaenopsis bellina	Leaf	½MS	3.0 mg l^{-1} (13.65 μM) TDZ	–	PLB	Khoddamzadeh et al. (2011)
Phalaenopsis gigantea	Leaf	New Dogashima medium	0.1 mg l^{-1} (0.46 μM) TDZ	1.0 mg l^{-1} NAA	PLB	Niknejad et al. (2011)
	PLB		–	–	Reg	
Cymbidium faberi	Seed	½MS	–	0.5 mg l^{-1} NAA	PLB	Tao et al. (2011)
	PLB		1.0 mg l^{-1} (4.55 μM) TDZ	0.5 mg l^{-1} NAA	Reg	
Cyrtopodium glutiniferum	Seed	MS	1 μM TDZ	–	PLB	Vogel and Macedo (2011)
	PLB		–	1 μM IAA	Reg	
Oncidium 'Gower Ramsey'	Root segments	½MS	3.0 mg l^{-1} (13.65 μM) TDZ	0.1 mg l^{-1} NAA	SE/PLB	Chen (2012)
	PLB		–	–	Reg	

Aranda Wan Chark Kuan 'Blue' x Vanda coerulea Grifft. ex. Lindl.	Leaf	MS	1.5 mg l^{-1} (6.83 μM) TDZ	–	PLB	Gantait et al. (2012)
	PLB		–	1 mg l^{-1} BA +0.5 mg l^{-1} IBA + 60 mg l^{-1} adenine sulphate	Reg	
Cymbidium giganteum	Pseudostem segments	MS	0.909 μM TDZ	–	PLB	Roy et al. (2012)
	PLB		–	–	Reg	
Renanthera Tom Thumb 'Qilin'	Leaf	Vacin and Went	1.0 mg l^{-1} (4.55 μM) TDZ	1.0 g l^{-1} peptone, + 10% coconut water	PLB	Wu et al. (2012)
	PLB	Vacin and Went	–	1.0 g l^{-1} peptone +1.0 mg l^{-1} NAA + 1.0 g l^{-1} AC	Reg	
Bletilla striata	Protocorms	VW	0.5 mg l^{-1} (2.28 μM) TDZ	–	PLB	Wang and Tian (2014)
Dendrobium thyrsiflorum	NS	MS	3 mg l^{-1} (13.65 μM) TDZ	–	PLB	Bhattacharyya et al. (2015)
	PLB		2 mg l^{-1} (9.1 μM) TDZ	0.5 mg l^{-1} NAA	Reg	
Dendrobium 'Gradita 31'	Shoot	½MS	0.3 mg l^{-1} (1.37 μM) TDZ	0.1 mg l^{-1} NAA + 15% CW	PLB	Winarto and Teixeira da Silva (2015)
	PLB	½ Rosasol	–	–	Reg	
Cibotium barometz	Juvenile sporophytes	½MS	1.0 mg l^{-1} (4.55 μM) TDZ	0.3 mg l^{-1} NAA	PLB	Yu et al. (2017)
	PLB	1/4 MS	–	0.2% AC	Reg	

2,4-D 2,4-dichlorophenoxy acetic acid, *AC* activated charcoal, *BA* N^6-benzyladenine, *FST* flower stem sections, *IAA* indole-3-acetic acid, *MS* Murashige and Skoog basal medium (Murashige and Skoog 1962), *NAA* α-naphthalene acetic acid, *PLB* protocorm-like bodies, *Reg* regeneration, *SDP* seed-derived protocorms, *SE* somatic embryo, *TDZ* thidiazuron, *VW* Vacin and Went medium (Vacin and Went 1949)

Sheelavanthmath et al. (2005) accounted that culture medium containing BA was better to TDZ for PLB induction, and 49.1 and 22.0 PLBs were obtained from protocorm and leaf explants on medium fortified with 1.0 and 2.0 μM BA, respectively. However, Winarto and Teixeira da Silva (2015) successfully produced PLBs by culturing shoot explants on $^{1}/_{2}$MS medium containing 1.0 mg l^{-1} TDZ plus 0.5 mg l^{-1} BA. Further, the use of coconut water at 15% (v/v) in the medium had significantly influenced the proliferation of PLBs.

13.4 Orchid PLB: Influence of TDZ on PLB Conversion

It is well known that TDZ induces effective organogenesis from PLBs in diverse orchid species. Dosage of TDZ in the culture medium stimulates the mode of PLB conversion either into complete plantlet (Fig. 13.1f) or into multiple shoots only. Table 13.1 also represents the overview of PLB-conversion into whole plant with or without the help of PGRs. In comparison to other PGRs, TDZ plays its role on PLB-conversion at much lower concentrations. For that reason, direct assessment amid TDZ and different other PGRs at equimolar dosages confuse statistical analyses. Amino purine cytokinins have parallel ranges of action (1–5 mg l^{-1}), but at this range TDZ results in callus formation as well as inhibition of shoot growth. Consequently, micropropagation experiments should be designed by considering TDZ at a range of much lower levels than the other amino purine cytokinins. In most of the reports, TDZ was not at all tested in PLB conversion medium. Instead, MS medium devoid of PGR was employed and accounted for its efficient organogenesis (Sheelavanthmath et al. 2005; Roy et al. 2007, 2012; Hong et al. 2010; Niknejad et al. 2011; Winarto and Teixeira da Silva 2015). Very limited number of reports are available that incorporated TDZ in the PLB conversion. Out of them, some researchers obtained complete plantlets with roots and shoots on the same TDZ-supplemented medium that was used for induction of PLBs (Chen et al. 2002; Malabadi et al. 2004; Chung et al. 2007, Cui et al. 2008). On the other hand, some found that supplementation of NAA into TDZ-containing medium resulted in enhanced shoot regeneration (Tao et al. 2011; Bhattacharyya et al. 2015). Malabadi et al. (2004) tested the effectiveness of TDZ on conversion of PLBs by culturing for different time periods (0, 2, 4, 6, 8, 10, 12 and 14 weeks) on 11.35 μM TDZ-supplemented medium. PLBs recorded 100% survival with maximum number (9.2 ± 2.02, 10.2 ± 3.06) of well-developed shoots when cultured for 6 and 8 weeks, respectively. However, the exposure to TDZ for more than 8 weeks caused fasciated or distorted shoots. Later on, to produce the adventitious shoots from PLB of *C. faberi*, Tao et al. (2011) cultured PLBs on the medium containing BA, TDZ and TDZ plus NAA, and the influence of these PGRs was compared. The shoot induction rate from PLB in the TDZ-fortified culture medium was up to 45.5% that was 2.5 times of BA-treated PLBs. Hong et al. (2010) reported an interesting outcome that TDZ had no significant effects on conversion of PLB to form shoots, but it retarded growth of shoot at 0.45–4.54 μM after 1 month of culture. Contrastingly, Gantait et al. (2012) found 1.0 mg l^{-1} BA along with 0.5 mg l^{-1} IBA and 60 mg l^{-1} adenine sulphate to be competent enough for PLB-conversion into complete plantlet. There are also reports where only auxins in the regeneration medium like NAA (Wu et al. 2012) or IAA (Vogel and Macedo 2011) was sufficient. Finally, to conclude, there is substantial variation and ambiguity regarding the composition of conversion media that ought to be worked upon and standardized in near future.

13.5 Conclusion and Future Prospect of the Use of TDZ in Orchid Organogenesis

Extensive advancement pertaining to in vitro regeneration of orchids during the last few decades has been made globally. The available literature provides considerable substantiation that TDZ is conceivably the most effective PGR presently obtainable for tissue culture of orchids (Fig. 13.2). TDZ was accounted to be superior in majority of the instances and also most effective at relatively lower concentrations, whereas the other cytokinins are only less effectual to induce PLB and regeneration as well. Majority of the reports investigating TDZ for PLB induction and subsequent plant regeneration in the orchids, normal development of plant devoid of any morphological abnormalities was observed (Guo et al. 2010; Gantait et al. 2012;

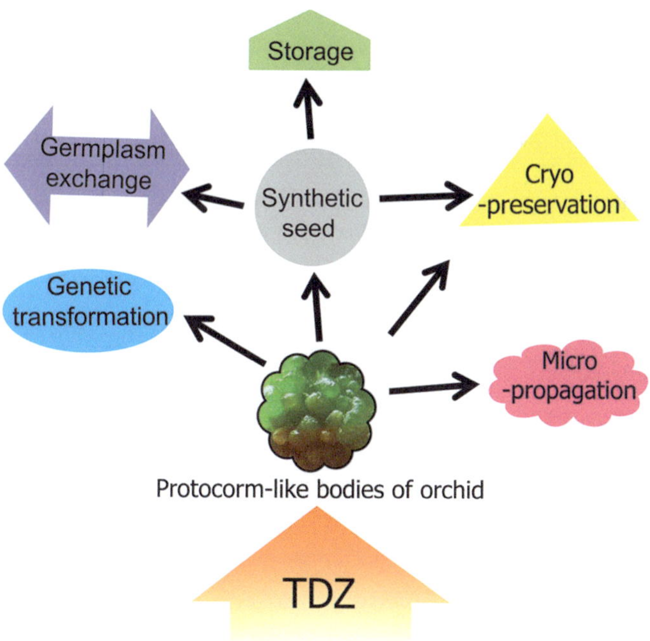

Fig. 13.2 Advancement pertaining to in vitro regeneration of orchids using TDZ as the most effective plant growth regulator during the last few decades

Roy et al. 2012; Winarto and Teixeira da Silva 2015). Conversely, in few other reports, disadvantageous effects of TDZ were also talked about. In spite of the high competence of TDZ, some investigations have found hindrance in growth, related with the use TDZ for long culture durations. Such unwanted morphologies observed are abnormal leaf shape, fascinated shoots as well as stunted shoots (Malabadi et al. 2004; Hong et al. 2010). The opposite effect may probably be due to higher cytokinin activity and perseverance of TDZ traces in the plant tissue. Hence, hindrance of growth morphology by TDZ might be normal with its highly active nature (Nayak et al. 1997a, b). To overcome the issues of shoot elongation, incorporation of auxin in the culture medium, at lower levels along with TDZ or by subculturing the clump of shoots to a medium containing other PGRs like BA and NAA would be helpful. Shoot elongation can be attained well by transferring the regenerated shoots to a secondary medium devoid of TDZ or with variable levels of other PGRs which is adopted by majority of the researchers. Hence, a primary medium should be used with the inclusion of TDZ, and after attaining sufficient number of shoots; those should be subcultured at a secondary medium containing other PGRs for further normal growth and elongation. On the whole, TDZ has exhibited its capability to induce callusing, PLB formation, and shoot proliferation, sometimes alone and sometimes in combinations with other PGRs. This compound has created a new dimension for the utilization of PGR in en masse clonal propagation and speed up the process of PLB induction that could facilitate a continual supply of healthy propagules for large-scale production in bioreactors, and synthetic seed technology, or cryopreservation for conservation. It would also provide a stable supply of PLB for genetic transformation-oriented research. Furthermore, transgenic orchid would have a better possibility of survival if enhanced shoot and root formation could be standardized by the inclusion of TDZ in culture medium.

Acknowledgements Authors acknowledge the e-library assistance from Bidhan Chandra Krishi Viswavidyalaya, Mohanpur, West Bengal, India. We further are thankful to the anonymous reviewers and the editor of this chapter for their critical comments and suggestions.

In compliance with ethical standards

Conflict of Interest: Declared none

References

Arditti J, Ernst R (1993) Micropropagation of orchids. Wiley, New York

Arndt FJ, Rusch R, Stilfried HV (1976) SN 49537, a new cotton defoliant. Plant Physiol 57:99

Bhattacharyya P, Kumaria S, Job N, Tandon P (2015) Phyto-molecular profiling and assessment of antioxidant activity within micropropagated plants of *Dendrobium thyrsiflorum*: a threatened, medicinal orchid. Plant Cell Tissue Organ Cult 122:535–550

Chang C, Chang WC (1998) Plant regeneration from callus culture of *Cymbidium ensifolium* var. misericors. Plant Cell Rep 17:251–255

Chen JT (2012) Induction of petal-bearing embryos from root-derived callus of *Oncidium* 'Gower Ramsey'. Acta Physiol Plant 34:1337–1343

Chen JT, Chang WC (2000) Plant regeneration via embryo and shoot bud formation from flower-stalk explants of *Oncidium* sweet sugar. Plant Cell Tissue Organ Cult 62:95–100
Chen JT, Chang WC (2001) Effects of auxins and cytokinins on direct somatic embryogenesis on leaf explants of *Oncidium* 'Gower Ramsey'. Plant Growth Regul 34:229–232
Chen JT, Chang WC (2002) Effects of tissue culture conditions and explant characteristics on direct somatic embryogenesis in *Oncidium* 'Gower Ramsey'. Plant Cell Tissue Organ Cult 69:41–44
Chen JT, Chang WC (2004) Induction of repetitive embryogenesis from seed-derived protocorms of *Phalaenopsis amabilis* var. *formosa* Shimadzu. In Vitro Cell Dev Biol Plant 40:290–293
Chen JT, Chang WC (2006) Direct somatic embryogenesis and plant regeneration from leaf explants of *Phalaenopsis amabilis*. Biol Plant 50:169–173
Chen Y, Piluek C (1995) Effects of thidiazuron and N^6-benzylaminopurine on shoot regeneration of *Phalaenopsis*. Plant Growth Regul 16:99–101
Chen JT, Chang C, Chang WC (1999) Direct somatic embryogenesis on leaf explants of *Oncidium* Gower Ramsey and subsequent plant regeneration. Plant Cell Rep 19:143–149
Chen YC, Chang C, Chang WC (2000) A reliable protocol for plant regeneration from callus culture of *Phalaenopsis*. In Vitro Cell Dev Biol Plant 36:420–423
Chen LR, Chen JT, Chang WC (2002) Efficient production of protocorm-like bodies and plant regeneration from flower stalk explants of the sympodial orchid *Epidendrum radicans*. In Vitro Cell Dev Biol Plant 38:441–445
Chung HH, Chen JT, Chang WC (2005) Cytokinins induce direct somatic embryogenesis of *Dendrobium Chiengmai* pink and subsequent plant regeneration. In Vitro Cell Dev Biol Plant 41:765–769
Chung HH, Chen JT, Chang WC (2007) Plant regeneration through direct somatic embryogenesis from leaf explants of *Dendrobium*. Biol Plant 51:346–350
Cui J, Liu J, Deng M, Chen J, Henny RJ (2008) Plant regeneration through protocorm-like bodies induced from nodal explants of *Syngonium podophyllum* 'White Butterfly'. Hortic Sci 43:2129–2133
Ernst R (1994) Effects of thidiazuron on *in vitro* propagation of *Phalaenopsis* and *Doritaenopsis* (Orchidaceae). Plant Cell Tissue Organ Cult 39:273–275
Gantait S, Sinniah UR (2012) Rapid micropropagation of monopodial orchid hybrid (*Aranda* Wan Chark Kuan 'Blue'× *Vanda coerulea* Grifft. ex. Lindl.) through direct induction of protocorm-like bodies from leaf segments. Plant Growth Regul 68:129–140
Gantait S, Sinniah UR, Mandal N, Das PK (2012) Direct induction of protocorm-like bodies from shoot tips, plantlet formation, and clonal fidelity analysis in *Anthurium andreanum* cv. CanCan. Plant Growth Regul 67:257–270
Gow WP, Chen JT, Chang WC (2009) Effects of genotype, light regime, explant position and orientation on direct somatic embryogenesis from leaf explants of *Phalaenopsis* orchids. Acta Physiol Plant 31:363
Gow WP, Chen JT, Chang WC (2010) Enhancement of direct somatic embryogenesis and plantlet growth from leaf explants of *Phalaenopsis* by adjusting culture period and explant length. Acta Physiol Plant 32:621–627
Guo WL, Chang YCA, Kao CY (2010) Protocorm-like bodies initiation from root tips of *Cyrtopodium paranaense* (Orchidaceae). Hortic Sci 45:1365–1368
Hong PI, Chen JT, Chang WC (2008) Plant regeneration via protocorm-like body formation and shoot multiplication from seed-derived callus of a maudiae type slipper orchid. Acta Physiol Plant 30:755–759
Hong PI, Chen JT, Chang WC (2010) Shoot development and plant regeneration from protocorm-like bodies of *Zygopetalum mackayi*. In Vitro Cell Dev Biol Plant 46:306–311
Huan LVT, Takamura T, Tanaka M (2004) Callus formation and plant regeneration from callus through somatic embryo structures in *Cymbidium* orchid. Plant Sci 166:1443–1449
Huang CH, Chung JP (2011) Efficient indirect induction of protocorm-like bodies and shoot proliferation using field-grown axillary buds of a *Lycaste* hybrid. Plant Cell Tissue Organ Cult 106:31–38

Hutchinson MJ, Saxena PK (1996) Acetylsalicylic acid enchances and synchronizes TDZ-induced somatic embryogenesis in geranium (*Pelargonium* X *horturum* Bailey) tissue cultures. Plant Cell Rep 15:512–515

Ishii Y, Takamura T, Goi M, Tanaka M (1998) Callus induction and somatic embryogenesis of *Phalaenopsis*. Plant Cell Rep 17:446–450

Jheng FY, Do YY, Liauh YW, Chung JP, Huang PL (2006) Enhancement of growth and regeneration efficiency from embryogenic callus cultures of *Oncidium* 'Gower Ramsey' by adjusting carbohydrate sources. Plant Sci 170:1133–1140

Jones TJ (2009) Maize tissue culture and transformation: the first 20 years. In: Kriz AL, Larkins BA (eds) Molecular genetic approaches to maize improvement: biotechnology in agriculture and forestry. Springer-Verlag, Berlin/Heidelberg, pp 7–27

Khoddamzadeh AA, Sinniah UR, Kadir MA, Kadzimin SB, Mahmood M, Sreeramanan S (2011) In vitro induction and proliferation of protocorm-like bodies (PLBs) from leaf segments of *Phalaenopsis bellina* (Rchb. f.) Christenson. Plant Growth Regul 65:381

Kuo HL, Chen JT, Chang WC (2005) Efficient plant regeneration through direct somatic embryogenesis from leaf explants of *Phalaenopsis* 'Little Steve'. In Vitro Cell Dev Biol Plant 41(4):453–456

Lee YI, Hsu ST, Yeung EC (2013) Orchid protocorm-like bodies are somatic embryos. Am J Bot 100:2121–2131

Lin YH, Chang C, Chang WC (2000) Plant regeneration from callus culture of a *Paphiopedilum* hybrid. Plant Cell Tissue Organ Cult 62:21–25

Lu MC (2004) High frequency plant regeneration from callus culture of *Pleione formosana* Hayata. Plant Cell Tissue Organ Cult 78:93–96

Malabadi RB, Mulgund GS, Nataraja K (2004) Efficient regeneration of *Vanda coerulea*, an endangered orchid using thidiazuron. Plant Cell Tissue Organ Cult 76:289–293

Mayer JLS, Stancato GC, Appezzato-Da-Glória B (2010) Direct regeneration of protocorm-like bodies (PLBs) from leaf apices of *Oncidium flexuosum* Sims (Orchidaceae). Plant Cell Tissue Organ Cult 103:411–416

Mok MC, Mok DWS, Armstrong DJ, Shudo K, Isogai Y, Okamoto T (1982) Cytokinin activity of N-phenyl-N′-1,2,3-thiadiazol-5-ylurea (Thidiazuron). Phytochemistry 21:1509–1511

Morel G (1960) Producing virus-free cymbidiums. Am Orchid Soc Bull 29:495–497

Murashige T, Skoog F (1962) A revised medium for rapid growth and bioassays with tobacco cultures. Physiol Plant 15:473–497

Murch SJ, Saxena PK (2001) Molecular fate of thidiazuron and its effects on auxin transport in hypocotyls tissues of *Pelargonium* x *hortorum* Bailey. Plant Growth Regul 35:269–275

Nayak NR, Patnaik S, Rath SP (1997a) Direct shoot regeneration from foliar explants of an epiphytic orchid, *Acampe praemorsa* (Roxb.) Blatter & McCain. Plant Cell Rep 16:583–587

Nayak NR, Rath SP, Patnaik S (1997b) In vitro propagation of three epiphytic orchids, *Cymbidium aloifolium* (L.) Sw., *Dendrobium aphyllum* (Roxb.) Fisch. And *Dendrobium moschatum* (Buch.-Ham.) Sw. through thidiazuron-induced high frequency shoot proliferation. Sci Hortic 71:243–250

Ng CY, Saleh NM (2011) In vitro propagation of *Paphiopedilum* orchid through formation of protocorm-like bodies. Plant Cell Tissue Organ Cult 105:193–202

Nhut DT, Van Le B, Tran Thanh Van K (2001) Manipulation of the morphogenetic pathways of *Lilium longiflorum* transverse thin cell layer explants by auxin and cytokinin. In Vitro Cell Dev Biol Plant 37:44–49

Niknejad A, Kadir MA, Kadzimin SB (2011) In vitro plant regeneration from protocorms-like bodies (PLBs) and callus of *Phalaenopsis gigantea* (Epidendroideae: Orchidaceae). Afr J Biotechnol 10:11808–11816

Palama TL, Menard P, Fock I, Choi YH, Bourdon E, Govinden-Soulange J, Bahut M, Payet B, Verpoorte R, Kodja H (2010) Shoot differentiation from protocorm callus cultures of *Vanilla planifolia* (Orchidaceae): proteomic and metabolic responses at early stage. BMC Plant Biol 10(1):82

Park SY, Yeung EC, Chakrabarty D, Paek KY (2002) An efficient direct induction of protocorm-like bodies from leaf subepidermal cells of *Doritaenopsis* hybrid using thin-section culture. Plant Cell Rep 21:46–51

Park SY, Murthy NH, Paek KY (2003) Protocorm-like body induction and subsequent plant regeneration from root tip cultures of *Doritaenopsis*. Plant Sci 164:919–923

Preece JE, Huetteman CA, Ashby WC, Roth PL (1991) Micro- and cutting propagation of silver maple. I. Results with adult and juvenile propagules. J Am Soc Hortic Sci 116:142–148

Roy J, Naha S, Majumdar M, Banerjee N (2007) Direct and callus-mediated protocorm-like body induction from shoot-tips of *Dendrobium chrysotoxum* Lindl. (Orchidaceae). Plant Cell Tissue Organ Cult 90:31–39

Roy AR, Sajeev S, Pattanayak A, Deka BC (2012) TDZ induced micropropagation in *Cymbidium giganteum* Wall. Ex Lindl. and assessment of genetic variation in the regenerated plants. Plant Growth Regul 68:435–445

Sheelavanthmath SS, Murthy HN, Hema BP, Hahn EJ, Paek KY (2005) High frequency of protocorm like bodies (PLBs) induction and plant regeneration from protocorm and leaf sections of *Aerides crispum*. Sci Hortic 106:395–401

Tao J, Yu L, Kong F, Zhao D (2011) Effects of plant growth regulators on in vitro propagation of *Cymbidium faberi* Rolfe. Afr J Biotechnol 10:15639–15646

Tokuhara K, Mii M (2003) Highly-efficient somatic embryogenesis from cell suspension cultures of *Phalaenopsis* orchids by adjusting carbohydrate sources. In Vitro Cell Dev Biol Plant 39:635–639

Vacin EF, Went FW (1949) Some pH changes in nutrient solutions. Bot Gaz 110:605–613

Vij SP, Sood A, Plaha KK (1984) Propagation of *Rhynchostylis retusa* Bl. (Orchidaceae) by direct organogenesis from leaf segment cultures. Bot Gaz 145:210–214

Visser C, Qureshi JA, Gill R, Saxena PK (1992) Morphoregulatory role of TDZ. Substitution of auxin and cytokinin requirement for the induction of somatic embryogenesis in geranium hypocotyl cultures. Plant Physiol 99:1704–1707

Vogel IN, Macedo AF (2011) Influence of IAA, TDZ, and light quality on asymbiotic germination, protocorm formation, and plantlet development of *Cyrtopodium glutiniferum* Raddi., a medicinal orchid. Plant Cell Tissue Organ Cult 104:147–155

Wang CX, Tian M (2014) Callus-mediated and direct protocorm-like body formation of *Bletilla striata* and assessment of clonal fidelity using ISSR markers. Acta Physiol Plant 36:2321–2330

Winarto B, Teixeira da Silva JA (2015) Use of coconut water and fertilizer for in vitro. In Vitro Cell Dev Biol Plant 51:303–314

Wu K, Zeng S, da Silva JAT, Chen Z, Zhang J, Yang Y, Duan J (2012) Efficient regeneration of *Renanthera* Tom Thumb 'Qilin' from leaf explants. Sci Hortic 135:194–201

Yu R, Zhang G, Li H, Cao H, Mo X, Gui M, Zhou X, Jiang Y, Li S, Wang J (2017) In vitro propagation of the endangered tree fern Cibotium barometz through formation of green globular bodies. Plant Cell Tissue Organ Cult 128:369–379

Zhao P, Wu F, Feng FS, Wang WJ (2008) Protocorm-like body (PLB) formation and plant regeneration from the callus culture of *Dendrodbium candidum* Wall ex Lindl. In Vitro Cell Dev Biol Plant 44:178–185

Zimmerman JL (1993) Somatic embryogenesis: a model for early development in higher plant. Plant Cell 5:1411–1423

Thidiazuron (TDZ): A Callus Minimizer for In Vitro Plant Production

14

Buhara Yücesan

Abstract

Thidiazuron (TDZ) seems to be quite suitable growth regulator for rapid and effective plant production in vitro. In many woody and herbaceous plants, lower TDZ concentrations induce less callus production around the explant. However, there is still insufficient number of studies how actually TDZ is effective on plant production in vitro and which biochemical and molecular genetic mechanism are involved in developmental physiology. Recent studies have shown that TDZ affects endogenous cytokinin and auxin production, and therefore morphogenetic identification of cells and tissues is downregulated by various genes acting on auxin regulation and transport as well as cytokinin response. These hormones that take a role in several metabolic pathways wherein more researches are required to identify the direct regulator and stress signaling factors during plant regeneration are induced by TDZ. A closer look on these subjects may bring about a better understanding on why urea derivative TDZ is more effective than adenine types of cytokinin on most species and how low concentration of TDZ has been found to be useful for micropropagation, especially in wood plants. Lastly, why does long exposure of TDZ have not been recommended and what are the possible reasons of hyperhydricity, abnormal shoot growth, and difficulty in rooting? The answers will be quite useful for management of culture conditions for the large-scale production and conservation of economical plants.

Keywords

TDZ · Thidiazuron · Callus · Growth Regulators · Morphogeny

B. Yücesan (✉)
Department of Seed Science and Technology, Faculty of Natural and Agricultural Sciences, Abant İzzet Baysal University, Bolu, Turkey
e-mail: buhara@ibu.edu.tr

N. Ahmad, M. Faisal (eds.), *Thidiazuron: From Urea Derivative to Plant Growth Regulator*, https://doi.org/10.1007/978-981-10-8004-3_14

14.1 Callus Formation in Plants

Callus is unorganized tissue mass and used widely in tissue culture systems. Exogenous application of auxin and cytokinin induces callus formation in many plant species. Generally, the ratio between auxin and cytokinin determines the fate of tissue morphogenesis. For instance, a high auxin/cytokinin ratio or vice versa ratio induces root or shoot regeneration, respectively. In this respect, other hormones, such as brassinosteroid as a novel growth regulator, abscisic acid, and gibberellic acid, are also effective for callus induction. Generally speaking, callus is either embryonic/organogenic which is capable of plant organ regeneration or non-embryonic/organogenic. The first type of callus might be shooty, rooty, or embryonic callus depending on the cultivation, while non-embryonic callus persists in being friable or compact depending on their tissue characteristics on respective culture media (Ikeuchi et al. 2013). Callus formation can also be induced by wounding plant tissue and pathogens in nature. In horticultural purposes, wound-induced callus tissue was induced from debarking of trees (Cline and Neely 1983). Stobbe et al. (2002) reported the potential use of wound-induced calli for new tissue or organs from various meristematic cell types including vascular and cortex cells since they are highly pluripotent. Upon to molecular and physiological properties of callus formation, wound-induced callus is raised by upregulation of cytokinins. This pattern is also caused by pathogen-related callus formation, namely, tumor induction as seen in many plant species infected by *Agrobacterium tumefaciens* (Nester et al. 1984). These bacteria cause an infection in wound sites promoting an outgrowth cell mass to produce galls therein (Barash and Manulis-Sasson 2007). Although tumor formation is induced by pathogens, these cell mass might be capable of producing an intact plant due to the presence of auxin and cytokinin released by pathogens. In this respect, tumor formation is particularly depending on pathogen type. For example, big vein viruses induce gall formation, and they give rise to an abnormal vascular tissue formation intervening the normal plant growth and development in host plants (Zhang et al. 2007). Similarly, other pathogenic organisms such as insects that cause gall formation (Mutun and Dinç 2011), parasitic protests, and nematodes that cause knot-root diseases in roots of host plants induce harmful damages in crops (Jammes et al. 2005). Additionally, callus formation of plants, particularly in *Brassica*, *Datura*, *Lilium*, and *Nicotiana*, might occur as a result of interspecific crosses. This naturally occurring callus tissue is also totipotent on medium-free growth medium when excised from hybrid plants (Ikeuchi et al. 2013 and references therein).

Plant tissue culture systems are widely based on hormone-applied approaches for plant propagation. Depending on the strategy for plant multiplication, with or without callus-mediated tissue culture protocols are used. Since our knowledge has been limited to understanding the complete genetic and epigenetic mechanisms acting on the morphogenesis, callus formation in tissue culture system is still the visible tip of iceberg with auxin and cytokinin ratio. Table 14.1 shows the efficacy of some selected growth regulators widely used for callus morphogeny of plants cultivated in vitro. On the other hand, in tissue culture systems, explant type, such as

Table 14.1 Effects of widely used growth regulators on callus formation (shooty, rooty, or morphogenic)

	BA		KIN		TDZ		NAA		IAA		IBA		2,4-D	
	LC	HC	LC	HC	LC	HC	LC	HC	LC	HC	LC	HC	LC	HC
Shooty callus	+	+++	+	++	+	++	+	–	–	–	–	–	+	+
Rooty callus	–	–	–	–	–	–	+	+++	–	+	–	+	+	++
Morphogenic callus	+	+++	+	++	+	++	+	+++	+	+	+	+	+	+++

+ signs represent degree of callus production as follows: +, slight; ++, mild; +++, massive; –: no callus

BA benzyl-adenine, *KIN* kinetin, *TDZ* thidiazuron, *NAA* naphtylene acetic acid, *IAA* indole-3-acetic acid, *IBA* indol-3-butyric acid, *2–4-D* diclorophenyl-fenoxy acetic acid

young embryonic axes, apical meristems, cotyledonary leaves, and immature zygotic embryos excised from seeds, is also effective for the maintenance of the callus morphogeny. Depending on the hormone concentration, duration of cultivation, and number of passages, various callus formations in plants have been largely reported for 50 years. Although callus cultures are useful for clonal propagation, they initially do not have active hormone biosynthesis (Duclercq et al. 2011). However, growth media conditions amended with exogenous growth regulators affect the cells to be habituated; thus these cells exhibit hormone-autotroph callus growth after long periods of callus culture (Kreis et al. 2015). However, in some cases morphogenic callus tissue might account for a somaclonal variation in respective tissues. Callus-mediated somaclonal variation is therefore not preferable for true-to-type production, but mutant selection for desired characters can be applicable.

14.2 Comparison of TDZ with Other Growth Regulators

TDZ is accepted as cytokinin-like synthetic hormone, and the use of TDZ in plant tissue culture is increasing rapidly in the last 20 years. With a closer look on cytokinins, they are a complex class of hormones (Mok et al. 1982; Huetteman and Preece 1993). There are six commonly known naturally occurring cytokinins, namely, zeatin, iP (isopentenyladenosine), and DHZ (dihydrozeatin riboside) with their ribosides ZR (zeatin riboside), IPR and DHZR (dihydrozeatin). Of those cytokinins, Z, ZR, IP, and IPA are in the conjugated form, and they can be oxidized from their side chains by plant tissues, whereas DHZ, DHZR, and BAP (6-benzyl-adenine) are conjugated without any side chain cleavage, in other words without an oxidation (Guo et al. 2011). It is noteworthy to say that in plant tissue culture systems, phenolic compounds mainly induce cytokinin oxidation, and this pattern most likely causes tissue necrosis that is an undesired issue for clonal propagation.

Cytokinins can be biologically inactive, and thus stable N-glycosylated on the purine ring or O-glycosylated on the N_6-substituted side chain, that may have a storage function as seen in Z and DHZ. Strikingly, cytokinins after being uptaken, very

Table 14.2 Comparison of TDZ and BAP on explant response either callus formation (CF) or shoot production (SP) in tea plant, *Camellia sinensis* (Adapted from Agarwal et al. 1992)

Concentrations	Explant response (%)	
	TDZ	BAP
0 μM	–	–
100 μM	94.0 ± 1.4 CF	–
10 μM	95.8 ± 1.2 CF	33.8 ± 1.0 SP
1 μM	96.7 ± 0.8 CF	12.8 ± 0.5 SP
100 nm	54.5 ± 1.2 SP	–
10 nm	49.2 ± 0.7 SP	–
1 nm	49.0 ± 0.9 SP	–
100 pM	29.0 ± 0.9 SP	–
10 pM	27.5 ± 0.6 SP	–
1 pM	28.2 ± 0.8 SP	–

low percentage remains in free form (not conjugated). TDZ is an exception and is conjugated only at a very low rate. It was extrapolated that phaseolus callus cultures on MS medium including radioactive-labeled TDZ taken up 60% of TDZ in a non-conjugated form (Mok and Mok 1985), whereas most purine-type cytokinins are considered to be some extent chemically unstable. Initially, effects of TDZ on morphogenesis were discovered in woody plants, therein high rate of shoot proliferation was induced at very low concentrations of TDZ as compared to the other PGR. In tea plant, 1 μM TDZ was effective with a frequency of 96.7% of explants forming callus, while the lower concentrations induce the shoot formation (Table 14.2; Agarwal et al. 1992). This is in accordance with many reports in which very low concentrations of TDZ induce axillary shoot formation in many woody plants, whereas BAP and other cytokinins are not effective (Huetemann and Preece 1993; Guo et al. 2011 and references therein; Yücesan et al. 2007). In terms of callus formation, it was effective at the high concentrations of TDZ; however, long exposure and overall callus overgrowth may result in necrosis, abnormal shoot growth, and difficulty in root formation (Lu 1993). Determination of critical TDZ concentration is important for the maintenance the fate of cell or tissue in a plant tissue culture system. In peanut (Gill and Ozias-Akins 1999), callus tissue on 1 or 5 μM TDZ continued proliferation, whereas 15 μM TDZ caused the callus to turn brown with inhibition of growth. In *Camellia sinensis*, the responsive explants were continuously grown on medium containing TDZ; an overall callus overgrowth and later necrosis were observed. In that, the general growth of the shoots during the earlier stages with respect to sturdy leaves, shoot length, and shoot diameter was also better in explants initiated on TDZ medium in comparison to BAP (Agarwal et al. 1992; see Table 14.2). However, it was necessary to remove TDZ in subsequent subcultures in order to increase the proliferation rates or the sturdiness of the shoots. This is probably due to the capacity of TDZ in stimulating endogenous cytokinin biosynthesis or in altering cytokinin metabolism. In this respect, Yücesan et al. (2015) reported the effect of TDZ on both shoot and root organogenesis in golden berry (*Cape gooseberry*) without any intervening callus formation on LS medium containing 2.2 μM TDZ producing 5.3 shoots per explant with some root formation (3.3 roots/shoot; Fig. 14.1a–f). As seen in golden berry, without any combination with an

Fig. 14.1 Comparative steps of in vitro plant regeneration with different plant growth regulators (PGR) of *Physalis peruviana* on LS medium (Linsmaier and Skoog 1965): (**a**) more root formation on PGR-free LS medium, (**b**) effects of kinetin (2.3 μM) on single shoot formation with some expanded leaves, (**c**) promising effect of TDZ (2.2 μM) on direct shoot regeneration, (**d**) highest shoot elongation with smaller leaves with1.4 μM of GA_3, (**e**) acclimatization process of the regenerants produced from TDZ-induced growth media, (**f**) successfully fruit development under ex vitro conditions 5 months after the culture initiation. Bar: **a** = 10 mm; **b** = 10 mm; **c** = 10 mm; **d** = 20 mm; **e** = 60 mm; **f** = 65 mm (Adapted from Yücesan et al. 2015)

auxin, TDZ might alter the auxin/cytokinin ratio in nodal explants favoring the axillary regeneration. Not only cytokinin-like effect but also root induction potential of TDZ may fulfill the auxin requirements of regeneration systems.

14.3 Effects of TDZ on Morphogeny

There are few reports on the molecular genetic mechanism upon how plant cells promote the regeneration competence after subsequent induction by thidiazuron. Before this, it is noteworthy to mention about the acquisition of regeneration competence induced by IAA. IAA is a major form of auxin and plays a critical role in plant physiology, i.e., auxin polarization, tissue morphogenesis, apical dominance and suppression of lateral branching, etc. Similarly, cytokinins participate in many aspects of growth and development. Very recently Kou et al. (2016) reported a novel study on protocorm-like body (PLB) model development system raised from callus-rhizoid cells of *Rosa canina* to attain how regeneration competence is induced by TDZ. In their study, PLBs are referred to as somatic embryogenesis, because the resemblance of protocorm is considered to be an extended state of the zygotic embryo derived from callus tissue. They observed in callus rhizoids that TDZ increased IAA (from 30 to 70 ng/g FW) and ZR (from 8 to 12 ng/g) concentration while decreased the level of iPA during the first 5 days of incubation. Their study extrapolates the accumulation of endogenous auxin such as IAA and cytokinins like zeatin in callus tissue. Therefore, the puzzle on the mechanism of how TDZ has been a topic of discussion for a long time will be clearly understood. Moreover, TDZ was effective to induce PLB morphogenesis from rhizoid tips, therein the cell fate was converted, and indirect shoot formation was established after treating the callus with TDZ. The molecular genetic mechanism of this morphogenetic response of TDZ was stimulated; thus auxin and cytokinin response were promoted by elevated level of *RcARF1*, *RcARF4*, *RcRR1*, *RcPIN2*, and *RcPIN3* expression, while expression of auxin transport protein gene RcPIN1 decreased. As a result, Kou et al. 2016 claimed that auxin efflux from rhizoid tips and its concentration gradient was affected by auxin transport accordingly. Therefore, effects of TDZ disturbing the balance of cytokinin and auxin around the explant provide an insight through the determination of cell fate. This study is in agreement with earlier studies in which how TDZ acted as a substitute for both auxin and CK during plant tissue culture (Visser et al. 1992; Casanova et al. 2004) and provides a new insight through TDZ incorporated with developmental pathways of embryonic callus tissues. In this respect, more researches are needed on how actually TDZ is effective on plant production in vitro and which biochemical and molecular genetic mechanisms are involved in developmental physiology.

References

Agarwal B, Singh U, Maitreyi B (1992) In vitro clonal propagation of tea (*Camellia sinensis*). Plant Cell Tissue Organ Cult 30:1–5

Barash I, Manulis-Sasson S (2007) Virulence mechanisms and host specificity of gall-forming *Pantoea agglomerans*. Trends Microbiol 15:538–545

Casanova E, Valdés AE, Fernández B et al (2004) Levels and immunolocalization of endogenous cytokinins in thidiazuron induced shoot organogenesis in carnation. J Plant Physiol 61:95–104. https://doi.org/10.1078/0176-1617-00957

Cline MN, Neely D (1983) The histology and histochemistry of wound-healing process in geranium cuttings. J Am Soc Hortic Sci 108:496–502

Duclercq J, Sangwan-Norreel B, Catterou M, Sangwan RS (2011) De novo shoot organogenesis: from art to science. Trends Plant Sci 16:597–606

Gill R, Ozias-Akins P (1999) Thidiazuron-induced highly morphogenic callus and high frequency regeneration of fertile peanut (*Arachis hypogaea* L.) plants. In Vitro Cell Dev Biol–Plant 35:455–450

Guo B, Abbasi BH, Zeb A, Xu LL, Wei YH (2011) Thidiazuron: a multi-dimensional plant growth regulator. Afr J Biotechnol 10(45):8984–9000

Huetteman CA, Preece JE (1993) Thidiazuron: a potent cytokinin for woody plant tissue culture. Plant Cell Tissue Organ Cult 33:105–119

Ikeuchi M, Sugimoto K, Iwase A (2013) Plant callus: mechanisms of induction and repression. Plant Cell 25(9):3159–3173

Kou Y, Yuan C, Zhao Q et al (2016) Thidiazuron triggers morphogenesis in *Rosa canina* L. Protocorm-like bodies by changing incipient cell fate. Front Plant Sci 7:557. https://doi.org/10.3389/fpls.2016.00557

Kreis W, Haug B, Yücesan B (2015) Somaclonal variation of cardenolide content in Heywood's foxglove, a source for the antiviral cardenolide glucoevatromonoside, regenerated from permanent shoot culture and callus. In Vitro Cell Dev Biol Plant 51:35–41

Linsmaier E, Skoog F (1965) Organic growth factor requirements of tobacco tissue cultures. Physiol Plant 18:100–127

Lu C (1993) The use of Thidiazuron in tissue culture. In Vitro Cell Dev Biol 29P(2):92–96. Retrieved from http://www.jstor.org/stable/4292979

Mok MC, Mok DWS (1985) The metabolism of [^{14}C]-thidiazuron in callus tissues of *Phaseolus lunatus*. Physiol Plant 65:427–432

Mok MC, Mok DWS, Armstrong DJ et al (1982) Cytokinin activity of N-phenyl-N′-1,2,3-thidiazol-5-yl urea (thidiazuron). Phytochemistry 21:1509–1511

Mutun S, Dinç S (2011) Contributions to the gallwasp (Hymenoptera: Cynipidae) fauna of Turkey with one new record. J Appl Biol Sci 5:83–85

Nester EW, Gordon MP, Amasino RM, Yanofsky MF (1984) Crown gall: a molecular and physiological analysis. Annu Rev Plant Physiol 35:387–413

Stobbe H, Schmitt U, Eckstein D, Dujesiefken D (2002) Developmental stages and fine structure of surface callus formed after debarking of living lime trees (*Tilia* sp.) Ann Bot (Lond) 89:773–782

Visser C, Qureshi JA, Gill R et al (1992) Morphoregulatory role of thidiazuron. Plant Physiol 99:1704–1707. https://doi.org/10.1104/pp.99.4.1704

Yücesan B, Türker AU, Gürel E (2007) TDZ-induced high-frequency plant regeneration through multiple shoot formation in witloof chicory (*Cichorium intybus* L). Plant Cell Tissue Organ Cult 91:243–250

Yücesan B, Mohammed A, Arslan M, Gürel E (2015) Clonal propagation and synthetic seed production from nodal segments of Cape gooseberry (*Physalis peruviana* L.), a tropical fruit plant. Turk J Agric For 39:797–806

Zhang HM, Yang J, Xin X, Chen JP, Adams M (2007) Molecular characterization of the genome segments S4, S6 and S7 of rice gall dwarf virus. Arch Virol 152:1593–1602

The Applications of TDZ in Medicinal Plant Tissue Culture

15

A. V. Deepa, M. Anju, and T. Dennis Thomas

Abstract

Thidiazuron (TDZ; N-phenyl-1, 2, 3-thidiazole-5ylurea) is a cytokinin-like compound routinely used for in vitro culture studies including shoot proliferation and regeneration in various plants. Recently, there is an increased use of TDZ in in vitro propagation of plants including medicinal and horticultural crops. TDZ was found to be more effective in micropropagation, multiple shoot induction, somatic embryogenesis, callus induction, and shoot organogenesis as compared to other cytokinins. In some systems the synergistic effect of TDZ with other cytokinin/auxin was found more effective than using alone. The optimum concentration of TDZ may vary based on the plant species, explants, and duration of exposure. In this review we have described the recent trends in the use of TDZ for micropropagation of some medicinal plants. The medicinal plants described in this chapter include *Picrorhiza kurroa*, *Aronia melanocarpa*, *Scutellaria ocmulgee*, *Salvia miltiorrhiza*, *Aphyllorchis montana*, *Gentiana decumbens*, *Aconitum balfourii*, *Pelargonium sidoides*, *Withania somnifera*, *Curculigo latifolia*, *Hedychium coronarium*, *Mentha arvensis*, *Clitoria ternatea*, *Saussurea involucrata*, *Jatropha curcas*, *Medicago sativa*, *Crocus sativus*, *Zingiber officinale*, *Cannabis sativa*, *Arnebia euchroma*, and *Azadirachta indica*.

Keywords

Thidiazuron · Medicinal plants · Plant tissue culture · Multiple shoot formation

---

A. V. Deepa · M. Anju · T. Dennis Thomas (✉)
Department of Plant Science, Riverside Transit Campus, Central University of Kerala, Kasaragod, Kerala, India
e-mail: den_thuruthiyil@cukerala.edu.in

N. Ahmad, M. Faisal (eds.), *Thidiazuron: From Urea Derivative to Plant Growth Regulator*, https://doi.org/10.1007/978-981-10-8004-3_15

15.1 Introduction

Phytohormones are organic compounds synthesized in various regions of plants and move to specific sites where they regulate plant growth and development. Auxins, gibberellins, ethylene, abscisic acid, and cytokinins are the major phytohormones responsible for the various functions in plant body. Among them auxin and cytokinin are unavoidable constituents of medium used in micropropagation of plants. IAA (indole-3-acetic acid) is a naturally occurring auxin which is widely used in plant tissue culture. In addition to IAA, synthetic auxins like IBA (indole-3-butyric acid), NAA (naphthalene acetic acid), and 2, 4-D (2, 4-dichlorophenoxyacetic acid) have significant role in morphogenesis of plants in vitro. Kinetin (KN), zeatin (ZT), BA (6-benzyladenine), or BAP (6-benzylaminopurine) are major cytokinins. Thidiazuron (TDZ; N-phenyl-1, 2, 3-thidiazole-5ylurea) is a cytokinin-like compound used for shoot proliferation and regeneration of plants. A study on the cytokinin metabolism in *Phaseolus* species using TDZ confirmed the cytokinin-like activity of TDZ (Mok et al. 1982). According to Mok et al. (1982), TDZ is an active cytokinin similar to highly active N-phenyl-N-4 pyridyl urea derivatives. It was synthesized by German Schering Corporation and was used initially as cotton defoliant (Murthy et al. 1998). Structurally TDZ is different from either auxin or cytokinin. It consists of two functional groups, phenyl and thidiazole groups, and the replacement of these groups results in the reduction of its activity (Fig. 15.1; Murthy et al. 1998). Many plants responded positively to TDZ, and the crucial role of TDZ in multiple shoot induction, callus organogenesis, shoot elongation, etc. had been well documented (Fig. 15.2a, b; Thomas 2003; Thomas and Puthur 2004; Thomas and Philip 2005; Cheruvathur et al. 2013; Cheruvathur and Thomas 2014). Usually low concentrations of TDZ induce shoot proliferation than many other cytokinins, while the higher level may inhibit shoot elongation (Huetteman and Preece 1993). The optimum concentration of TDZ may vary based on the plant species, explants, and duration of exposure. TDZ at 0.11 mg/l was found effective in shoot induction in *Withania somnifera* while 2.0 mg/l in *Jatropha curcas* (Fatima et al. 2015; Kumar and Reddy 2012). The TDZ-treated explants with shoot buds subcultured on plant growth regulator (PGR)-free medium resulted in highest rate of shoot proliferation (Siddique and Anis 2006). A combination of TDZ with other phytohormones could be more effective than used alone (Guo et al. 2011). The combination of TDZ with NAA was superior over the combination of TDZ with IAA and TDZ with adenine in promoting shoot regeneration in cabbage (Gambhir and Srivastava 2015). TDZ

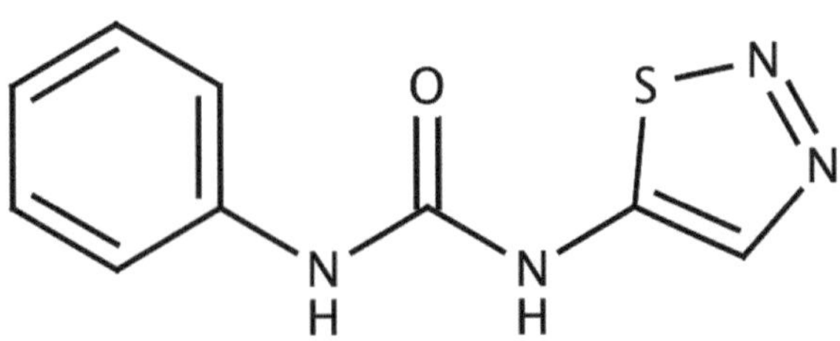

Fig. 15.1 Chemical structure of thidiazuron (TDZ, N-phenyl-1, 2, 3-thidiazole-5ylurea)

Fig. 15.2 TDZ-induced morphogenesis. (**a**) TDZ (1.5 mg/l)-induced callus organogenesis in *Bacopa monnieri* (unpublished work of Linta and Thomas) (**b**) Multiple shoot induction in *Elephantopus scaber* from nodal segments on MS medium supplemented with 2.0 mg/l TDZ (Abraham and Thomas, unpublished work)

can significantly influence the metabolism of other PGRs. It can increase auxin accumulation and translocation (Casanova et al. 2004), cytokinin content (Kodja et al. 2015), and ABA concentration (Murch and Saxena 1997). TDZ-induced ACO (1-aminocyclopropane-1-carboxylate oxidase) production in *Medicago sativa* resulted in a production of ethylene (Feng et al. 2012). Recently, TDZ has been effectively used to induce various morphogenic responses in numerous plant systems including medicinal plants (Table 15.1). The present review is an attempt to analyze the effect of TDZ on micropropagation of various selected medicinal plants.

15.1.1 *Picrorhiza kurroa*

Picrorhiza kurroa (common name, kutki) is an endangered herbaceous medicinal plant belonging to the family Scrophulariaceae. It is a native of Western Himalaya,

Table 15.1 Some recent reports on TDZ-induced morphogenesis in various medicinal plants

Name of the plant	Explant[a]	Type of medium and plant growth regulators	Response[a]	References
Picrorhiza kurroa	N	Pretreatment in TDZ (0.11 mg/l)	MSh, RT	Patial et al. (2016)
Aronia melanocarpa	N	MS + TDZ (0.5 mg/l), MS with TDZ (0.5 mg/l) + NAA (0.1 mg/l)	MSh	Sivanesan et al. (2016)
Scutellaria ocmulgee	Leaf and shoot TCL	MS + maltose + TDZ (5.5 mg/l) + IAA (0.22 mg/l)	MSh	Vaidya et al. (2016)
Salvia miltiorrhiza	L	MS + TDZ (0.5 mg/l)	MSh	Tsai et al. (2015)
Salvia miltiorrhiza	L	MS + TDZ (0.1 mg/l) + 2, 4-D (5.0 mg/l)	C	Tsai et al. (2015)
Aphyllorchis montana	SD	BM-TM + TDZ (1.5 mg/l) + NAA (0.24 mg/l)	MSh	Mahendran and Bai (2015)
Gentiana decumbens	PT	MS salts (without NH_4NO_3) + Vitamins + Glucose (30 g/l) + glutamine (3.0 g/l) + NAA (2 mg/l) + TDZ (0.1 mg/l)	C	Tomiczak et al. (2015)
Aconitum balfourii	L, P	MS + TDZ (0.5 mg/l) + NAA (1.0 mg/l)	C	Gondval et al. (2016)
Pelargonium sidoides	L	MS + TDZ (0.22–1.1 mg/l) + picloram (0.484 mg/l)	C	Kumar et al. (2015)
Withania somnifera	N	MS + TDZ (0.11 mg/l)	MSh	Fatima et al. (2015)
Curculigo latifolia	ST	MS + TDZ (0.5 mg/l) + IBA (0.25 mg/l)	MSh	Babaei et al. (2014)
Curculigo latifolia	ST	MS + TDZ (2.0 mg/l) + IBA (0–0.5 mg/l)	Scalp	Babaei et al. (2014)
Hedychium coronarium	RB	MS + TDZ (1.0 mg/l)	MSh	Verma et al. (2014)
Mentha arvensis	N	MS + TDZ (3.3 mg/l)	MSh	Faisal et al. (2014)
Clitoria ternatea	CN	MS + TDZ (0.022 mg/l)	MSh	Mukhtar et al. (2012)
Clitoria ternatea	N	MS + TDZ (0.22 mg/l)	MSh	Mukhtar et al. (2012)
Saussurea involucrata	L	MS + TDZ (0.011 mg/l)	MSh	Guo et al. (2012)
Jatropha curcas	CP	MS + TDZ (0.5 mg/l)	MSh	Kumar and Reddy (2012)
Medicago sativa	P	MS + TDZ (0.2 mg/l) + $CoCl_2$ (6.5 mg/l)	SE	Feng et al. (2012)

(continued)

Table 15.1 (continued)

Name of the plant	Explant[a]	Type of medium and plant growth regulators	Response[a]	References
Crocus sativus	CR	MS + TDZ (0.99 mg/l)	C	Sharifi et al. (2010)
Zingiber officinale	AS	MS + TDZ (1.0 mg/l) + IBA (1.0 mg/l)	MSh, R	Lincy and Sasikumar (2010)
Cannabis sativa	N	MS + TDZ (0.11 mg/l)	MSh	Lata et al. (2009)
Arnebia euchroma	Ct and HC	MS + TDZ (1.0 mg/l)	MSh	Jiang et al. (2005)
Azadirachta indica	R, L, N	MS + TDZ (0.5–0.99 mg/l) + 2, 4-D (0.11 mg/l)	SE	Akula et al. (2003)

[a]*AS* aerial stem, *BM-TM* terrestrial orchid medium, *C* callus, *CN* cotyledonary node, *CP* cotyledonary petiole, *CR* corm, *Ct* cotyledon, *HC* hypocotyl, *L* leaf, *P* petiole, *MR* multiple root, *MSh* multiple shoot, *N* node, *PT* protoplast, *RB* rhizome bud, *R* root, *SD* seed, *SE* somatic embryo, *ST* shoot tip, *TCL* thin cell layer

whose rhizome and roots are enriched with several glycosides like kutkoside, picroside I, and picroside II (Ansari et al. 1988), which are responsible for its medicinal properties such as anti-inflammatory (Turaskar et al. 2013), hepatoprotective (Sinha et al. 2011), and anti-asthmatic (Sehgal et al. 2013) activities.

An efficient in vitro plant regeneration protocol for *P. kurroa* using TDZ has been standardized by Patial et al. (2016). The effect of TDZ on proliferation and survival of the plant has been tested. Nodal explants (2.0–2.5 cm) were collected from in vitro-cultured *P. kurroa* and pretreated with 0.055–0.22 mg/l TDZ for 7–30 days. Then they were transferred to MS medium supplemented with either 0.473 mg/l KN, 0.26 mg/l BAP, or without PGRs. Rate of shoot multiplication was influenced by both concentration and duration of TDZ treatment. Within 4 or 5 days, the base of the node began to swell irrespective of the TDZ concentration. After 15 days, shoot initiation was noticed in 0.25 and 0.11 mg/l TDZ. But increasing the duration to 30 days had deleterious effect on the growth and elongation of shoot, resulting in the formation of a rosette-like shape.

Rooting hormones like IAA and NAA are usually used for root induction. But here, 15 days pretreatment with 0.11 mg/l TDZ followed by culturing on MS medium devoid of PGRs resulted in highest number of roots (13.00) when compared to MS medium supplemented with IAA or NAA.

15.1.2 *Aronia melanocarpa*

A. melanocarpa (common name, chokeberry) belongs to the family Rosaceae. *A. melanocarpa* contains phenolic compounds which is responsible for its antioxidant activity (Oszmiański and Wojdylo 2005). It has other health-promoting properties such as antimutagenic, anticancer, cardioprotective, hepatoprotective, gastroprotective, antidiabetic, anti-inflammatory, and antiviral (Sivanesan et al. 2016; Kokotkiewicz et al. 2010; Do and Hwang 2014).

Sivanesan et al. (2016) demonstrated the effect of TDZ on shoot multiplication of *A. melanocarpa* from the nodal explants. The explants failed to produce shoot on MS medium without PGRs, while the MS medium fortified with TDZ alone or in combination with TDZ- and NAA-induced shoot formation. The shoot formation was concentration dependent. The lower concentrations of TDZ (0.25–0.5 mg/l) promoted shoot induction, whereas higher concentrations (1.0–2.0 mg/l) reduced the mean shoot length and number of shoots per explant (Sivanesan et al. 2016). The optimum concentration of TDZ for *A. melanocarpa* was 0.5 mg/l. In this concentration 82.2% shoot multiplication was observed with a mean number 10.4 shoots per explant (Sivanesan et al. 2016). Higher concentration such as 2.0 mg/l TDZ resulted in hyperhydric shoots.

The combination of TDZ and NAA was most effective than TDZ alone in shoot formation from nodal explants of *A. melanocarpa*. On this combination, the highest percentage of shoot multiplication was 98.9%, and mean number of shoots per explant was 19.8, and mean shoot length was 3.8 cm. Highest result was observed when the MS medium is fortified with 0.5 mg/l TDZ and 0.1 mg/l NAA. Higher concentration of TDZ, i.e., 2.0 mg/l and 0.1 mg/l NAA, caused hyperhydric shoots (Sivanesan et al. 2016). Subculture medium without TDZ was good for further shoot formation and maintenance of morphogenetic potential of nodal explants (Sivanesan et al. 2016).

15.1.3 *Scutellaria ocmulgee*

Scutellaria ocmulgee (common name, skullcap) is a herbaceous plant, belonging to the family Lamiaceae. It is a rare and threatened medicinal plant with ornamental values, which requires immediate conservation. *S. ocmulgee* crude extract and its constituent flavonoids are reported to have antitumor properties (Patel et al. 2013).

An efficient protocol for shoot regeneration of *S. ocmulgee* from leaf and shoot thin cell layer (TCL) explants was developed for the first time by Vaidya et al. (2016). Among the various media and PGR combinations tried, maximum number of shoot was observed from leaf transverse TCL explants grown on MS medium supplemented with 5.5 mg/l TDZ, 0.175 mg/l IAA, and maltose. Among the PGRs used, media with higher concentration of TDZ gave best results, though concentration of TDZ (5.5 mg/l) was significantly lower compared to other studies (Vaidya et al. 2016).

15.1.4 *Salvia miltiorrhiza*

Salvia miltiorrhiza (common name, Chinese sage) is a perennial herbaceous plant belonging to the family Lamiaceae. It is a native of China and Japan and hence used in their traditional medicine to treat cardiovascular diseases (Wang et al. 2013).

Chemical constituents isolated from dried roots of this plant include cryptotanshinone, dihydrotanshinone, tanshinone I, tanshinone IIA, and salvianolic acid B (Pan et al. 2001; Ai and Li 1988).

Tsai et al. (2016) established an efficient protocol for mass propagation of *S. miltiorrhiza* using TDZ from leaf explants via direct organogenesis and indirect organogenesis. Initial explants were 2 cm^2 leaf segments collected from in vitro raised plants. Explants were incubated on MS medium supplemented with various combinations of TDZ (0.1–1.0 mg/l) and 2, 4-D (1.0–10.0 mg/l). Direct rhizogenesis was obtained from explants cultured on basal medium with an average numbers of 39.00 roots per explant. Direct shoot formation was observed on MS medium supplemented with 0.1, 0.5, and 1.0 mg/l TDZ. Among these various concentrations, the highest number of shoots was obtained with 0.5 mg/l of TDZ.

When the leaf explants of *S. miltiorrhiza* were cultured on MS medium supplemented with TDZ and 2, 4-D, all the explants developed calli. Maximum callus proliferation was observed on 5.0 mg/l 2, 4-D and 0.1 mg/l TDZ. Multiplication and maintenance of the calli were best done with the combinations of 0.1 mg/l TDZ plus 5.0 mg/l 2, 4-D; 0.1 mg/l TDZ plus 10.0 mg/l 2, 4-D; and 5.0 mg/l 2, 4-D alone. Since there was no morphogenesis during subculturing, Tsai et al. (2016) were able to maintain 16 callus lines. On subculturing the calli on PGR-free MS medium, a few of them developed shoots, and all of them developed roots. Highest number of shoots was formed on 1.0 mg/l 2, 4-D plus 0.5 mg/l TDZ. When the calli were transferred to MS medium containing varying concentrations of TDZ, maximum number of shoots were obtained with 1.0 mg/l TDZ.

15.1.5 *Aphyllorchis montana*

Aphyllorchis montana is a saprophytic achlorophyllous orchid. It is a terrestrial mycoheterotrophic orchid species that is categorized under the data deficient category of orchids in India (McKendrick et al. 2002). As it is a rare medicinal orchid in the Western Ghats, it seeks immediate attention and conservation (Sinu et al. 2012). Plant extract of *A. montana* has been employed to cure cold, cough, and anemia and also to strengthen the vitality (Prajapati et al. 2003). Traditional healers used *A. montana* to cure diabetes mellitus. Hypoglycemic activity of ethanolic extracts of *A. montana* was proven by Pentela et al. (2012). It is the main ingredient of "chyawanprash" – an ayurvedic tonic (Sreenu et al. 2013).

The germination of *A. montana* seed requires the help of specific fungi and has poor seed viability, resulting in low germination rate (Warcup 1973). A protocol for plant regeneration from asymbiotic seed germination was standardized by Mahendran and Bai (2015). Immature capsules of *A. montana* were collected and sterilized. The seeds were inoculated in several different growth media, all with 2% sucrose and 0.8% agar. Among the various media tried, BM-I terrestrial orchid medium (BM-TM; Van Waes and Debergh 1986) was found to be the best for seed germination (79% cultures respond). For the multiplication of protocorms, BM-TM medium was supplemented with various cytokinins at different concentrations, alone or in combination with NAA (0.24 mg/l). The seed-derived protocorms produced multiple seedlings in the presence of cytokinins. TDZ was most effective among all the cytokinins tested. The number of seedlings derived from protocorms was maximum on a combination of 1.5 mg/l TDZ and 0.24 mg/l NAA.

15.1.6 *Gentiana decumbens*

Gentiana decumbens is a herbaceous medicinal plant belonging to the family Gentianaceae. Pharmacological studies have revealed its antioxidant activity (Myagmar and Aniya 2000). It produces steroid glucosides and flavonoids (Dungerdorj et al. 2006). It is used in traditional medicine to cure liver diseases (Kletter et al. 2008) and stomach disorders (Qureshi et al. 2007).

Tomiczak et al. (2015) reported plant regeneration from protoplasts of differentiated green leaf mesophyll cells of *G. decumbens.* Protoplast was isolated from young leaves of in vitro-cultured *G. decumbens.* Purified protoplasts were cultured in PCM (protoplast culture medium; Tomiczak et al. 2015) with 0.5% mannitol and 0.8% agarose. Of the two cytokinins (i.e., BAP and TDZ) tested with PCM, TDZ-induced optimum cell division (plating efficiency) than BAP. The microcalli formed from protoplast culture were transferred to callus proliferation medium (CPM) and incubated in the dark. Increased callus multiplication was obtained on CPM with 2.0 mg/l NAA and 0.2 mg/l TDZ.

15.1.7 *Aconitum balfourii*

Aconitum balfourii (common name, mitha) is a perennial herb, endemic to Himalayan Alpine. It is a highly valuable, rare, and threatened medicinal plant, belonging to Ranunculaceae family. Its tuberous roots are rich source of alkaloids like aconite, pseudoaconitine, balfourine, and bikhaconite (Sharma and Gaur 2012). Analgesic, anti-inflammatory, antirheumatic, and vermifuge drugs are produced from the root of this plant (Sharma and Gaur 2012).

An efficient protocol for in vitro plantlet regeneration of *A. balfourii* via callus-mediated organogenesis using TDZ has been established by Gondval et al. (2016). Leaf and petiole segments were used as explants. Surface-sterilized explants were cultured on MS medium containing different combinations of TDZ (0.5–2.0 mg/l) and NAA (1.0–4.0 mg/l). For both petiole and leaf, callus induction was best on MS medium supplemented with 0.5 mg/l TDZ and 1.0 mg/l NAA. While lower concentration or the absence of TDZ exhibited reduced callus induction, increased concentration of TDZ caused the formation of brown-colored callus with reduced growth. The calli were subcultured on MS medium supplemented with varying concentrations of TDZ (0.5–2.0 mg/l). Maximum shoots were obtained on MS medium supplied with 0.5 and 1.5 mg/l TDZ for leaf and petiole, respectively. No shoot induction was observed, both in the absence and higher concentration of TDZ (Gondval et al. 2016).

15.1.8 *Pelargonium sidoides*

Pelargonium sidoides is a herbaceous medicinal plant of the family Geraniaceae and is extensively used as traditional herbal medicine in many countries. It is effective in treatment of tuberculosis, dysentery, diarrhea, cough, fever, and bronchitis (Kumar et al. 2015). It consists of various phytochemicals such as oxygenated coumarins, gallic acid, flavonoids, and hydroxycinnamic acid derivatives which are responsible for its phytochemical activities (Kumar et al. 2015; Colling et al. 2010).

Kumar et al. (2015) studied the effect of various concentrations of TDZ in combination with picloram on callus induction and evaluated the phenolic composition in *P. sidoides*. Callus induction in *P. sidoides* was high on MS medium supplemented with a combination of TDZ and picloram than control. Hundred percent callus formation was obtained in 0.22 mg/l TDZ + 0.484 mg/l picloram, 0.55 mg/l TDZ + 0.484 mg/l picloram, and 1.1 mg/l TDZ + 0.484 mg/l picloram (Kumar et al. 2015).

P. sidoides contains hydroxycinnamic acid and hydroxybenzoic acid. Picloram-TDZ combination had a significant role in the concentration of hydroxybenzoic derivatives such as P-hydroxybenzoic acid, protocatechuic acid, vanillic acid, and salicylic acid. Highest concentration of this hydroxybenzoic acid was obtained in 0.484 mg/l picloram and 0.55 mg/l TDZ. There was also influence in hydroxycinnamic derivatives by TDZ and picloram. Picloram (0.484 mg/l) and TDZ (0.55 mg/l, 1.1 mg/l) promoted the production of highest concentration caffeic acid, ferulic acid, and coumaric acid. However, catechin was absent in picloram-TDZ-treated plants (Kumar et al. 2015).

15.1.9 *Withania somnifera*

Withania somnifera is a multipurpose medicinal agent belonging to the family Solanaceae. The significant phytochemicals in *W. somnifera* include alkaloids, somniferin, steroidal lactones, and saponins. It can be used to improve blood circulation since it is rich in iron (Umadevi 2012).

MS basal medium devoid of TDZ was not effective to induce morphogenic response and shoot induction from nodal segments of *W. somnifera* (Fatima et al. 2015). Of the various concentrations of TDZ employed along with MS medium, 0.11 mg/l TDZ was most effective and induced highest number of shoots (20 ± 0.40) with an average shoot length of 4.04 ± 0.24 cm after 4 weeks (Fatima et al. 2015). Period of exposure to TDZ was also an important factor in shoot induction. From this study, it is clear that continuous exposure and higher concentration of TDZ resulted in loss of regenerative potential and stunted shoots (Fatima et al. 2015).

15.1.10 *Curculigo latifolia*

Curculigo latifolia (common name, lembah) is a monocotyledonous perennial herb. It is a native of Malaysia, belonging to the family Hypoxidaceae. Yamashita et al.

(1990) discovered and isolated curculin from *C. latifolia*. Curculin is a sweet protein that has both sweet taste and taste-modifying activity. It can change sour taste into sweet taste (Barre et al. 1997). Neoculin is another sweet protein isolated from *C. latifolia* (Koizumi et al. 2007). *C. latifolia* is also a valuable medicinal plant with anticancerous (Ismail et al. 2010) and antidiabetic (Kant 2005) activities. Fadlalla et al. (2007) reported that *C. latifolia* has inhibitory effects on hepatitis B virus.

A micropropagation protocol for *C. latifolia* via shoot regeneration from shoot tip explants using TDZ and IBA has been standardized by Babaei et al. (2014). Shoot tips were surface sterilized and cultured on MS medium containing various concentrations of TDZ (0.5–2.0 mg/l) and IBA (0.25–0.5 mg/l). While in control (MS without PGR), meristem elongation with a single plantlet occurred; the presence of TDZ prevented the apical dormancy and promoted the formation of several axillary and adventitious shoots. Regenerated shoots were developed within 14 weeks of culture. Though varying combination of TDZ- and IBA-induced shoots from shoot tip explants, highest number of shoots (83.3%) and increased shoot length were observed on MS medium containing 0.5 mg/l TDZ and 0.25 mg/l IBA (i.e., lower PGR concentrations). From this it is evident that lower concentrations of TDZ induced shooting.

In a study carried out by Babaei et al. (2014), shoot tip explants were incubated in basal medium supplemented with different concentrations of TDZ (0.5–2.0 mg/l) and IBA (0.25–0.5 mg/l). Scalp (proliferating meristem cultures) induction at the base of the shoot tip explant was found more prominent in high TDZ concentration (2.0 mg/l). IBA had no significant role in inducing scalp, but the synergistic effect of TDZ and IBA had significant role in increasing the size of the scalp. The scalps that were initially white in color turned yellowish once segregated from the explant.

15.1.11 *Hedychium coronarium*

Hedychium coronarium (common name, white ginger lily) is a herbaceous perennial flowering plant of the family Zingiberaceae. Though it is grown as an ornamental plant, it is also an important aromatic medicinal plant. This is an endangered plant that has been red listed (Singh and Singh 2009). It has an underground fleshy rhizome that has several nodes and buds. The rhizome, flower, leaf, and stem produce important alkaloids. *H. coronarium* has antidiabetic, anti-arthritic, and anticancerous activities (Jain et al. 2003a, b). Previous studies have shown the presence of diterpenes – coronarin A, B, C, and D and isocoronarin D, E, and F in *H. coronarium* (Kunnumakkara et al. 2008).

Micropropagation of *H. coronarium* from rhizome bud explants using TDZ has been standardized by Verma and Bansal (2014). Rhizomes were surface sterilized and cut into small pieces each with a single bud. For shoot bud induction, the explants' outer sheaths were removed and placed on MS medium supplemented with various concentrations (0.1–5.0 mg/l) of TDZ, BAP, and KN. Highest number of multiple shoots (14.21 ± 0.09) were obtained on 1.0 mg/l TDZ-containing medium. Explants with multiple shoots were transferred to fresh MS medium with different concentrations of TDZ (0.1–5.0 mg/l), BAP (0.1–5.0 mg/l), and KN (0.1–5.0 mg/l). Maximum shoot elongation (12.89 ± 0.07 cm) was obtained with 1.0 mg/l BAP. Normal rooting was obtained in shoots regenerated from medium containing TDZ, when transferred to rooting medium.

15.1.12 *Mentha arvensis*

Mentha arvensis is a herbaceous perennial plant commonly known as "wild mint" having active principle menthol responsible for dental care. *M. arvensis* showed antimicrobial activity against *Proteus mirabilis*, *Candida albicans*, *Escherichia coli*, and *Staphylococcus aureus* (Mickienė et al. 2011). Dar et al. (2014) demonstrated antioxidant activity of methanolic extract of *M. arvensis* by using iron chelating, radical scavenging, reducing power, nitrous oxide scavenging, and H_2O_2 scavenging methods.

Shoot regeneration capacity of *M. arvensis* from nodal explants in the presence of TDZ has been demonstrated by Faisal et al. (2014). MS medium without the pretreatment with TDZ failed to produce shoots, while the nodal explants pretreated with TDZ in liquid MS medium followed by transfer on PGR-free MS medium showed shoot induction and multiplication. Of the various pretreatment experiments, the explants pretreated with 3.3 mg/l TDZ were the best. Here, the percent shoot formation was 92.6, number of shoots per explants was 23.7, and average length was 5.2 cm (Faisal et al. 2014).

15.1.13 *Clitoria ternatea*

Clitoria ternatea (common name, Asian pigeonwing) is a rare perennial herbaceous plant belonging to the family Fabaceae (Panday et al. 1993). Chemical studies have revealed the presence of various terpenoids, flavanol glycosides, anthocyanins, steroids, and a peptide called clitoides (Mukherjee et al. 2008). In traditional medicine, the plant is used to treat neurological disorders (Jain et al. 2003a, b), stress, and depression and for enhanced memory (Mukherjee et al. 2008). It is also used to cure skin diseases, eye infection, ulcers, and urinary infection (Jain et al. 2003a, b).

Mukhtar et al. (2012) carried out an in vitro performance comparison of cotyledonary node explants and nodal explants of *C. ternatea* at various concentrations of TDZ. Both explants were cultured on MS medium supplemented with various concentrations of TDZ (0.011–0.55 mg/l). Maximum shoot regeneration in cotyledonary node explants was obtained at 0.022 mg/l TDZ, while nodal explants gave the best result with 0.22 mg/l TDZ. Subculturing for 4 weeks in basal MS medium gave root induction in shoots originated from both explants.

15.1.14 *Saussurea involucrata*

Saussurea involucrata (common name, snow lotus) is an endangered Chinese medicinal plant (Guo et al. 2007) that belongs to Asteraceae family. In Chinese traditional medicine, *S. involucrata* is used for the treatment of cough, cold, rheumatoid arthritis, dysmenorrhea, and stomachache and to promote blood circulation (Chik et al. 2015). Studies have demonstrated its anti-inflammatory, analgesic, cardiotonic, anticancerous, abortifacient, and anti-fatigue activities (Jia et al. 2005).

Guo et al. (2012) developed an efficient protocol for in vitro micropropagation of *S. involucrata*. Leaf explants were cultured on MS medium supplemented with different concentrations of TDZ (0.022–4.4 mg/l). Among which, MS medium with 0.11 mg/l TDZ gave maximum number of shoots (15.6 ± 1.4 shoots/explant) within 28 days of culture. In another study, Guo et al. (2016) found that the TDZ-induced shoot morphogenesis is a complex process that involves increased levels of GA_3 and ZT and also the accumulation of H_2O_2 in explants.

15.1.15 *Jatropha curcas*

Jatropha curcas is an important medicinal plant belonging to the Euphorbiaceae family. The leaves, stem, bark, latex, seeds, and roots of *J. curcas* are used to cure various diseases such as malaria, dysentery, wound, eczema, and skin diseases (Prasad et al. 2012). This plant has been used in the treatment for fever, rheumatism, diarrhea, asthma, and piles (Upadhyay et al. 2007). It also has insecticidal, antitumor, and anticancerous activities (Agbogidi et al. 2013).

Kumar and Reddy (2012) developed an efficient method for the in vitro regeneration of plants by direct shoot induction from cotyledonary petiole of *J. curcas*. The highest number of shoot buds (13.76 per explants) and percentage of shoot bud formation (66.97%) were observed in a medium containing 2.0 mg/l TDZ. However, this concentration leads to the inhibition of further shoot proliferation and elongation due to compact shoot induction (Kumar and Reddy 2012). Optimum concentration of TDZ for shoot bud induction (percentage of shoot induction 51.19% and number of shoot buds 9.75) was 0.5 mg/l (Kumar and Reddy 2012). Prolonged TDZ treatment is hazardous to this plant. The extended TDZ treatment caused necrosis of callus from the 6th day onward, and the treatment for a duration of 11 days resulted in the death of callus (Aishwariya et al. 2015). The explant orientation on the medium also plays a crucial role in shoot induction. Percent shoot bud formation ranged from 5.88 to 66.97% in the horizontal position and 5.17–62.97% for vertical position of explants, while the shoot bud number per explant varied from 2.21 to 13.76 for horizontal position and 2.01–13.01 for vertical position of explants in the presence of TDZ with concentration ranged from 0.048 to 2.156 mg/l. Higher percentage of shoot bud induction and higher number of shoot buds were observed in horizontal position (Kumar and Reddy 2012). Therefore, besides TDZ the orientation of explants on medium is also an important factor for shoot induction. Other significant factors which determine the response are source of explants and genotypes. In vitro explants produced higher percentage of shoot induction and higher number of shoots than in vivo explants. The genotype CSMCRI-JC-3 performed well at various concentrations of TDZ than other genotypes (Kumar and Reddy 2012).

15.1.16 *Medicago sativa*

Medicago sativa is a perennial herb belonging to Fabaceae family. It is one of the most popular medicinal plants which is widely used to cure kidney pain and cough, to improve memory, and as antidiabetic, anti-inflammatory, and antimicrobial traditionally (Doss et al. 2011). The major phytochemicals present in this plant include alkaloids, phytoestrogens, coumarins, saponins, phytosterols, and triterpenes (Doss et al. 2011).

The effect of TDZ alone and in combination with $CoCl_2$ (cobalt (II) chloride) on the embryogenic competency of *M. sativa* has been studied by Feng et al. (2012). They cloned full-length ACO (1-aminocyclopropane-1-carboxylate oxidase) cDNA from the *Medicago* callus and analyzed the gene expression during somatic embryogenesis under treatment with TDZ alone or in combination with $CoCl_2$. ACO is an important enzyme in the biosynthesis of ethylene, and $CoCl_2$ is its inhibitor. TDZ promoted the ACO gene expression, while TDZ in combination with $CoCl_2$ reduced ACO expression (Feng et al. 2012). According to Feng et al. (2012), ethylene mediated by ACO could enhance the inhibitory effect of TDZ on the embryogenic competence of *Medicago* callus. Therefore, inhibitory effect of TDZ on embryogenic competence has been mediated by ACO (Feng et al. 2012).

15.1.17 *Crocus sativus*

Crocus sativus (saffron) is a herbaceous medicinal plant with purple-colored flowers, and this plant is normally propagated via corms. The stigma is the medicinally important and most valuable part of this plant. Chemical studies of *C. sativus* revealed the presence of carotenoids like crocetin and its glycosidic forms like glucoside, digentiobioside (crocin), gentiobioside, and diglucoside (Bhargava 2011). Among these, crocetin is mostly responsible for its pharmacological activity (Fernandez and Pandalai 2004). The antibacterial, antifungal, antiseptic, aphrodisiac, antispasmodic, and expectorant properties of stigmas of *C. sativus* had been well documented (Bhargava 2011).

An efficient micropropagation protocol for *C. sativus* from corm explants was standardized by Sharifi et al. (2010). Meristematic region of healthy corms were used as explants. A comparison was made between the effects of BA and TDZ. The explants were first inoculated on MS or B5 medium supplemented with either TDZ (0.25, 1.0, or 2.0 mg/l) or BA (0.5, 2.0, or 4.0 mg/l) at different concentrations. Maximum (100%) callus induction was obtained on MS medium supplemented with 1.0 mg/l TDZ and minimum by 2.0 mg/l BA. When the callus was transferred to TDZ-containing media, shoot buds were formed with an intermediate globular embryo-like callus structure. This study shows that using TDZ alone at a concentration below 2.2 mg/l induces shoots directly from callus, instead of forming somatic embryos (Sharifi et al. 2010).

15.1.18 *Zingiber officinale*

Zingiber officinale commonly known as "ginger" is a tropical, herbaceous medicinal plant belonging to the family Zingiberaceae. The fleshy rhizome is used as spice to flavor food. Ginger is widely used in traditional medicine for the treatment of sore throats, crams, constipation, indigestion, vomiting, fever, arthritis, rheumatism, and helminthiasis (Ali et al. 2008). Pharmacological studies revealed the antioxidant, anti-inflammatory (Ali et al. 2008), antidiabetic, hypolipidemic (Al-Amin et al. 2006), analgesic, and hypoglycemic (Ojewole 2006) activities of *Z. officinale.*

Lincy and Sasikumar (2010) standardized an efficient protocol for direct regeneration of plantlets from aerial stems of *Z. officinale* using TDZ. Two varieties of *Z. officinale*, namely, Jamaica and Varada, were used for the study. Surface-sterilized aerial stem explants were trimmed to 1.0–1.5 cm pieces and transferred to half-strength MS medium containing TDZ alone and in combination with IBA, BAP, and GA_3 at varying concentrations. Among different combinations, maximum shoots and roots were obtained on MS medium supplemented with 1.0 mg/l TDZ and 1.0 mg/l IBA, except for Jamaica, which gave highest number of roots with 1.0 mg/l TDZ and 0.5 mg/l IBA. While Jamaica gave an average number of 11.1 shoots per explants, Varada produced 14.6 shoots per explants. TDZ alone resulted in poor root and shoot regeneration (Lincy and Sasikumar 2010).

15.1.19 *Cannabis sativa*

Cannabis sativa is an annual herb included in Cannabinaceae family originated in Central Asia. The major phytochemical in *C. sativa* is cannabinoids which are responsible for the pharmacological effects. Therefore, *Cannabis* is widely used as medicinal plant and is administrated to patients suffering from rheumatism and epilepsy (Ali et al. 2012). *C. sativa* has been utilized as hallucinogenic, hypnotic, and anti-inflammatory agent in traditional medicine also (Ali et al. 2012).

Thidiazuron is the most effective cytokinin-derived hormone as compared to other cytokinins such as BA and KN for shoot proliferation of *C. sativa* (Lata et al. 2009). TDZ also showed higher efficacy than ZT in shoot proliferation of *C. sativa* from cotyledons (Chaohua et al. 2016). In 0.11 mg/l TDZ, 100% cultures responded with the highest number of shoots (average 13 shoots) per culture (Lata et al. 2009). Even though shoot proliferation was successful, the rooting efficiency was poor in other PGRs. However, Lata et al. (2009) obtained better rooting in TDZ-containing media as compared to BA. In another study 51.7% induction frequency and three shoots per explants were obtained in optimum concentration 0.4 mg/l TDZ and 0.2 mg/l NAA (Chaohua et al. 2016).

15.1.20 *Arnebia euchroma*

Arnebia euchroma is a critically endangered medicinal plant commonly seen in Himalayan region. This plant has been included in the family Boraginaceae. It showed various phytochemical activities such as antimicrobial, anti-inflammatory, antibacterial, antifungal, and wound healing properties (Ashkani-Esfahani et al. 2012; Manjkhola et al. 2005).

To study the effect of TDZ on embryo growth into seedlings of *A. euchroma*, Jiang et al. (2005) excised the embryos and inoculated on medium containing various concentrations of TDZ (0.1–4.0 mg/l). In this system, although TDZ promoted growth and proliferation, the seedlings showed certain abnormalities. The abnormal leaves with lengthened hypocotyls, enlarged cotyledons, and inhibited root growth were the immediate effects of TDZ in the medium (Jiang et al. 2005).

Cotyledon and hypocotyl explants were used for shoot induction. Among the different concentrations of TDZ (0.1–4.0 mg/l), the optimum concentration for the formation of shoots was 1.0 mg/l regardless of explants. As the concentration of TDZ increased above 1.0 mg/l, the number of regenerated shoots decreased. The medium without TDZ did not give good result which indicates that the TDZ has significant role in shoot formation in explants of *A. euchroma*. Comparatively, cotyledonary tissue gave better regeneration capacity than hypocotyl explants (Jiang et al. 2005). The optimum duration for pretreatment with TDZ was 12 days which gave approximately 6.6 shoots per explant. However, the longer exposure was not effective (Jiang et al. 2005). Therefore, the duration of pretreatment should be considered during experiment.

15.1.21 *Azadirachta indica*

In traditional medicine *Azadirachta indica* has significant place and is considered as sacred gift of nature. This plant belongs to the family Meliaceae. *A. indica* shows antibacterial, antiviral, anti-inflammatory, antioxidant, anticarcinogenic activities, and it is also effective in the treatment of sexually transmitted diseases, skin diseases, digestive disorders, and parasitic diseases (Kumar and Navaratnam 2013).

TDZ has significant role in inducing somatic embryogenesis in *A. indica*. Akula et al. (2003) demonstrated somatic embryogenesis in neem by using three vegetative tissues such as root, node, and leaf as explants collected from four selected clones. In their study PGR-free MS medium was suitable for induction of somatic embryos from nodal and root cuttings, while the application of BA or 2, 4-D either singly or in combination inhibited the somatic embryogenesis. TDZ (0.51–0.99 mg/l) and 2, 4-D (0.11 mg/l) have significant role in promoting somatic embryos via callus phase from leaf explants. Approximately 15% somatic embryos from nodal and root explants from PGR-free medium developed into whole plantlets after a series of developmental phases, while somatic embryos from leaf explants showed poor (nearly <5%) conversion rate (Akula et al. 2003).

The induction of somatic embryos from hypocotyls, epicotyls, cotyledons, cotyledonary node, and leaves of seedlings of *A. indica* by using TDZ had been reported (Gairi and Rashid 2004). Reduction in root length and its branching, reduction in leaf size and intensification in its greening, and suppressions of hypocotyls elongation were obtained in low concentration (0.022 mg/l) of TDZ, and more suppression was noticed at a higher concentration (0.11 mg/l TDZ). The length of hypocotyls and epicotyls was strongly reduced at the highest concentration (0.22 mg/l TDZ). Direct differentiation of somatic embryos was high in the presence of 0.22 mg/l TDZ on hypocotyls, epicotyls, cotyledons, cotyledonary node, and leaves of seedlings of *A. indica* (Gairi and Rashid 2004).

References

Agbogidi OM, Akparobi SO, Eruotor PG (2013) Health and environmental benefits of *Jatropha curcas* Linn. Unique Res J Agric Sci 1:76–79

Ai CB, Li LN (1988) Stereo structure of salvianolic acid B and isolation of salvianolic acid C from *Salvia miltiorrhiza*. J Nat Prod 51:145–149

Aishwariya V, Ramrao RK, Kokiladevi E, Arul L, Sudhakar D, Kumar KK, Balasubramanian P (2015) Impact of TDZ (Thidiazuron) pulse treatment in single and multiple shoot formation in calli of *Jatropha curcas* L. Int JAdv Res 3:879–884

Akula C, Akula A, Drew R (2003) Somatic embryogenesis in clonal neem, *Azadirachta indica* A. Juss. and analysis for in vitro azadirachtin production. In Vitro Cell Dev Biol-Plant 39:304–310

Al-Amin ZM, Thomson M, Al-Qattan KK, Peltonen-Shalaby R, Ali M (2006) Anti-diabetic and hypolipidaemic properties of ginger (*Zingiber officinale*) in streptozotocin-induced diabetic rats. Br J Nutr 96:660–666

Ali BH, Blunden G, Tanira MO, Nemmar A (2008) Some phytochemical, pharmacological and toxicological properties of ginger (*Zingiber officinale* Roscoe): a review of recent research. Food Chem Toxicol 46:409–420

Ali EM, Almagboul AZ, Khogali SM, Gergeir UM (2012) Antimicrobial activity of *Cannabis sativa* L. Chin Med 3:61–64

Ansari RA, Aswal BS, Chander R, Dhawan BN, Garg NK, Kapoor NK, Kulshreshtha DK, Mehdi H, Mehrotra BN, Patnaik GK (1988) Hepatoprotective activity of kutkin-the iridoid glycoside mixture of *Picrorhiza kurroa*. Ind J Med Res 87:401–404

Ashkani-Esfahani S, Imanieh MH, Khoshneviszadeh M, Meshksar A, Noorafshan A, Geramizadeh B, Ebrahimi S, Handjani F, Tanideh N (2012) The healing effect of *Arnebia euchroma* in second degree burn wounds in rat as an animal model. Iran Red Crescent Med J 14:70–74

Babaei N, Psyquay Abdullah NA, Saleh G, Lee Abdullah T (2014) An efficient in vitro plantlet regeneration from shoot tip cultures of *Curculigo latifolia*, a medicinal plant. Sci World J2014:1–9

Barre A, Van Damme EJ, Peumans WJ, Rougé P (1997) Curculin, a sweet-tasting and taste-modifying protein, is a non-functional mannose-binding lectin. Plant Mol Biol 33:691–698

Bhargava V (2011) Medicinal uses and pharmacological properties of *Crocus sativus* Linn (Saffron). Int J Pharm Pharma Sci 3:22–26

Casanova E, Valdés AE, Fernández B, Moysset L, Trillas MI (2004) Levels and immunolocalization of endogenous cytokinins in thidiazuron-induced shoot organogenesis in carnation. J Plant Physiol 161:95–104

Chaohua C, Gonggu Z, Lining Z, Chunsheng G, Qing T, Jianhua C, Jianguang S (2016) A rapid shoot regeneration protocol from the cotyledons of hemp (*Cannabis sativa* L.) Ind Crop Prod 83:61–65

Cheruvathur MK, Thomas TD (2014) High frequency multiple shoot induction from nodal segments and rhinacanthin production in the medicinal shrub *Rhinacanthus nasutus* (L.) Kurz. Plant Growth Regul 74:47–54

Cheruvathur MK, Kumar GK, Thomas TD (2013) Somatic embryogenesis and synthetic seed production in *Rhinacanthus nasutus* (L.) Kurz. Plant Cell Tissue Organ Cult 113:63–71

Chik WI, Zhu L, Fan LL, Yi T, Zhu GY, Gou XJ, Yu ZL (2015) *Saussurea involucrata*: a review of the botany, phytochemistry and ethnopharmacology of a rare traditional herbal medicine. J Ethnopharmacol 172:44–60

Colling J, Groenewald JH, Makunga NP (2010) Genetic alterations for increased coumarin production lead to metabolic changes in the medicinally important *Pelargonium sidoides* DC (Geraniaceae). Metabol Eng 12:561–572

Dar MA, Masoodi MH, Wali AF, Mir MA, Shapoo NS (2014) Antioxidant potential of methanol root extract of *Mentha arvensis* L. from Kashmir region. J Appl Pharm Sci 4:50–57

Do Thi N, Hwang ES (2014) Bioactive compound contents and antioxidant activity in Aronia (*Aronia melanocarpa*) leaves collected at different growth stages. Prev Nutr Food Sci 19:204–212

Doss A, Parivuguna V, Vijayasanthi M, Surendran S (2011) Antibacterial evaluation and phytochemical analysis of *Medicago sativa* L. against some microbial pathogens. Indian J Sci Technol 4:550–552

Dungerdorj D, Tsetsegmaa S, Purevsuren S, Bayasgalan B (2006) On the phytochemical investigation of the genus *Gentiana* used for the treatment of liver diseases in Mongolian traditional medicine. Postępy Fitoterapii 2:71–74

Fadlalla AM, Ratnam W, Heng LY (2007) Genetic effects of arsenic and heavy metals pollutants on *Curculigo latifolia* (Lumbah). J Biol Sci 7:1155–1162

Faisal M, Alatar AA, Hegazy AK, Alharbi SA, El-Sheikh M, Okla MK (2014) Thidiazuron induced in vitro multiplication of *Mentha arvensis* and evaluation of genetic stability by flow cytometry and molecular markers. Ind Crop Prod 62:100–106

Fatima N, Ahmad N, Ahmad I, Anis M (2015) Interactive effects of growth regulators, carbon sources, ph on plant regeneration and assessment of genetic fidelity using single primer amplification reaction (SPARS) techniques in *Withania somnifera* L. Appl Biochem Biotechnol 177:118–136

Feng BH, Wu B, Zhang CR, Huang X, Chen YF, Huang XL (2012) Cloning and expression of 1-aminocyclopropane-1-carboxylate oxidase cDNA induced by thidiazuron during somatic embryogenesis of alfalfa (*Medicago sativa*). J Plant Physiol 169:176–182

Fernandez JA, Pandalai SG (2004) Biology, biotechnology and biomedicine of saffron. Rec Res Dev Plant Sci 2:127–159

Gairi A, Rashid A (2004) Direct differentiation of somatic embryos on different regions of intact seedlings of *Azadirachta* in response to thidiazuron. J Plant Physiol 161:1073–1077

Gambhir G, Srivastava DK (2015) Thidiazuron induces high frequency shoot regeneration in leaf and petiole explants of cabbage (*Brassica oleracea* L. Var. Capitata). J Biotechnol Biomater 5:1–6

Gondval M, Chaturvedi P, Gaur AK (2016) Thidiazuron-induced high frequency establishment of callus cultures and plantlet regeneration in *Aconitum balfourii* Stapf.: an endangered medicinal herb of North-West Himalayas. Indian J Biotechnol 15:251–255

Guo B, Gao M, Liu CZ (2007) In vitro propagation of an endangered medicinal plant *Saussurea involucrata* Kar.et Kir. Plant Cell Rep 26:261–265

Guo B, Abbasi BH, Zeb A, Xu LL, Wei YH (2011) Thidiazuron: a multi-dimensional plant growth regulator. Afr J Biotechnol 10:8984–9000

Guo B, Stiles AR, Liu CZ (2012) Thidiazuron enhances shoot organogenesis from leaf explants of *Saussurea involucrata* Kar et. Kir. In Vitro Cell Dev Biol Plant 48:609–612

Guo B, He W, Zhao Y, Wu Y, Fu Y, Guo J, Wei Y (2016) Changes in endogenous hormones and H_2O_2 burst during shoot organogenesis in TDZ-treated *Saussurea involucrata* explants. Plant Cell Tissue Organ Cult 128:1–8

Huetteman CA, Preece JE (1993) Thidiazuron: a potent cytokinin for woody plant tissue culture. Plant Cell Tissue Organ Cult 33:105–119

Ismail MF, Abdullah NAP, Saleh G, Ismail M (2010) Anthesis and flower visitors in *Curculigo latifolia* Dryand (Hypoxidaceae). J Biol Life Sci 1:13–15

Jain MN, Dhal CC, Shroff RH, Bhutada RH, Somani RS, Kasture VS, Kasture SB (2003a) *Clitoria ternatea* and the CNS. Pharmacol Biochem Behav 75:529–536

Jain SP, Singh J, Singh SC (2003b) Rare and endangered medicinal and aromatic plants of Madhya Pradesh. J Econ Tax Bot 27:925–932

Jia JM, Wu CF, Liu W, Yu H, Hao Y, Zheng JH, Ji YR (2005) Anti-inflammatory and analgesic activities of the tissue culture of *Saussurea involucrata*. Biol Pharm Bull 28:1612–1614

Jiang BO, Yang YG, Guo YM, Guo ZC, Chen YZ (2005) Thidiazuron-induced in vitro shoot organogenesis of the medicinal plant *Arnebia euchroma* (Royle) Johnst. In Vitro Cell Dev Biol-Plant 41:677–681

Kant R (2005) Sweet proteins-potential replacement for artificial low calorie sweeteners. Nutrition J4:5

Kletter C, Glasl S, Thalhammer T, Narantuya S (2008) Traditional Mongolian medicine-a potential for drug discovery. Sci Pharm 76:49–63

Kodja H, Noirot M, Khoyratty SS, Limbada H, Verpoorte R, Palama TL (2015) Biochemical characterization of embryogenic calli of *Vanilla planifolia* in response to two years of thidiazuron treatment. Plant Physiol Biochem 96:337–344

Koizumi A, Nakajima KI, Asakura T, Morita Y, Ito K, Shmizu-Ibuka A, Abe K (2007) Taste-modifying sweet protein, neoculin, is received at human T1R3 amino terminal domain. Biochem Biophys Res Commun 358:585–589

Kokotkiewicz A, Jaremicz Z, Luczkiewicz M (2010) *Aronia* plants: a review of traditional use, biological activities and perspectives for modern medicine. J Med Food 13:255–269

Kumar VS, Navaratnam V (2013) Neem (*Azadirachta indica*): prehistory to contemporary medicinal uses to humankind. Asian Pac J Trop Biomed 3:505–514

Kumar N, Reddy MP (2012) Thidiazuron (TDZ) induced plant regeneration from cotyledonary petiole explants of elite genotypes of *Jatropha curcas*: a candidate biodiesel plant. Ind Crop Prod 39:62–68

Kumar V, Moyo M, Gruz J, Šubrtová M, Van Staden J (2015) Phenolic acid profiles and antioxidant potential of *Pelargonium sidoides* callus cultures. Ind Crop Prod 77:402–408

Kunnumakkara AB, Ichikawah AP, Mohan KCJ, Hema PS, Nair MS, Aggarwal BB (2008) Coronarin D, a labdane diterpene, inhibits both constitutive and inducible nuclear factor-B κ pathway activation, leading to potentiation of apoptosis, inhibition of invasion and suppression of osteoclastogenesis. Mol Cancer Ther 7:3308–3316

Lata H, Chandra S, Khan I, ElSohly MA (2009) Thidiazuron-induced high-frequency direct shoot organogenesis of *Cannabis sativa* L. In Vitro Cell Dev Biol Plant 45:12–19

Lincy A, Sasikumar B (2010) Enhanced adventitious shoot regeneration from aerial stem explants of ginger using TDZ and its histological studies. Turk J Bot 34:21–29

Mahendran G, Bai VN (2015) An efficient in vitro propagation, antioxidant and antimicrobial activities of *Aphyllorchis montana* Rchb f. Plant Biosyst 150:1087–1095

Manjkhola S, Dhar U, Joshi M (2005) Organogenesis, embryogenesis and synthetic seed production in *Arnebia euchroma*- a critically endangered medicinal plant of the Himalaya. In Vitro Cell Dev Biol-Plant 41:244–248

McKendrick SL, Leake JR, Taylor DL, Read D (2002) Symbiotic germination and development of the mycoheterotrophic orchid *Neottia nidus*-Avis in nature and its requirement for locally distributed *Sebacina* spp. New Phytol 154:233–247

Mickienė R, Ragažinskienė O, Bakutis B (2011) Antimicrobial activity of *Mentha arvensis* L. and *Zingiber officinale* R. essential oils. Biologija 57:92–97

Mok MC, Mok DWS, Armstrong DJ, Shudo K, Isogai Y, Okamoto T (1982) Cytokinin activity of N-phenyl-N′-1, 2, 3-thiadiazol-5-ylurea (thidiazuron). Phytochemistry 21:1509–1511
Mukherjee PK, Kumar V, Kumar NS, Heinrich M (2008) The Ayurvedic medicine *Clitoria ternatea*-from traditional use to scientific assessment. J Ethnopharmacol 120:291–301
Mukhtar S, Ahmad N, Khan MI, Anis M, Aref IM (2012) Influencing micropropagation in *Clitoria ternatea* L. through the manipulation of TDZ levels and use of different explant types. Physiol Mol Biol Plants 18:381–386
Murch SJ, Saxena PK (1997) Modulation of mineral and free fatty acid profiles during thidiazuron mediated somatic embryogenesis in peanuts (*Arachis hypogaea* L.) J Plant Physiol 151:358–361
Murthy BNS, Murch SJ, Saxena PK (1998) Thidiazuron: a potent regulator of in vitro plant morphogenesis. In Vitro Cell Dev Biol Plant 34:267–275
Myagmar BE, Aniya Y (2000) Free radical scavenging action of medicinal herbs from Mongolia. Phytomedicine 7:221–229
Ojewole JA (2006) Analgesic, anti-inflammatory and hypoglycemic effects of ethanol extract of *Zingiber officinale* (Roscoe) rhizomes (Zingiberaceae) in mice and rats. Phytother Res 20:764–772
Oszmiański J, Wojdylo A (2005) *Aronia melanocarpa* phenolics and their antioxidant activity. Eur Food Res Technol 221:809–813
Pan X, Niu G, Liu H (2001) Microwave-assisted extraction of tanshinones from *Salvia miltiorrhiza* Bunge with analysis by high-performance liquid chromatography. J Chromatogr A 922:371–375
Panday NK, Tewari KC, Tewari RN, Joshi GC, Pande VN, Pandey G (1993) Medicinal plants of Kumaon Himalaya: strategies for conservation. In: Dhar U (ed) Himalayan biodiversity conservation strategies, no. 3. Himavikas, Nanital, pp 293–302
Patel PS, Joshee N, Rimando AM, Parajuli P (2013) Anti-cancer scopes and associated mechanisms of *Scutellaria* extract and flavonoid wogonin. Curr Cancer Ther Rev 9:34–42
Patial V, Sharma M, Bhattacharya A (2016) Potential of thidiazuron in improved micropropagation of *Picrorhiza kurroa* – an endangered medicinal herb of alpine Himalaya. Plant Biosyst 151:729–736
Pentela B, Thalla S, Tharangini K, Geethanjali J, Govinda RT, Venkata Lakshmi D (2012) Hypoglycemic activity of ethanolic extract of *Aphyllorchis montana* on Alloxan induce diabetes in rats. Int J Chem Pharm Sci 3:27–29
Prajapati NS, Purohit SS, Sharma AK, Kumar T (2003) A handbook of medicinal plants. A complete source book. Agrobios, Jodhpur, pp 110–115
Prasad DR, Izam A, Khan MMR (2012) *Jatropha curcas*: plant of medical benefits. J Med Plant Res 6:2691–2699
Qureshi RA, Gilani SA, Ashraf M (2007) Ethnobotanical studies with special reference to plant phenology at Sudhan Gali and Ganga Chotti Hills (District Bagh, AK). Electron J Environ Agric Food Chem 6:2207–2215
Sehgal R, Chauhan A, Gilhotra UK, Gilhotra A (2013) In vitro and in vivo evaluation of anti-asthmatic activity of *Picrorhiza kurroa* plant. Int J Pharm Sci Res 4:3440–3443
Sharifi G, Ebrahimzadeh H, Ghareyazie B, Karimi M (2010) Globular embryo-like structures and highly efficient thidiazuron-induced multiple shoot formation in saffron (*Crocus sativus* L.) In Vitro Cell Dev Biol Plant 46:274–280
Sharma E, Gaur AK (2012) *Aconitum balfourii* Stapf: a rare medicinal herb from Himalayan alpine. J Med Plant Res 6:3810–3817
Siddique I, Anis M (2006) Thidiazuron induced high frequency shoot bud formation and plant regeneration from cotyledonary node explants of *Capsicum annuum* L. Indian J Biotechnol 5:303–308
Singh NR, Singh MS (2009) Wild medicinal plants of Manipur included in the red list. Asian Agrihist 13:221–225

Sinha S, Bhat J, Joshi M, Sinkar V, Ghaskadbi S (2011) Hepatoprotective activity of *Picrorhiza kurroa* Royle Ex. Benth extract against alcohol cytotoxicity in mouse liver slice culture. Int J Green Pharm 5:244–253

Sinu PA, Sinu N, Chandrashekara K (2012) Ecology and population structure of a terrestrial mycoheterotrophic orchid, *Aphyllorchis montana* Rchb. f. (Orchidaceae) in Soppinabetta forests of the Western Ghats, India. J Threat Taxa 4:2915–2919

Sivanesan I, Saini RK, Kim DH (2016) Bioactive compounds in hyperhydric and normal micropropagated shoots of *Aronia melanocarpa* (Michx.) Elliott. Ind Crop Prod 83:31–38

Sreenu T, Venkata Ramana K, Jyothibasu T, Delhiraj N (2013) Hypoglycemic activity of ethanolic extract of *Aphyllorchis montana* induced by streptozocin in rats. Int J Pharm Tech Res 5:320–323

Thomas TD (2003) Thidiazuron induced multiple shoot induction and plant regeneration from cotyledonary explants of mulberry. Biol Plant 46:529–533

Thomas TD, Philip B (2005) Thidiazuron-induced high-frequency shoot organogenesis from leaf-derived callus of a medicinal climber, *Tylophora indica* (Burm. F.) Merrill. In Vitro Cell Dev Biol Plant 41:124–128

Thomas TD, Puthur JT (2004) Thidiazuron induced high frequency shoot organogenesis in callus from *Kigelia pinnata* L. Bot Bull Acad Sinica 45:307–313

Tomiczak K, Mikuła A, Sliwinska E, Rybczyński JJ (2015) Autotetraploid plant regeneration by indirect somatic embryogenesis from leaf mesophyll protoplasts of diploid *Gentiana decumbens* Lf. In Vitro Cell Dev Biol Plant 51:350–359

Tsai KL, Chen EG, Chen JT (2016) Thidiazuron-induced efficient propagation of *Salvia miltiorrhiza* through in vitro organogenesis and medicinal constituents of regenerated plants. Acta Physiol Plant 38:29

Turaskar A, More S, Sheikh R, Gadhpayle J, Bongade SI (2013) Inhibitory potential of *Picrorhiza kurroa* Royle Ex. Benth extracts on phenylhydrazine induced reticulocytosis in rats. Asian J Pharm Clin Res 6:215–216

Umadevi M (2012) Traditional and medicinal uses of *Withania somnifera*. The Pharma Innov 1:102–110

Upadhyay B, Roy S, Kumar A (2007) Traditional uses of medicinal plants among the rural communities of Churu district in the Thar Desert, India. J Ethnopharmacol 113:387–399

Vaidya BN, Jackson CL, Perry ZD, Dhekney SA, Joshee N (2016) Agrobacterium-mediated transformation of thin cell layer explants of *Scutellaria ocmulgee* small: a rare plant with anti-tumor properties. Plant Cell Tissue Organ Cult 127:57–69

Van Waes JM, Debergh PC (1986) In vitro germination of some western European orchids. Physiol Plant 67:253–261

Verma M, Bansal YK (2014) Effect of a potent cytokinin thidiazuron (TDZ) on in vitro regeneration of *Hedychium coronarium* J Koenig-A valuable medicinal plant. Int J Rec Biotech 2:38–44

Wang J, Xiong X, Feng B (2013) Cardiovascular effects of salvianolic acid B. Evid Based Complement Alternat Med 2013:1–16

Warcup JH (1973) Symbiotic germination of some Australian terrestrial orchids. New Phytol 72:387–392

Yamashita H, Theerasilp S, Aiuchi T, Nakaya K, Nakamura Y, Kurihara Y (1990) Purification and complete amino acid sequence of a new type of sweet protein with taste-modifying activity, curculin. J Biolchem 265:15770–15775

Factors Affecting Thidiazuron-Induced Direct Somatic Embryogenesis of *Phalaenopsis aphrodite*

16

Wee-Peng Gow, Hsiao-Hang Chung, Jen-Tsung Chen, and Wei-Chin Chang

Abstract

The effects of NaH_2PO_4, sucrose, activated charcoal, polyvinylpyrrolidone (PVP), and strength of MS medium were studied to optimize thidiazuron (TDZ)-induced direct somatic embryogenesis from leaf explants of *Phalaenopsis aphrodite* subsp. *formosana*. The results showed that full- and quarter-strength macroelements of MS medium were not suitable for direct embryo induction from leaf explants. Thus, a half-strength macroelement and full-strength microelements of MS nutrients plus full-strength of MS vitamins, 170 mg l^{-1} NaH_2PO_4, 1 g l^{-1} peptone, 3 mg l^{-1} TDZ, and 20 g l^{-1} sucrose are proposed as a suitably modified medium. In addition, PVP at 0.25 g l^{-1} significantly promoted direct embryogenesis on the cut ends of the explants, but activated charcoal at 0.5–1 g l^{-1} was inhibitory.

Keywords

Activated charcoal · Direct somatic embryogenesis · Embryogenic competence · Polyvinylpyrrolidone

W.-P. Gow · W.-C. Chang
Institute of Plant and Microbial Biology, Academia Sinica, Taipei, Taiwan, Republic of China

H.-H. Chung
Department of Horticulture, National Ilan University, Yilan City, Taiwan, Republic of China

J.-T. Chen (✉)
Department of Life Sciences, National University of Kaohsiung, Kaohsiung, Taiwan, Republic of China
e-mail: jentsung@nuk.edu.tw

N. Ahmad, M. Faisal (eds.), *Thidiazuron: From Urea Derivative to Plant Growth Regulator*, https://doi.org/10.1007/978-981-10-8004-3_16

16.1 Introduction

Phalaenopsis orchids are popular in international flower markets and have high commercial value as cut flower and potted plant production. Conventional in vitro culture protocols had been developed for propagation of this genus mainly via protocorm-like body formation, shoot multiplication, and callus culture (Tanaka et al. 1975; Arditti and Ernst 1993; Tokuhara and Mii 1993, 2001; Ernst 1994; Chen and Piluek 1995: Duan et al. 1996; Ishii et al. 1998: Islam and Ichihashi 1999: Chen et al. 2000: Park et al. 2000, 2002). Recently, more efficient regeneration systems through direct somatic embryogenesis had been developed using leaf cultures (Kuo et al. 2005; Chen and Chang 2006; Gow et al. 2008, 2009). However, further systematic investigations on medium composition and physiological status are needed to optimize the protocol for practical use in regenerating transgenic plants or mass propagation of this orchid. The aim of this present report is to study the effects of NaH_2PO_4, sucrose and strength of MS medium, activated charcoal, and polyvinylpyrrolidone on direct somatic embryogenesis using the leaf culture system of *Phalaenopsis aphrodite.*

16.2 Materials and Methods

16.2.1 Plant Materials

In vitro grown seedlings of *Phalaenopsis aphrodite* Rchb.f. subsp. *formosana* Christenson (formerly also referred to as *Phalaenopsis amabilis*) were purchased from Taiwan Sugar Corporation (TSC), Chiayi, Taiwan. The plants were maintained on a plant growth regulator (PGR)-free half-strength MS (Murashige and Skoog 1962) medium in 250 ml flasks with a 2-month-interval subculture period and for two times of subculture. All of the cultures were incubated under a 16/8-h (light/dark) photoperiod at photosynthetic photon flux density of 32 $\mu molm^{-2}\ s^{-1}$ (daylight fluorescent tubes FL-30D/29, 40 W, China Electric Co., Taipei, Taiwan) and temperature of 26 ± 1 °C. The seedlings with three to five leaves and two to four roots were used as donor plants.

16.2.2 Somatic Embryo Induction (in Darkness)

The basal medium for somatic embryo induction was a modified MS medium containing half-strength macroelements and full-strength microelements and supplemented with [mg l^{-1}]: myoinositol (100), niacin (0.5), pyridoxine HCl (0.5), thiamine HCl (0.1), glycine (2.0), peptone (1000), NaH_2PO_4 (170), sucrose (20,000), thidiazuron (3.0), and Gelrite (2500). The pH of variants of the medium was adjusted to 5.2 with 1M KOH or HCl prior to autoclaving at 121 °C for 15 min. Leaf tip segments (about 1 cm in length) taken from the donor plants were used to induce direct somatic embryogenesis on different variants of the medium. The leaf explants were placed adaxial side up on the culture medium and were incubated in 90×15 mm^2 Petri dishes under darkness for 2 months in an incubator at temperature of 26±1 °C. Modification of medium composition including sucrose (0, 10, 20, 30, and 40 g l^{-1}), NaH_2PO_4 (0, 42.5, 85, and 170 mg l^{-1}), MS medium strength (full-strength macro- and microelements, half-strength macroelements and full-strength microelements as half-strength, and quarter-strength macroelements and full-strength microelements as quarter-strength), activated charcoal (0, 0.5, 1.0, and 2 g l^{-1}), and polyvinylpyrrolidone (PVP; 0, 0.1, 0.25, and 0.5 g l^{-1}) was used to test their effects on direct somatic embryo formation.

16.2.3 Somatic Embryo Development (in Light Condition)

Leaf-derived embryos were transferred onto a PGRs-free half-strength MS medium in 250 ml flasks under a light condition with 16/8-h (light/dark) photoperiod at photosynthetic photon flux density of 32 μmolm^{-2} s^{-1} and temperature of 26±1 °C for 45 days.

16.2.4 Histological Analysis

Tissues for histological observations were fixed in FAA (95% ethyl alcohol, glacial acetic acid, formaldehyde, water, 10: 1: 2: 7), dehydrated in a tertiary-butyl-alcohol series, embedded in paraffin wax, sectioned at 10 μm thickness, and stained with 0.5% safranin-O and 0.1% fast green (Jensen 1962).

16.2.5 Scanning Electron Microscopy (EM) Observations

Samples for scanning EM were fixed in 2.5% glutaraldehyde in 0.1 M phosphate buffer (pH 7.0) for 4 h at 4 °C and then dehydrated in ethanol (Dawns 1971), dried using a critical point dryer (HCP-2, Hitachi), and coated with gold in an ion coater (IB-2, Giko Engineering Co.). A scanning EM (DSM-950, Carl Zeiss) was used for examination and photography of the samples.

16.2.6 Data Analysis

The percentage of explants forming somatic embryos was recorded as those formed from entire explants or different parts of the explants (LT, leaf tips; Ad, adaxial sides; Ab, abaxial sides; CE, cut ends). The number of embryos formed from each responding explant was counted under a stereomicroscope (SZH, Olympus, Tokyo, Japan) at the protocorm stage. Data were scored after 60 days of culture. Five replicates (dishes) each with four leaf explants were provided for each treatment. The data expressed as percentages were transformed using arc sine prior to ANOVA and then converted back to the original scale (Compton 1994). All means were compared by following Duncan's multiple range test (Duncan 1955). Significant differences between means were presented at the level of $p \leq 0.05$.

16.3 Results and Discussion

16.3.1 The Morphogenetic Pathway of Direct Embryogenesis

When leaf explants of *P. aphrodite* were cultured on TDZ-containing half-strength MS medium supplemented with 20 g l^{-1} sucrose, pale yellow-green globular embryos were obtained after 45 days of culture in darkness (Table 16.1; Fig. 16.1a). These embryos subsequently turned green, enlarged, and developed scale leaves 45 days after transfer onto PGR-free half-strength MS medium in light condition (Fig. 16.1b). The histological study revealed that the epidermal cells had undergone a process of dedifferentiation and gained mitotic ability to form meristematic cells (Fig. 16.2a). Subsequently, the meristematic cells gave rise to form somatic embryos without the intervening of callus tissues (Fig. 16.2b). These leaf-derived embryos developed and consist of scale leaves and the vascular tissue on the parent explants (Fig. 16.2c). Scanning EM observation revealed that the subepidermal cells were also able to divide into meristematic cells, thus forming protuberances through the epidermis (Fig. 16.3a). Single-state embryos formed on surfaces of explants with scattering dedifferentiated leaf cells (Fig. 16.3b). When a mass of leaf cells were induced to dedifferentiate, it became easier to form multiple-state of embryos (Fig. 16.3c). In addition, asynchronous formation of embryos was frequently found on the explants (Fig. 16.3c). The foliar embryos had the ability to form secondary embryos from their anterior end when they were still on the parent explants (Fig. 16.3d).

Table 16.1 Effect of sucrose on direct somatic embryogenesis from leaf explants of *P. aphrodite*

Sucrose (g l^{-1})	% of explant with embryogenesis	% of explant with browning	% of each part of explant with embryogenesis				No. of embryos per responding explant
			CE	Ad	Ab	LT	
0	0 d	65 a	0 c	0 b	0 b	0 b	0
10	40 ab	35 ab	25 b	30 a	0 b	5 b	7.5
20	65 a	15 b	55 a	25 a	15 a	40 a	7.8
30	10 cd	50 a	0 c	10 ab	0 b	0 b	10.5
40	5 bc	65 a	10 c	10 ab	0 b	0 b	11.0

Means within a column followed by the same letter are not significantly different according Duncan's multiple range test at $P \leq 0.05$ (Duncan 1955)

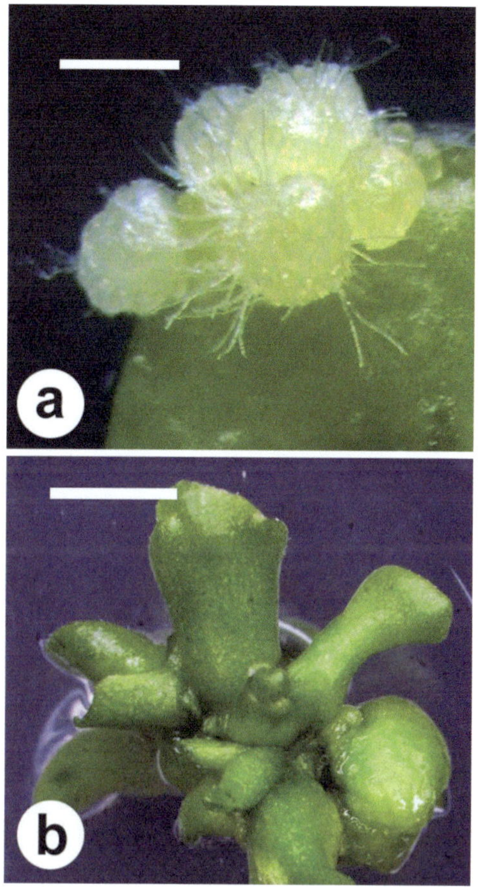

Fig. 16.1 Direct somatic embryogenesis from leaf explants of *P. aphrodite*. (**a**) Somatic embryos with absorbing hairs formed on a leaf explant after 45 days of culture on half-strength MS medium with 3 mg l^{-1} TDZ in darkness (scale bar = 1.5 mm). (**b**) Green, enlarged embryos with developing leaves after 45 days of culture after transfer the somatic embryos shown in (A) to PGR-free half-strength MS medium in light (scale bar = 4 mm)

16.3.2 Effect of Sucrose

In *Oncidium* orchid tissue culture, concentration of sucrose significantly affected somatic embryogenesis from leaf explants (Chen and Chang 2002, Su et al. 2006). Without sucrose, leaf explants of *P. aphrodite* failed to form embryos with necrosis after 2 months of culture on TDZ-containing medium (Table 16.1). Sucrose at 20 g l^{-1} gave the most suitable results with highest percentage of explants with embryogenesis from the entire explant and lowest percentage of explants with browning (Table 16.1). Except for the adaxial side, sucrose at 20 g l^{-1} gave significantly higher percentage of explants with embryogenesis when compared with other concentrations on the leaf parts (Table 16.1). Higher concentrations of sucrose resulted in lower embryogenic responses, higher browning rates, but higher number of embryos per responding explant (Table 16.1).

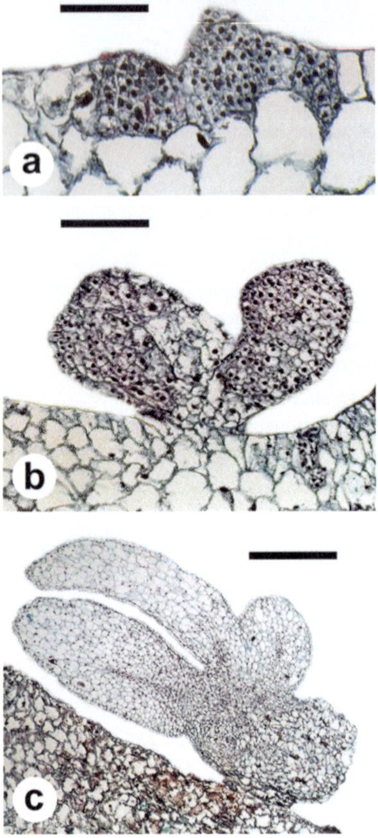

Fig. 16.2 Histology of direct somatic embryogenesis from leaf explants of *P. aphrodite*. (**a**) Embryogenic cells originated from the epidermal layer of a leaf explant after 30 days of culture on half-strength MS medium with 3 mg l^{-1} TDZ in darkness (scale bar 350 μm). (**b**) Globular embryos formed after 40 days of culture on half-strength MS medium with 3 mg l^{-1} TDZ in darkness (scale bar = 500 μm). (**c**) An embryo developed vascular tissues after 30 days of culture after transfer the somatic embryos shown in Fig. 16.1a to a PGR-free half-strength MS medium in light (scale bar = 1.5 mm)

16.3.3 Effect of NaH_2PO_4

Phosphate plays an important role in plant growth and development, and the process of somatic embryogenesis may be greatly influenced by phosphate (Pedroso and Pais 1995). NaH_2PO_4 was usually supplemented in media as an additive phosphate source for in vitro culture in *Dendrobium*, *Epidendrum*, *Oncidium*, and *Paphiopedilum* (Chen et al. 1999, 2000, 2002, 2004; Chung et al. 2005, 2007). In *Oncidium*, NaH_2PO_4 was found to be effective in induction of direct embryogenesis from leaf cultures, and the optimal concentrations was between 85 and 170 mg l^{-1} (Chen and Chang 2002). NaH_2PO_4 at 170 mg l^{-1} resulted in the highest efficiency of direct embryogenesis with 65% of explants forming an average of 7.8 embryos per responding explants. By contrast, other concentrations of NaH_2PO_4 had no significant effects on direct embryogenesis. Except for adaxial and abaxial sides, NaH_2PO_4 at 170 mg l^{-1} gave significantly higher percentage of explants with embryogenesis when compared with other concentrations on the leaf locations (Table 16.2).

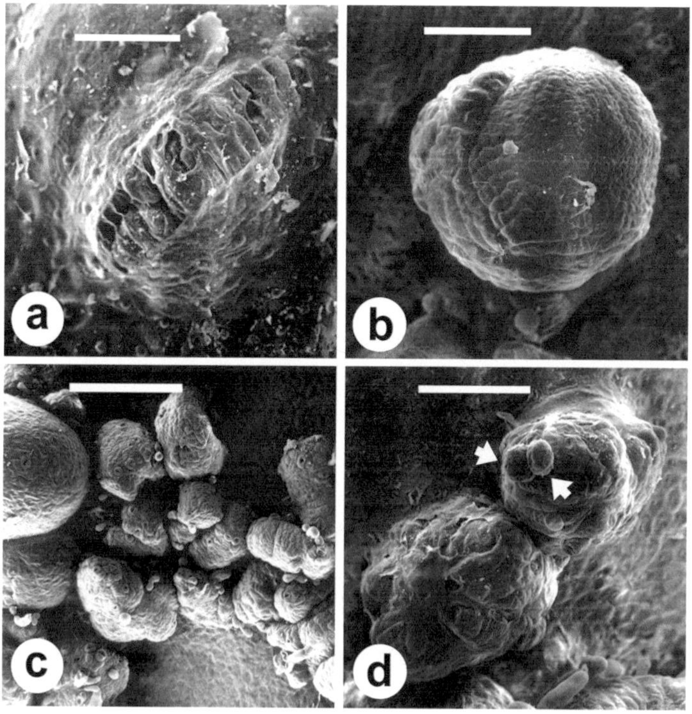

Fig. 16.3 Scanning electron microscopic observation on direct somatic embryogenesis from leaf explants of *P. aphrodite*. (**a**) An early event of direct embryogenesis from subepidermal cells (scale bar = 100 μm). (**b**) A globular embryo (scale bar = 200 μm). (**c**) Embryos with scale leaves formed on a leaf explant (scale bar = 350 μm). (**d**) Secondary embryos (arrow) formed on a primary embryo (scale bar = 200 μm)

Table 16.2 Effect of NaH_2PO_4 on direct somatic embryogenesis from leaf explants of *P. aphrodite*

NaH_2PO_4 (mg l^{-1})	% of explant with embryogenesis	% of explant with browning	% of each part of explant with embryogenesis				No. of embryos per responding explant
			CE	Ad	Ab	LT	
0	10 b	50 a	0 b	10 a	5 ab	5 b	13.5
42.5	25 b	30 a	10 b	25 a	0 b	10 b	13.5
85	20 b	30 a	10 b	10 a	0 b	10 b	4.3
170	65 a	15 a	55 a	25 a	15 a	40 a	7.8

Means within a column followed by the same letter are not significantly different according Duncan's multiple range test at $P \leq 0.05$ (Duncan 1955)

Table 16.3 Effect of MS medium strength on direct somatic embryogenesis from leaf explants of *P. aphrodite*

Strength of MS medium	% of explant with embryogenesis	% of explant with browning	% of each part of explant with embryogenesis				No. of embryos per responding explant
			CE	Ad	Ab	LT	
Full	30 b	40 a	15 b	25 a	0 b	0 b	5.2
1/2	65 a	15 a	55 a	5 ab	15 a	40 a	7.8
1/4	20 b	40 a	20 b	0 b	0 b	0 b	10.0

Means within a column followed by the same letter are not significantly different according Duncan's multiple range test at $P \leq 0.05$ (Duncan 1955)

16.3.4 Effect of Medium Strength

High concentration of nutrients in culture medium seems did not favor in vitro culture of *Phalaenopsis* (Arditti and Ernst 1993). In the present study, the result showed that half-strength MS medium was the most suitable one for induction of direct embryo formation from leaf explants of *P. aphrodite* (Table 16.3). Both full-strength and quarter-strength MS gave lower embryogenic responses and higher browning rates (Table 16.3). Indeed, half-strength MS medium was used as basal medium for in vitro culture of *Dendrobium*, *Epidendrum*, and *Oncidium* (Chen et al. 1999, 2000, 2002; Chung et al. 2005, 2007). Except for the adaxial side, half-strength MS medium gave significantly higher percentage of explants with embryogenesis when compared with other strength on the leaf locations (Table 16.3).

16.3.5 Effect of Activated Charcoal

Activated charcoal was usually used in conventional in vitro culture medium of *Phalaenopsis* to reduce the toxic effect of phenolic compounds secreted by explants (Arditti and Ernst 1993). However, in the present study, the application of activated charcoal gave a negative effect on direct embryo induction from leaf explants of *P. aphrodite* (Table 16.4). Activated charcoal doses of 0.5, 1.0, and 2.0 g l^{-1} were all totally inhibitory and likely could be related to the obtained explant browning rates between 55% and 80% (Table 16.4). Suggestion is that the activated charcoal may absorb TDZ or reduce its activity to induce embryogenesis.

16.3.6 Effect of Polyvinylpyrrolidone

Polyvinylpyrrolidone (PVP) is soluble in water and binds to polar molecules exceptionally well, owing to its polarity. In plant tissue culture media, PVP adsorb not only toxic exudates (phenolics) but also growth regulators and nutrients (Bhat and

Table 16.4 Effect of activated charcoal on direct somatic embryogenesis from leaf explants of *P. aphrodite*

Activated charcoal (g l^{-1})	% of explant with embryogenesis	% of explant with browning	% of each part of explant with embryogenesis				No. of embryos per responding explant
			CE	Ad	Ab	LT	
0	65 a	15 b	55 a	25 a	15 a	40 a	7.8
0.5	0 b	55 ab	0 b	0 b	0 b	0 b	0
1	0 b	70 a	0 b	0 b	0 b	0 b	0
2	0 b	85 a	0 b	0 b	0 b	0 b	0

Means within a column followed by the same letter are not significantly different according Duncan's multiple range test at $P \leq 0.05$ (Duncan 1955)

Table 16.5 Effect of PVP on direct somatic embryogenesis from leaf explants of *P. aphrodite*

PVP (g l^{-1})	% of explant with embryogenesis	% of each part of explant with embryogenesis				No. of embryos per responding explant
		CE	Ad	Ab	LT	
0	45 a	25 b	20 a	0 a	25 a	7.8
0.1	50 a	40 ab	30 a	5 a	25 a	10.7
0.25	60 a	60 a	35 a	10 a	45 a	15.5
0.5	50 a	50 a	25 a	5 a	25 a	10.6

Means within a column followed by the same letter are not significantly different according Duncan's multiple range test at $P \leq 0.05$ (Duncan 1955)

Chandel 1991). In the present study, the use of PVP significantly enhanced direct embryo induction from cut ends of explants (Table 16.5). In *Dioscorea alata* L., the exudate from the cut end of the explant was responsible for browning of the culture medium (Bhat and Chandel 1991). Therefore, the suggestion is that PVP may absorb the toxic exudate(s) from cut ends and this way promoted the somatic embryogenesis. According to experimental results, a suitable concentration of PVP would be 0.25 g l^{-1} (Table 16.5).

According to the present results, a modified MS medium with 1/2-strength macroelements, full-strength microelements and vitamins, 170 mg l^{-1} NaH_2PO_4, 0.25 g l^{-1} PVP, and 20 g l^{-1} sucrose could be proposed as a suitable medium for direct somatic embryogenesis in *Phalaenopsis aphrodite* subsp. *formosana*.

References

Arditti J, Ernst R (1993) Micropropagation of orchids, vol 2. Wiley, New York. 467–520pp

Bhat SR, Chandel KPS (1991) A novel technique to overcome browning in tissue culture. Plant Cell Rep 10:358–361

Chen JT, Chang WC (2002) Effects of tissue culture conditions and explant characteristics on direct somatic embryogenesis in *Oncidium* 'Gower Ramsey'. Plant Cell Tissue Organ Cult 69:41–44

Chen JT, Chang WC (2006) Direct somatic embryogenesis and plant regeneration from leaf explants of *Phalaenopsis amabilis*. Biol Plant 50:169–173

Chen Y, Piluek C (1995) Effects of thidiazuron and N^6-benzylaminopurine on shoot regeneration of *Phalaenopsis*. Plant Growth Regul 16:99–101

Chen JT, Chang C, Chang WC (1999) Direct somatic embryogenesis on leaf explants of *Oncidium* 'Gower Ramsey' and subsequent plant regeneration. Plant Cell Rep 19:143–149

Chen YC, Chang C, Chang WC (2000) A reliable protocol for plant regeneration from callus culture of *Phalaenopsis*. In Vitro Cell Dev Biol-Plant 36:420–423

Chen LR, Chen JT, Chang WC (2002) Efficient production of protocorm-like bodies and plant regeneration from flower stalk explants of the sympodial orchid *Epidendrum radicans*. In Vitro Cell Dev Biol-Plant 38:441–445

Chen TY, Chen JT, Chang WC (2004) Plant regeneration through shoot bud formation from leaf explants of *Paphiopedilum* orchids. Plant Cell Tissue Organ Cult 76:11–15

Chung HH, Chen JT, Chang WC (2005) Cytokinins induce direct somatic embryogenesis of *Dendrobium* Chiengmai Pink and subsequent plant regeneration. In Vitro Cell Dev Biol-Plant 41:765–769

Chung HH, Chen JT, Chang WC (2007) Plant regeneration through direct somatic embryogenesis from leaf explants of *Dendrobium*. Biol Plant 51:346–350

Compton ME (1994) Statistical methods suitable for the analysis of plant tissue culture data. Plant Cell Tissue Organ Cult 37:217–242

Dawns CJ (1971) Biological techniques in electron microscopy. Barnes and Noble, New York, 193pp

Duan JX, Chen H, Yazawa S (1996) In vitro propagation of *Phalaenopsis* via culture of cytokinins-induced nodes. J Plant Growth Regul 15:133–137

Duncan DB (1955) Multiple range and multiple F test. Biometrics 11:1–42

Ernst R (1994) Effects of thidiazuron on in vitro propagation of *Phalaenopsis* and *Doritaenopsis* (Orchidaceae). Plant Cell Tissue Organ Cult 39:273–275

Gow WP, Chen JT, Chang WC (2008) Influence of growth regulators on direct embryo formation from leaf explants of *Phalaenopsis* orchids. Acta Physiol Plant 30:507–512

Gow WP, Chen JT, Chang WC (2009) Effects of genotype, light regime, explant position and orientation on direct embryo formation from leaf explants of *Phalaenopsis* orchids. Acta Physiol Plant 31:363–369

Ishii Y, Takamura T, Goi M, Tanaka M (1998) Callus induction and somatic embryogenesis of *Phalaenopsis*. Plant Cell Rep 17:446–450

Islam MO, Ichihashi S (1999) Effects of sucrose, maltose and sorbitol on callus growth of *Phalaenopsis*, *Doritaenopsis* and *Neofinetia*. J Jpn Soc Hortic Sci 68:1124–1131

Jensen WA (1962) Botanical histochemistry. Freeman, San Francisco. 408pp

Kuo HL, Chen JT, Chang WC (2005) Efficient plant regeneration through direct somatic embryogenesis from leaf explants of *Phalaenopsis* 'Little Steve'. In Vitro Cell Dev Biol-Plant 41:453–456

Murashige T, Skoog F (1962) A revised medium for rapid growth and bioassays with tobacco tissue cultures. Physiol Plant 15:495–497

Park SY, Murthy HN, Paek KY (2000) Mass multiplication of protocorm-like bodies using bioreactor system and subsequent plant regeneration in *Phalaenopsis*. Plant Cell Tissue Organ Cult 63:67–72

Park SY, Murthy HN, Paek KY (2002) Rapid propagation of *Phalaenopsis* from floral stalk-derived leaves. In Vitro Cell Dev Biol-Plant 38:168–172

Pedroso MC, Pais MS (1995) Factors controlling somatic embryogenesis. Plant Cell Tissue Organ Cult 43:147–154

Su YJ, Chen JT, Chang WC (2006) Efficient and repetitive production of leaf-derived embryos of *Oncidium*. Biol Plant 50:107–110

Tanaka M, Hasegawa A, Goi M (1975) Studies on the clonal propagation of monopodial orchids by tissue culture. I. Formation of protocorm-like bodies from leaf tissues in *Phalaenopsis* and *Vanda*. J Jpn Soc Hortic Sci 44:47–58

Tokuhara K, Mii M (1993) Micropropagation of *Phalaenopsis* and *Doritaenopsis* by culturing shoot tips of flower stalk buds. Plant Cell Rep 13:7–11

Tokuhara K, Mii M (2001) Induction of embryogenic callus and cell suspension culture from shoot tips excised from flower stalk buds of *Phalaenopsis* (Orchidaceae). In Vitro Cell Dev Biol-Plant 37:457–461

TDZ Induced Diverse In Vitro Responses in Some Economically Important Plants

17

Parvatam Giridhar, Sridevi Vaddadi, Pradeep Matam, and S. V. Shreelakshmi

Abstract

A fleeting look at the diverse responses exhibited by a variety of plants against a urea derived, multifunctional plant growth regulator thidiazuron (*N*-phenyl-*N*′-1,2,3-thidiazol-5-ylurea (TDZ)) reveals its significance in plant tissue culture. Few decades ago this substituted phenylurea compound was in use rarely for inducing tissue culture responses in woody plant species, but later became one of the most sought-after plant growth regulators for induction of in vitro responses even in non-woody plant species. One notable feature of this growth regulator is its cytokinin- and auxin-type responses reported in excised tissues of plants, which is otherwise familiar as a defoliant for the mechanical harvest of cotton bolls. Often morphogenetic responses such as organogenesis via direct or indirect way, somatic embryogenesis, callus cell culture proliferations, etc. and physiological changes at cellular level were demonstrated upon administering TDZ in vitro. Various biochemical core mechanisms for such triggered responses of TDZ have been reviewed. Under this context, a glance at TDZ induced in vitro responses in a woody plant species, such as coffee, vanilla orchid, annatto dye-yielding tropical plant *Bixa orellana*, banana, a natural non-calorie sweetener *Stevia rebaudiana* and natural flavour 2H4MB producing *Decalepis hamiltonii*, will evoke interest among researchers in this area, as the outcome facilitates further advances in this area and also helpful for better utilization of TDZ for biotechnological improvement of economically important plant species.

Keywords

Callus · Cell proliferation · Embryogenesis · Plant regeneration · Thidiazuron

P. Giridhar (✉) · S. Vaddadi · P. Matam · S. V. Shreelakshmi
Plant Cell Biotechnology Department, CSIR-Central Food Technological Research Institute, Mysore 570020, India
e-mail: giridharp@cftri.res.in

N. Ahmad, M. Faisal (eds.), *Thidiazuron: From Urea Derivative to Plant Growth Regulator*, https://doi.org/10.1007/978-981-10-8004-3_17

17.1 Introduction

Thidiazuron (TDZ) is a heterocyclic substituted phenylurea compound initially employed as a cotton defoliant (Arndt et al. 1976) and later emerged as one of the potential plant growth regulators. In view of its intrinsic cytokinin-like activity, sometimes it works better than adenine-type cytokinins (Mok et al. 1982). An arrays of biological responses were reported in response to TDZ administration to different plant species, which include physiological and biochemical changes.

It is opined that TDZ promotes the conversion of cytokinin ribonucleotides to the biologically more active ribonucleosides (Capelle et al. 1983). Various other functions of TDZ include the following: it indirectly influences endogenous levels of growth hormones in treated plant cells (Suttle 1985); is able to stimulate cell division, cotyledonary tissue expansion and the synthesis of endogenous purine cytokinins; or inhibits their degradation (Thomas and Katterman 1986).

The role of TDZ in plant morphogenesis, mechanism of action, influence of other hormones, involvement of enzymes and ions, etc. is reviewed (Guo et al. 2011). TDZ is found to provoke a stress response in tissues upon concentration-dependent exposure. This causes a change in stress-associated metabolites such as proline, abscisic acid and 4-aminobutyrate and associated storage of numerous mineral ions (Murch and Saxena 1997). Since TDZ has been reported to induce micropropagation of various plant species, suggesting the possibility of a novel mode of its action by modulation of endogenous growth regulators in somatic embryos and callus tissue, in this article we highlight the potential influence of TDZ on in vitro propagation methods of some economically important plants.

17.2 Banana

Organized growth of banana tissue in vitro is limited to embryo culture and shoot tip culture, especially embryo culture is an important aid for classical breeding in banana since the germination frequency of seed is extremely low. A wide range of cytokinins [BA, 2iP, purine, zeatin and kinetin] including TDZ were explored for their influence on shoot tips freshly excised from in vitro rooted plantlets. Among these, TDZ was reported to outperform BA, kinetin, zeatin and 2iP even at low concentrations (Arinaitwe et al. 2000), while at relatively higher concentrations (above 8 μM), it induces scalps in bananas (Sadik et al. 2007). The use of MS medium with 1 μM TDZ and 1 μM IAA followed for shoot multiplication from excised shoot tips of *Musa* spp., followed by rooting with activated charcoal (AC), was investigated (Gubbuk and Pekmezci 2006). Multiplication rate of shoots decreases after few subculturing cycles. The addition of paclobutrazol (PP333) to TDZ-containing medium resulted in further increase of proliferation of buds when compared with the medium containing only TDZ (Lee 2005). Similarly, an efficient medium culture for clonal mass propagation was established for the propagation of two banana (*M. acuminate* L.) cultivars Cavendish Dwarf and Valery using the MS medium in combination with 0.5 mg/L TDZ and 2 mg/L IAA and also somaclonal variation was reported (Farahani et al. 2008; Sheidai et al. 2010). Application of more active cytokinins with carry-over effect such as TDZ sustained bud proliferation in bananas for some time (Makara et al. 2010). Direct shoot and cormlet regeneration was reported from whole male inflorescence on MS medium supplemented with optimal levels of BA, TDZ and KIN produced the maximum number of shoots in the presence of varying concentrations of TDZ (0.45–13.5 μM) irrespective of their genotypes was reported for the four cultivars, namely, *M. acuminata* cv. Matti, *M. acuminata* cv. Sannachenkadali, *M. acuminata* cv. Chingan and *M. acuminata* cv. Njalipoovan also shoot proliferation in *M. cavendishii* (Smitha et al. 2014, 2015; Wijerathna and Kumarihami 2016). But elongation of shoots was inhibited at higher concentration of TDZ (9.1 μM), and clumps of small globular buds were reported at the base of shoots and genetic variation (Shirani et al. 2009, 2010). Anuradha et al. (2014) opined that BA alone was not sufficient to induce high proliferation rates and inclusion of TDZ was essential for shoot recovery in five wild species of *Musa* (*M. acuminata, M. balbisiana, M. basjoo, M. jackeyi* and *M. textilis*) and their subsequent cryopreservation. Theodosy and Juliana (2014) provided the evidence that in vivo shoot multiplication rates would be good through sucker growth of banana cv. *Mzuzu, Bukoba* and *Mtwike* and the same can be increased by dipping for 12 h deshealthed corms in TDZ solution at 2.0 mg/L. Recently, Kassim et al. (2015) optimized TDZ and N-(2-chloro-4-pyridyl)-N′-phenylurea (4-CPPU) increase embryogenic response from scalps of EA-AAA banana which can enhance their genetic transformation. Similarly, by using MS medium supplemented with 0.1 μM TDZ, the highest mean number of microshoots, rooting and their length were reported in cooking banana cultivar (INJAGI) (Jane et al. 2015).

17.3 *Bixa orellana*

B. orellana (Bixaceae) produce a unique reddish-orange dye known as annatto dye with bixin as the major apocarotenoid of it. This annatto dye is the most sought-after natural colourant in food industry, especially for colouring milk products, etc. Although many tissue culture methods were suggested for this economically important plant, which includes in vitro shoot multiplication, somatic embryogenesis and callus formation from leaf and shoot tip explants, no efficient direct organogenesis methods were prevailing before 2004. By using in vitro established seedling parts as explants, direct organogenesis in the presence of TDZ was investigated (Parimalan et al. 2008). Noteworthy organogenesis response from hypocotyl segments, rooted hypocotyls, and cotyledonary leaf explants on MS medium supplemented with TDZ (2.0 mg/L) and 0.25% coconut water (CW) or N6-benzyladenine (BA) (7.0 mg/L) and α-naphthalene acetic acid (NAA) (0.1 mg /L) was accomplished for *B. orellana* (pink flowers variety). A maximum of 6–8 and 20–22 shoots per explant, with a shoot length of 10–12 mm, were obtained on media containing TDZ at 0.5 mg/L and 2.0 mg/L, respectively.

De Paiva Neto et al. (2003) observed the presence of distorted leafy structures that fail to develop into normal shoots when TDZ alone was used in culture medium. But in the presence of CW (0.25%), TDZ (2.0 mg/L) supports multiple shoot bud growth. So, the major highlight of this study was that thidiazuron in combination with coconut water promoted higher organogenic response in rooted hypocotyls. Similarly, in a recent study, multiple shoots (5–6 in number) were reported from cotyledonary node explants of *B. orellana* on MS medium supplemented with 4 mg/L TDZ (Vijayasekhar et al. 2012).

17.4 *Capsicum* spp.

Capsicum (chilli peppers) belongs to Solanaceae is a highly versatile plant system investigated for various tissue culture studies. However, it is one of the highly recalcitrant plants in view of its poor responses and also genotypic dependence. A glance at literature reveals various inherent problems associated with in vitro studies of *Capsicum* such as rosette shoots, ill-defined shoot buds formation and abnormal morphogenetic nature, etc. (Kehie et al. 2012). Rafael and Neftali (1996) reported plant regeneration based on shoot formation from wounded hypocotyls with TDZ (1 μg/L), for *C. annuum* L., wherein the percentage of explants with buds and shoots was at 39.8%. Similarly the effects of different concentrations of TDZ were optimized for in vitro micropropagation involving culturing of nodal segments in MS media (Ju et al. 2010; Otroshy et al. 2011). TDZ at 3.4 μM induced seven to eight shoots (from *C. chinense* Jacq. cv.) per explant that developed into healthy plants (Nancy et al. 2005). An in vitro regeneration protocol was developed by Kehie et al. (2012) for *C. chinense* Jacq. cv., wherein up to 13 shoots were induced with bud-forming capacity (BFC) index of 10.8, by culturing nodal segments in MS medium supplemented with 18.16 μM TDZ, followed by 35.52 μM BA. Recently, Mythili et al. (2017) reported the presence of coconut water in the elongation media enhanced the regeneration of well-developed shoots from differentiating explants on TDZ. High regeneration frequency was obtained from different *C. annuum* varieties, when explants were cultured on TDZ (2 mg/L). Significantly high regeneration response was noticed with up to 12 shoots per explant with elongated shoots on TDZ (2 mg/L)-containing medium. Venkataiah et al. (2003) reported TDZ-mediated organogenesis in ten pepper cultivars, and the extent of the response varies with the genotype. Specifically, out of ten genotypes tested, *C. annuum* cv. CA 960, G4 and X-235 produced maximum number of adventitious shoots.

17.5 *Coffea* spp.

Coffea which belongs to Rubiaceae is a woody plant, and it is not so easy to achieve in vitro propagation. Especially the presence of high content of phenolics in leaves hinders callogenesis from leaf explants. During the last three decades, successful in vitro shoot multiplication and direct and indirect somatic embryogenesis methods have been reported for this commercially important crop (Sridevi and Giridhar 2008). However the response is mostly specific to a species or variety. Direct organogenesis and direct somatic embryogenesis are always advantageous, as they cut short the entire in vitro propagation period. Under this context, the role of TDZ in induction of direct somatic embryogenesis was demonstrated successfully for commercially important *Coffea arabica* L. and *Coffea canephora* P ex Fr. (robusta). Giridhar et al. (2004) used segments taken from cotyledon leaf, first leaf and stalk of regenerated plantlets and able to obtain clusters of somatic embryos directly from cut portions of explants on 9.08 μM TDZ containing MS medium supplemented with 2% sucrose within 8 weeks which aid mass propagation through direct regeneration. This rapid and reproducible effect of TDZ was subsequently used to culture transformed in vitro plants of *C. canephora* (Giridhar et al. 2004). Later Ibrahim et al. (2013) reported direct and indirect somatic embryogenesis in *C. arabica*, wherein the combination of 2.26 μM 2,4-D+4.54 and 9.08 μM TDZ induced direct somatic embryogenesis from explants, while that of 4.52 μM and 9.04 μM 2,4-D + 9.08 μM TDZ induced indirect somatic embryogenesis. Similarly, influence of TDZ on secondary metabolites cafestol and kahweol of somatic embryos of robusta coffee was demonstrated (Sridevi and Giridhar 2014). TDZ (2.27 μM) in combination with IAA (2.85 μM) and 5–10% coconut water in half-strength MS basal medium drastically reduced the levels of cafestol (caf) and kahweol (kah) in somatic embryos of *C. canephora* (Sridevi and Giridhar 2014).

17.6 *Decalepis hamiltonii*

Lack of systematic cultivation methodologies for *D. hamiltonii* and poor viability of the seeds after a couple of months of their maturity and dispersal are two reasons for its limited availability. Adding to its endemic nature, now it is an endangered plant. These factors have led to the development of several in vitro propagation protocols for its conservation. Vedavathy (2004) reported 100% germination of immature zygotic embryos when placed on MS medium supplemented with GA_3 (0.05 ppm), BA (1.0 ppm) and TDZ (1.0 ppm). Giridhar et al. (2005) compared the influence of different cytokinins (2iP, BA, Kn, TDZ and Zeatin) on shoot multiplication through shoot tips. Maximum numbers of multiple shoots (6.5) were noticed on MS medium supplemented with 4.9 μM 2iP. Further elongation of shoots and adventitious shoot formation were obtained on MS medium with 2.5 μM 2iP and 0.3 μM GA_3. Saini and Giridhar (2012) found that exposing seeds to 0.3% H_2O_2 resulted in effective (94%) seed germination in *D. hamiltonii*. Similarly, 100% of immature zygotic embryo germination on MS medium supplemented with GA_3 (0.05 ppm), 6-benzylaminopurine (BA, 1.0 ppm) and TDZ (1.0 ppm) was also reported in the same study.

17.7 *Stevia*

Incorporation of TDZ in culture medium appears to be a promising step towards multiple shoots induction from shoot tip explants of *Stevia rebaudiana*. To achieve the same, inoculation of explants in two different modes was studied, i.e. normal method and reverse polarity (upside down) on culture medium (Giridhar et al. 2010), wherein 13.65 μM TDZ induced up to 12 shoots from shoot tip explants on modified MS medium. However, only three to four shoots were induced in normal mode of inoculation. In another study, the poor response of TDZ for multiple shoot induction from shoot tip explants especially when cultured on medium containing TDZ and NAA at 1.0 and 0.1 mg/L was reported (Zayova et al. 2013) for *S. rebaudiana* collected from different regions of Bulgaria. Later, TDZ-induced high frequency plant regeneration through direct shoot organogenesis in this plant was demonstrated (Lata et al. 2013) by using nodal explants (96%), wherein up to 60 shoots per explant were claimed with an average shoot length of 6 cm. HPLC analysis of in vitro raised plant leaves showed significant presence of steviolbioside, dulcoside A, rebaudioside A and rebaudioside C (Giridhar et al. 2010; Lata et al. 2013). Recently, a two-stage culture procedure using TDZ for efficient micropropagation of *S. rebaudiana* (Singh and Dwivedi 2014) was reported. In this study, the best response was observed in the presence of 0.5 mg/L TDZ in half-strength MS medium, followed by transfer of induced shoots onto MS medium containing 0.01 mg/L TDZ to get up to 11 long shoots with a shoot length of 7–8 cm. For the first time, in vitro shoot formation from root explants of *S. rebaudiana* was reported by Ghauri et al. (2013) on medium comprising TDZ, especially TDZ at 1.25 mg/L evoked 40.5% response from explants with a shoot length of 2–3 cm. However TDZ at lower concentration (0.5 mg) and 1.25 mg BA combination, best response was noticed (89.5%) with 3.25 cm long shoots. Apart from this TDZ was used to obtain regeneration from encapsulated nodal explants of *S. rebaudiana* (Lata et al. 2014).

17.8 *Vanilla planifolia*

Vanilla planifolia, native to Mexico and Central America, now cultivated in other parts of the tropics, is the source of natural vanillin produced in its beans upon curing. Being a succulent orchid, tissue culture protocol optimization is tricky. Successful plant regeneration from shoot and seed-derived callus was reported in vanilla (Nirmal Babu et al. 1997; Minoo 2002; Minoo and Nirmal Babu 2009; Giridhar 2007). To achieve mass multiplication, large number of shoot buds induction from nodal explants is very vital. The combination of 4.54 μM TDZ and 10% coconut water was used to produce multiple shoots and bulbous shoot buds. These shoot buds developed into plantlets and successful hardening and field transfer accomplished (Giridhar and Ravishankar 2004c). Palama et al. (2010) established a novel protocol for the regeneration of vanilla plants in the presence of IAA (0.5 mg/L) and TDZ (0.5 mg/L) and investigated the biochemical and molecular mechanisms that trigger shoot organogenesis from embryogenic callus that initially induced from in vitro raised seedlings of *V. planifolia* on MS basal medium containing TDZ (0.5 mg/L). Periodical subculturing and maintenance of primary callus on MS medium containing IAA (0.5 mg/L) and TDZ (0.3 mg/L) for 6 months induced embryogenic callus, and the same further leads to shoot organogenesis in 15 and 20 days upon transfer to MS basal medium with NAA (0.5 mg/L). Kodja et al. (2015) reported that the presence of TDZ in medium for over 2 years hampers adventitious shoot formation from protocorm-like bodies (PLBs). Similarly, two protocols of thin cell layer (TCL) system of *V. planifolia* cultures such as cross section and longitudinal section were reported by Jing et al. (2014), among that longitudinal section TCL system on medium containing BA (1.0 mg/L) and TDZ (1.0 mg/L) supported shoot induction of *V. planifolia*.

17.9 Conclusion

Thidiazuron is considered as a better option whenever other conventionally used cytokinins such as kinetin, BA and 2iP fail to show the desired response for in vitro propagation of plants. In this chapter we have summarized the potential benefits of TDZ incorporation in culture medium for various in vitro responses in some economically important plants (Table 17.1). Although TDZ is considered as a good choice for woody plant species, nowadays it is very commonly used for inducing varied in vitro responses in diverse plants. As evident from the respective examples afore mentioned, in most of the cases, TDZ alone is capable of evoking significant response for callusing, shoot bud proliferation or organogenesis. Rarely TDZ exhibit a synergistic effect with other cytokinins or auxins such as IAA, NAA or IBA and worked well based on the medium used, explant type or other factors. However, being a synthetic cytokinin and also a nucleotide-based growth regulator, abnormal morphological appearances in cultures are not uncommon. Such abnormal responses could be prevented by employing TDZ along with optimized levels of coconut water. In general, for most of the cytokinins, the response is genotype specific and also for TDZ, the same is true. From these observations, it is also evident that thidiazuron alone has been found to substitute for both auxin and cytokinin requirements in culture media and induce better organogenesis or somatic embryogenesis which is well documented by others (Saxena et al. 1992; Gill et al. 1993). Not much is known about the cascade of events takes place during the induction of direct organogenesis or somatic embryogenesis that appears to follow the TDZ treatment. Probably, reprogramming and expression of morphologically competent cells are involved, and such regulatory signals could control plant morphogenesis with particular interest of the phenomenon of direct organogenesis or callusing or somatic embryo formation. Another reason for the promising response of TDZ in tissue culture is due to the accumulation of the biosynthetic precursors of both auxins and cytokinins which is well demonstrated by others by measuring alterations in the growth regulator levels; hence, the effect is due to de novo synthesis rather than reduced catabolism. Now the efficacy of this synthetic cytokinin in effecting regeneration in a wide range of plant species has been a proven fact. However, it is

Table 17.1 Approaches on TDZ-induced in vitro propagation for various economical plants

Plant	Explant	Response	Hormones	Reference
Musa spp. (banana) cultivars	Shoot tips	Shoot proliferation	MS+ 16.8 μM TDZ	Arinaitwe et al. (2000)
	Scalps	Embryogenic callus	MS+ 8 μM TDZ	Sadik et al. (2007)
	Shoot tips	Rapid clonal propagation	MS+ 1 μM TDZ and 1 μM IAA	Gubbuk and Pekmezci (2006)
			MS+ 0.5 mg/L TDZ and 2 mg/L IAA	Farahani et al. (2008), Sheidai et al. (2010)
		Globular buds	MS+ 9.1 μM TDZ	Shirani et al. (2009, 2010)
		Multiple shooting	MS+ 10 μM BA+ 1 μM IAA+ 1 μM TDZ	Anuradha et al. (2014)
	Corms	In vivo shoot multiplication	2 mg/L TDZ solution	Theodosy and Juliana (2014)
	Scalps	Embryogenic response	MS+ TDZ and 4-CPPU	Sadik et al. (2007)
	Shoot tips	Microshoots, rooting and their length	MS+ 0.1 μM TDZ	Jane et al. (2015)
Bixa orellana	Hypocotyl segments, rooted hypocotyls and cotyledonary leaf	Organogenesis	MS+ 2.0 mg/L TDZ and 0.25% coconut water	De Paiva Neto et al. (2003)
	Cotyledonary node	Multiple shoots	MS+ 4 mg/L TDZ	Vijayasekhar et al. (2012)
Capsicum. annuum L.	Buds and shoots	Shoot formation from wounded hypocotyls	MS+ 1 μg/L TDZ	Rafael and Neftali (1996)
	Nodal segments	Plantlet regeneration	MS+ TDZ	Ju et al. (2010), Otroshy et al. (2011)
Capsicum chinense Jacq. cv.	Shoots	Shoot regeneration	MS+ 3.4 μM TDZ	Nancy et al. (2005)
	Nodal segments	Bud-forming capacity (BFC)	MS+ 18.16 μM TDZ+ 35.52 μM BAP	Kehie et al. (2012)

(continued)

Table 17.1 (continued)

Plant	Explant	Response	Hormones	Reference
Coffea arabica L. and *Coffea canephora* P ex Fr.	Leaf explant	Somatic embryos	MS+ 2% sucrose+ 9.08 μM TDZ	Giridhar et al. (2004) and Sridevi and Giridhar (2014)
		Callus production	MS+ 2% sucrose+ 9.08 μM TDZ+ 10% coconut water	
	In vitro stalk leaf explant	Somatic embryos	MS+ 2% sucrose+ 9.08 μM TDZ	
		Callus production	MS+ 2% sucrose+ 9.08 μM TDZ+ 10% coconut water	
	In vivo cotyledonary leaf explants	Somatic embryos	MS+ 2% sucrose+ 9.08 μM TDZ	
		Callus production		
	In vivo hypocotyl explants	Somatic embryos	MS+ 3% sucrose+ 9.08 μM TDZ	
		Callus production	MS+ 2% sucrose+ 9.08 μM TDZ	
	Leaf explants	Callus, direct and indirect embryogenesis	MS+ 3% sucrose+ 9.04 μM 2,4 D+ 9.08 μM TDZ	Ibrahim et al. (2013)
Decalepis hamiltonii	Seeds	100% Germination	MS+ GA_3 (0.05 ppm), BA (1.0 ppm) and TDZ (1.0 ppm)	Vedavathy (2004), Saini and Giridhar (2012)
	Shoot tips	Multiple shooting	MS with different concentration of 2iP, BA, Kn, TDZ and Zeatin	Giridhar et al. (2005)

(continued)

Table 17.1 (continued)

Plant	Explant	Response	Hormones	Reference
Stevia rebaudiana	Shoot tip	Normal method and reverse polarity (upside down) of shoots	MS+ 13.65 uM TDZ	Giridhar et al. (2010)
		Multiple shoot induction	MS+ 1 mg/L TDZ+ 0.1 mg/l NAA	Zayova et al. (2013)
	Nodal explants	Direct shoot organogenesis	MS + TDZ	Lata et al. (2013)
	Induced shoots	Shoot length	MS+ 0.01 mg/L TDZ	Singh and Dwivedi (2014)
			MS+ 0.5 mg/L TDZ and 1.25 mg/L BA	Ghauri et al. (2013)
	Encapsulated nodal explants	Regeneration	MS+ TDZ	Lata et al. (2014)
Vanilla planifolia	Nodal explants	Multiple shoot buds	MS+ 4.54 μM TDZ and 10% coconut water	Giridhar and Ravishankar (2004c)
	Seedling explant-based callus	Embryogenic callus	IAA (0.5 mg/L) and TDZ (0.3 mg/L)	Palama et al. (2010)
	Embryogenic callus	Multiple shoots	IAA (0.5 mg/L) and TDZ (0.5 mg/L)	
	Protocorm-like bodies	Adventitious shoot inhibition	TDZ (0.5 mg/L)	Kodja et al. (2015)
	Thin cell layers	Shoot induction	BA (1.0 mg/L) and TDZ (1.0 mg/L)	Jing et al. (2014)

necessary to have an insight into the interaction of TDZ and other factors involved in morphogenesis of plants.

References

Anuradha A, Smriti V, Namrata S, Priyanka V, Meena DPS, Tyagi RK (2014) Cryoconservation of some wild species of *Musa* L. Indian J Genet 74(4):665–669

Arinaitwe G, Rubaihayo PR, Magambo MJS (2000) Proliferation rate effects of cytokinins on banana (*Musa spp.*) cultivars. Sci Hortic 86:13–21

Arndt F, Rusch R, Stllfried HV (1976) SN 49537 a new cotton defoliant. Plant Physiol 57:99

Capelle SC, Mok DWS, Kirchner SC (1983) Effects of thidiazuron on cytokinin autonomy and the metabolism of Nt-(/x 2-isopentenyl) [8-14C] adenosine in callus tissues of *Phaseolus lunatus* L. Plant Physiol 73:796–802

De Paiva Neto VB, Ribeiro da Mota T, Otoni WC (2003) Direct organogenesis from hypocotyl-derived explants of annatto (*Bixa orellana*). Plant Cell Tissue Organ Cult 75:159–167

Farahani F, Aminpoor H, Sheidai M, Noormohammadi Z, Mazinani MH (2008) An improved system for in vitro propagation of banana (*Musa acuminate L.*) cultivars. Asian J Plant Sci 7(1):116–118

Ghauri EG, Afridi MDS, Marwat AG, Rahman I, Akram MD (2013) Micropropagation of *Stevia rebaudiana* Bertoni through root explants. Pak J Bot 45(4):1411–1416

Gill R, Gerrath JM, Saxena PK (1993) Hïgh-frequency direct somatic embryogenesis in thin layer cultures of hybrid seed geranium (*Pelargonium x hortorum*). Can J Bot 71:408–413

Giridhar P (2007) Micropropagation of vanilla: a foremost orchid spice crop of humid tropics. J Ind Bot Soc 86:51–53

Giridhar P, Ravishankar GA (2004) Efficient micropropagation of *Vanilla planifolia* Andr. under the influence of thidiazuron, zeatin and coconut water. Indian J Biotechnol 3:113–118

Giridhar P, Kumar V, Indu EP, Ravishankar GA, Chandrasekar A (2004) Thidiazuron induced somatic embryogenesis in *Coffea arabica* and *C. canephora* P ex Fr. Acta Bot Croat 63:25–33

Giridhar P, Gururaj B, Ravishankar GA (2005) In vitro shoot multiplication through shoot tip cultures of *Decalepis hamiltonii* Wight & Arn., a threatened plant endemic to southern India. In Vitro Cell Dev Biol Plant 41:77–80

Giridhar P, Sowmya KS, Ramakrishna A, Ravishankar GA (2010) Rapid clonal propagation and Stevioside profiles of *Stevia rebaudiana* Bertoni. Int J Plant Dev Biol 4(1):47–52

Gubbuk H, Pekmezci M (2006) In vitro propagation of banana (*Musa* spp.) using thidiazuron and activated charcoal. Acta Agric Scand Sect B-Soil Plant Sci 56:65–69

Guo B, Abbasi BH, Zeb A, Xu LL, Wei YH (2011) Thidiazuron: a multi-dimensional plant growth regulator. Afr J Biotechnol 10(45):8984–9000

Ibrahim MSD, Hartati SR, Rubiyo, Agus P, Sudarsono (2013) Direct and indirect somatic embryogenesis on Arabica coffee (*Coffea Arabica*). Indones J Agric Sci 14(2):79–86

Jane K, Felicien N, Bancy W, Ntirugulirwa B, Amini M, Peter YS, Njenga PK, Theodore A (2015) In vitro propagation of two elite cooking banana cultivars- FHIA 17 and INJAGI. Int J Biotech Mol Bio Res 6:40–47

Jing GF, Razak WNWAB, Rahaman ZA, Subramaniam S (2014) The effect of thin cell layer system in *Vanilla planifolia* in vitro culture. Curr Bot 5:22–25

Ju YS, Sivanesan I, Geon An C, Jeong BR (2010) Adventitious shoot regeneration from leaf explants of miniature paprika (*Capsicum annuum*) 'Hivita Red' and 'Hivita Yellow'. African J Biotech 9(19):2768–2773

Kassim S, Geofrey A, Patrick RR, Settumba BM (2015) TDZ and 4-CPPU induce embryogenic response on scalps of recalcitrant East African highland banana. J Agri Sci 7(8):1–11

Kehie M, Kumaria S, Tandon P (2012) In vitro plantlet regeneration from nodal segments and shoot tips of *Capsicum chinense* Jacq. cv. Naga King chili. 3 Biotechnol 2:31–35

Kodja H, Michel N, Shahnoo SK, Hafsah L, Robert V, Palama TL (2015) Biochemical characterization of embryogenic calli of *Vanilla planifolia* in response to two years of thidiazuron treatment. Plant Physio Biochem:1–36

Lata H, Chandra S, Wang YH, Raman V, Khan IA (2013) TDZ-induced high frequency plant regeneration through direct shoot organogenesis in *Stevia rebaudiana* Bertoni: an important medicinal plant and a natural sweetener. Am J Plant Sci 4:117–128

Lata H, Chandra S, Wang YH, Elsohly MA, Khan IA (2014) In vitro germplasm conservation of elite *Stevia rebaudiana* bertoni. Acta Hort (ISHS) 1039:303–308

Lee SW (2005) Thidiazuron in the improvement of banana micropropagation. Acta Hortic 692:67–74

Makara AM, Rubaihayo PR, Magambo MJS (2010) Carry-over effect of Thidiazuron on banana in vitro proliferation at different culture cycles and light incubation conditions. African J Biotech 9:3079–3085

Minoo D (2002) Seedling and somaclonal variation and their characterization in *Vanilla*. PhD thesis, Calicut University, Kerala

Minoo D, Nirmal Babu K (2009) Micropropagation and in vitro conservation of vanilla (*Vanilla planifolia* Andrews). In: Jain SM, Saxena PK (eds) Springer protocols, methods in molecular

biology 547, protocols for in vitro cultures and secondary metabolite analysis of aromatic and medicinal plants. The Humana Press/Springer, New York, pp 129–138
Mok MC, Mok DWS, Armstrong DJ, Shudo K, ISO Y, Okamoto T (1982) Cytokinin activity of N-phenyl-N-1, 2,3 -thidiazol-5-ylurea (thidiazuron). Phytochemistry 21:1509–1511
Murch SJ, Saxena PK (1997) Modulation of mineral and fatty acid profiles during thidiazuron mediated somatic embryogenesis in peanuts (*Arachis hypogaea* L.) J Plant Physiol 151:358–361
Mythili JB, Rajeev PR, Vinay G, Nayeem A (2017) Synergistic effect of silver nitrate and coconut water on shoot differentiation and plant regeneration from cultured cotyledons of *Capsicum annum* L. Indian J Exp Bio 55:184–190
Nancy SB, Adriana CF, Felipe BP, Montalvo-Peniche MC, Zapata-Castillo PY, Anabel SR, Amilcar ZC, Omar GA, Miranda- Ham ML (2005) Regeneration of Habanero Pepper (*Capsicum chinense* Jacq.) via organogenesis. Hort Sci 40(6):1829–1831
Nirmal Babu K, Ravindran PN, Peter KV (1997) Protocols for micropropagation of spices and aromatic crops. Indian Inst Spices Res Calicut 35
Otroshy M, Moradi K, Khayam Nekouei M (2011) The effect of different cytokines in propagation of *Capsicum annuum* L. by in vitro nodal cutting. Trakia J Sci 9:21–30
Palama TL, Menard P, Isabelle F, Choi YH, Bourdon E, Joyce GS, Muriel B, Bertrand P, Robert V, Kodja H (2010) Shoot differentiation from protocorm callus cultures of *Vanilla planifolia* (Orchidaceae): proteomic and metabolic responses at early stage. BMC Plant Biol 10:82
Parimalan R, Giridhar P, Ravishankar GA (2008) Mass multiplication of *Bixa orellana* L. through tissue culture for commercial propagation. Ind Crop Prod 28:122–127
Rafael RM, Neftali OA (1996) An improved and reliable chili pepper *(Capsicum annuum* L.) plant regeneration method. Plant Cell Rep 16:226–231
Sadik K, Rubaihayo PR, Magambo MJS, Pillay M (2007) Generation of cell suspensions of East African highland bananas through scalps. Afr J Biotechnol 6:1352–1357
Saini RK, Giridhar P (2012) Effect of temperature, chemical treatment, hydration and plant growth regulators on germination of seeds and immature zygotic embryo of swallow root (*Decalepis hamiltonii*). Research J Agri Sci 3:40–44
Saxena PK, Malik KA, Gill LG (1992) Induction by Thidiazuron of somatic embryogenesis in intact seedlings of peanut. Planta 187:421–424
Sheidai M, Aminpoor H, Noormohammadi Z, Farahani F (2010) RAPD analysis of somaclonal variation in banana (*Musa accuminata* L.) cultivar Cavendish Dwarf. Geneconserve 9:1–10
Shirani S, Mahdavi F, Maziah M (2009) Morphological abnormality among regenerated shoots of banana and plantain (*Musa* spp.) after in vitro multiplication with TDZ and BA from excised shoot-tips. Afr J Biotechnol 8(21):5755–5761
Shirani S, Sariah M, Zakaria W, Maziah M (2010) Study of genetic and phenotypic variability among somaclones induced by BA and TDZ in micropropagated shoot tips of banana (*Musa* spp.) using RAPD markers. J Agri Sci 2(3):49–60
Singh P, Dwivedi P (2014) Two-stage culture procedure using thidiazuron for efficient micropropagation of *Stevia rebaudiana*, an anti-diabetic medicinal herb. 3 Biotech 4(4):431–437
Smitha PD, Binoy KR, Ashalatha SN (2014) Effect of TDZ on direct shoot regeneration from whole male inflorescence of four diploid banana cultivars from South India. Plant Sci Int 1(1):24–32
Smitha PD, Binoy KR, Ashalatha SN (2015) Plantlet production through development of cormlet from bract meristem of four diploid banana cultivars from south India. Int J Curr Sci and Tech 3(6):32–36
Sridevi V, Giridhar P (2008) Recent trends in coffee biotechnology towards quality improvement – a review. Ind J Bot Res 4:5–12
Sridevi V, Giridhar P (2014) Establishment of somaclonal variants of Robusta coffee with reduced levels of cafestol and kahweol. In Vitro Cell Dev Biol Plant 50:618–626
Suttle JC (1985) Involvement of ethylene in the action of the cotton defoliant thidiazuron. Plant Physiol 78:272–276
Theodosy JM, Juliana M (2014) Effect of thidiazuron on in vivo shoot proliferation of popular banana (*Musa* spp. L) cultivars in Tanzania. J App Biosci 81:7214–7220

Thomas JC, Katterman FR (1986) Cytokinin activity induced by thidiazuron. Plant Physiol 81:681–683

Vedavathy S (2004) *Decalepis hamiltonii* Wight & Arn.-an endangered source of indigenous health drink. Nat Product Rad 3:22–23

Venkataiah P, Christopher T, Subhash K (2003) Thidiazuron induced high frequency adventitious shoot formation and plant regeneration in *Capsicum annuum* L. J Plant Biotechnol 5:245–250

Vijaya Sekhar VE, Krishna Satya A, Sudhakar P, Sambasivarao KRS (2012) Improved in vitro shoot multiplication of *Bixa orellana* under the influence of phytohormones particularly thidiazuron. J Pharm Res 5(2):1144–1147

Wijerathna YMAM, Kumarihami HMPC (2016) Effects of different hormonal concentrations and culture medium on multiplication and rooting of stage II banana (*Musa cavendishii*). Not Sci Biol 8(1):69–72

Zayova E, Stancheva I, Geneva M, Petrova M, Dimitrova L (2013) Antioxidant activity of in vitro propagated *Stevia rebaudiana* Bertoni plants of different origins. Turkish J of Bio 37:106–113

18 Regulation of Morphogenesis and Improvement in Shoot Multiplication in *Vitex* Species Using Thidiazuron

Naseem Ahmad, Mohammad Faisal, Sheikh Altaf Hussain, and Mohammad Anis

Abstract

Thidiazuron (TDZ), a phenyl urea derivative, possesses strong cytokinin-like activity with a wide range of morphogenic responses in all plant species including the woody ones, where other cytokinins are least effective. The current communication reviews the effect of various concentrations and exposure durations of TDZ in in vitro morphogenesis in selected *Vitex* species. Four-week exposed nodal explants at 5 μM TDZ before transfer to secondary medium gave maximum shoot multiplication rate in *V. negundo*, while in *V. trifolia* 1-week exposed explants at 5.0 μM TDZ showed maximum regeneration frequency and shoot multiplication after 8 weeks of transfer to secondary medium. Best somatic embryogenesis has also been reported in leaf explants of *V. doniana* on TDZ-enriched medium with amino acid tryptophan. Best rhizogenesis of these TDZ-induced shootlets were reported on IBA in *V. negundo* and on NAA in *V. trifolia*. A 10-minute pulse treatment to excised shootlets at 500 μM IBA gave the maximum ex vitro rooting in *V. negundo*, while best in vitro rooting in *V. trifolia* was reported on ½ MS medium supplemented with 0.5 μM NAA. Rooted plantlets were acclimatized successfully in growth chamber and then transferred to field conditions with high survival rate without any morphological variation.

Keywords

Nodal segment · Thidiazuron · Somatic embryogenesis · In vitro rooting

N. Ahmad (✉) · S. A. Hussain · M. Anis
Plant Biotechnology Laboratory, Department of Botany, Aligarh Muslim University, Aligarh, UP, India

M. Faisal
Department of Botany & Microbiology, College of Science, King Saud University, Riyadh, Saudi Arabia

N. Ahmad, M. Faisal (eds.), *Thidiazuron: From Urea Derivative to Plant Growth Regulator*, https://doi.org/10.1007/978-981-10-8004-3_18

Abbreviations

BA	Benzyl adenine
IBA	Indole-3-butyric acid
MS	Murashige and Skoog medium
NAA	α-Napthalene acetic acid
TDZ	Thidiazuron

18.1 Introduction

Vitex, the largest genus in Verbenaceae, consists of 250 species; among them 14 are found in India (Ganapaty and Vidyadhar 2005). These deciduous shrubs are widely distributed in tropics and warm temperate regions of both hemispheres of the world. Most of these species like *V. negundo*, *V. trifolia and V. doniana* are medicinally very important plants (Anonymous 2003). Due to their great medicinal and economical importance, these species have been indiscriminately harvested from the wild because of high industrial demands. The propagation through traditional conventional methods like stem cuttings and suckers is least efficient, and the seed viability is poor with low germination and poor seedling survival rates. Natural population is decreasing day by day, and therefore there is an immediate need to develop the reproducible micropropagation methods for mass multiplication of these valuable shrubs for germplasm conservation and ecorestoration.

Plant tissue culture technique paved the way for mass multiplication of the many prized medicinal and aromatic plants in short time span. The in vitro propagation of these selected species from various explants would be beneficial in accelerating large-scale mass production improvement and conservation of these overexploited plant species (Anis and Ahmad 2016). Thidiazuron (TDZ) is a phenylurea compound with a diverse array of morphogenic responses from shoot bud induction to shoot proliferation and from callus formation to somatic embryogenesis in many plant species. TDZ may act as endogenous plant growth regulator having both auxin- and cytokinin-like activity with other functional roles in modification of cell membranes, energy levels and nutrient uptake or in nutrient assimilation. This morphoregulatory potential of TDZ has led to well-established micropropagation systems in various woody plant species where other cytokinins are least effective. Short exposure of TDZ is sufficient to stimulate best regeneration system, while prolonged exposure has various negative effects on growing cultures. In this chapter the role of TDZ in inducing morphogenetic effects was reviewed in selected *Vitex* species.

18.2 *Vitex trifolia*

Vitex trifolia is an aromatic hardy coastal shrub with pale grey bark growing up to 5 m in a tropical climate having close resemblance and distribution with *V. negundo* and found in large parts of India, distributed from foot of Himalayas from southward to the Andamans. Whole plant is used in herbal medicine, but leaves are most important possessing maximum number of phytochemicals. The leaves contain aucubin, agnuside, casticin, orientin, isoorientinluteolin-7-glycoside, camphene, alfa pinene, terpinyl acetate, flavones, beta-sitosterol, camphene, etc. The leaf extract is used to cure headaches, rheumatic pains, muscle sprains, tuberculosis and post-childbirth complications and possesses anticancerous activity. This decoction is also used to treat the swollen testicles and to treat beriberi, whole leaf bath is highly recommended. The flowers are used to treat fever, vomiting and severe thirst. The fruit contains the important alkaloid vitricine, and the decoction of fruits is used to treat amenorrhoea, common cold, headache, watery eyes and mastitis and to lower blood pressure. The inner bark is chewed and swallowed as a remedy for dysentery. The root extract is used as febrifuge and diuretic and to cure liver disorders. Propagation of this plant through seeds and seedlings is very difficult; thus, this plant is raised from stem cuttings and offsets requiring the rainy season. Under such limitations of vegetative propagation where efficiency of generation cycle is low and time consuming, plant tissue culture technique is the only solution to ease down the propagation cycle with maximum efficiency and mass multiplication, without any seasonal constraints.

Ahmed and Anis (2012) reported shoot regeneration in *Vitex trifolia*, via adventitious shoot proliferation using nodal explants. The properly disinfected explants were inoculated either on MS (Murashige and Skoog 1962) alone or MS supplemented with different concentrations (0.5, 1.0, 2.5, 5.0, 7.5 or 10 μM) of TDZ. All the tested concentrations of TDZ facilitated shoot bud induction after 1 week of incubation. No morphogenic response was observed on MS medium devoid of TDZ even after 4 weeks of incubation, served as control. MS + 5.0 μM TDZ facilitated maximum regeneration potential and shoot induction. Due to negative effects of prolonged exposure of TDZ, the responsive cultures were transferred to secondary medium either devoid of any PGR or enriched with different concentrations of BA alone or in combination with NAA. 1 to 10day exposure was given to nodal segments which were collected from different months, and it was found that 7-day TDZ-exposed explants collected from mid-September to November yielded maximum 93% regeneration frequency with maximum (23.0 ± 0.2) mean shoot number and (3.4 ± 0.3 cm) mean shoot length on (MS + 1.0 μM BA +0.5 μM NAA) after 8 weeks of transfer. Further the growing cultures were subcultured onto the same fresh medium, and it was observed that all growth parameters increased up to fourth subculture with maximum stability on fifth subculture, beyond which all growth factors started declining. Healthy microshoots (4–5 cm) were excised and transferred to MS or 1/2MS rooting medium supplemented with varied concentrations

(0.1, 0.5, 1.0 or 2.0 μM) of NAA. After 4 weeks of incubation, 87% rooting response was observed on MS medium supplemented with (0.5 μM) NAA with maximum (4.4 ± 0.4) mean root number and (1.8 ± 0.09 cm) mean root length.

For acclimatization healthy rooted plantlets were transferred to soilrite-filled plastic cups, watered and covered with polybags to ensure and retain high humidity. After 4 weeks of acclimatization, plantlets with four or more fully expanded leaves were transferred to pots containing different planting substrates like soilrite, garden soil, vermiculite, or 1:1 mixture of vermiculite and garden soil. Among the tested planting substrates, best growth and maximum (92.2%) survival rate was observed on 1:1 ratio of vermiculite and garden soil. All the regenerated plantlets were morphologically similar with mother plant without any detectable variation.

18.3 *Vitex negundo* L.

Vitex negundo is a woody aromatic medicinal shrub growing on altitude up to 1500 m in outer Himalayas, distributed commonly in various parts of India and mainly found near tropical forests, moist places, riverbank sides and in open waste lands. It is widely grown as hedge plant along fields and road sides. Whole plant is used in herbal medicine but the leaves and roots are most potent. Nishindine and hydrocotylene are the two main alkaloids present in leaves. The other important phytochemicals present include agnuside, aucubin, casticin, gluconontol, glucosides, iso-orientin, tannic acid, etc. Leaf extract is used to cure the rheumatic pains, sprains, foetid discharge and fever. The insecticidal, larvicidal, antibacterial and anticancerous properties are also reported (https://easyayurveda.com/2014/07/27/nirgundi-vitex-negundo-uses-dose-side-effects-research/). Further leaf decoction is used as a tonic against vermifuge and hair disorders. Flowers are used to care the intestinal and liver disorders. Fruit juice is prescribed in headache and to treat watery eyes. Boiled seeds are used as heath supplement. Rhizome is used for treating leprosy. The root extract possesses febrifugal, diuretic and anthelmintic properties and is also used to cure the diseases like dysentery and dyspepsia. Thus, this plant is most commonly used in Indian system of herbal and indigenous medicine (Anonymous 2003). Indiscriminate collection from wild with limited cultivation, coupled with ever-increasing market and industrial demands has strained its natural population. Conventional propagation through stem cuttings and suckers is least efficient and age dependent. Regeneration through seeds is also problematic due to poor germination and low survival rates of seedlings. Other factors include transportation and seasonal constraints (Sahoo and Chand 1998). Thus, it was important to establish a reproducible and efficient protocol for this valuable medicinal shrub. Plant tissue culture technique offers a valuable tool for mass multiplication and germplasm conservation of rare, endangered and aromatic medicinal plants by applying various plant growth regulators like BA, Kin, 2iP, TDZ, etc. TDZ potent cytokinin, was earlier tested for this species by Sahoo and Chand (1998), but they did not got satisfactory results; protocol was modified by Ahmad and Anis (2007), for better multiplication of this plant species.

Ahmad and Anis (2007) reported enhanced shoot multiplication from nodal explants. Nodal explants were collected from the plants maintained in the net house of the Department of Botany, Aligarh Muslim University Aligarh, India. Sterilized nodal segments were inoculated on MS medium which served as control and failed to produce any morphogenic response even after 8 weeks of incubation. However, shoot bud induction was facilitated on all supplementations (0.1, 0.5, 1.0, 2.5, 5.0 or 10 μM) of TDZ to MS medium after 1 week of incubation. Among the tested concentration, 1.0 μM TDZ was found most responsive (98%) for producing maximum (24.8 ± 0.96) mean shoot number with mean shoot length (5.50 ± 0.41) after 8 weeks of incubation. Different (2, 4, 6 or 8 weeks) exposure treatments were given to explants, and a 4-week exposure proved to be optimal for better response and regeneration. However, continuously growing cultures on the same fresh medium beyond 4 weeks showed some deleterious effects. To minimize these harmful effects, these responsive cultures were transferred on MS supplemented with (1.0 μM) BA + (0.5 μM) NAA for better growth and multiplications. These cultures were subcultured on the similar fresh medium; all growth factors like shoot number and shoot length increased up to fourth subculture, beyond which a decreasing trend was observed. This efficient protocol yielded the highest number of shoots (with better shoot length up to the fourth subculture than its earlier report; Sahoo and Chand 1998) where a low mean shoot number (1.26) was observed after 4 weeks with increase in shoot number up to second subculture. For stimulating ex vitro rhizogenesis, healthy shoots of length more than 4 cm were excised, and a 10-min pulse treatment of different (100, 200, 300, 500 or 1000 μM) higher concentrations of IBA was given. Of all the tested treatments, a maximum response of 97% was observed with the highest mean root number (13.6 ± 0.67) and mean root length (3.3 ± 0.45 cm) at 500 μM pulse treated per IBA shootlet after 4 weeks of transfer to soilrite. Healthy plantlets with six to seven fully expanded leaves were transferred to field conditions with 95% survival rate. In this 12-week micropropagation cycle, all plantlets regenerated from nodal explants grew and flowered normally without any detectable morphological variation with donor mother plant.

18.4 *Vitex doniana*

Vitex doniana is the most abundant and widespread indigenous species of Africa, and on different evidences, it has been transferred to family Lamiaceae (Wagstaff et al. 1998; Orwa et al. 2009). Leaves and fruits are eaten raw or after processing by rural people. All plant parts are used in traditional medicine to cure and relieve the diseases like anaemia, chickenpox, diabetes, fever, dysentery gonorrhoea, jaundice, leprosy and rachitis and also used as tonic to enhance milk production in lactating mothers. Further it is used for the treatment of kidney, liver and infertility disorders (Louppe et al. 2008; Orwa et al. 2009). Due to its widespread medicinal, nutritional and economic uses, it has been chosen as a model species to be domesticated in

Benin (Codjia et al. 2003; Dadjo et al. 2012). The conventional methods of propagation of this plant are weak and least efficient (Thies 1995, Sanoussi et al. (2012). The other factors like increasing population, growing industrial demands, overexploitation and unimproved propagation methods have led to sharp decrease in its natural population (Achigan-Dako et al. 2010). Thus, there is an immense need to generate the alternate propagation methods through plant tissue culture technique to meet the said challenges. This is the most promising mass multiplication technique to protect, propagate and conserve the germplasm of valuable medicinal herb from a single cell, tissue or organ. Among the in vitro methods, regeneration through single cell (somatic embryogenesis) is the most feasible one for large-scale production of transgenic plants in short duration (Saiprasad 2001).

Dadjo et al. (2015) reported the effect of TDZ on somatic embryogenesis in leaf explant of *Vitex doniana*. They evaluated the influence of TDZ and different amino acids on induction and regeneration of somatic embryos from the leaf explants of *V. doniana*, as no earlier tissue culture report on this plant was available. Healthy young leaves were collected from 2-year-old plants maintained in the greenhouse and were washed, disinfected, excised and cultured on MS medium supplemented with TDZ and other amino acids at varying concentrations. Among the tested concentration, MS + 0.11 mg/l thidiazuron + 2% sucrose +100 mg/l myo-inositol + tryptophan at 30.6 mg/l proved to be optimal for production of maximum (6.5) number of somatic embryos per explant. Further addition of 8.45 mg/l $AgNO_3$ to medium produced the same number of somatic embryos in 90% embryogenic cultures. Induction of callus starts from cut ends of the leaves with the formation of globular embryos after 4 weeks and cotyledonary stage embryos after 2 months of incubation. Among the tested amino acids, threefold increase was observed in proline-supplemented medium while fivefold increase on tryptophan-supplemented medium than the control. Thus, these differential results for induction and regeneration of somatic embryos verify different requirements of specific amino acids in *V. doniana*.

18.5 Conclusion

In conclusion these established protocols showed the positive effect of TDZ on increasing the adventitious organogenesis from nodal or leaf explants of selected *Vitex* species. The present communication reviews efficiency of TDZ in vitro propagation protocols of three *Vitex* species through effective bud breaking and shoot multiplication or by somatic embryogenesis without any morphological abnormalities. Increase in shoot multiplication was observed in TDZ-exposed explants when transferred to secondary medium supplemented with BA and NAA. Shoot multiplication rates increased up to fourth subculture with maximum stability at fifth subculture. For in vitro rhizogenesis, IBA was found best for maximum response with highest mean root number. The efficiency of these established micropropagated protocols can be utilized for maximum clonal production of rare and endangered medicinal plants for their germplasm conservation and ecorestoration and to meet the industrial demands. These tissue cultural approaches can be employed all over the year for large-scale production of planting material by bypassing the limitations of conventional methods. Biotechnology has made significant contributions for understanding the biological processes which permits the manipulation of in vitro morphogenesis and investigations on various physiological, biochemical and molecular aspects to address the recalcitrance of vegetative methods.

References

Achigan-Dako EG, Pasquini MW, Assogba-Komlan F, NDanikou S, Yedomonhan H, Dansi A, Ambrose-Oji B (2010) Traditional vegetables in Benin: diversity, distribution, ecology, agronomy and utilisation. Institut National des Researches Agricoles du Benin, Benin

Ahmad N, Anis M (2007) Rapid clonal multiplication of a woody tree, *Vitex negundo* L. through axillary shoot proliferation. Agrofor Syst 71:195–200

Ahmed MR, Anis M (2012) Role of TDZ in the quick regeneration of multiple shoots from nodal explants of *Vitex trifolia* L. –an important medicinal plant. Appl Biochem Biotechnol 168:957–966

Anis M, Ahmad N (2016) Plant tissue culture: propagation conservation and improvement. Springer Nature, Singapore

Anonymous (2003) The wealth of India, raw materials, vol I. Publication and information Directorate, CSIR, New Delhi, pp 522–524

Codjia JTC, Assogbadjo AE, Ekue MRM (2003) Diversite et valorisation au niveau local des resource vegetables foresteries alimentairis du Benin. Chaiers Agric 12:321–331

Dadjo C, Assogbadjo AE, Fandohan B, Glele Kakai R, Charkeredza S, Houehanou TD, Van Damme P, Sinsin B (2012) Uses and management of black plum (*Vitex doniana* Sweet) in Southern Benin. Fruits 67:239–248

Dadjo C, Kahia J, Muthuri C, Diby L, Kouame C, Njenga K (2015) Induction and regeneration of somatic embryos from *Vitex doniana* (Lamiaceae) leaf explants. Int J Biotechnol Mol Biol Res 6:28–23

Ganapaty S, Vidyadhar KN (2005) Phytoconstituents and biological activities of *Vitex*: a review. J Nat Rem 5:75–95

Louppe D, Oteng-Amoako AA, Brink M (2008) Timbers 1. Plant resources of tropical Africa. PROTA Foundation. Backhuys Publishers. CTA, Wageningen

Murashige T, Skoog F (1962) A revised medium for rapid growth and bioassay with tobacco tissue cultures. Plant Physiol 15:473–497

Orwa C, Mutua A, Kindt R, Jamnadass R, Anthony S (2009) Agro forest database: a tree reference and selection guide version 4.0

Sahoo Y, Chand PK (1998) Micropropagation of *Vitex negundo* L. a woody aromatic medicinal shrub, through high frequency axillary shoot proliferation shoot. Plant Cell Rep 18:301–307

Saiprasad GVS (2001) Artificial seeds and their applications. Resonance 6:39–47

Sanoussi A, Ahoton LE, Odjo T (2012) Propagation of black plum (*Vitex doniana* Sweet) using stem and root cuttings in the ecological conditions of South Benin. Tropicultura 2:107–112

Thies E (1995) Principaux ligneux agro foreateries de la Guinee, Zone de transition. Guinee Bissau, Guinee, Cote d, Ivoire, Ghana, Togo, Benin, Nigeria, Cameroun. Schriftenreihe der, GTZ No 253

Wagstaff SJ, Hickerson L, Spangler R, Reeves PA, Olmstead RG (1998) Phylogeny in labiatae S. L., inferred from Cp dna sequences. Plant Syst Evol 209(3–4):265–274

19 TDZ-Induced Regeneration in *Stevia rebaudiana* Bertoni: An Important Natural Sweetener

Arpan Modi and Nitish Kumar

Abstract

A substituted phenylurea, thidiazuron (TDZ), plays a vital role as cytokinins in the regeneration of large number of plant species including stevia. There are several reports that are available on regeneration of stevia using TDZ. Apart from inducing regeneration with very high frequency, TDZ is also proved to be non-mutagenic. Stevia is of great importance due to its steviol glycosides (SGs) which are natural sweeteners used by the food industry as well as having medicinal purposes. This chapter is reviewed focusing on the impact of TDZ on shoot regeneration of stevia.

Keywords

Micropropagation · Regeneration · Stevia · Stevioside · TDZ

19.1 Introduction

Stevia rebaudiana Bertoni ($2n = 22$), belonging to Asteraceae family, is a traditional medicinal herb containing sweet diterpene glycosides generally known as steviol glycosides. Major steviol glycosides are stevioside and rebaudioside A (Tadhaniet al. 2007). Stevioside, a diterpenoid glycoside, is a natural sweetener, predominantly found in the leaves of the plant. It is widely used for several years as a sweetener in

A. Modi
Department of Plant Sciences, Agricultural Research Organization Volcani Center, Rishon LeZiyyon, Israel

N. Kumar (✉)
Centre for Biological Sciences (Biotechnology), Central University of South Bihar, Patna, Bihar, India
e-mail: nitish@cub.ac.in

N. Ahmad, M. Faisal (eds.), *Thidiazuron: From Urea Derivative to Plant Growth Regulator*, https://doi.org/10.1007/978-981-10-8004-3_19

South America, Japan, and China and in different countries of Europe. Stevioside resembles the structure of gibberellins and is also synthesized through the same pathway as gibberellins (Genus 2003). Stevia is a highly self-incompatible plant, and a plant produced by seed generally shows a wide variation in the stevioside content (Tamura et al. 1984; Nakamura and Tamura 1985). In general the percentage of seed germination is quite low in stevia and requires long time to establish seedlings (Kawatani et al. 1977). Recent studies have displayed a high degree of variability in sweetening level and composition due to the heterogenous populations obtained through propagation by seed (Miyagawa and Fujioka 1986; Anbazhagan et al. 2010). Vegetative propagation is also limited by the lower number of individuals that can be obtained simultaneously from a single plant due to pathogen accumulation in the tissues (Sakaguchi and Kan 1982; Debnath 2008; Mishra et al. 2010). Therefore, clonal propagation is ideal to reduce some of these problems, and micropropagation using plant tissue culture technology offers an approach that is capable of producing large numbers of genetically similar disease-free stevia plants in a short period of time and limited space.

Besides the development of a novel micropropagation system, an efficient growth regulator capable of inducing high organogenesis is necessary. Thidiazuron (TDZ: *N*-phenyl-*N*-[1,2,3-thiadiazol-5-yl]urea), a non-purine cytokinin compound, has been shown to exhibit a stronger effect than N^6-benzyladenine (BA) on in vitro morphogenesis of a large number of plants (Mithila et al. 2003; Ganeshan et al. 2006; Jones et al. 2007; Shan et al. 2000; Kumar and Reddy 2012). TDZ was synthesized by German Schering Corporation for defoliation of cotton (Arndt et al. 1976). Originally, TDZ was classified as a type of cytokinin that induces many responses that were similar to the responses induced by natural cytokinins. It was proved that, TDZ alone, unlike traditional phytohormones, fulfilled the requirements of various regenerative responses of many different plant species. Recently, the morphoregulatory potential of TDZ has led to its application in plant tissue culture for the development of feasible morphogenetic systems. TDZ emerged as an effective bioregulant in cell and tissue cultures in wide array of plant species (Mithila et al. 2003; Ganeshan et al. 2006; Jones et al. 2007; Shan et al. 2000; Kumar and Reddy 2012). Moreover, TDZ triggers a basic survival mechanism in plant tissues that includes asexual reproduction for species survival.

19.2 Micropropagation of Stevia

Conventional method of propagation for this plant might not be useful for plant as well as stevioside production as the seeds of stevia have very low germination percentage. On the other hand, vegetative propagation is limited with less numbers of cuttings produced from whole plant with less survival rate. Tissue culture techniques are best alternative for crop production to crop improvement. Since last 25 years research on micropropagation of stevia plant has been conducted which includes all the methods of in vitro techniques (direct and indirect organogenesis). Summary of the micropropagation work on stevia is described in Table 19.1.

Table 19.1 Standardized protocol for micropropagation stevia

Sr. no.	Explant	Composition	Result	References
1.	Callus	MS + 0.1 mgl^{-1} BA + 3% sucrose + 0.8% agar (pH 5.8)	Bud formation initiation	Ferreira and Handro (1988)
2.	Callus	MS + 1 mgl^{-1} BAP + 3% sucrose (pH 5.8)	Bud formation initiation	Swanson et al. (1992)
3.	Shoot tip	MS + 2 mgl^{-1} BA + 3% sucrose + 0.7% agar (pH 5.8)	8.5 ± 0.9 shoots per explants	Sivaram and Mukundan (2003)
4.	Node	MS + 2 mgl^{-1} IAA + 0.5 mgl^{-1} Kn + 3% sucrose + 0.3% Phytagel (pH 5.8)	23.4 ± 2.1 shoots per explants	Hwang (2006)
5.	Node	MS + 1 mgl^{-1} IAA + 10 mgl^{-1} Kn + 30 mgl^{-1} ADS + 3% sucrose + 0.9% agar (pH 5.8)	>10 shoots per explants	Mitra and Pal (2006)
6.	Node	MS + 1.5 mgl^{-1} BA + 0.5 mgl^{-1} Kn + 3% sucrose + 0.7% agar (pH 5.7)	85.33% growth rate with 8.75 no. of shoots/culture	Ahmed et al. (2007)
7.	Shoot tip	MS + 1 mgl^{-1} BA + 3% sucrose + 0.6% agar (pH 5.7–5.8)	3.1 number of shoots per culture with 8.35 cm length of shoots	Hosain et al. (2008)
8.	Shoot tip	MS + 3% sucrose + 0.5% agar (pH 5.7)	6.63 shoots per explants	Ibrahim et al. (2008)
9.	Leaf	MS + 2 mgl^{-1} BA + 1 mgl^{-1} Kn + 3% sucrose + 0.75% agar (pH 5.7)	4.33 number of shoots per explants with average shoot length of 3.73 cm	Sreedhar et al. (2008)
10.	Callus	MS + 1 mgl^{-1} BA + 0.2 mgl^{-1} NAA + 3% sucrose + 0.9% agar (pH 5.6 ± 0.2)	14 number of shoots per explants with average shoot length of 5.6 cm	Janarthanam et al. (2009)
11.	Callus	MS + 1.8 mgl^{-1} BA + 0.12 mgl^{-1} NAA + 3.5% sucrose + 0.7% agar (pH 5.7–5.8)	2.17 shoots per culture with an average length of shoots of 3.22 cm	Moktaduzzman and Rahman (2009)
12.	Node	MS + 3.5 mgl^{-1} BA + 3% sucrose + 0.45% agar agar (pH 5.7)	83.2 number of shoots per explants	Sairkar et al. (2009)
13.	Callus	MS + 5 mgl^{-1} BA + 1 mgl^{-1} NAA + 3% sucrose + 0.45% agar agar (pH 5.7)	3.8 plantlet formation per calli	Sairkar et al. (2009)
14.	Shoot tip	MS + 1 mgl^{-1} BA + 0.5 mgl^{-1} IAA + 3% sucrose + 0.8% agar (pH 5.8)	16.20 number of shoots per explants	Anbazhagan et al. (2010)
15.	Shoot tip	MS + 2 mgl^{-1} Kin + 3% sucrose + 0.8% agar (pH 5.6)	11.33 number of shoots with an average length of 11 0.13 cm	Das et al. (2010)
16.	Node	MS + 0.5 mgl^{-1} Kn + 0.5 mgl^{-1} NAA + 3% sucrose + 0.8% agar (pH 5.7)	8.2 number of shoots with an average length of 6.8 cm	Kalpana et al. (2010)
17.	Node	Modified MS + 1% sucrose + 0.7% agar (pH 6.2)	5.41 number of shoots with an average shoot length of 4.76 cm	Modi et al. (2012)

Efficient protocols have been developed from explants, viz., callus, leaf, node, and shoot tips. Although the protocol looks very simple (Modi et al. 2012), there is an obvious impact of several growth promoter in micropropagation of stevia.

19.3 Regeneration of Stevia Plantlets Using TDZ

Direct regeneration of plant, i.e., without callus stage, is a valuable consistent technique of micropropagation. Development of plants through direct regeneration system may show high genetic uniformity than those developed through callus intervening regeneration system. Numerous research reported on direct organogenesis of shoot buds using TDZ in culture medium from leaf, axillary nodes, shoot tips, and roots (Giridhar et al. 2010; Ghauri et al. 2013; Singh and Dwivedi 2014; Lata et al. 2015; Ahmed et al. 2016). Summary of the regeneration of stevia plantlets using TDZ is described in Table 19.2.

In experiments conducted with TDZ on stevia, interesting findings were observed by Giridhar et al. (2010) in which shoot tip explants were inoculated on medium with two different positions viz., horizontal and inverted (tip of the shoot goes downwards in the medium) and the medium contained different concentrations of TDZ (4.53-18.16 μM). They observed multiple shoot formation in both modes of explant position, and surprisingly, normally placed explants showed 3–4 shoots per explant, whereas 11–12 shoots per explant were observed in inverted explants. Not only on nodal explants but also on root explants that the TDZ showed regeneration of shoots. Incorporation of 13.62 μM TDZ in modified MS medium induced 11–12 multiple shoots from shoot tip explant when inoculated in reverse polarity, i.e., shoot tip in downward direction (inverted mode).

Ghauri et al. (2013) established shoots on leaves of exotic stevia plants with an average shoot length of 2.75 cm in a medium containing 0.75 mgl^{-1} TDZ. Moreover,

Table 19.2 Standardized protocol for in vitro regeneration of stevia plantlets using TDZ

Sr. no.	Explant	Composition	Result	References
1.	Shoot tip	MS + 13.62 μM TDZ	11–12 shoots per explants	Giridhar et al. (2010)
2.	Leaf	MS + 0.5 mgl^{-1} TDZ + 1.25 mgl^{-1} BAP	40 shoots per explant	Ghauri et al. (2013)
3.	Nodal explant	½ MS + 0.01 mgl^{-1} TDZ	11 shoots per explants	Singh and Dwivedi (2014)
4.	Nodal segment	MS + 1 μM TDZ	60 shoots per explant	Lata et al. (2013)
5.	Nodal segment	MS + 1.0 mgl^{-1} BAP + 0.5 mgl^{-1} TDZ + 0.05 mgl^{-1}NAA	3.38 shoots per explants	Soliman et al. (2014)
6.	Shoot tip and nodal segment	MS + 2 mgl^{-1} BAP + 1 mgl^{-1} TDZ	3.6 shoots per explants	Pawar et al. (2015)
7.	Shoot tip	MS + 1.5 mgl^{-1} TDZ	14.5 shoots per explant	Ahmed et al. (2016)

TDZ (0.5 mgl^{-1}) along with 1.25 mgl^{-1} BA showed significantly higher numbers of shoots per culture.

The role of TDZ has been well characterized in shoot development of stevia (Singh and Dwivedi 2014). They established a culture of stevia on ½ MS medium containing cytokinins, viz., BA, Kn, and TDZ. The protocol had been developed with two stage procedures. Shoots from the best four media were selected for shoot proliferation, i.e., 0.2 mgl^{-1} BA, 0.2 mgl^{-1} Kn, 0.2 mgl^{-1} BA + 0.2 mgl^{-1} Kn, and 0.5 mgl^{-1} TDZ. They were again transferred to either the same media or to the ½ MS or to the ½ MS + 0.01 mgl^{-1}TDZ. It was observed that the medium containing ½ MS + 0.01 mgl^{-1}TDZ showed significantly higher numbers of shoots, increased shoot length, and higher numbers of leaves. Among all, shoots transferred from 0.5 mgl^{-1} TDZ to 0.01 mgl^{-1} TDZ-containing medium showed maximum numbers of shoots (11), highest shoot length (7.17 cm), and highest numbers of leaves (61).

Lata et al. (2013) studied the effect of various cytokinins on shoot formation in stevia. Various concentrations of BAP, Kn, and TDZ ranging from 1.0 to 9.0 μM were added to the medium, and they observed 60 shoots per culture with an average shoot length of 6 cm. Moreover, it was also observed that the average number of shoots was decreased with increased concentration of TDZ, whereas in BA- and Kn-treated plants the same value increased with increased concentration which implies the efficiency of TDZ at very low concentration. Further, Lata et al. (2015) studied the effect of TDZ on germination of synthetic seeds prepared from nodal segment of stevia. After 4 months of storage, 88% of the population germinated when placed on MS medium containing TDZ (0.2 mgl^{-1}). Regenerated plants were then placed on root-inducing medium and subsequently in polyhouse for hardening. Well-developed plantlets were analyzed for clonal fidelity using photosynthesis rate, CO_2 and water vapor exchange rate, water use efficiency, stevioside, and rebaudioside A content. Among all the parameters tested, the result between tissue culture raised plants and mother plants is comparable, and the difference between them is found at par ($p < 0.05$). Now, it is obvious that TDZ-induced regeneration does not lead to somaclonal variations, as, from physiological processes to biochemical constituents, the results were found at par (Lata et al. 2015).

Soliman et al. (2014) attempted in vitro propagation of stevia through multiple shoot regeneration from nodal segments cultured on Murashige and Skoog (MS) medium supplemented with various concentrations of BAP, NAA, IAA, and TDZ. The maximum number of axillary shoots per explant (3.38) and the highest shoot length (2.92 cm) were observed with MS medium supplemented with 1.0 mgl^{-1} BAP + 0.5 mgl^{-1} TDZ and 0.05 mgl^{-1}NAA.

Pawar et al. (2015) observed the regeneration of bud from shoot tip and nodal segment on basal MS medium containing BAP along with Kin, TDZ, and NAA at different concentration. They found 3.3–3.6 shoots per explant on MS medium supplemented with 2 mgl^{-1}BAP and 1 mgl^{-1} TDZ.

Ahmed et al. (2016) established the micropropagation protocol for stevia in which they tried different levels of cytokinins (BA, Kn, and TDZ) for shoot proliferation. However, they observed that more numbers of shoots were formed in the medium containing TDZ (1.5 mgl^{-1}) than any medium having alone BA or Kn or their combination. Moreover, they also observed that days to shoot initiation were higher in TDZ-treated explants than BA or Kn. Formation of multiple shoots in TDZ-containing medium implied its significance in commercial plant production. Though 1.5 mgl^{-1} TDZ gave the best number of shoots (14.5), 1.0 mgl^{-1} TDZ gave the best performance.

19.4 Somatic Embryogenesis of Stevia Plantlets Using TDZ

Somatic embryos, which are bipolar structures, arise from individual cells and have no vascular connection with the maternal tissue of the explants. Embryos may develop directly from somatic cells (direct embryogenesis), or development of recognizable embryogenic structures is preceded by numerous, organized, non-embryogenic mitotic cycles (indirect embryogenesis). Somatic embryogenesis has a great potential for clonal multiplication. Under controlled environmental conditions, somatic embryos germinate readily, similar to their seedling counterpart. Like other crops, the TDZ also helped in embryogenesis process in stevia. Banerjee and Sarkar (2010) established somatic embryos formed from leaf explants. TDZ with the concentration of 0.2 mgl^{-1} in the medium supplemented with 2 mgl^{-1} 2,4-D and 0.2 mgl^{-1} BAP showed maximum callus formation (96%) leading to somatic embryo formation, whereas in control (without TDZ) no callus formation was observed. Thus the efficacy of TDZ for somatic embryo formation, like other crops, may be justified for stevia also.

19.5 Conclusion

Regeneration, either in vivo or ex vivo, needs stable regulation of cell proliferation and organogenesis, which are assumed to confine to plant hormones, auxin, and cytokinins. TDZ is believed to be the best synthetic cytokinin present for the regeneration of numerous plant species including stevia. TDZ improved greatly the ex vivo generation and multiplication of species recalcitrant to propagation. In stevia regeneration, TDZ is much effective in concentrations 10–1000 times less than other phytohormones. However, in spite of the popularity of TDZ as a phytohormone and more than 30 years of research work, the exact biological role of TDZ is still a mystery.

References

Ahmed SR, Howlader MMS, Sutradhar P, Yasmin S (2016) An efficient protocol for in vitro regeneration of *Stevia rebaudiana*. Asian J Med Biol Res 2:95–106

Anbazhagan M, Kalpana M, Rajendran R, Natrajan V, Dhanavel D (2010) In vitro production of *Stevia rebaudiana* Bertoni. Emir J Food Agric 22:216–222

Arndt FR, Rusch R, Stillfried HV, Hanisch B, Martin WC (1976) A new cotton defoliant. Plant Physiol 57:S–99

Banerjee M, Sarkar P (2010) Somatic embryogenesis in *Stevia rebaudiana* Bertoni using different concentration of growth hormones. Int J Plant Sci 5:284–289
Das A, Gantait S, Mandal N (2010) Micropropagation of an elite medicinal plant: *Stevia rebaudiana*. Int J Agric Res 6:40–48
Debnath M (2008) Clonal propagation and antimicrobial activity of an endemic medicinal plant *Stevia rebaudiana*. J Med Plant Res 2:45–51
Ferreira CM, Handro W (1988) Micropropagation of *Stevia rebaudiana* through leaf explants from adult plants. Planta Med 54:157–160
Ganeshan S, Chodaparambil SV, Baga M, Fowler DB, Hucl P, Rossnagel BG, Chibbar RN (2006) In vitro regeneration of cereals based on multiple shoot induction from mature embryos in response to thidiazuron. Plant Cell Tissue Organ Cult 85:63–73
Genus JMC (2003) Molecule of interest stevioside. Phytochemistry 64:913–921
Ghauri E, Afridi MS, Marwat GA, Rahman I, Akram M (2013) Micropropagation of *Stevia rebaudiana* Bertoni from root explants. Pak J Bot 45:1411–1416
Giridhar P, Sowmya KS, Ramakrishna A, Ravishankar GA (2010) Rapid clonal propagation and stevioside profiles of *Stevia rebaudiana* Bertoni. Int J Plant Dev Biol 4:47–52
Hossain MA, Shamimkabir AHM, Jahan TA, Hasan MN (2008) Micropropagation of *Stevia*. Int J Sustain Crop Prod 3:1–9
Hwang SJ (2006) Rapid in vitro propagation and enhanced stevioside accumulation in *Stevia rebaudiana* Bert. J Plant Biol 49:267–270
Ibrahim IA, Nasr MI, Mohammed BR, El-Zefzafi MM (2008) Plant growth regulators affecting *in vitro* cultivation of *Stevia rebaudiana*. Sugar Tech 10:254–259
Janarthanam B, Gopalakrishnan M, Sai GL, Sekar A (2009) Plant regeneration from leaf derived callus of *Stevia rebaudiana* Bertoni. Plant Tissue Cult Biotech 19:133–141
Jones MP, Yi Z, Murch SJ, Saxena PK (2007) Thidiazuron-induced regeneration of *Echinacea purpurea* L.: micropogation in solid and liquid culture systems. Plant Cell Rep 26:13–19
Kalpana M, Anbazhagan M, Natarajan V, Dhanavel D (2010) Improved micropropagation method for the enhancement of biomass in *Stevia rebaudiana* Bertoni. Recent Res Sci Technol 2:8–13
Kawatani T, Kaneki Y, Tanabe T (1977) On the cultivation of Kaa-hee (*Stevia rebaudiana* (Bert.)). Jpn J Trop Agric 20:137–142
Kumar N, Reddy MP (2012) Thidiazuron (TDZ) induced plant regeneration from cotyledonary petiole explants of elite genotypes of *Jatropha curcas*: a candidate biodiesel plant. Ind Crop Prod 39:62–68
Lata H, Chandra S, Wang YH, Raman V, Khan IA (2013) TDZ-induced high frequency plant regeneration through direct shoot organogenesis in *Stevia rebaudiana* Bertoni: an important medicinal plant and a natural sweetener. Am J Plant Sci 4:117–128
Lata H, Chandra S, Wang YH, ElSohly MA, Khan IA (2015) Polyhouse cultivation of in vitro raised elite *Stevia rebaudiana* Bertoni: an assessment of biochemical and photosynthetic characteristics. Int J Trop Agric 33:2381–2389
Mishra PK, Singh R, Kumar U, Prakash V (2010) *Stevia rebaudiana*- a magical sweetener. Glob J Biotechnol Biochem 5:62–74
Mithila J, Hall JC, Victor JMR, Saxena PK (2003) Thidiazuron induces shoot organogenesis at low concentrations and somatic embryogenesis at high concentrations on leaf and petiole explants of African violet (*Saintpaulia ionantha* Wendl.) Plant Cell Rep 21:408–414
Mitra A, Pal A (2006) In vitro regeneration of *Stevia rebaudiana* (Bert.) from the nodal explants. J Plant Biochem Biot 16:59–62
Miyagawa H, Fujioka N (1986) Studies on the tissue culture of *Stevia rebaudiana* and its components: II. Induction of shoot primordia. Planta Med 52:321–323
Modi AR, Patil G, Kumar N, Singh AS, Subhash N (2012) A simple and efficient in vitro mass multiplication procedure for *Stevia rebaudiana* Bertoni and analysis of genetic fidelity of in vitro raised plants through RAPD. Sugar Tech 14:391–397
Moktaduzzman M, Rahman SMM (2009) Regeneration of *Stevia rebaudiana* and analysis of somaclonal variation by RAPD. Biotechnology 8:449–455

Nakamura S, Tamura Y (1985) Variation in the main glycosides of Stevia (*Stevia rebaudiana* Bertoni). Jpn J Trop Agric 29:109–116

Pawar SV, Khandagale VG, Jambhale VM, Jadhav AS, Pawar BD (2015) In vitro regeneration studies in *stevia* through nodal segment and shoot tip. Bioscan 10:1007–1010

Sairkar P, Chandravanshi MK, Shukla NP, Mehrotra NN (2009) Mass production of an economically important medicinal plant *Stevia rebaudiana* using in vitro propagation techniques. J Med Plant Res 3:266–270

Sakaguchi M, Kan T (1982) Japanese researches on *Stevia rebaudiana*. Ci Cult 34:235–248

Shan XY, Li DS, Qu RD (2000) Thidiazuron promotes in vitro regeneration of wheat and barley. In Vitro Cell Dev Biol Plant 36:207–210

Singh P, Dwivedi P (2014) Two-stage culture procedure using thidiazuron for efficient micropropagation of *Stevia rebaudiana*, an anti-diabetic medicinal herb. 3 Biotech 4:431–437

Sivaram L, Mukundan U (2003) In vitro culture studies on *Stevia rebaudiana*. In Vitro Cell Dev Biol Plant 39:520–523

Soliman HIA, Metwali EMR, Almaghrabi OAH (2014) Micropropagation of *Stevia rebaudiana* Bertoni and assessment of genetic stability of in vitro regenerated plants using inter simple sequence repeat (ISSR) marker. Plant Biotechnol 31:249–256

Sreedhar RV, Venkatachalam L, Thimmaraju R, Bhagyalalakhmi N, Naeayan MS, Ravishankar GA (2008) Direct organogenesis from leaf explants of *Stevia rebaudiana* and cultivation in bioreactor. Biol Plant 52:355–360

Swanson SM, Mahady GB, Beecher CWW (1992) Stevioside biosynthesis by callus, root, shoot and rooted-shoot cultures in vitro. Plant Cell Tissue Organ Cult 28:151–157

Tadhani MB, Patel VH, Subhash R (2007) In vitro antioxidant activities of *Stevia rebaudiana* leaves and callus. J Food Compos Anal 27:323–329

Tamura Y, Nakamura S, Fukui H, Tabata M (1984) Comparison of *Stevia* plants grown from seeds, cuttings and stem tip cultures for growth and sweet diterpene glycosides. Plant Cell Rep 3:180–182

Shoot Organogenesis of Aloe Plants with Emphasis on TDZ

20

Thayamini H. Seran and Naseem Ahmad

Abstract

Aloe species particularly *Aloe vera* (L.) is a potential crop in pharmaceutical, cosmetic and food industries. As a result of increasing demand for aloe, large-scale plant production is needed for commercial cultivation. Hence, in vitro plant propagation is necessary for its commercial production due to male sterility and limited number of conventionally produced propagules. In aloe in vitro propagation, adenine-type cytokinins, particularly BAP, are mostly used for shoot organogenesis. According to the literature survey, TDZ at low concentration is required for a short period to induce multiple shoot buds than BAP. However, TDZ is not commonly used in aloe plant production although TDZ has been used in plant tissue culture of the other medicinal plant and horticultural crop species. Hence, it could be incorporated into culture medium even with other cytokinins, normally BAP, at appropriate concentration. Addition of auxins particularly IAA and NAA may enhance the initiation of shoots. Thus, TDZ would utilize in aloe in vitro clonal propagation technique which is useful in mass plant production, genetic transformation and germplasm conservation for international germplasm exchange.

Keywords

Aloe plant · BAP · In vitro culture · Plant propagation · TDZ

T. H. Seran (✉)
Department of Crop Science, Faculty of Agriculture, Eastern University, Chenkalady, Sri Lanka
e-mail: thayaminis@esn.ac.lk

N. Ahmad
Department of Botany, Aligarh Muslim University, Aligarh, UP, India

N. Ahmad, M. Faisal (eds.), *Thidiazuron: From Urea Derivative to Plant Growth Regulator*, https://doi.org/10.1007/978-981-10-8004-3_20

20.1 Aloe Plant and Its Use

Plants belonging to genus *Aloe* are widely distributed in the world (Steenkamp and Stewart 2007), and there are many identified aloe species belonging to Aloaceae family. Among them, *Aloe vera* L. (syn *A. barbadensis* Mill.) is the most popular species of the genus *Aloe* and commercially grown in many tropical climate countries for its medicinal properties (Park and Lee 2006). It is a succulent herbal plant and can grow in any soil preferably dry, sandy and well-drained soil (Fuentes et al. 2000). The aloe plants are commonly used in skin care and cosmetics and as nutraceuticals (Gordon and David 2001). The two major valuable substances from leaf source of *Aloe vera* are gel and latex (Ni and Tizard 2004). Aloe gel is widely used as a folk remedy for gastrointestinal medical problem, skin injuries and burns and also as a constituent in health foods and cosmetics (Capasso et al. 1998). Aloe juice is useful in smooth functioning of the body mechanism (Saroj et al. 2004). In aloe plants, there are several potentially active components including mono- and polysaccharides, anthraquinones, saponins, enzymes, amino acids, vitamins, salicylic acid, lignin and steroids (Campestrini et al. 2006). Polysaccharide in aloe plants has the ability to control blood sugar, stimulate the production of body's own antioxidant as well as reduce cholesterol (Jones and Aloecorp 2005). *Aloe vera* emodin, an anthraquinone, has the ability to suppress or inhibit the growth of malignant cancer cells (Thomson 2004).

20.2 Need for In Vitro Plant Propagation

Aloe vera L. is one of the economically important crops, and it is a succulent plant which can be propagated by using seeds, root suckers and rhizome cuttings. But, aloe is mostly propagated by root suckers as seed propagation is not efficient due to the high occurrence of male sterility and also great interval between male and female flowering time (Liu et al. 2011); further plants raised from seeds are normally heterozygous. As planting materials, 3–4-month-old plants having three to four leaves are normally used. Generally, limited number of lateral shoots per plant is produced in a year in conventional method. Marfori and Malasa (2005) mentioned that five to ten suckers per plant are formed in a year. As a result, it is difficult to obtain larger number of planting materials for commercial cultivation through conventional vegetative propagation. Alternatively, rapid mass propagation of uniform healthy plants is possible using in vitro technique in aloe which is widely distributed worldwide (Steenkamp and Stewart 2007). Hosseini and Parsa (2007) mentioned that the conventional vegetative propagation in aloe is not sufficient to meet its demand for growing plants continuously and it is an important plant for medical, ornamental and cosmetic uses.

There are several reports stating the potential of in vitro propagation of aloe during the past decades (Aggarwal and Barna 2004; Liao et al. 2004; Baksha et al. 2005; Marfori and Malasa 2005; Hosseini and Parsa 2007; Hashemabadi and Kaviani 2008; Bhandari et al. 2010; Nayanakantha et al. 2010; Haque and Ghosh 2013; Daneshvar et al. 2013; Abdi et al. 2013; Dwivedi et al. 2014; Sahoo and Rout

2014; Lobine et al. 2015). In vitro propagation technique has been used to produce significantly larger number of genetically identical healthy plants from a single mother plant within a short period over conventional method of propagation. In addition, plants can be produced throughout the year, and plantlets can be conserved under in vitro conditions. Further, somatic variants would be developed through indirect organogenesis. Moreover, genetic transformation of aloe varieties needs efficient in vitro culture techniques for obtaining competent explants for plant transformation and regeneration (Velcheva et al. 2005).

20.3 In Vitro Shoot Organogenesis

In vitro organogenesis in plants involves formation of plant organ either shoots or roots directly from cultured explants or indirectly from callus under in vitro conditions. They are unipolar structures and physically attached to the explants or callus. Rapid in vitro plant propagation of *Aloe vera* L. (syn *Aloe barbadensis* Mill) and other aloe species has been reported by various researchers (Roy and Sarkar 1991; Abrie and Staden 2001; Baksha et al. 2005; Saroha et al. 2005; Hashemabadi and Kaviani 2008; Nayanakantha et al. 2010; Abdi et al. 2013; Dwivedi et al. 2014; Lavakumaran and Seran 2014; Lobine et al. 2015). Plantlets were effectively produced from shoot tip, rhizomatous stem and leaf segment of aloe. For successful plant production of *Aloe vera*, most of the researchers performed two-step process using one medium for shoot formation and the other medium for rooting of shoots to produce plantlets. Hashemabadi and Kaviani (2008) and Dwivedi et al. (2014) recommended one-step method to produce plantlets directly from shoot tips and used MS (Murashige and Skoog 1962) medium containing 6-benzylaminopurine (BAP) with α-naphthaleneacetic acid (NAA) or BAP alone, respectively, for the shoot induction, shoot proliferation and also rooting of shoots. In contrary, Abdi et al. (2013) demonstrated three-step procedure, i.e. shoot induction in MS medium with BAP and NAA, shoot proliferation in MS medium with BAP and rooting of proliferated shoots in B5 (Gamborg et al. 1968) medium with NAA. It may be due to the genetic variation of *Aloe vera*.

Roy and Sarkar (1991) reported the rapid propagation of *Aloe vera* through callus stage where callus was formed in stem segments excised from young axillary shoots. Garro-Monge et al. (2008) and Choudhary et al. (2011) regenerated the shoots successfully from callus derived from zygotic embryos and shoot tips, respectively. Callus is an unorganized cell mass which can redifferentiate to form plant organs or somatic embryos. Garro-Monge et al. (2008) stated that the maximum number of shoots was formed from embryogenic calluses derived from zygotic embryos in medium containing 2 mg/L BAP and 0.05 mg/L 2,4-dichlorophenoxy acetic acid (2,4-D); however, somatic embryos were not developed. It may be that appropriate concentrations of phytohormones are very important to regulate the pathway of plant regeneration. In most of the previous studies, BAP alone or in combination with NAA has been widely used in shoot organogenesis of aloe plants; however, thidiazuron (TDZ) has not been generally used in aloe species. Huetteman and Preece (1993) suggested that TDZ is also an active cytokinin for shoot induction in plant tissue culture. It has equal or more effect than BAP, and it requires at one-tenth of the optimum BAP concentration (Tawfik and Mohamed 2005). In vitro shoot proliferation and subsequent rooting can be automated in bioreactors. Plant propagation technique through in vitro culture is useful in rapid plant production, genetic improvement and germplasm conservation of aloe plants.

20.3.1 Establishment of Culture

Several factors have effect on successful micropropagation of *Aloe* species under in vitro conditions. They are disinfestation procedure, explant source, size and physiological stage of explants, phenolic compounds in explants, genotypes, culture medium and other aspects. Aseptic cultures are needed to establish cultures for morphogenesis of plant tissues.

20.3.2 Explant Disinfection

An explant is a piece of plant tissue or organ which is used as initial source of in vitro culture. They are thoroughly surface sterilized with a disinfectant solution to remove surface contaminants prior to culturing. Most common contaminants are fungi and bacteria which emit metabolic wastes that are noxious to the cultured tissues. Normally plant parts are collected from the healthy mother plants, and explants are excised from them and thereafter washed under running tap water for about 1 h. Subsequently they are dipped in 70% alcohol for 30 s and surface sterilized for 20–30 min by using disinfectant solution with two or three drops of Tween 20 as a wetting agent. They are then washed three times with sterilized distilled water. Exposure time and selection of disinfectant are crucial to eradicate contaminates without damaging plant tissues. The exposure time of sterilizing chemicals depends on source, type and size of explants to be sterilized. Generally, sterilizing chemicals used in micropropagation of aloe plants are ethyl alcohol, sodium hypochlorite, calcium hypochlorite and mercuric chloride.

Sodium hypochlorite (NaOCl) is commonly used as disinfectant solution for sterilizing many plant types in plant tissue culture, and 1.4% NaOCl solution for 1 min is an effective disinfectant agent from leaf explants of *A. crasna* and *A. sinensis* from greenhouse source (Okudera and Ito 2009). In *Aloe vera*, shoot tips were surface sterilized with 2% (w/v) NaOCl by Hashemabadi and Kaviani (2008) and 20% Clorox™ (NaOCl – 5.25% active ingredient) by Lavakumaran and Seran (2014) for 30 min to excise 1.0–1.5-cm-long segments. Blume and Neis (2005) indicated that combined application of NaOCl and ultrasound treatment increased the effectiveness of explant sterilization. Further, it is demonstrated by Qi et al. (2005) who treated the seeds effectively by using mercuric chloride ($HgCl_2$) at 0.1–0.2% for 10–15 min in *A. agallocha* and *A. sinensis*. Choudhary et al. (2011) revealed that shoot tips were immersed in 1% Bavistin for 30 min and then treated with 0.1% (w/v) $HgCl_2$ solution for 20 min to take 1.0–2.0-cm-long segments in *Aloe vera*. This is also stated by Dwivedi et al. (2014) but treated with 0.15% Bavistin for 5 min and then surface sterilized with 0.1% $HgCl_2$ solution for 5 min. Thus, disinfectant solution can be chosen depending on explant type and source to effectively sterilize explants for in vitro culture.

20.3.3 Explant Type

Healthy plant materials which are to be propagated are imperative to produce healthiest plants. The larger the size of explants, the more is the chance of contamination. As a result, the size of explants should be optimum. Juvenile tissues are generally quick to respond as compared to older tissues. Type, size and physiological stage of explants to be used are crucial in the initiation of culture and subsequently the expression of morphogenic potential of the cultured explants. Every living cell has a potential to produce complete plantlet, but not all plant cells have the equal capability to express totipotency (Sasikumar et al. 2009). The commonly used explants for direct in vitro propagation of *Aloe vera* L. (syn. *Aloe barbadensis* Mill) are shoot tips and lateral buds (Saroha et al. 2005; Baksha et al. 2005; Marfori and Malasa 2005; Hashemabadi and Kaviani 2008; Bhandari et al. 2010; Nayanakantha et al. 2010; Abdi et al. 2013; Daneshvar et al. 2013; Lavanya and Seran 2013; Dwivedi et al. 2014), whereas rhizomatous stem and leaf segment are also used as explants by some researchers (Roy and Sarkar 1991; Haque and Ghosh 2013; Sahoo and Rout 2014). Shoot tip explants showed quick response to induce shoot buds over leaf base explants (Lavanya and Seran 2013); further shoot tips having leaf base were better than those without leaf base (Abdi et al. 2013). Long et al. (2007) declared that stem segment was the best explant for callus formation from aloe tissue. *Aloe vera* which is a succulent plant exudes phenolic substance from cut surface into the culture medium, and this compound leads to reduce the survival of explants (Roy and Sarkar 1991). In general, polyvinylpyrrolidone (PVP), ascorbic acid, citric acid and charcoal are incorporated into the culture medium to reduce the phenolic compound. Nayanakantha et al. (2010) stated that inclusion of 1 g/L PVP was not effective to control phenolic browning of explants; hence they used MS medium containing 4 mg/L BAP + 0.2 mg/L NAA + 1 g/L PVP + 10 mg/L citric acid + 0.5 g/L activated charcoal to obtain more adventitious buds in *Aloe vera*.

20.3.4 Culture Initiation

Successful plant propagation of aloe using in vitro techniques mainly depends on selection of suitable explant, composition of basal medium as well as combination and concentrations of plant growth regulators. Intact or segment of shoot tip which has meristematic tissue is commonly used as an explant for shoot regeneration of *Aloe vera* (Table 20.1). Explant browning is a major obstacle for the establishment of aloe culture (Roy and Sarkar 1991; Abrie and Staden 2001). It can be eliminated or minimized by the addition of an antioxidant into the culture medium or other possible ways for reducing phenolase activity. Roy and Sarkar (1991) reported that incorporation of PVP in the culture medium assuaged the explant browning. This is in conformity with the finding of Bhatt and Dhar (2004) who stated that pretreatment of mature explants with PVP was effective to control phenol exudation in *Myrica esculenta*. Conversely, Nayanakantha et al. (2010) noticed that inclusion of PVP at a rate of 1 g/L was not effective to control phenolic browning of explants. This may be due to genetic variation.

MS medium is commonly used in micropropagation of aloe plants (Table 20.1). The medium has been solidified with 0.6–0.8% agar (Baksha et al. 2005; Nayanakantha et al. 2010; Abdi et al. 2013; Sahoo and Rout 2014), and as a carbon source, 3% sucrose has been generally incorporated into the medium in aloe organogenesis (Marfori and Malasa 2005; Hashemabadi and Kaviani 2008; Nayanakantha et al. 2010; Choudhary et al. 2011; Abdi et al. 2013; Dwivedi et al. 2014). Organogenesis implies initiation of shoots and roots directly or indirectly from in vitro cultured explants. Plant growth regular is an important factor for morphogenesis. Depending on the genotypes, appropriate level of phytohormones should be determined. In general, cytokinins are used for shoot regeneration. Many reports showed that cytokinin is requisite at optimal concentration to stimulate multiple shoots in many plant species; however, incorporation of auxin into the culture medium containing cytokinin enhances shoot regeneration when added at low concentration.

Selection of proper cytokinin and/or auxin at appropriate concentration is vital in clonal propagation of aloe plants. In general, a high ratio of cytokinin/auxin encourages shoot formation, while low ratio induces root formation. BAP alone (Marfori and Malasa 2005; Haque and Ghosh 2013; Dwivedi et al. 2014) or with auxins is often used in initiation of aloe shoots (Baksha et al. 2005; Hashemabadi and Kaviani 2008; Bhandari et al. 2010; Nayanakantha et al. 2010; Abdi et al. 2013; Sahoo and Rout 2014) as shown in Table 20.1. Liao et al. (2004) examined the effects of BA, NAA and sucrose on shoot bud initiation and suggested semi-solid MS medium containing 2.0 mg/L BA, 0.3 mg/L NAA, 30 g/L sucrose and 0.6 g/L PVP is best for bud induction. The medium containing kinetin and BA alone could not stimulate shoots (Abdi et al. 2013). This is attributed by Lavakumaran and Seran (2014) who expressed that MS medium containing 3 mg/L BAP or 1 mg/L TDZ in combination with 0.5 mg/L NAA exhibited more response for shoot bud induction than the other treatments from the cultured shoot tip explants. It is also reported by Sahoo and Rout (2014) who exhibited that BA was effective for shoot bud development from leaf explants than kinetin and adenine sulphate.

Shoots can also be regenerated indirectly from the cultured explants, i.e. through callus stage. Generally, medium with auxin alone or equal concentrations of cytokinin and auxin stimulates callus from in vitro explants. This is in harmony with Daneshvar et al. (2013) who showed that most of explants produced callus in medium without cytokinin. Further, Choudhary et al. (2011) pointed out that MS medium supplemented with 0.5 mg/L kinetin and 0.5 mg/L 2,4-D induced better callus when tested in different concentrations of phytohormones; nevertheless shoots were developed better in 1 mg/L BAP and 0.5 mg/L NAA medium from

Table 20.1 Adenine-type cytokinins used in shoot induction and multiplication of *Aloe vera* L. (syn *Aloe barbadensis* Mill)

Explants	Cytokinin/s tested	Best culture medium	In vitro response	References
Shoot tips	BAP (0.5–4.0 mg/L)	MS + 2.0 mg/L BAP + 0.5 mg/L NAA	Shoot bud induction and multiple shoot formation	Baksha et al. (2005)
Shoot tips	BAP, kinetin (0.0–2.0 mg/L)	MS + 1.0 mg/L BAP	Shoot bud induction and multiple shoot formation	Marfori and Malasa (2005)
Embryogenic callus derived from zygotic embryos	BAP (0.5–4.0 mg/L)	MS + 2.5 mg/L BAP + 0.05 mg/L 2,4 D	Shoot formation	Garro-Monge et al. (2008)
Shoot tips	BAP (0.5–2.0 mg/L)	MS + 0.5 mg/L BAP + 0.5 mg/L NAA	Shoot bud initiation, shoot proliferation and rooting	Hashemabadi and Kaviani (2008)
Shoot tips	BAP, kinetin (0.2–1.0 mg/L)	MS + 1.0 mg/L BAP + 0.2 mg/L IBA	Shoot bud initiation and shoot proliferation	Bhandari et al. (2010)
Lateral shoot (suckers)	BAP (0.0–8.0 mg/L)	MS + 4.0 mg/L BAP + 0.2 mg/L NAA	Shoot bud induction and proliferation	Nayanakantha et al. (2010)
Callus derived from shoot tips	BAP (0.5–1.0 mg/L)	MS + 1.0 mg/L BAP + 0.5 mg/L NAA	Shoot formation	Choudhary et al. (2011)
Shoot tips	BAP, kinetin (0.0–1.0 mg/L)	MS + 1.0 mg/L BAP + 0.2 mg/L IBA	Shoot bud initiation and shoot proliferation	Kumar et al. (2011)
Shoot tips	BAP (0.5–3.0 mg/L)	MS + 2.5 mg/L BAP + 0.15 mg/L NAA	Shoot proliferation	Daneshvar et al. (2013)
	BAP, kinetin (0.5–1.0 mg/L)	MS + 1.0 mg/L BAP + 1.0 mg/L kinetin + 0.15 mg/L NAA	Shoot proliferation	
Shoot tips	BAP, kinetin (0.0–4.0 mg/L)	MS + 4.0 mg/L BAP + 0.2 mg/L NAA	Shoot bud induction	Abdi et al. (2013)
		MS + 4.0 mg/L BAP	Shoot proliferation	
Rhizomatous stem (8 mm long)	BAP, kinetin (1.0–4.0 mg/L)	MS + 2.5 mg/L BAP	Shoot bud induction and shoot multiplication	Haque and Ghosh (2013)
Shoot tips	BAP, kinetin (50 μg/L–400 μg/L)	MS + 50.0 μg/L BAP	Shoot bud initiation, shoot proliferation and rooting	Dwivedi et al. (2014)
Leaf segments (~0.5 cm^2)	BAP (0.0–3.0 mg/L), kinetin (0.0–3.0 mg/L)	MS + 2.0 mg/L BA +0.5 mg/L NAA + 40 mg/L adenine sulphate	Shoot bud initiation and shoot proliferation	Sahoo and Rout (2014)

callus that was derived from shoot tips. In contrary, You (2001) reported that BAP was the vital phytohormone for callus induction on MS basal media, followed by IBA, 2,4-D and NAA. It is also supported by Velcheva et al. (2005) who stated that MS medium with 3 mg/L BAP and 2 mg/L IAA produced callus from young seedlings. Garro-Monge et al. (2008) revealed that a high number of shoots were formed from embryogenic calluses which were derived from zygotic embryos cultured in medium containing 2 mg/L BAP and 0.05 mg/L 2,4-D. In shoot organogenesis of aloe plants, the cultures are normally incubated at temperature ranging from 23 to 27 °C under white fluorescent light with 16-h photoperiod (Baksha et al. 2005; Nayanakantha et al. 2010; Abdi et al. 2013). Light intensity of 2000–2500 lux and 70% relative humidity are often provided (Hashemabadi and Kaviani 2008; Lavakumaran and Seran 2014) for shoot regeneration.

20.3.5 Adenine-Type Cytokinins in Shoot Regeneration

The plant growth regulars particularly cytokinins are considered as an important component in the culture medium for shoot regeneration of aloe plants (Abrie and Staden 2001; Aggarwal and Barna 2004; Liao et al. 2004; Velcheva et al. 2005; Debiasi et al. 2007). There are two types of cytokinins which are adenine-type (6-benzylaminopurine, kinetin and zeatin) and phenylurea-type cytokinins (diphenylurea and thidiazuron) as stated by Aina et al. (2012). Adenine-type cytokinins have been widely used in micropropagation of aloe plants by several researchers. Among them, BAP alone or in combination with auxins is usually employed in aloe propagation under in vitro conditions by the researchers (Table 20.1). Further, some researchers stated that BAP has better response than kinetin for aloe shoot bud induction and proliferation (Marfori and Malasa 2005; Bhandari et al. 2010; Kumar et al. 2011; Haque and Ghosh 2013; Abdi et al. 2013; Sahoo and Rout 2014; Dwivedi et al. 2014). It is also reported by Singh et al. (2009) who mentioned that the response of BAP was more than the other cytokinins on micropropagation of *Aloe* species.

Some researchers exhibited that BAP alone was effective for shoot proliferation of *Aloe vera* L. (syn *Aloe barbadensis* Mill) at the different concentrations 50.0 μg/L (Dwivedi et al. 2014), 1.0 mg/L (Marfori and Malasa 2005), 2.0 mg/L (Tanabe and Horiuchi 2006), 2.5 mg/L (Haque and Ghosh 2013) and 4.0 mg/L (Abdi et al. 2013). Although shoot multiplication was achieved more in auxin-free medium, shoot initiation was more prominent in BAP with NAA (Abdi et al. 2013). On the other hand, a combination of both auxin and cytokinin is essential to enhance shoot proliferation of aloe plants (Roy and Sarkar 1991; Aggarwal and Barna 2004; Liao et al. 2004; Velcheva et al. 2005; Baksha et al. 2005; Saroha et al. 2005; Hashemabadi and Kaviani 2008; Nayanakantha et al. 2010; Daneshvar et al. 2013). MS medium containing 2.0 mg/L BAP and 0.18 mg/L IAA produced multiple shoots in *Aloe vera* in 12 days after inoculation (Saroha et al. 2005). It is also reported by Aggarwal and Barna (2004), Bhandari et al. (2010) and Kumar et al. (2011) who stated that MS medium supplemented with 1.0 mg/L BA and 0.2 mg/L IBA (indole-4-butyric acid) gave the highest shoot multiplication.

Liao et al. (2004) achieved shoot multiplication from shoot tip explants by using MS medium supplemented with 2.0 mg/L BA and 0.3 mg/L NAA. Similarly MS medium containing BA and NAA was the best medium for direct shoot regeneration in aloe plants (Baksha et al. 2005; Hashemabadi and Kaviani 2008; Nayanakantha et al. 2010). Moreover, Hashemabadi and Kaviani (2008) recommended 0.5 mg/L BA and 0.5 mg/L NAA combination is best for shoot proliferation and also rooting; further they also suggested that suitable ratio of BA to NAA was 1:1 or 2:1 for shoot bud initiation and their proliferation of *Aloe vera*. In contrast, Wu (2000) and Daneshvar et al. (2013) showed that suitable ratio of cytokinin to auxin was 10:1 for shoot multiplication of *A. arborescence* and *A. vera*, respectively. According to the available information, MS media containing 0.5–3.0 mg/L BAP with 0.0–0.5 mg/L NAA or IBA have been commonly used in direct shoot organogenesis of aloe plants.

The research study showed that the highest shoot proliferation was attained in MS medium supplemented with a combination of 2.0 mg/L BA, 0.5 mg/L kinetin and 0.2 mg/L NAA (Ahmed et al. 2007). However, Haque and Ghosh (2013) noticed the effective shoot multiplication in BAP at 2.5 mg/L when compared to kinetin at 4.0 mg/L and further noted decrease in shoot production with an increased BAP strength above the optimum level. Baksha et al. (2005) stated that there was no increase in shoot proliferation while increase concentration of BAP (4.0 mg/L) with or without NAA (0.5 mg/L). These findings revealed that the quality and quantity of phytohormones as well as genotypes of explants are more crucial for successful morphogenetic response and shoot proliferation.

In addition to BAP, adenine sulphate has been used by some researchers (Bhandari et al. 2010; Kalimuthu et al. 2010; Sahoo and Rout 2014) to enhance shoot proliferation of aloe plants. Though MS medium supplemented with 1.5 mg/L BAP and 50 mg/L adenine sulphate produced high shoot proliferation (Kalimuthu et al. 2010), inclusion of 50–150 mg/L adenine sulphate into the culture medium containing 0.2 mg/L BA and 0.2 mg/L IBA has no remarkable effect on shoot multiplication (Bhandari et al. 2010). Adenine sulphate generally performs as a precursor for natural cytokinin synthesis. Moreover, MS medium supplemented with 4 mg/L BAP + 0.2 mg/L NAA + 1 g/L PVP gave better shoot induction and elongation (Nayanakantha et al. 2010). In MS medium containing 1 mg/L BA and 0.2 mg/L IBA, the highest shoot multiplication was achieved by Aggarwal and Barna (2004) and Kumar et al. (2011) who further reported that citric acid at 10 mg/L and liquid medium with the same composition enhanced the shoot multiplication. According to their report, liquid medium is better than solid medium for shoot proliferation in *A. vera* and reduces the cost of production for commercial plant propagation. It is also reported by Adelberg and Adelberg (2012) who recommended liquid medium with lesser concentration (about 6 μM) of BA for the production of larger in vitro plantlets of *A. barbadensis* over those in solid medium without affecting the multiplication rate. The cultured tissue in liquid medium and composition of mineral salts in basal medium also determine the degree of shoot bud induction and their proliferation. Abdi et al. (2013) showed the evidence of highest rate of shoot induction from shoot tips cultured in MS basal medium with BA and NAA than that in B5 basal medium containing BA and kinetin with NAA. Debiasi et al. (2007) stated that the use of MS medium is best for in vitro shoot multiplication.

20.3.6 Thidiazuron in Shoot Organogenesis

Thidiazuron (TDZ) is a phenylurea-based cytokinin which was initially tested by Mok et al. (1982) to make use of it in plant tissue culture. During the past decades, many reports indicated that application of TDZ enhances shoot bud initiation and shoot proliferation in annual and perennial herbs. It boosts cell division and shoot differentiation (Nathiya et al. 2013). And also it increases cell elongation and breaks apical dominance (Guo et al. 2011). Generally, responses of TDZ are similar to the responses stimulated by the commonly used adenine-based cytokinins in plant regeneration under in vitro conditions. TDZ has a powerful role in physiological activity of plant tissues during its in vitro culture, and it is more effective than other plant growth regulators as it requires low concentration (Guo et al. 2011). Genkov and Iordanka (1995) mentioned that TDZ has a potent cytokinin-like activity at 50–100 times lower concentrations than BAP; thus, it is inexpensive because it needs very less amount. Yildirim and Turker (2009) showed that TDZ was more productive in shoot organogenesis of meadowsweet as compared to BA and kinetin. TDZ exhibited higher in vitro response than BAP in shoot regeneration from leaf and petiole explants of *Begonia tuberhybrida* (Nada et al. 2011) and shoot bud induction from shoot tips of *Aloe vera* (Lavakumaran and Seran 2014). This effect is consistent with the finding of Thinesh and Seran (2015) in friable callus production from stem explants of *Hylocereus undatus* (dragon fruit) for plant regeneration through organogenesis or somatic embryogenesis. The stronger effect of TDZ is also confirmed by Wang et al. (2009) who revealed that when evaluating shoot bud formation in BA (1.0–5.0 mg/L), kinetin (1.0–5.0 mg/L) and TDZ (0.1–0.5 mg/L with NAA (0.05–0.5 mg/L), 0.2 mg/L TDZ and 0.1 mg/L NAA gave better axillary bud induction in *Cannabis sativa* (hemp). Moreover, they pointed out that regenerated plantlets in the medium supplemented with TDZ were more dense and vigorous. Nathiya et al. (2013) stated that BAP had better multiple shoot initiation than TDZ in *Withania somnifera* (ashwagandha). This finding may be attributed to genotypic variation.

Endogenous plant growth regulators may be changed directly or indirectly by TDZ which has an efficient role in plant cell and tissue culture (Guo et al. 2011). TDZ is a vital component in tissue culture of woody plant and herbaceous crop species (Huetteman and Preece 1993). According to available literature, TDZ is not widely used in micropropagation of aloe species although it has a significant effect on shoot proliferation of other plant species (Table 20.2). Addition of TDZ to the culture medium induced viable adventitious shoot regeneration in many plant species (Mok et al. 2000; Uranbey 2005; Faisal et al. 2005; Peddabonia et al. 2006; Yildirim and Turker 2009; Nada et al. 2011; Nathiya et al. 2013; Lobine et al. 2015). Morphogenesis of in vitro cultured explants to TDZ mainly depends on the type of explants, genotypes, concentration of TDZ added to the culture medium and also endogenous plant growth regulars available in the cultured explants. The low concentration of TDZ is normally used for axillary shoot proliferation in culture (Mok et al. 2000; Singh and Syamal 2001; Guo et al. 2011). Lobine et al. (2015) reported that the highest protocorm-like structures were formed in the medium containing a very low level (0.01 mg/L) of TDZ. On the other hand, they noticed the necrotic tissues when these structures were maintained in the same medium for a long period. Petiole explants which were excised from seedlings pretreated with 1 mg/L TDZ for 5 weeks produced considerably more shoots in BAP and NAA medium than those taken from seedlings pretreated with 3 or 5 mg/L TDZ (Gürel et al. 2003). Short exposure to TDZ stimulates various morphogenetic responses (Guo et al. 2011); meanwhile TDZ inhibits shoot elongation (Ledbetter and Preece 2004). As a result, the cultured explants in TDZ would be transferred to the basal medium (Guo et al. 2011) or medium with other commonly used cytokinins (Debnath 2005; Lavakumaran and Seran 2014) to reproduce normal shoots.

TDZ and BAP have diverse roles in plant development during the activity of shoot meristems and shoot formation (Kyozuka 2007). Saroha et al. (2005) expressed that when vertically excised shoot tips of *Aloe vera* were placed on MS basal media with various growth regulators for shoot regeneration, MS medium supplemented with BAP at 2 mg/L gave better response within 15 days, while BAP at 10 mg/L exhibited poor shoot proliferation. Further, they revealed that medium containing TDZ at 1 mg/L slightly increased shoot formation. This finding is in harmony with Lavakumaran and Seran (2014), and also microshoot buds were transferred to BAP (3 mg/L) from TDZ (1 mg/L) for shoot elongation, and further they observed abnormal structures at the high concentration (3 mg/L) of TDZ medium. Liao et al. (2004) revealed that the degree of shoot bud proliferation decreases in higher concentrations of NAA or BA or kinetin medium. A high concentration of cytokinin leads to cytogenetic instability (Qu et al. 2000). Optimum concentration and appropriate type of cytokinins are principally required to expose the morphogenetic response for shoot regeneration under in vitro conditions. Combined application of TDZ and BA showed efficient shoot initiation in *Aloe arborescens* (Velcheva et al. 2005). It is further supported by Nathiya et al. (2013) in *Withania somnifera* that MS media containing 0.5 mg/L BAP and 0.05 mg/L TDZ produced the highest shoot induction than media supplemented with BAP and TDZ alone, whereas Nada et al. (2011) highlighted that the number of shoots per explants of *Begonia tuberhybrida* was decreased in BAP and TDZ medium.

Not only TDZ and BAP have a stronger effect on shoot bud induction and proliferation, but also TDZ in combination with auxins possesses better results in some plant species (Table 20.2). Inclusion of auxins, particularly IAA, NAA or IBA, into the culture medium containing TDZ exhibited greater multiple shoot induction in aloe plants (Lincy and Sasikumar 2010; Lavakumaran and Seran 2014; Lobine et al. 2015) and other plants (Kusula et al. 2008; Wang et al. 2009; Nada et al. 2011). Lavakumaran and Seran (2014) stated that TDZ at low concentration (1 mg/L) with NAA (0.5 mg/L) had better shoot bud induction in *Aloe vera* as appeared in BAP at 3 mg/L and NAA at 0.5 mg/L. It is also observed in endemic threatened Mascarene *Aloes*, namely, *Aloe lomatophylloides*, *A. macra*, *A. purpurea* and *A. tormentorii*, by Lobine et al. (2015) who noted more protocorm-like bodies in MS medium with TDZ and NAA. This is in conformity with Lincy and Sasikumar (2010) who reported that TDZ and IBA alone or in combination with other hormones resulted in enhanced multiple shoot induction in Chinese aloe. Similarly, it has been noticed in

Table 20.2 TDZ used for in vitro shoot induction and proliferation of some plant species

Plant species (explants)	PGR tested	Best culture medium	Growth response	References
Aloe vera L. (shoot tip segments)	BAP alone or with TDZ	MS + 2.0 mg/L BAP + 1.0 mg/L TDZ	Shoot formation	Saroha et al. (2005)
Aloe lomatophylloides Balf.f	BAP, TDZ alone or with NAA	MS + 0.01 mg/L TDZ + 0.01 mg/L NAA	Adventitious shoot proliferation	Lobine et al. (2015)
A. macra Haw				
A. purpurea Lam (0.5-cm-long hypocotyls with radicles)				
Begonia tuberhybrida (1 cm^2 mature leaves)	BAP alone or with NAA and/or TDZ	MS + 2.0 mg/L TDZ + 1.0 mg/L NAA	Shoot bud formation	Nada et al. (2011)
Begonia tuberhybrida (1-cm-long petioles)		MS + 2.0 mg/L TDZ + 0.5 mg/L NAA		
Capsicum annum L. (1 cm^2 cotyledon and leaf explants)	BAP, TDZ, kinetin alone or with IAA	MS + 9.0 μM TDZ + 2.8 μM IAA	Shoot regeneration and elongation	Kusula et al. (2008)
Solanum tuberosum L. (1.0-cm-long shoot apices)	TDZ	MS + 10^{-8} M TDZ	Plantlet formation	Sajid and Aftab (2009)
Cannabis sativa L. (shoot tips)	BAP, TDZ, kinetin with NAA	MS + 0.2 mg/L DZ + 0.1 mg/L NAA	Auxiliary bud multiplication	Wang et al. (2009)
Calendula officinalis (apical explants 1.0–1.2 cm)	BAP, TDZ alone or with IAA	MS + 0.8 mg/L BAP or 0.2 mg/L TDZ	Shoot proliferation	Victório et al. (2012)
Withania somnifera (shoot tips)	BAP, TDZ alone or in combination	MS + 0.5 mg/L BAP + 0.05 mg/L TDZ	Multiple shoot formation	Nathiya et al. (2013)
Curculigo latifolia (4.0-mm-long shoot tips)	TDZ with IBA	MS + 0.5 mg/L TDZ + 0.25 mg/L IBA	Shoot regeneration	Babaei et al. (2014)
Achillea millefolium (terminal segment of rhizome)	TDZ	MS + 0.75 mg/L TDZ	Shoot multiplication	Alvarenga et al. (2015)

chilli pepper (Kusula et al. 2008), hemp (Wang et al. 2009) and *Begonia* (Nada et al. 2011) by the application of TDZ in combination with IAA or NAA. Moreover, Victório et al. (2012) found that BAP and IAA or TDZ and IAA did not remarkably affect shoot responses. In petiole explants of *Begonia*, the highest number of shoots per explant was obtained in 2.0 mg/L TDZ and 0.5 mg/L NAA, but the number of shoots produced was significantly decreased in BAP alone or in combination with NAA and/or TDZ (Nada et al. 2011).

In ginger, low concentration of BAP with NAA persuaded the shoot induction slowly but favours for shoot elongation from multiple shoots, whereas comparatively high concentration of BAP with NAA resulted in stunted multiple shoot buds from rhizome explants (Sathyagowri and Seran 2011, 2013). It is also reported by Hiremath (2006) who observed that high concentration of cytokinins exhibited the greater number of multiple shoots and reduction in shoot length. Ivanova (2009) stated that medium containing TDZ at 5 μM and 15 μM formed shoot buds which did not develop into shoots in *Aloe polyphylla*. However, Nathiya et al. (2013) noticed the shoots in medium containing TDZ which were very short and firm in *Withania somnifera*; further they observed that BAP medium produced less shoot primordia and well-differentiated shoot length than TDZ medium. It is also supported by Victório et al. (2012) who revealed that shoot elongation of *C. officinalis* was suppressed by the application of TDZ though it gave maximum shoots in low concentrations. TDZ at 0.5 mg/L stimulates a highest number of shoots by indirect organogenesis in blueberry (Cappelletti et al. 2016). According to the available reports, TDZ is used to induce multiple shoot buds in some plant species, but adenine-type cytokinins, particularly BAP, are needed for the production of elongated shoots in organogenesis under in vitro conditions where TDZ functions as an effective bioregulant (Matand and Prakash 2007).

20.4 In Vitro Rooting

The in vitro proliferated shoots are rooted to form plantlets. For in vitro rooting, the uniform healthy shoots are selected and normally transferred to culture medium supplemented with auxin alone or in combination with reduced level of cytokinin. The requirement of plant growth regulators for root initiation and their subsequent development mainly depends on plant species and phytohormones added in medium for shoot proliferation. In the previous reports, auxin alone particularly NAA or IBA has been used by many researchers for in vitro rooting of aloe shoots (Baksha et al. 2005; Marfori and Malasa 2005; Nayanakantha et al. 2010; Daneshvar et al. 2013; Abdi et al. 2013; Sahoo and Rout 2014). Daneshvar et al. (2013) exhibited the root induction within 3 weeks after culturing in medium containing NAA or IBA (0.25–0.5 mg/L) and further described that the quality of developed roots in 1.0 mg/L IBA is better than in 1.0 mg/L NAA. Baksha et al. (2005) indicated that in higher auxin level (1.0–1.5 mg/L), the quality of developed roots was poor. The best auxin for root induction (Baksha et al. 2005) and root formation (Marfori and Malasa 2005) was NAA than IAA and IBA. An optimum concentration of NAA added to MS medium is 0.10 mg/L for rooting (Marfori and Malasa 2005).

The rooted shoots have been attained not only in auxin alone medium but also in combination with cytokinin. Hashemabadi1 and Kaviani (2008) stated that rooting of the proliferated shoots was better in MS medium containing 0.5 mg/L BA plus 0.5 mg/L NAA or 1 mg/L BA plus 0.5 mg/L NAA. In contrast, Nayanakantha et al. (2010) reported that shoots in MS medium with BAP alone or in combination with NAA did not produce roots. The reason for the result obtained may be due to relatively high ratio of cytokinin/auxin used by them. In addition to the phytohormones, the quantity of mineral salts in the medium is a vital role in rooting of shoots. Within 3–4 weeks of culture, the high percentage (<75–95%) of rooting was attained in half MS medium with 0.5 mg/L NAA (Sahoo and Rout 2014; Baksha et al. 2005) but 100% rooting in MS medium with 0.2 mg/L NAA (Nayanakantha et al. 2010) and B5 medium with 2.0 mg/L NAA (Abdi et al. 2013).

Some researchers reported rooting of aloe shoots in MS medium without growth regulators (Aggarwal and Barna 2004; Bhandari et al. 2010; Nayanakantha et al. 2010; Haque and Ghosh 2013; Lavakumaran and Seran, 2014) or with cytokinin alone (Hashemabadi1 and Kaviani 2008; Dwivedi et al. 2014). The rooting percentage was 90% in MS medium supplemented with 0.5 g/L activated charcoal (Nayanakantha et al. 2010). Plant regeneration, i.e. shoot regeneration and root formation, was achieved using only MS medium supplemented with 2.5 mg/L BAP (Hashemabadi and Kaviani 2008) or with 50 μg/L BAP (Dwivedi et al. 2014) in micropropagation of *Aloe vera*. Leaf gel of *Aloe vera* (AVG) as organic supplement (20.0%) was also incorporated to one-third strength of MS medium for successful rooting of regenerated shoots from nodal segment of rhizomatous stem by Haque and Ghosh (2013). Sometimes, the regenerated shoots are dipped into auxin solution and then directly transferred to the auxin-free medium or the soil to develop roots. And also the proliferated shoots or plantlets in culture should be prepared for acclimatization to increase the survival rate during transplanting.

20.5 Acclimatization

Successful transfer of the in vitro plantlets to the soil under ex vitro condition is one of the most important stages in plant propagation. Normally, roots formed in the culture are non-functional, and in vitro plantlets are very sensitive to ambient environmental conditions; therefore, hardening of plants is necessary to adapt them to ex vitro conditions and also to produce normal roots. There is also a possibility to acclimatize the non-rooted shoots in the soil under the controlled environment. In general, the rooted shoots are collected from the in vitro culture and washed the agar. Later they are transplanted to the suitable soil mixture in small pots and placed under the controlled environmental conditions in greenhouse. Subsequently, plants are gradually exposed to reduce the relative humidity, and the acclimatized plants are then transferred to the field. Haque and Ghosh (2013) and Sahoo and Rout (2014) mentioned that in vitro plantlets of aloe plants were successfully hardened in the greenhouse. The various potting mixtures such as mixture of soil and farmyard manure (Aggarwal and Barna 2004), coconut peat and perlite (Hashemabadi and Kaviani 2008) and sand/soil/cow dung (1:1:1) (Sahoo and Rout 2014) have been used by the researchers for hardening plantlets in the greenhouse. The acclimatized plants were transplanted in soil and vermin compost (3:1) mixture with 100% survival rate (Haque and Ghosh 2013), and they were successfully transferred in the field (Marfori and Malasa 2005).

References

Abdi G, Hedayat M, Modarresi M (2013) In vitro micropropagation of *Aloe vera* -impacts of plant growth regulators, media and type of explants. J Biol Environ Sci 7(19):19–24

Abrie A, Staden JV (2001) Micropropagation of endangered *Aloe polyphylla*. Plant Growth Regul 33(1):19–23

Adelberg J, Adelberg JN (2012) Effects of cytokinin on multiplication and rooting of *Aloe barbadensis* during micropropagation on agar and liquid medium. J Med Active Plant 1(1):1–5

Aggarwal D, Barna KS (2004) Tissue culture propagation of elite plant of *Aloe vera* Linn. J Plant Biochem Biotech 13(1):77–79

Ahmed S, Kabir AH, Ahmed MB, Razvy MA, Ganesan S (2007) Development of rapid micropropagation method of *Aloe vera* L. Sjemenarstvo 24(2):121–128

Aina O, Quesenberry K, Gallo M (2012) Thidiazuron-induced tissue culture regeneration from quartered-seed explants of Arachis paraguariensis. Crop Sci 52(3):1076–1083

Alvarenga ICA, Silva ST, Bertolucci SKV, Pinto JEBP, Pacheco FV (2015) Application of Thidiazuron (TDZ) for in vitro multiplication of yarrow (*Achillea millefolium* L.) and profile of volatile compounds. Aust J Crop Sci 9(10):948–953

Babaei N, Abdullah NAP, Saleh G, Abdullah TL (2014) An efficient in vitro plantlet regeneration from shoot tip cultures of *Curculigo latifolia*, a medicinal plant. Sci World J. https://doi.org/10.1155/2014/275028

Baksha R, Jahan MAA, Khatun R, Munshi JL (2005) Micropropagation of *Aloe barbadensis* Mill through in vitro culture of shoot tip explants. Plant Tissue Cult Biotech 15(2):121–126

Bhandari AK, Negi JS, Bisht VK, Bharti MK (2010) In vitro propagation of *Aloe vera*-A plant with medicinal properties. Nat Sci 8(8):174–176

Bhatt ID, Dhar U (2004) Factors controlling micropropagation of *Myrica esculenta* buch.-Ham.ex D. Don. A high value wild edible of Kumaun Himalaya. Afr J Biotech 3(10):534–540

Blume T, Neis U (2005) Improving chlorine disinfection of wastewater by ultrasound application. Water Sci Tech 52(10–11):139–144

Campestrini LH, Kuhnen S, Lemos PMM, Bach DB, Dias PF, Maraschin M (2006) Cloning protocol of *Aloe vera* as a study-case for "tailor-made" biotechnology to small farmers. J Technol Manag Innov 1(5):76–79

Capasso F, Borrelli F, Capasso R (1998) *Aloe* and its therapeutics use. Phytother Res 12(S1):121–127

Cappelletti R, Sabbadini S, Mezzetti B (2016) The use of TDZ for the efficient *in vitro* regeneration and organogenesis of strawberry and blueberry cultivars. Sci Hortic 207:117–124. https://doi.org/10.1016/j.scienta.2016.05.016

Choudhary AK, Ray AK, Jha S, Mishra IN (2011) Callus formation, shoot initiation and in vitro culture of *Aloe vera*. Biotechnol Bioinformatics Bioeng 1(4):551–553

Daneshvar MH, Moallemi N, Zadeh NA (2013) The effect of different media on shoot proliferation from the shoot tip of *Aloe vera* L. Jundishapur J Nat Pharm Prod 8(2):93–97

Debiasi C, Silva CG, Pescador R (2007) Micropropagation of *Aloe vera* L. Rev Bras Plant Med Botucatu 9(1):36–43

Debnath SC (2005) Strawberry sepal: another explant for thidiazuron induced adventitious shoot regeneration. In Vitro Cell Dev Biol Plant 41(5):671–676

Dwivedi NK, Indiradevi A, Suma A, Nair RA (2014) Rapid in vitro micropropagation of non-bitter vegetable type *Aloe vera* L. (IC333202). Res Plant Biol 4(4):39–43

Faisal M, Ahmad N, Anis M (2005) Shoot multiplication in *Rauvolfia tetraphylla* L using thidiazuron. Plant Cell Tissue Organ Cult 80(2):187–190

Fuentes SRL, Calheiros MBP, Manetti-Filho J, Vieira LGE (2000) The effects of silver nitrate and different carbohydrate sources on somatic embryogenesis in *Coffea canephora*. Plant Cell Tissue Organ Cult 60(1):5–13

Gamborg OL, Miller RA, Ojima K (1968) Nutrient requirements of suspension cultures of soybean root cells. Exp Cell Res 50(1):151–158

Garro-Monge G, Gatica-Arias AM, Valdez-Melara M (2008) Somatic embryogenesis, plant regeneration and acemannan detection in aloe (*Aloe barbadensis* Mill.) Agron Costarric 32(2):41–52
Genkov T, Iordanka I (1995) Effect of cytokinin-active phenylurea derivatives on shoot multiplication, peroxidase and superoxide dismutase activities of in vitro cultured carnation. Bulg J Plant Physiol 21(1):73–83
Gordon MC, David JN (2001) Natural product drug discovery in the next millennium. Pharm Biol 39(1):8–17
Guo B, Abbasi BH, Zeb A, Xu LL, Weib YH (2011) Thidiazuron: a multi-dimensional plant growth regulator. Afr J Biotech 10(45):8984–9000
Gürel S, Topal E, Gürel E (2003) The effect of pretreating seedlings with TDZ on direct shoot regeneration from petiole explants of sugar beet (*Beta vulgaris* L.) Asia Pac J Mol Biol Biotechnol 11(1):57–62
Haque SKM, Ghosh B (2013) High frequency microcloning of *Aloe vera* and their true-to-type conformity by molecular cytogenetic assessment of two years old field growing regenerated plants. Bot Stud 54:46. http://www.as-botanicalstudies.com/content/54/1/46
Hashemabadi D, Kaviani B (2008) Rapid micropropagation of *Aloe vera* L. via shoot multiplication. Afr J Biotech 7(12):1899–1902
Hiremath RC (2006) Micro propagation of ginger (*Zingiber officinale* Rosc.). M.Sc. thesis, College of Agriculture, Dharwad University of Agricultural Sciences, Dharwad, India
Hosseini R, Parsa M (2007) Micropropagation of *Aloe vera L.* grown in South Iran. Pak J Biol Sci 10(7):1134–1137
Huetteman CA, Preece JE (1993) Thidiazuron: a potent cytokinin for woody plant tissue culture. Plant Cell Tissue Organ Cult 33(2):105–119
Ivanova MV (2009) Regulation of hyperhydricity in *Aloe polyphylla* propagated in vitro. Ph.D dissertation, School of Biological and Conservation Sciences, University of KwaZulu-Natal, South Africa
Jones K, Aloecorp (2005) The antidiabetic activity of *Aloe vera*. Cosmet Sci Technol 2:34–35
Kalimuthu K, Vijayakumar S, Senthilkumar RR, Sureshkumar M (2010) Micropropagation of *Aloe vera* Linn. A medicinal plant. Int J Biotechnol Biochem 6(3):405–410
Kumar M, Singh S, Singh S (2011) In vitro morphogenesis of a medicinal plant–*Aloe vera L.* Asian J Plant Sci Res 1(1):31–40
Kusula K, Prasad S, Umate P, Gadidasu K, Abbagani S (2008) Efficient TDZ and IAA assisted plant regeneration from cotyledon and leaf explants of *Capsicum annum* L. – one step protocol for shoot bud differentiation and elongation. Int J Plant Dev Biol 2(2):114–117
Kyozuka J (2007) Control of shoot and root meristems function by cytokinin. Curr Opin Plant Biol 10(5):442–446
Lavakumaran L, Seran TH (2014) Effect of 6-benzyl-aminopurine and thidiazuron on *in vitro* shoot organogenesis of *Aloe vera* (L.) Burm. f. Chil J Agric Res 74(4):497–501
Lavanya L, Seran TH (2013) Microshoot formation from shoot tip of *Aloe vera* (L.) under in vitro conditions. In: Proceeding of 2nd international symposium on minor fruits and medicinal plants for better lives, University of Ruhuna, Sri Lanka, 20 December 2013
Ledbetter DI, Preece JE (2004) Thidiazuron stimulates adventitious shoot production from *Hydrangea quercifolia* Bartr. leaf explants. Sci Hortic 101(01–02):121–126
Liao Z, Chen M, Tan F, Sun X, Tang K (2004) Micropropagation of endangered Chinese aloe. Plant Cell Tissue Organ Cult 76(1):83–86
Lincy A, Sasikumar B (2010) Enhanced adventitious shoot regeneration from aerial stem explants of ginger using TDZ and its histological studies. Turk J Bot 34(1):21–29
Liu X, Li J, Zhang Y, Li L, He D (2011) Biological research advancement in *Aloe*. J Med Plant Res 5(7):1046–1052
Lobine D, Soulange JG, Sanmukhiya MR, Lavergne C (2015) A tissue culture strategy towards the rescue of endangered mascarene aloes. ARPN J Agric Biol Sci 10(1):28–38
Long WX, Yang XB, Qi MY, Li DH, Guo T (2007) A review of research in tissue culture of *Aloe*. Subtrop Plant Sci 36(1):70–74

Marfori EC, Malasa AB (2005) Tissue culture for the rapid clonal propagation of *Aloe barbadensis* Mill. Philipp Agric Sci 88(1):167–170

Matand K, Prakash CS (2007) Evaluation of peanut genotypes for in vitro plant regeneration using thidiazuron. J Biotechnol 130(2):202–207

Mok MC, Mok DWS, Armstrong DJ, Shudo K, Isogai Y, Okamoto T (1982) Cytokinin activity of N-phenyl-N′-1,2,3-thidiazol-5-yl urea (thidiazuron). Phytochemistry 21(7):1509–1511

Mok MC, Martin RC, Mok DWS (2000) Cytokinins: biosynthesis, metabolism and perception. In Vitro Cell Dev Biol Plant 36(2):102–107

Murashige T, Skoog F (1962) A revised medium for rapid growth and bioassays with tobacco tissue cultures. Physiol Plant 15(3):473–497

Nada S, Chennareddy S, Goldman S (2011) Direct shoot bud differentiation and plantlet regeneration from leaf and petiole explants of *Begonia tuberhybrida*. HortSci 46(5):759–764

Nathiya S, Pradeepa D, Devasena T, Senthil K (2013) Studies on the effect of sucrose, light and hormones on micropropagation and in vitro flowering of *Withania somnifera* var.jawahar-20. J Anim Plant Sci 23(5):1391–1397

Nayanakantha NMC, Singh BR, Kumar A (2010) Improved culture medium for micropropagation of *Aloe vera* L. Trop Agric Res Ext 13(4):87–93

Ni Y, Tizard IR (2004) Analytical methodology: the gel analysis of Aloe pulp and its derivatives. In: Reynolds T (ed) Aloes: the genus *Aloe*, 2nd edn. CRC Press, Boca Raton, pp 111–126

Okudera Y, Ito M (2009) Production of agarwood fragrant constituent in Aquilaria calli and cell suspension culture. Plant Biotech 26(3):307–315

Park YI, Lee SK (2006) New perspectives on aloe, 1st edn. Springer, New York

Peddabonia V, Thamidala C, Karampuri S (2006) In vitro shoot multiplication and plant regeneration in four *Capsicum* species using thidiazuron. Sci Hortic 107(2):117–122

Qi SY, He ML, Lin LD, Zhang CH, Hu LJ, Zhang HZ (2005) Production of 2-(2-phenylethyl) chromones in cell suspension cultures of *Aquilaria sinensis*. Plant Cell Tissue Organ Cult 83(2):217–221

Qu L, Polashock J, Vorsa N (2000) A high efficient in vitro cranberry regeneration system using leaf explants. HortSci 35(5):948–952

Roy SC, Sarkar A (1991) In vitro regeneration and micropropagation of *Aloe vera* (L.) Sci Hortic 47(1–2):107–113

Sahoo S, Rout GR (2014) Plant regeneration from leaf explants of *Aloe barbadensis* Mill. and genetic fidelity assessment through DNA markers. Physiol Mol Biol Plants 20(2):235–240

Sajid ZA, Aftab F (2009) Effect of thidiazuron (TDZ) on in vitro micropropagation of *Solanum tuberosum* L. cvs. desiree and cardinal. Pak J Bot 41(4):1811–1815

Saroha V, Yadav RC, Yadav NR (2005) Remove from marked Records High frequency shoot regeneration in *Aloe vera* using BAP and TDZ combinations of growth regulators. In: ICAE national symposium on biotechnological interventions for improvement of horticultural crops: issues and strategies, Vellanikkara, Kerala, India, 10–12 January 2005

Saroj PL, Dhandar DG, Singh RS (2004) Indian aloe, 1st edn. Central Institute for Arid Horticulture, Bikaner, pp 6–10

Sasikumar S, Raveendar S, Premkumar A, Ignacimuthu S, Agastian P (2009) Micropropagation of *Baliospermum montanum* (Willd.) Muell. Ara.-A threatened medicinal plant. Indian J Biotechnol 8(2):223–226

Sathyagowri S, Seran TH (2011) In vitro plant regeneration of ginger (*Zingiber officinale* Rosc.) with emphasis on initial culture establishment. Int J Med Aroma Plant 1(3):195–202

Sathyagowri S, Seran TH (2013) A note on morphological studies of shoot bud initiation from ginger (*Zingiber officinales* Rosc) rhizome cultured in vitro conditions. Sudan J Agric Res 22:101–106

Singh SK, Syamal MM (2001) A short pre-culture soak in thidiazuron or forchlorfenuron improves axillary shoot proliferation in rose micropropagation. Sci Hortic 91(1–2):169–177

Singh M, Rathore MS, Panwar D, Rathore JS, Dagla HR, Shekhawat NS (2009) Micropropagation of selected genotype of *Aloe vera* L–an ancient plant for modern industry. J Sustain For 28(8):935–950

Steenkamp V, Stewart MJ (2007) Medicinal applications and toxicological activities of aloe products. Pharm Biol 45(5):411–420

Tanabe MJ, Horiuchi K (2006) *Aloe barbadensis* Miller ex vitro autotrophic culture. J Hawahan Pac Agric 13:55–59

Tawfik AA, Mohamed MF (2005) Organogenic response of *Salvia officinalis* L. to dark preconditioning, thidiazuron and benzyladenine. Plant Growth Regul Soc Am Q 33(4):125–133

Thinesh A, Seran TH (2015) In vitro callogenesis from bud and stem explants of dragon fruit (*Hylocereus undatus*). Asian J Pharma Sci Tech 5(4):253–256

Thomson PDR (2004) Herbal medicines, 3rd edn. Thomson PDR, Montvale

Uranbey S (2005) Thidiazuron induced adventitious shoot regeneration in *Hyoscyamus niger*. Biol Plant 49(3):427–430

Velcheva M, Faltin Z, Vandi A, Eshdat Y, Perl A (2005) Regeneration of *Aloe arborescens* via organogenesis from young inflorescences. Plant Cell Tissue Organ Cult 83(3):293–301

Victório CP, Lage CLS, Sato A (2012) Tissue culture techniques in the proliferation of shoots and roots of *Calendula officinalis*. Rev Ciênc Agron 43(3):539–545

Wang R, He LS, Xia B, Tong JF, Li N, Peng F (2009) A micropropagation system for cloning of hemp (*Cannabis sativa* L.) by shoot tip culture. Pak J Bot 41(2):603–608

Wu HZ (2000) Tissue culture of *Aloe arborescens* Mill. Acta Hortic Sin 27(2):151–152

Yildirim AB, Turker AU (2009) In vitro adventitious shoot regeneration of the medicinal plant meadowsweet (*Filipendula ulmaria* (L.) Maxim). In Vitro Cell Dev Biol 45(2):135–144

You JH (2001) Advances in the study on tissue culture of Aloe. Chin Wild Plant Resour 20(2):10–11

21 Morphogenic Potential of Different Explants of Broccoli (*Brassica oleracea* L. var. *italica*): Important "Nutrient-Rich" Vegetable, Using Thidiazuron

Pankaj Kumar, Karuna Dhiman,
and Dinesh Kumar Srivastava

Abstract

Broccoli is a high-valued nutritionally rich important cole vegetable crop. But its quality and quantity are largely affected by various pests, diseases, and environmental stresses. Traditional plant breeding technologies alone are not sufficient to control its massive losses because of no natural genetic variations present in its germplasm. Genetic manipulation can supplement the conventional agricultural practices for broccoli improvement. The objective of the present study was to evaluate the effectiveness of thidiazuron (TDZ) in in vitro high-frequency plant regeneration studies in broccoli using hypocotyl, cotyledon, petiole, and leaf explants. Highly efficient and reproducible regeneration protocol has been optimized in broccoli cv. Solan green head. Explants were cultured on Murashige-Skoog (MS) medium, containing different TDZ concentrations and combinations such as TDZ alone, TDZ + adenine, TDZ + NAA (naphthalene acetic acid), and TDZ + IAA (indole acetic acid). Total 36 combinations were used; the maximum shoot regeneration response was observed in hypocotyl explants (95.92%) on MS medium supplemented with 2.0 μM TDZ + 0.5 μM IAA followed by petiole explants (91.55%) on MS medium with 2.0 μM TDZ + 0.107 μM NAA. Leaf explants gave (89.25%) shoot regeneration response on MS medium containing 1.0 μM TDZ + 0.107 μM NAA and cotyledon explants producing multiple shoots (88.88%) on MS medium supplemented with 2.0 μM TDZ + 0.59 mM adenine. After completed shoot regeneration, high-frequency (100%) root regeneration response was observed on rooting medium containing 0.10 mg/l NAA. For hardening, in vitro regenerated plantlets were transferred to pots containing cocopeat

P. Kumar
National Institute of Plant Genome Research, New Delhi, India

K. Dhiman · D. K. Srivastava (✉)
Department of Biotechnology, Dr. Yashwant Singh Parmar University of Horticulture and Forestry, Solan, Himachal Pradesh, India

N. Ahmad, M. Faisal (eds.), *Thidiazuron: From Urea Derivative to Plant Growth Regulator*, https://doi.org/10.1007/978-981-10-8004-3_21

and successfully acclimatized. Well developed in vitro plant regeneration protocol has been optimized in broccoli cv. Solan green head using potent cytokinin thidiazuron.

Keywords
Thidiazuron · Explant · Organogenesis · Rooting · Acclimatization

21.1 Introduction

Broccoli (*Brassica oleracea* L. var. *italica*) is nutritionally rich, an economically important, cole vegetable crop of the family Brassicaceae ($2n = 18$), especially rich in vitamin A, vitamin K, calcium, antioxidants, β-carotene, riboflavin, and iron content (Vallejo et al. 2003; Abdel-Wahhab and Aly 2003). The anticancer properties of broccoli are contributed by anticancerous compound sulforaphane (Zhang et al. 1992; Kumar and Srivastava 2016a, b; Kumar et al. 2017a, b) and high selenium content (Finley et al. 2001; Finley 2003). Quality and quantity of broccoli produced are largely affected by different pests, diseases, and environmental stresses during commercial cultivation of broccoli because of its limited gene pool (Cao and Earle 2003; Viswakarma et al. 2004; Parmar et al. 2017). Therefore, genetic manipulation is an appropriate method for broccoli genetic improvement.

Broccoli cv. Solan green head is used in crop breeding program due to its early maturity as well as high yield potential. The cultivar was developed by the Department of Vegetable Science, Dr. Yashwant Singh Parmar University of Horticulture and Forestry, Nauni, Solan, Himachal Pradesh, India. However, the broccoli cv. Solan green head is severely affected by insect pests, such as diamondback moth, cabbage looper, beetles, and aphids (Kumar and Srivastava 2015a; Kumar et al. 2015b). The diamondback moth is considered to be the major pest of the crucifers worldwide and has become resistant to all major categories of insecticides (Tabashnik et al. 1991). According to Kumar and Srivastava (2015a) and Kumar et al. (2015b), plant tissue culture applications along with plant genetic engineering can be used to add agronomically important target characteristics to broccoli cultivars. Establishment of reproducible, highly efficient in vitro plant regeneration system is a key step for genetic improvement of broccoli via genetic transformation procedure.

A stable, reliable, and efficient plant regeneration system is the basis of genetic manipulation studies of any plant species. Various researchers have reported plant regeneration studies in broccoli by using different explants such as anther (Chang et al. 1996), peduncle (Christey and Earle 1991), protoplasts (Kaur et al. 2006), hypocotyl (Ravanfar et al. 2009; Huang et al. 2011; Kumar and Srivastava 2015a; Kumar et al. 2015a), leaf tissue (Farzinebrahimi et al. 2012; Kumar and Srivastava 2015a), cotyledon (Qin et al. 2006; Ravanfar et al. 2011, 2014; Kumar and Srivastava 2015a, b), and petiole (Kumar et al. 2015a, b).

Thidiazuron (*N*′-phenyl-*N*′-1,2,3-thidiazol-5-ylurea, TDZ) is heterocyclic phenyl urea that has tremendous application as a potent plant growth regulator (PGR) having biological properties similar to cytokinin (Mok et al. 1982), for in vitro regeneration studies in different crops such as rubus (Fiola et al. 1990), geranium (Visser et al. 1992), peanut (Murthy et al. 1995), mustard (Guo et al. 2005), and broccoli (Ravanfar et al. 2014). Ravanfar et al. (2014), Kumar and Srivastava (2015a), and Kumar et al. (2015b) reported that TDZ in an appropriate ratio with IAA and NAA increased shoot regeneration from hypocotyl and cotyledon explants of broccoli. However, according to Ravanfar et al. (2009, 2011), BAP showed maximum shoot regeneration response from hypocotyl and cotyledon explants with or without NAA in broccoli cv. Green Marvel. This varied response of plant growth regulators in evoking morphogenic response is either due to genotypic differences or plants reacting differently during in vitro organogenesis. Different factors such as source and age of explant, genotype, culture conditions, physical environment, and concentration of plant growth regulators used affect the adventitious shoot regeneration frequency (Achar 2002; Huang et al. 2011).

The effects of source of explants, i.e., donor seedlings, on shoot regeneration from hypocotyl, cotyledon, and leaf have been reported in various species (Srivastava et al. 1989, 1991a, b; Sharma et al. 2014). In most *Brassica* species, efficient shoot regeneration depended on explant age (Rani et al. 2013). Young explants gave higher regeneration response than older explants (Chakrabarty et al. 2002; Sharma et al. 2012, 2014). In the present study, we report on the high-frequency shoot regeneration ability of different explants of broccoli (*Brassica oleracea* L. var. *italica* cv. Solan green head) using TDZ.

21.2 Materials and Methods

21.2.1 Plant Material

The certified seeds of broccoli cv. Solan green head were obtained from the Department of Vegetable Science of Dr. Yashwant Singh Parmar University of Horticulture and Forestry, Nauni, Solan (HP). The seeds of broccoli were thoroughly washed under running tap water and teepol and were soaked for half an hour. Seeds were surface sterilized with 0.1% mercuric chloride ($HgCl_2$) for 1.5 min and then thoroughly washed (three to four times) with sterilized distilled water under laminar air flow chamber. The seeds were then inoculated on MS half strength basal medium containing 0.5% sucrose. The pH of the medium was adjusted to 5.8 before sterilization. The culture tubes inoculated with seeds were then covered with carbon paper in order to maintain the dark condition for seed germination and kept in culture room at 26 ± 2 °C temperature.

21.2.2 Source of Explants

Cotyledon and hypocotyl explants were excised from 10- to 12-day-old in vitro grown seedlings, and petiole and leaf explants were excised from 18 to 20 days in vivo grown seedlings of broccoli. Leaf and petiole explants were surface sterilized with 0.1% mercuric chloride ($HgCl_2$) for 1.5 min and then thoroughly washed (three to four times) with sterilized distilled water under laminar air flow chamber. All the four explants were used for plant regeneration studies in broccoli.

21.3 Nutrient Medium

21.3.1 Murashige and Skoog's Medium

Nutrient medium and concentration of plant growth regulators for the optimum growth and differentiation of cells and tissues in vitro may vary with species or cultivar. Even the tissues from different parts of the same plant may have different nutritional requirements, and different concentrations of plant growth regulators may be required by them for optimum plant growth and differentiation. MS salts (macro and micro), vitamins supplemented with 100 mg/l mesoinositol, 3% sucrose, and 0.8% agar-agar were used as basal medium (Murashige and Skoog 1962). Different concentrations and combinations of TDZ alone, TDZ and adenine, TDZ and IAA, and TDZ and NAA were used in MS medium for in vitro plant regeneration experiment. The pH of the medium was adjusted to 5.8 with the help of 0.1N NaOH and 0.1N HCl before adding agar-agar. The medium was poured in culture vessels and sterilized by autoclaving at 15 lbs/in.^{2} for 15–20 min.

21.4 Sterilization of Glasswares and Instruments

All the glasswares were cleaned in a solution of 10% teepol with test tube brush and rinsed in tap water. Finally they were rinsed in double distilled water and dried in hot dry oven for 3–4 h. Glasswares and instruments such as flasks, petri dishes, beakers, test tubes, scalpels, forceps, filter assembly, etc. were wrapped with paper/plugged with cotton plug and sterilized at 15 lbs/in.^{2} for 15–20 min.

21.5 Culture Conditions

All the aseptic manipulations were carried out under laminar air flow chamber. Laminar flow was presterilized with ultraviolet light for 15 min, and sterilized instruments were used for inoculation. Defined precautions were taken to maintain aseptic conditions inside the cabinet. After inoculation of explants, all the culture vessels were kept in culture room at 26 ± 2 °C under 16 h of photoperiod.

21.6 Plant Regeneration in Broccoli cv. Solan Green Head

21.6.1 Shoot Regeneration from Hypocotyl and Cotyledon Explants

Hypocotyl and cotyledon explants were excised from aseptically grown 10–12-day-old seedlings and cut into 0.5–1.0 cm pieces and inoculated on MS medium with varying concentrations and combinations of TDZ, TDZ + NAA, TDZ + adenine, and TDZ + IAA (Table 21.1) to obtain efficient shoot regeneration. For every combination of five flasks with five explants, each was inoculated and all the experiments were repeated thrice. Observations were taken at 7-day interval till shoot regeneration. Explants were evaluated for percentage shoot regeneration with mean number of shoot per explant.

21.6.2 Shoot Regeneration from Leaf and Petiole Explants

Leaf and petiole explants were excised from glasshouse-grown 18–20-day-old seedlings and surface sterilized. The explants were cut into small pieces (0.5–1 cm), and their surface was gently tapped with the scalpel blade to injure them. These explants were cultured on MS medium containing different combinations and concentrations of TDZ, TDZ + NAA, TDZ + adenine, and TDZ + IAA (Table 21.1) to obtain highly efficient shoot regeneration. For every combination of five flasks with five explants, each was inoculated and all the experiments were repeated thrice. Observations were taken at 7-day interval till shoot regeneration. Explants were evaluated for percentage shoot regeneration with mean number of shoot per explant.

Table 21.1 Different concentrations and combinations of TDZ (thidiazuron) alone, TDZ and adenine, TDZ and NAA, and TDZ and IAA used in medium (MS) for efficient shoot regeneration from hypocotyl, cotyledon, petiole, and leaf explants of broccoli cv. Solan green head

Medium (MS) with different concentrations and combinations			
TDZ (μM)	TDZ (μM) and adenine (mM)	TDZ (μM) and NAA (μM)	TDZ (μM) and IAA (μM)
0.25	0.25 + 0.59	0.25 + 0.107	0.25 + 0.5
0.5	0.5 + 0.59	0.5 + 0.107	0.5 + 0.5
0.75	0.75 + 0.59	0.75 + 0.107	0.75 + 0.5
1.0	1.0 + 0.59	1.0 + 0.107	1.0 + 0.5
1.25	1.25 + 0.59	1.25 + 0.107	1.25 + 0.5
1.5	1.5 + 0.59	1.5 + 0.107	1.5 + 0.5
1.75	1.75 + 0.59	1.75 + 0.107	1.75 + 0.5
2.0	2.0 + 0.59	2.0 + 0.107	2.0 + 0.5
2.5	2.5 + 0.59	2.5 + 0.107	2.5 + 0.5

21.6.3 Root Regeneration from In Vitro Developed Shoots and Development of Complete Plantlets

For rooting, the regenerated shoots obtained from cultured explants (hypocotyl, cotyledon, leaf, and petiole) were transferred to root regeneration media containing MS medium with different concentrations of IAA, NAA, and IBA for root induction to get complete plantlets.

21.7 Acclimatization of In Vitro Regenerated Plantlets

21.7.1 Washing of Roots

After proper in vitro plantlet regeneration with well-developed root and shoot system, the plantlets were taken out of the tubes carefully so that there would be no damage to the root system. For removing adhering medium, roots were washed gently under running tap water. After that, plantlets were kept in running tap water for a few minutes so that they do not wilt after transfer to soil. The plantlets were treated with 0.5% Bavistin for 3–4 min.

21.7.2 Planting in Cocopeat Mixture

In vitro regenerated plantlets were transferred to the cocopeat mixture which was sterilized by autoclaving at 15 lbs/in.^{2} for 30 min at 121 °C (after sterilization of the cocopeat mixture, the pots were filled with this mixture). After transfer, the plantlets were watered with 0.5% Bavistin solution and covered with polythene bags. The plantlets were watered daily in order to maintain relative humidity.

21.7.3 Planting in Potting Mixture

The plantlets from cocopeat were transferred to pots, which contain presterilized potting mixture {consisting of sand + soil + FYM (1:1:1)}. The potted plantlets were kept under varying conditions of light intensity and humidity and observed for growth/survival.

21.8 Statistical Analysis

The data recorded for the different parameters were subjected to "completely randomized design." Each experiment was repeated thrice and each treatment consisted of at least 30 explants. The statistical analysis based on mean values per treatment was made using analysis of variance of CRD (Gomez and Gomez 1984).

21.9 Results

21.9.1 Efficient Shoot Regeneration from Hypocotyl and Cotyledon Explants Using Thidiazuron

The hypocotyl and cotyledon explants were cultured on supplemented MS medium with varying concentrations of TDZ alone, TDZ + adenine, TDZ + NAA, and TDZ + IAA for multiple shoot regeneration response. During the initial days of culture, the explants began to expand and swell. After 1 week of culturing, cotyledon explants showed initial expansion, whereas hypocotyl explants showed swelling at their margins. After 2 weeks, callus initiation occurred at the cut ends in hypocotyl explants, whereas in cotyledon explants, callus initiation was observed in between second and third weeks of culturing. In case of cotyledon explants, shoot initiation was observed after 6 weeks of culturing, whereas in hypocotyl explants, shoot initiation was recorded between 27 and 30 days. After 7–8 weeks, significant differences were recorded among the treatments for percentage of explants forming shoots. Multiple shoot regeneration response (95.92%) with mean number of shoots (4.28) from the treatment containing 2.0 μM TDZ and 0.5 μM IAA was observed in case of cultured hypocotyl explants. Meanwhile, for cotyledon explants, maximum percent shoot regeneration (88.88%), with average number of shoots (3.46) per explants, was recorded from the treatment containing 2.0 μM TDZ and 0.59 mM adenine (Kumar and Srivastava 2015a).

Kumar and Srivastava (2015a) reported that total of 36 combinations of plant growth regulators with different concentrations and combinations were used for efficient shoot regeneration studies. Out of nine concentrations of TDZ used, maximum percent shoot regeneration (81.11%) with average number of shoots (4.22) per explants was observed on MS medium with 1.5 μM TDZ, in case of hypocotyl explants, whereas in cotyledon explants, average number of shoots (1.51) per explant with 53.33% shoot regeneration was observed on MS medium with 0.5 μM TDZ. From nine different combinations of TDZ and adenine, the hypocotyl explants showed high-frequency shoot regeneration (91.10%) with highest mean number of shoots (4.15) with 0.5 μM TDZ and 0.59 mM adenine treatment combination. However, in the case of cotyledon explants, maximum shoot regeneration response was (88.88%) with highest mean number of shoots (3.46) with 2.0 μM TDZ and 0.59 mM adenine treatment combination. Out of nine different concentrations and combinations TDZ + IAA, hypocotyl explants gave maximum shoot induction (95.92%) with average number of shoots (4.28) from treatment 2.0 μM TDZ and 0.5 μM IAA (Table 21.2). Whereas, in cotyledon explants, callus initiation was observed after 2 weeks of culturing and callus proliferation after 25–28 days of culturing. There was no shoot or root differentiation was recorded from callus even after 8 weeks of culturing. From nine different concentrations and combinations TDZ + NAA, maximum percent shoot regeneration (51.10%) with mean number of shoots (1.94) per explant was recorded from the treatment 0.5 μM TDZ and 0.107 μM NAA; in case of hypocotyl explants and for cotyledon explants, 83.33% shoot regeneration with 2.86 mean number of shoots per explant was observed.

Table 21.2 Best medium for efficient shoot regeneration from hypocotyl, cotyledon, petiole, and leaf explants of broccoli cv. Solan green head

Explant used	MS medium	Mean number of shoots/ explant	Percent shoot regeneration
Hypocotyl	2.0 μM TDZ + 0.5 μM IAA	4.28 ± 0.169	95.92 ± 0.37
Cotyledon	2.0 μM TDZ + 0.59 mM adenine	3.46 ± 0.234	88.88 ± 1.78
Leaf	1.0 μM TDZ + 0.107 μM NAA	3.34 ± 0.041	89.25 ± 0.373
Petiole	2.0 μM TDZ + 0.107 μM NAA	4.22 ± 0.06	91.55 ± 0.924

Source: Kumar and Srivastava (2015a) and Kumar et al. (2015b)

21.9.2 Efficient Shoot Bud Differentiation from Leaf and Petiole Explants Using Thidiazuron

In the present study, completely green, young, and tender petiole and leaf explants were used from in vivo grown seedlings for multiple shoot induction. After 1 week of culturing, explants began to expand, and callus proliferation was observed after 3 weeks of culturing for both (petiole and leaf) explants. Significant differences were observed among the treatments for percentage of explants forming shoots after 7–8 weeks of culturing. Maximum shoot regeneration response was achieved using nine different concentrations and combinations of TDZ + NAA for both (petiole and leaf) explants. Petiole explants gave highest percent shoot regeneration (91.55%), with highest mean number of shoots (4.22) per explant from 2.0 μM TDZ + 0.107 μM NAA treatment combination (Table 21.2), whereas in case of leaf explants, maximum percent shoot regeneration (89.25%) with average number of shoot (3.34) per explant was resulted from the treatment containing 1.0 μM TDZ and 0.107 μM NAA (Kumar et al. 2015b).

A total of 36 combinations of plant growth regulators with varying concentrations for shoot regeneration have been reported (Kumar et al. 2015b). From nine different concentrations of TDZ used, maximum percent shoot regeneration (72.59%) with average number of shoots (2.19) per explant was recorded from the treatment 1.75 μM TDZ in cultured leaf explants, whereas in the case of petiole explants, the highest mean number of shoots per explant was 1.57 with 70.74% shoot regeneration and was recorded from the treatment 2.0 μM TDZ. From nine different combinations of TDZ and adenine, the petiole explants showed maximum shoot regeneration response (77.29%) with average number of shoots (2.66) from 1.5 μM TDZ + 0.59 mM adenine treatment combination. However, in the case of leaf explants, multiple shoot induction response was 67.40%, with average number of shoots (1.71) per explant from 2.5 μM TDZ + 0.59 mM adenine treatment combination. Kumar et al. (2015b) reported significant differences in the shoot regeneration potential for nine different combinations of TDZ + IAA used. In cultured leaf explants, medium supplemented with different concentrations of TDZ + IAA

(1.25, 1.5, 1.75, 2.0, 2.5 μM) did not produce any shoots. Callus initiation was observed after 3 weeks of culturing and callus proliferation occurred after 4 weeks of culturing, but there was no shoot or root differentiation from callus even after 8 weeks of culturing. However, petiole explants gave multiple shoot regeneration response (85.18%), with highest mean number of shoots (2.72) per explant from 1.75 μM TDZ and 0.5 μM IAA treatment combination.

21.10 Rooting and Acclimatization

For root induction, three different types of auxins were tried in medium (MS) from in vitro developed shoots. The presence of 0.10 mg/l NAA in nutrient medium promoted the early root regeneration, and high-frequency (100%) root regeneration was recorded. The in vitro regenerated complete plantlets were successfully acclimatized on cocopeat, and 80% survival of plants was observed during acclimatization. A protocol for plant regeneration from cotyledon, hypocotyl, petiole, and leaf explants has been optimized in broccoli cv. Solan green head.

21.11 Discussion

In the present investigation, 10–12-day-old aseptically in vitro grown seedlings (for cotyledon and hypocotyl explants) and 18–20-day-old in vivo glasshouse-grown seedlings (for petiole and leaf explants) were used for plant regeneration studies. Among earlier findings, the effects of age of donor seedlings (explants source) on shoot induction from different explants such as cotyledon, hypocotyl, petiole, and leaf have been studied in various species (Srivastava et al. 1991a, b; Kumar and Srivastava 2016a, b). Similarly in *Brassica* species, also the effect of age of donor explants has been reported (Chen and Hou 2008; Deng-Xia et al. 2011; Sharma et al. 2014; Gambhir 2014; Gaur 2015). Dong and Jia (1991) have reported that young tender 5-day-old cotyledons were more responsive for multiple shoot induction, whereas in cotyledonary explants from seedlings older than 7 days, shoot regeneration frequency dropped sharply. Eisner et al. (1992) obtained best result with 5- to 7-day-old hypocotyls. Choi et al. (1996) reported that cotyledon explants from 10-day-old seedlings were not responsive and also reported that explants became more competent as the age of the source seedlings increased up to 8 days. Kumar and Srivastava (2015a, 2015b) and Kumar et al. (2015a, b) have reported high-frequency shoot regeneration with mean number of shoots per explant using completely green fully expanded hypocotyl and cotyledon explants which were greenish in color and turgid in nature and young tender completely green petiole and leaf explants. A promising explanation is that young explants are physiologically and biochemically more active as they have less rigid cell wall and are easily affected by the environmental factors such as exogenous plant growth regulators.

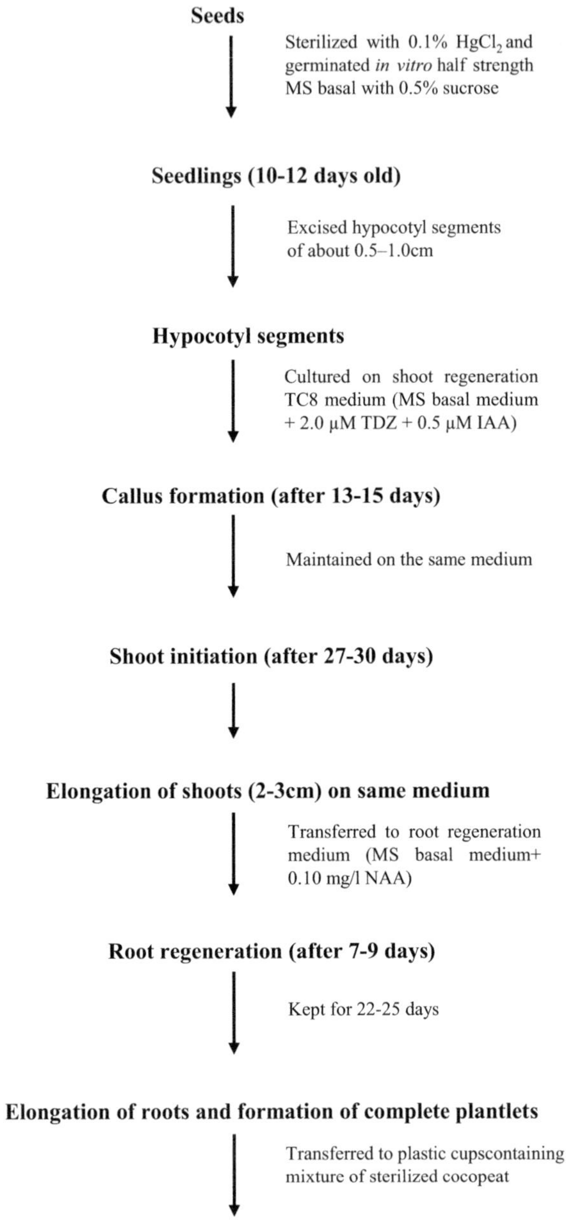

Fig. 21.1 Flowchart depicting optimized protocol for in vitro plant regeneration from hypocotyl explants in broccoli cv. Solan green head

In this study, cotyledon, hypocotyl, petiole, and leaf were used as explants, but hypocotyl was found best for shoot regeneration and multiplication as compared to other explants (Fig. 21.1).

Petrova and Antonova (1996), Sharma and Srivastava (2013), and Sharma et al. (2014) have reported that hypocotyls were more responsive than cotyledons, whereas

Arora et al. (1996) found that cotyledons produced shoots in much closer time than hypocotyls. Qin et al. (2006) have observed cotyledon was more responsive in comparison to hypocotyl explants, whereas Ravanfar et al. (2009) reported that hypocotyl promising explants for shoot regeneration in broccoli. Chen and Hou (2008) reported that the regeneration frequency of cotyledon and petiole was the highest among different explants types. Cheng et al. (2001) observed that among different explants tested, young internodal segments and hypocotyls from in vitro shoots were the most regenerative. According to Lim et al. (1998), hypocotyl explants revealed better shoot regeneration response as compared to petiole and cotyledon explants. Pavlovic et al. (2010) have observed highest percentage of shoot formation in hypocotyl explants as compared to root and cotyledon explants in all *Brassica oleracea* L. varieties. Seedling hypocotyls were preferred for regeneration of plants by Ravanfar et al. (2009, 2011), Yang et al. (2010), Huang et al. (2011), Sharma et al. (2012, 2014), Kumar and Srivastava (2015a, b), and Kumar et al. (2015a, b).

TDZ (thidiazuron) was used in MS medium for efficient shoot regeneration studies. Thidiazuron, a heterocyclic phenyl urea (*N′*-phenyl-*N′*-1,2,3-thidiazol-5-ylurea, TDZ), has gained significance as effective plant growth regulator (PGR) for in vitro plant regeneration studies in large number of crops (Guo et al. 2005; Ravanfar et al. 2014; Kumar et al. 2015a, b). Thidiazuron biological properties were reported qualitatively similar to cytokinin (Mok et al. 1982) and effective in in vitro plant regeneration studies in different recalcitrant species (Liu et al. 2003; Mithila et al. 2003). The efficacy of TDZ with different concentrations and combinations was tested for enhancing shoot regeneration frequency from broccoli cv. Solan green head explants (cotyledon, hypocotyl, petiole, and leaf). Different explants responded in vitro with different regeneration frequencies. Different published reports for in vitro regeneration in *Brassica* species whether they include genetic transformation step or not are based on TDZ and found to be very efficient for enhancing the frequency of shoot regeneration (Christey et al. 1997; Henzi et al. 2000; Cheng et al. 2001; Lu et al. 2003; Jonoubi et al. 2005; Chen and Hou 2008; Song et al. 2012; Kumar and Srivastava 2015a; Kumar et al. 2015b).

In the present investigation, TDZ was reported to be superior in promoting shoot regeneration frequency from cotyledon, hypocotyl, petiole, and leaf explants in broccoli. Maximum (95.92%) percent shoot regeneration from hypocotyl explants followed by petiole (91.55%), leaf (89.25%), and cotyledon (88.88%) was optimized on different media supplemented with TDZ during the present study. The obtained results were in accordance with results obtained by Cheng et al. (2001) and Lu et al. (2003). Hypocotyl explants gave better result on MS medium containing TDZ as compared to BA. The regeneration frequency reached 98.80% on MS medium containing 0.25 mg/l TDZ + 0.5 mg/l NAA + 5.0 mg/l $AgNO_3$. Whereas Chen and Hou (2008) reported the percent regeneration of petiole with cotyledon explants was highest in MS medium containing 0.5 mg/l TDZ + 0.5 mg/l NAA + 7.5 mg/l $AgNO_3$, Ravanfar et al. (2014) observed that in cotyledon explants, TDZ with NAA increased shoot formation in broccoli cv. Green Dragon King. In present studies, cotyledon and leaf explants showed less or no shoot regeneration, when supplemented with different concentrations of TDZ and IAA. Similar results were reported by Gambhir (2014) and Gaur (2015) in cabbage and cauliflower.

Media and its constituents play significant role in the in vitro plant regeneration studies in different plant species. Most commonly used medium in plant regeneration studies is MS medium (Murashige and skoog 1962); however, other different kinds of media had been used by various researchers. Linsmaier and Skoog (LS) media were used by Christey and Earle (1991) for regeneration from peduncle explants and

Fig. 21.2 (**a**) High-frequency in vitro shoot regeneration from hypocotyl explants. (**b**) Multiple shoot induction response from cotyledon explants. (**c**) Efficient shoot regeneration from leaf explants. (**d**) Multiple shoot regeneration from petiole explants. (**e**) In vitro regenerated plantlets with well-developed root and shoot system. (**f**) Successful acclimatization of regenerated plantlets (*Source*: Kumar and Srivastava 2015a; Kumar et al. 2015b)

shoot regeneration frequency in their study varied from 41% to 98%. Effect of MS medium was compared with mMS (supplemented with organic supplement PG-96) and found effective than that of simple MS medium by Qin et al. (2006). In our study MS medium was used for carrying out the plant regeneration studies and was found effective in inducing high percent shoot regeneration (Fig. 21.2).

Three different auxins (IBA, NAA, and IAA) were tested for the efficient root regeneration from in vitro developed shoots. Among the three auxins, after 3 weeks of culturing, the 100% root regeneration response with well-developed healthy and vigorous roots was observed on MS medium with 0.10 mg/l NAA. Ravanfar et al. (2009) and Sharma et al. (2014) reported that rooting medium with IBA was most promising for root regeneration. However, it was also reported that root induction was highly genotype-dependent rather than supplemented auxins (NAA, IAA, or IBA) (Arinson et al. 1990; Vandemoortele et al. 1999). After rooting, plantlets were successfully acclimatized and recorded with 80% survival rate. The grown plants were morphologically uniform, and no obvious variation in appearance was observed.

21.12 Conclusion

The present study showed that TDZ was efficient in high-frequency plantlet regeneration responses from broccoli cultivar "Solan green head" explants. Finally it is concluded that hypocotyl was better explants than petiole, cotyledon, and leaf as they showed high percent shoot regeneration with mean number of shoots per explant and high multiplication rate. Shoot regeneration and root regeneration were comparatively early in hypocotyls as compared to other three explants. This standardized protocol can be satisfactorily exploited for genetic engineering purposes using different agronomically important traits.

Acknowledgment We are thankful to the Head of Department of Vegetable Science, Dr. Yashwant Singh Parmar University of Horticulture and Forestry, Solan, for providing certified seeds of broccoli cv. Solan green head. The main author (P.K.) gratefully acknowledges the DST INSPIRE JRF/SRF fellowship award of the Department of Science & Technology, Innovation in Science Pursuit for Inspired Research (INSPIRE), New Delhi, India.

References

Abdel-wahhab MA, Aly SE (2003) Antioxidants and radical scavenging properties of vegetable extracts in rats fed aflatoxin-contaminated diet. J Agric Food Chem 51:2409–2414

Achar PN (2002) Study of factors affecting embryo yields from anther culture of cabbage. Plant Cell Tissue Organ Cult 69:183–188

Arinson PG, Donaldson P, Jackson A, Semple C, Keller W (1990) Genotype-specific response of cultured broccoli (*Brassica oleracea* L. var. *italica*) anthers to cytokinins. Plant Cell Tissue Organ Cult 20:217–228

Arora N, Yadav NR, Chowdhury JB (1996) Efficient plant regeneration in cauliflower (*Brassica oleracea* var. *botrytis*). Cruciferae Newslett 18:26–27

Cao J, Earle ED (2003) Transgene expression in broccoli (Brassica oleracea var. italica) clones propagated in vitro via leaf explants. Plant Cell Rep 21:789–796

Chakrabarty R, Viswakarma N, Bhat SR, Kirti PB, Singh BD, Chopra VL (2002) *Agrobacterium*-mediated transformation of cauliflower: optimization of protocol and development of *Bt*-transgenic cauliflower. J Biosci 27:495–502

Chang YM, Liou PC, Hsiao CH (1996) Anther culture of cabbage (*Brassica oleracea* L. var. *capitata*) and broccoli (*B. oleracea* L. var. *italica*). J Agric Res China 45:35–46

Chen MM, Hou XL (2008) Optimization of regeneration in vitro of non-heading Chinese cabbage. J Nanjing Agric Univ 3:217–223

Cheng PK, Lakshmanan P, Swarup S (2001) High frequency direct shoot regeneration and continuous production of rapid-cycling *Brassica oleracea* in vitro. In Vitro Cell Dev Biol Plant 37:592–598

Choi PS, Soh WY, Liu JR (1996) Somatic embryogenesis and plant regeneration in cotyledonary explant cultures of Chinese cabbage. Plant Cell Tissue Organ Cult 44:253–256

Christey MC, Earle ED (1991) Regeneration of Brassica oleracea from peduncle explants. Hort Sci 26:1069–1072

Christey MC, Sinclair BK, Braun RH, Wyke L (1997) Regeneration of transgenic vegetable brassicas (*B. oleracea* and *B. campestris*) via *Ri*-mediated transformation. Plant Cell Rep 16(9):587–593

Deng-Xia Y, Lei C, Yu-Mei L, Mu Z, Yang-Yong Z, Zhi-Yuan F, Li-Mei Y (2011) Transformation of cabbage (*Brassica oleracea* L. var. *capitata*) with *Bt cry1Ba3* gene for control of diamondback moth. Agric Sci China 10:1693–1700

Dong JZ, Jia SR (1991) High efficiency plant regeneration from cotyledons of watermelon (*Citrullus vulgaris* Schrad). Plant Cell Rep 9:858–863

Eisner GI, Mar YI, Ya SMF (1992) Optimization of conditions for *in vitro* plant regeneration for cabbage transformations. Soviet Agric Sci 11(12):15–19

Farzinebrahimi R, Taha RM, Fadainasab M, Mokhtar S (2012) In vitro plant regeneration, antioxidant and antibacterial studies on broccoli, *Brassica oleracea* var. *italica*. Pak J Bot 44:2117–2122

Finley JW (2003) Reduction of cancer risk by consumption of selenium-enriched plants: enrichment of broccoli with selenium increases the anticarcinogenic properties of broccoli. J Med Food 6:19–26

Finley JW, Ip C, Lisk DJ, Davis CD, Hintze KG, Whanger PD (2001) Cancer-protective properties of high-selenium broccoli. J Agric Food Chem 49:2679–2683

Fiola JA, Hassan MA, Swartz HJ, Bors RH, McNicol R (1990) Effect of thidiazuron, light fluency rates and kanamycin on in vitro shoot organogenesis from excised rubus cotyledons and leaves. Plant Cell Tissue Organ Cult 20:223–228

Gambhir G (2014) Studies on *Agrobacterium*-mediated insect resistance gene transfer in cabbage (*Brassica oleracea* L. var. *capitata*) and molecular analysis of regenerated plantlets. Ph. D. thesis, Dr. Y.S. Parmar University of Horticulture and Forestry, Nauni, Solan (H.P.), India

Gaur A (2015) Studies on *Agrobacterium*-mediated insect resistance gene [*cry1A*(*a*)] transfer in cauliflower (*Brassica oleracea* L. var. *botrytis*). Ph.D. thesis, Dr. Y.S. Parmar University of Horticulture and Forestry, Nauni, Solan (H.P.), India

Gomez KA, Gomez AA (1984) Statistical procedures for agricultural research. Wiley, New York

Guo DP, Zhu ZJ, Hu XX, Zheng SJ (2005) Effect of cytokinins on shoot regeneration from cotyledon and leaf segment of stem mustard (*Brassica juncea* var. *tsatsai*). Plant Cell Tissue Organ Cult 83:123–127

Henzi MX, Christey MC, McNeil DL (2000) Factors that influence *Agrobacterium rhizogenes*-mediated transformation of broccoli (*Brassica oleracea* L. var. *italica*). Plant Cell Rep 19:994–999

Huang K, Jiashu C, Xiaolin Y, Wanzhi Y, Gang L, Xiang X (2005) Plant male sterility induced by antigene *CYP86MF* in *Brassica oleracea* L. var. *italica*. Agricult Sci China 4(11):806–810

Huang K, Qiuyun W, Juncleng L, Zheng J (2011) Optimization of plant regeneration from broccoli. Afr J Biotechnol 10:4081–4085

Jonoubi P, Mousavi A, Majd A, Salmanian AH, Jalali Javaran A, Daneshian J (2005) Efficient regeneration of *Brassica napus* L. hypocotyls and genetic transformation by *Agrobacterium tumefaciens*. Biol Plant 49(2):175–180

Kaur N, Vyvadilova M, Klima M, Bechyne M (2006) A simple procedure for mesophyll protoplast culture and plant regeneration in *Brassica oleracea* L. and *Brassica napus* L. Czech J Genet Plant Breed 3:103–110

Kumar P, Srivastava DK (2015a) Effect of potent cytokinin thidiazuron (TDZ) on in vitro morphogenic potential of broccoli (*Brassica oleracea* L. var. *italica*), an important vegetable crop. Indian J Plant Physiol 20(4):317–323

Kumar P, Srivastava DK (2015b) High frequency organogenesis in hypocotyl, cotyledon, leaf and petiole explants of broccoli (*Brassica oleracea* L. var. *italica*), an important vegetable crop. Physiol Mol Biol Plants 21(2):279–285

Kumar P, Srivastava DK (2016a) Biotechnological advancement in genetic improvement of broccoli (*Brassica oleracea* L. var. *italica*), an important vegetable crop: a review. Biotechnol Lett 38(7):1049–1063

Kumar P, Srivastava DK (2016b) Biotechnological application in in vitro plant regeneration studies of Broccoli (*Brassica oleracea* l. var. *italica*), an important vegetable crop: a review. Biotechnol Lett 38(4):561–571

Kumar P, Gambhir G, Gaur A, Srivastava DK (2015a) Molecular analysis of genetic stability in in vitro regenerated plants of broccoli (*Brassica oleracea* L. var. *italica*). Curr Sci 109(8):1470–1475

Kumar P, Gaur A, Srivastava DK (2015b) Morphogenic response of leaf and petiole explants of broccoli using thidiazuron. J Crop Improv 29:432–446

Kumar P, Gaur A, Srivastava DK (2017a) *Agrobacterium* – mediated insect resistance gene (cry1Aa) transfer studies pertaining to antibiotic sensitivity on cultured tissues of broccoli (*Brassica oleracea* L. var. *italica*): an important vegetable crop. Int J Veg Sci 23:523–535

Kumar P, Shaunak I, Thakur AK, Srivastava DK (2017b) Health promising medicinal molecules in vegetable crops. J Gen Genome 1:102

Lim HT, You YS, Park ET, Thomas G (1998) High plant regeneration, genetic stability of regenerants and genetic transformation of herbicide resistance gene (*bar*) in Chinese cabbage (*Brassica campestris* ssp. *pekinensis*). Acta Hortic 459:199–208

Liu CZ, Murch SJ, Demerdash MEL, Saxena PK (2003) Regeneration of the Egyptian medicinal plant *Artemisia judaica* L. Plant Cell Rep 21:525–530

Lu YE, Li HX, Ye ZB (2003) Effects of two kinds of cytokinins on shoot regeneration from cotyledonary explants of chinese cabbage. Plant Sci J 21(4):361–364

Mithila J, Hall JC, Victor JMR, Saxena PK (2003) Thidiazuron induces shoot organogenesis at low concentration and somatic embryogenesis at high concentration on leaf and petiole explants of African violet (*Saintpaulia ionantha* W. Endl.) Plant Cell Rep 21:408–414

Mok MC, Mok DWS, Amstrong DJ, Shudo K, Isogai Y, Okamoto T (1982) Cytokinin activity of N-Phenyl- n-1, 2, 3-thidiazol-5-yl urea (thidiazuron). Phytochemistry 21:1509–1511

Murashige T, Skoog F (1962) A revised medium for rapid growth and bioassays with tobacco tissue cultures. Physiol Plant 15:473–497

Murthy BNS, Murch SJ, Saxena PK (1995) Thidiazuron-induced somatic embryogenesis in intact seedlings of peanut (*Arachis hypogaea* L.): endogenous growth regulator levels and significance of cotyledons. Physiol Plant 94:268–276

Parmar N, Singh KH, Sharma P, Singh L, Kumar P, Nanjundan J, Khan YJ, Chauhan DK, Thakur AK (2017) Genetic engineering strategies for biotic and abiotic stress tolerance and quality enhancement in horticultural crops: a comprehensive review. 3 Biotech 7:239

Pavlovic S, Vinterhalter B, Mitic N, Adzic S, Pavlovic N, Zdravkovic M, Vinterhalter D (2010) In vitro shoot regeneration from seedling explants in *Brassica* vegetables: red cabbage, broccoli, savoy cabbage and cauliflower. Arch Biol Sci Belgrade 62(2):337–345

Petrova S, Antonova G (1996) Plant regeneration from seedling explants of cabbage. Cruciferae-Newslett 18:31

Qin Y, Li HL, Guo YD (2006) High frequency embryogenesis, regeneration of broccoli (*Brassica oleracea* var. *italica*) and analysis of genetic stability by RAPD. Sci Hortic 111:203–208

Rani T, Yadav RC, Yadav NR, Rani A, Singh D (2013) Genetic transformation in oilseed brassicas: a review. Indian J Agric Sci 83:367–373

Ravanfar SA, Aziz MA, Kadir MA, Rashid AA, Sirchi MHT (2009) Plant regeneration of *Brassica oleracea* var. *italica* (broccoli) cv. Green marvel was affected by plant growth regulators. Afr J Biotechnol 8:2523–2528

Ravanfar SA, Aziz MA, Kadir MA, Rashid AA, Haddadi F (2011) In vitro shoot regeneration and acclimatization of *Brassica oleracea* var. *italica* cv. Green marvel. Afr J Biotechnol 10:5614–5619

Ravanfar SA, Aziz MA, Rashid AA, Shahida S (2014) In vitro adventitious shoot regeneration from cotyledon explant of *Brassica oleracea* subsp. *italica* and *Brassica oleracea* subsp. *capitata* using TDZ and NAA. Pak J Bot 46:329–335

Sharma C, Srivastava DK (2013) Efficient *Agrobacterium*- mediated genetic transformation of tomato using petiole explants. Crop Improv 40:44–49

Sharma S, Sharma C, Srivastava DK (2012) Plant regeneration genetic transformation and gene expression in in vitro tissues of cauliflower (*Brassica oleracea* L. var *botrytis*). Bioinfolet 9:760–764

Sharma S, Gambhir G, Srivastava DK (2014) High frequency organogenesis in cotyledon and hypocotyls explants of cabbage (*Brassica oleracea* L. var. *capitata*). Nat Acad Sci Lett 37:5–12

Song GQ, Walworth AE, Hancock JF (2012) *Agrobacterium tumefaciens*-mediated transformation of rutabaga (*Brassica napus* var. *napobrassica*) cultivar "American Purple Top Yellow". In Vitro Cell Dev Biol Plant 48:383–389

Srivastava DK, Andrianov VM, Piruzian ES (1989) Tissue culture and plant regeneration of watermelon (*Citrullus vulgaris* Schrad cv. *Melitopolski*). Plant Cell Rep 8:300–302

Srivastava DK, Andrianov VM, Piruzian ES (1991a) Regeneration and genetic transformation studies in watermelon (*Citrullus vulgaris* L. cv. *Melitopolski*). In: Parkash J, Pierika RLM (eds) Horticulture – new technologies and applications. Kluwer Academic Publishers, Dordrecht, pp 127–130

Srivastava DK, Kolgonova TV, Mett VL, Piruzian ES (1991b) Genetic transformation of cotton (*Gossypium hirsutum* L. cv. 108-F). Acta Hortic 289:263–264

Suri SS, Saini ARK, Ramawat KG (2005) High frequency regeneration and Agrobacterium tumefaciens- mediated transformation of broccoli (*Brassica oleracea* var. *italica*). European J Hort Sci 70(2):71–78

Tabashnik BE, Finson N, Johnson MW (1991) Managing resistance to *Bacillus thuringiensis:* lessons from the diamondback moth (Lepidoptera: Plutellidae). J Econ Entomol 84:49–55

Vallejo FC, Garcia-viguera C, Tomas-barberan FA (2003) Changes in broccoli (*Brassica oleracea* var. *italica*) health promoting compounds with inflorescence development. J Agric Food Chem 51:3776–3782

Vandemoortele JL, Billard JP, Boucaud J, Gaspar T (1999) Evidence for an interaction between basal medium and plant growth regulators during adventitious or axillary shoot formation of cauliflower. In Vitro Cell Dev Biol Plant 35:13–17

Visser C, Qureshi JA, Gill R, Saxena PK (1992) Morphoregulatory role of thidiazuron: substitution of auxin and cytokinin requirement for the induction of somatic embryogenesis in geranium hypocotyls. Plant Physiol 99:1704–1707

Viswakarma N, Bhattacharya RC, Chakrabarty R (2004) Insect resistance of transgenic broccoli expressing a synthetic *cryIA(b)* gene. J Hortic Sci Biotechnol 79:182–188

Yang JL, Seong ES, Kim MJ, Ghimire BK, Kang WH, Yu CY, Li CH (2010) Direct somatic embryogenesis from pericycle cells of broccoli (*Brassica oleracea* L. var. *italica*) root explants. Plant Cell Tissue Organ Cult 100:49–58

Zhang YS, Talalay P, Cho CG, Posner G (1992) A major inducer of anticarcinogenic protective enzymes from broccoli: isolation and elucidation of structure. Proc Natl Acad Sci U S A 89:2399–2403

Check for updates

Thidiazuron: A Potent Phytohormone for In Vitro Regeneration

22

Sujatha Govindaraj

Abstract

Thidiazuron ($C_9H_8N_4OS$) is one of the most effective substituted phenylureas that has been examined for cytokinin-like activity in plant tissue cultures. A wide range of physiological responses were ascertained in response to TDZ application in several plant species. Apart from its cytokinin-like activity, TDZ has been steered to modulate the endogenous auxin levels. However, it remains to be resolved whether it possesses an auxin activity or if it is concerned with auxin metabolism. It induces numerous morphogenic responses, starting from tissue proliferation to induction of shoot buds and somatic embryos. It has been shown to promote shoot regeneration expeditiously than that of other cytokinins, and organized centers of growth are attained at much lower concentrations. Other prospects embody modification in cell membrane, energy levels, nutrient absorption, transport, assimilation, etc. TDZ exhibits the distinctive property of mimicking both auxin and cytokinin effects on growth and differentiation of cultured explants, though structurally it is different from either auxins or purine-based cytokinins. The effectiveness of TDZ as an inductive chemical for ontogenesis is not restricted to tissue culture systems, and the regeneration is settled in vivo further. During this review, many recently revealed studies on characterization of TDZ-induced in vitro regeneration are bestowed and mentioned.

Keywords

Thidiazuron · In vitro morphogenesis · Callusing · Somatic embryos · Metabolism

S. Govindaraj (✉)
Department of Botany, Periyar E. V. R. College (Autonomous), Tiruchirappalli, Tamil Nadu, India

N. Ahmad, M. Faisal (eds.), *Thidiazuron: From Urea Derivative to Plant Growth Regulator*, https://doi.org/10.1007/978-981-10-8004-3_22

22.1 Introduction

Plant tissue culture technique is the culture of plant cells, tissues, or organs under specific in vitro conditions to create a great number of true-to-type plants in short time using assorted starting plant material through phases of explant selection and preparation, culture establishment, regeneration, and acclimatization of the plantlets to ex vitro conditions (George 2008). The technology is progressing in applications for clonal propagation of medicinal, horticultural, agronomic crops, and forest trees. Many factors influence in vitro response of plants including the selected explant to be cultured; physiological state of the explant, juvenile or mature state; genotype; the health status of the explant; and culture media (Lee 2004; Kane 2005). The chosen explant for in vitro studies needs physiological adjustment to the culture conditions so as to achieve enhanced clonal multiplication and for the cultivated plant to accomplish physiological stability, and recurrent subculture to fresh media is necessary as medium nutrients get exhausted over time (Lee 2004; Kozai and Xiao 2006). The capacity to regenerate the entire plant from cultured somatic cells, tissue, or organ has been known for several decades; however, the problem of how the cultures differentiate into a whole plant and various physiological and anatomical features of the regenerated plants and during transfer to field conditions is still being studied by many research groups (Skoog and Miller 1957; Pospisilova et al. 1999; Vogel 2005; Jariteh et al. 2015). Manipulation of the in vitro development of plants is of paramount and applied interest as it proffers a model to portray developmental stages at genomic and proteomic levels and also offers potential to rejuvenate plants for increased propagation (Lee 2004; Moyo et al. 2015). A profound understanding of the in vitro plant development, the morphophysiology, as well as stress physiology mechanism and potential for acclimatization to ex vitro environment are of significance in foreseeing and enhancing the survival rate of plantlets during the in vitro culture conditions and acclimatization stages (Pospisilova et al. 1999; Cassells and Curry 2001; Moyo et al. 2015).

The two primary morphogenic pathways leading to whole plant regeneration – which is a prerequisite for most plant breeding, genetic, and transgenic applications of in vitro biology – involve either somatic embryogenesis or shoot organogenesis followed by root organogenesis. Both developmental pathways can occur either directly without a callus intermediate stage, termed adventitious, or indirectly following an unorganized callus stage, termed de novo (Gamborg and Phillips 1995). Few plant species have been shown to regenerate by both organogenic and somatic embryogenic pathways, but many plant species can regenerate by one or the other of these pathways (Phillips 2004). Plant cells can be maintained for extended periods in the apparent absence of all known plant hormones; it seems safe to conclude that no hormone is essential just to maintain the viability of plant cells (Davies 1995). However, the auxins and cytokinins are very important for proper growth and maintenance of culture. Indole-3-acetic acid (IAA) is a major naturally occurring auxin, which is widely reported in plant tissue culture and morphogenesis. In addition to the natural auxin, a whole host of synthetic auxins are known. The most widely used are α-naphthalene acetic acid (NAA) and 2,4-dichlorophenoxyacetic acid (2,4-D). The natural cytokinins are a series of adenine molecules modified by the addition of 5-carbon side chains of the sixth position. About 50 years ago, Skoog and Miller (1957) described the controlled organ regeneration in plants; however, developmental biologists were surprised by the unbelievable capacity of plant tissues to regenerate the whole plants (concept of totipotency). The capacity of cultured plant tissues and cells to undergo morphogenesis, resulting in the formation of discrete organs or whole plants, has provided opportunities for numerous applications of in vitro plant biology in studies of basic botany, biochemistry, propagation, breeding, and development of transgenic crops.

Thidiazuron (TDZ, 1-phenyl-3-(1,2,3-thiadiazol-5-yl) urea) with molecular formula $C_9H_8N_4OS$ and molecular mass 220.25 g/mol is a light yellow crystalline chemical that is sparingly soluble in water, but highly soluble in ethanol and at varying levels in other organic solvents such as acetone, benzene, DMSO, etc. (Table 22.1, Fig. 22.1). Ethanol is the preferred solvent in using TDZ for in vitro studies. TDZ was manufactured by the German Schering Corporation for defoliation of cotton (*Gossypium hirsutum*) (Arndt et al. 1976). The defoliating property of

Table 22.1 Physical and chemical properties of thidiazuron

Character	Description
Color	Light yellow crystals
Odor	Odorless
Trade name	DROPP
IUPAC name	1-Phenyl-3-(1,2,3-thiadiazol-5-yl) urea
Molecular formula	$C_9H_8N_4OS$
Molecular weight	220.25 g/mol
Melting point	210.5–212.5 °C
Vapor pressure	2.30 × 10–11 mmHg at 25 °C
pH	6.50 at 20 °C
Dissociation constant	pKa = 8.86
Storage	Dry conditions
Purity	≥98%
CAS number	51707-55-2

Source: Pub Chem; URL: https://pubchem.ncbi.nlm.nih.gov

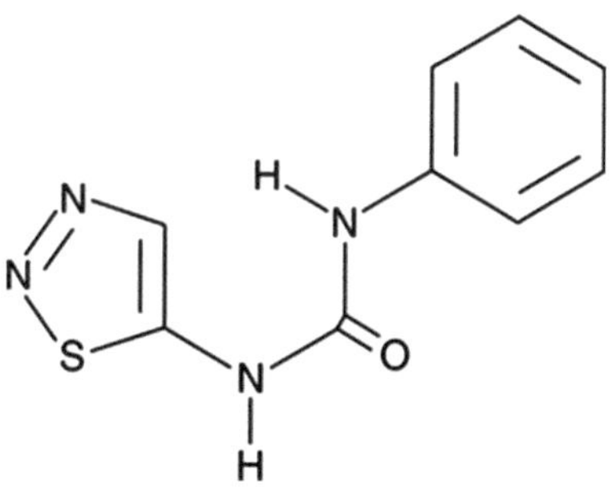

Fig. 22.1 Structure of thidiazuron

TDZ is restricted to a few species belonging to the Malvaceae family (Grossmann 1991; Zubkova et al. 1991). Thidiazuron induces abscission of cotton leaves, characteristically without the breakdown of chlorophyll or any alteration in the leaf water potential (Grossmann 1991).This chemical has now emerged as a highly efficacious bioregulant of morphogenesis in tissue culture of a diverse array of plant species, and the culture responses range from induction of callusing to the formation of somatic embryos. There are two functional groups in TDZ molecule, viz., phenyl and thiadiazol groups, and replacement of any of these groups with other ring structures results in the reduction in activity. A wide range of physiological responses were observed in response to TDZ application in different plant species. Examples of the diversity of physiological effects mediated by TDZ include enhanced seed germination in lettuce (Baskakov et al. 1981), substitution of chilling requirement for seed germination in *Pyrus* sp. (Lin et al. 1994), accelerated bud break in apple (Wang et al. 1986), stimulation of sprouting in potato (Ji and Wang 1988), cotyledon growth in pumpkin (Burkhanova et al. 1984), formation of branched trichomes and stomata on floral organs (Venglat and Sawhney 1994), and increased cluster and berry weight in grapes (Reynolds et al. 1992). TDZ has revealed both auxin- and cytokinin-like effects, though, chemically, it is totally different from frequently used auxins and cytokinins. A number of physiological and biochemical responses in cells are likely to be influenced by TDZ, but these may or may not be directly related to the induction of morphogenesis. Reports showed that TDZ may modify endogenous plant growth regulators, either directly or indirectly, and produce reactions in cell/tissue necessary for its division/regeneration. Other possibilities include the modifications in cell membrane, energy levels, nutrient absorption, transport, assimilation, etc. (Guo et al. 2011). The precise mechanism of action of TDZ is explained by two hypotheses: It is possible that TDZ directly promotes growth due to its own biological activity in a fashion similar to that of N^6-substituted cytokinins, or it may induce the synthesis and (or) accumulation of endogenous cytokinins (Mok and Mok 1985). The latter notion is based on the effects of the high ability of TDZ in inducing cytokinin-dependent shoot regeneration and modulation of endogenous levels of cytokinins.

In vitro propagation of plants is widely used, quickly obtaining a large number of identical plants, with phytochemical and sanitary quality (Sivanesan et al. 2010). For the success of an in vitro culture, protocols specific to each species are necessary, using different culture media, salt concentrations, and plant growth regulators. Thidiazuron (TDZ) is widely used in tissue culture and promotes cell division and elongation (Murthy et al. 1998). It operates in the regeneration and proliferation of meristems and, in combination with other regulators, can be used for the formation and maintenance of callus (Kokotkiewicz et al. 2012). It was proved that TDZ, unlike traditional phytohormones, individually fulfilled the requirements of various regenerative responses of many different plant species. The morpho-regulatory potential of TDZ has led to its application in plant tissue culture for the development of feasible morphogenetic systems. High intrinsic activity (economic factor) coupled with stability against heat (ease of use) and enzymes renders TDZ a choice chemical for establishing regenerable tissue culture systems (Mok and Mok 1985).

TDZ has been successfully used for the propagation of several plants where the increase in the number and length of shoots is observed when there is an increase in concentrations of the regulator (Ahmed and Anis 2012; Grabkowska et al. 2014). Further advantage observed in the use of TDZ on proliferation of plants is the maintenance of genetic stability, important for obtaining plants *true-to-type* and germplasm conservation (Faisal et al. 2014). In this review, several recently published studies on characterization of TDZ-induced in vitro regeneration are presented and discussed.

22.2 TDZ-Mediated Organogenesis

22.2.1 *Stevia rebaudiana*

Stevia rebaudiana Bertoni, a member of Asteraceae family which is indigenous to certain regions of South America-Brazil and Paraguay, is one of the important antidiabetic medicinal herbs. The compounds in its leaves, stevioside and rebaudioside, taste about 300 times sweeter than sucrose. It is used as sweetening agent and has enormous commercial importance. Its other medicinal uses include regulating blood sugar, preventing hypertension and tooth decay, and treatment of skin disorders. *Stevia* also has healing effect on blemishes, wound cuts, and scratches, besides being helpful in weight and blood pressure management. Conventional propagation in this plant is restricted due to the poor seed viability coupled with very low germination rate. Role of vegetative propagation method is also limited as specific habitat conditions are mandatory to grow the plants in addition to low acclimatization rate in soil. A suitable alternative method to prepare sufficient amount of plants within short time duration is the use of in vitro cultures. An efficient high frequency plant regeneration protocol through direct organogenesis was successfully developed for *Stevia rebaudiana* Bertoni (Lata et al. 2013). Nodal segments containing axillary buds were used as an explant and inoculated on MS (Murashige and Skoog 1962) medium containing 3% (w/v) sucrose, 0.8% (w/v) agar supplemented with various concentrations of benzyladenine (BA), kinetin (KN), and thidiazuron (TDZ) ranging from 1.0 to 9.0 μM. Maximum multiple shoots (96%) were obtained in MS medium supplemented with 1.0 μM TDZ with an average of 60 shoots per culture, having an average shoot length of 6.0 cm, whereas BA (9.0 μM) produced a maximum of 32 shoots with an average shoot length of 2 cm, and KN (9.0 μM) produced a maximum of 22 shoots with an average shoot length of 2 cm (Fig. 22.2). It is evident that TDZ at very low concentrations was able to produce the maximum number of shoots than BA and KN in *Stevia rebaudiana*. Several reports have stated that TDZ results in better shoot regeneration than any other cytokinins (Lata et al. 2013). According to Capelle et al. (1983), TDZ directly promotes growth due to its own biological activities similar to that of an N-substituted cytokinin, or it may induce the synthesis and accumulation of an endogenous cytokinin.

Fig. 22.2 Micropropagation of *Stevia rebaudiana* using nodal segment. (**a**) Mother plant, (**b–c**) in vitro shoot multiplication, (**d–e**) rooting, (**f**) fully rooted plant under acclimatization, (**g–i**) hardened tissue culture-raised plant, and (**j**) mature tissue culture-raised plants (*Source*: Lata et al. 2013; Article DOI – https://doi.org/10.4236/ajps.2013.41016)

22.2.2 *Withania somnifera*

Withania somnifera (L.) Dunal, Ashwagandha (Solanaceae), is cosmopolitan throughout the drier regions of India up to an altitude of 2000 m in range. It is useful in treating iatrogenic malnutrition in kids. In Ayurveda, the roots are prescribed for feminine disorders, bronchitis, arthritis, rheumatism, inflammation, central nervous system disorders, skin diseases, etc. The shortage of correct cultivation practices, the loss of habitats, and the illegal, indiscriminate collection of this plant from its natural habit create a heavy threat to its existence in the wild. Moreover, propagation through seeds is troublesome attributable to low germination percentage. Therefore, a protocol was standardized for in vitro regeneration of *Withania somnifera* on TDZ-supplemented medium (Fathima and Anis 2011). The nodal explants of *Withania somnifera* placed on MS basal medium lacking TDZ failed to show any morphogenetic response and did not produce shoots even after 6 weeks of incubation. On the other hand, MS basal medium supplemented with varied concentrations of TDZ (0.0–10.0 μM) showed swelling of explants, followed by differentiation of shoot bud primordia with different regeneration frequencies. Of the numerous concentrations of TDZ tested, 0.5 μM was found to be foremost effective in inducing highest percentage regeneration (98%) with the maximum number of shoots (23.8) and shoot length (4.83 cm) after 4 weeks of culture (Fathima and Anis 2011). TDZ has been used to induce shoot regeneration in several plants including *Psoralea corylifolia* L. (Faisal and Anis 2006) and *Cyamopsis tetragonoloba* L. (Ahmad and

Anis 2007). This behavior is believed to be due to the efficiency of TDZ to enhance the biosynthesis and accumulation of endogenous adenine-type cytokinins (Huettman and Preece 1993), thus creating TDZ as an effective cytokinin for the stimulation of shoot buds. In distinction, higher concentration of TDZ (1.0 μM and above) suppressed shoot formation during the same week culture period.

22.2.3 *Pluchea lanceolata*

Pluchea lanceolata is a perennial herb belonging to the family Asteraceae grown in warm climatic regions of India and known as Rasana. This plant is prized for its anti-arthritic and anti-inflammatory activity. A protocol was standardized for micropropagation of *P. lanceolata* by using nodal explants (Kher et al. 2014). The ability of the nodal explants for the bud break varied depending on the plant growth regulators and their concentration. Nodal explants were inoculated onto MS (Murashige and Skoog 1962) medium supplemented with 6-benzylaminopurine (BAP), kinetin (Kin), thidiazuron (TDZ), and 2iP (2-isopentenyladenine) at varied concentrations (0.0, 0.5, 1.0, 1.5, and 2.0 mg dm-3). Shoots developed with all the concentrations of cytokinins investigated and bud break occurred after 7–8 days of culture. Nodal explants cultured on MS medium augmented with 0.5 mg dm^{-3} thidiazuron (TDZ) exhibited the highest multiplication rate (9.7 shoots/explant). It was observed that in the cultures where shoot number was higher, the shoot length remained shorter.

22.2.4 *Cannabis sativa*

Hemp (*Cannabis sativa* L.) belongs to the Cannabaceae family. It is an annual herb that has been cultivated for the value of its fiber and more recently for paper manufacturing, oil extraction, and medicinal or drug preparation. Aseptic shoot tips were introduced to MS medium supplemented with different types of cytokinin for auxiliary bud induction. The effect of different concentrations of BA, KN, and TDZ on shooting response in the shoot tips of hemp was investigated. Among the three cytokinins tested, TDZ (0.2 mg l^{-1}) was found to provide the best bud induction, inducing an average of 3.22 buds with the thickest stem (Wang et al. 2009). Furthermore, the type of cytokinin in the medium also affected plantlet morphology, with the plantlets grown in TDZ-containing medium being more compact and vigorous. The suitability of TDZ for in vitro auxiliary shoot propagation has been well established in many woody plant tissue culture (Carl and John 1993) and also determinated in many herbage plants (Donna and John 2004).

22.2.5 *Gossypium hirsutum*

Most cotton *Gossypium hirsutum* (Malvaceae) genotypes of commercial interest present problems of in vitro regeneration. Aiming at improving regeneration rate, meristems and caulinar apices (region with about 5 mm length, immediately below the meristematic region) of IAC 22 and COKER 312 cultivars were extracted from plants with two or three primordia leaves and grown in Murashige and Skoog (MS) medium containing thidiazuron (TDZ), with concentrations ranging between 0.02 and 5.0 μM, in 3- to 7-day periods, with naphthalene acetic acid (NAA) and gibberellic acid (GA_3). The number of shoots from meristems was higher in 0.02 μM TDZ (5.80 shoots/explant) concentrations. In the case of caulinar apices, best results were obtained with 0.5 and 1.0 μM of TDZ (4.08 shoots/explant) (Caramori et al. 2001).

22.2.6 *Curculigo latifolia*

The monocotyledonous plant, *Curculigo latifolia*, commonly known as lemba, is a perennial herb belonging to the Hypoxidaceae family and was thought to be natively from Malaysia. The plant is known for its sweet proteins, namely, curculin and neoculin, that have been proven to be 500–9000 times sweeter than sucrose by weight. Curculin which is a good low-calorie sweetener is absorbed by the human body and has a great potential for low-calorie sweetener-based industries. Besides the industrial and economic importance of *C. latifolia*, it is also considered as a valuable medicinal plant in having anticancer properties and antidiabetic properties and inhibiting hepatitis B virus. A procedure was developed for in vitro propagation of *Curculigo latifolia* through shoot tip culture (Babaei et al. 2014). Direct regeneration and indirect scalp induction of *Curculigo latifolia* were obtained from shoot tips grown on MS medium supplemented with different concentrations and combinations of thidiazuron and indole-3-butyric acid. Maximum response for direct regeneration in terms of percentage of explants producing shoot, shoot number (7.52 shoots/explant), and shoot length (2.71 cm) was obtained on MS medium supplemented with combination of thidiazuron (0.5 mg l^{-1}) and indole-3-butyric acid (0.25 mg l^{-1}) after both 10 and 14 weeks of cultures. Indole-3-butyric acid in combination with thidiazuron exhibited a synergistic effect on shoot regeneration. The shoot tips were able to induce maximum scalp from basal end of explants on the medium with 2 mg l^{-1} thidiazuron. Cultures showed that shoot number, shoot length, and scalp size increased significantly after 14 weeks of culture.

22.2.7 *Guizotia abyssinica*

Guizotia abyssinica Cass. belonging to family Asteraceae is an herbaceous crop with a lot of industrial as well as medicinal importance. The plant oil is good absorbent of fragrance of flowers used as base oil by perfume industry. The plant is used by the various tribal communities of India in the treatment of rheumatism, arthritis, microbial infections applied to treat burns, used for birth control, and treatment of syphilis. A simple, efficient, and reproducible regeneration protocol for in vitro propagation of *G. abyssinica* was established (Baghel and Bansal 2015). Different explants, viz., apical and axillary buds, leaf, and internode, were selected for in vitro regeneration study to observe the effect of different concentrations of TDZ. Among all the four explants used, apical bud proved best in terms of shoot regeneration and multiplication. The variations in the regeneration potential of explants are attributable to the differences in the physiological and genetic makeup of cells. The best multiple shoot regeneration (4.44 ± 0.1) was observed on TDZ (0.45 μM)-supplemented medium. The reason for the better efficacy of apical bud is probably the presence of meristematic shoot bud at the already grown shoot tip. But in case of nodal explants, new shoot buds from the nodes are needed to be induced through purely hormonal control, or in other words, new shoot buds develop after inoculation in the absence of apical dominance. The role of TDZ in inducing regeneration is attributed to the ability of TDZ in enhancing the synthesis of adenine-type cytokinins. These results corroborate the fact that TDZ is an effective plant growth regulator for induction of shoot bud regeneration. The possible reason for the higher activity of individual TDZ treatment might be its high stability due to its resistance to cytokinin oxidase. TDZ-induced regeneration is linked to accumulation and transport of certain endogenous signals such as auxins or the related compounds like melatonin and serotonin (Jones et al. 2007).

22.2.8 *Artemisia vulgaris*

Artemisia vulgaris L. (mugwort) belongs to the family Asteraceae and is a tall aromatic perennial herb that grows in the hilly district of India in areas up to 2400 m in elevation. In traditional medicine, this plant is widely used for the treatment of diabetes, and extracts of the whole plant are used for epilepsy and in combination for psychoneurosis, depression, irritability, insomnia, and anxiety stress. Mugwort is commonly used in traditional European medicine as a choleretic and for amenorrhea and dysmenorrhea. The essential oil of the plant was reported to exhibit 90% mosquito repellency against *Aedes aegypti*, a mosquito that transmits dengue and yellow fever. An in vitro propagation system for *Artemisia vulgaris* has been developed (Sujatha and Ranjitha Kumari 2007). Hypocotyl segments (8–12 mm) excised from 10-day-old in vitro-grown seedlings were inoculated vertically on MS medium containing 3% (w/v) sucrose, 0.7% (w/v) agar supplemented with different concentrations of BA (6-benzyladenine) (0.44–13.32 μM), and TDZ (*N*-phenyl-*N*′-(1,2,3-thiadiazol-yl) urea) (0.23–11.35 μM) individually for multiple shoot induction (Fig. 22.3). The incubation of hypocotyl explants on MS media supplemented with BA (4.44 μM) or TDZ (4.54 μM) resulted in organogenic frequencies of 98.6% and 99.7%. The best organogenic response, including adventitious shoot number and elongation, was obtained when hypocotyl segments were cultured onto MS medium supplemented with 4.54 μM TDZ. Up to 28 shoots were formed per explant for an optimal duration of exposure of 48 days, and the maximum shoot length recorded was 9.8 cm. The media containing high concentrations of TDZ (>4.54 μM) decreased shooting frequency and shoot elongation.

Fig. 22.3 Shoot multiplication in *Artemisia vulgaris* L. (**a**) Shoot initiation from hypocotyl segments, (**b**) shoot multiplication, (**c**) rooting of shoots, (**d**) hardened plants in plastic cups (*Source*: Sujatha and Ranjitha Kumari 2007; Article DOI https://doi.org/10.1007/s11816-007-0028-1)

22.2.9 *Capsicum annuum*

Capsicum annuum termed as "hot pepper" forms an important economical crop of the family Solanaceae. It acts on the circulation and the digestion and is used to treat a wide range of complaints from arthritis and chilblains to colic and diarrhea. An efficient protocol for rapid in vitro propagation of *Capsicum annuum* cv. Pusa Jwala through multiple shoot bud formation from cotyledonary node explants of 15-day-old aseptic seedlings has been developed (Siddique and Anis 2006). The morphogenetic responses of cotyledonary node explant to TDZ alone or in combination with IAA were studied. Explants cultured onto a growth regulator-free MS medium failed to produce shoots even after 4 weeks of culture. When MS medium was supplemented with different concentrations of TDZ (0.1–10.0 μM), multiple shoots emerged from cotyledonary node explants after 15 days of culture. TDZ in combination with IAA at different concentrations induced more shoots per explant compared to TDZ alone in the investigation. The explants cultured on a medium containing TDZ (1.0 μM) produced a maximum of 11.6 shoot buds per explant with 72.67% regeneration. Optimum shoot differentiation was observed in a media containing TDZ (1.5 μM) and IAA (0.5 μM) after 4 weeks of culture. In this medium, the highest regeneration frequency (82.33%) and highest number of shoot buds per explant (19.0 ± 0.67) were achieved. At higher levels of TDZ, the number of shoot buds and regeneration frequency was reduced considerably, which may possibly be due to excessive callus growth while its specific concentration supports the maximum shoot bud induction.

22.2.10 *Rauvolfia serpentina*

Rauvolfia serpentina (L.) Benth. ex Kurz. (Apocynaceae) is an endangered medicinal plant recognized worldwide. Due to the presence of indole alkaloids and its usage as an antihypertensive drug, the demand for this plant has increased manifold in the global pharmaceutical industry. An efficient system for in vitro propagation of the endangered medicinal plant *Rauvolfia serpentina* has been developed (Alatar 2015). In vitro proliferation of shoots is usually promoted by incorporating growth regulators into the culture medium. TDZ is being selected for in vitro propagation of many plant species because of its tremendous ability to stimulate shoot proliferation. Compared to most of the other active compounds added to the media, extremely low concentrations of TDZ stimulate axillary shoot proliferation of many plant species. Proliferation of shoots was achieved from nodal segment explants, excised from field grown plants on Murashige and Skoog (MS) medium supplemented with thidiazuron (TDZ) (0.1–2.5 μ mol l^{-1}) although with low regeneration response and few number of shoots per explant. Greater number of shoots was achieved from nodal explants pretreated with higher concentrations of TDZ (5–100 μ mol l^{-1}) in liquid MS medium for different time periods (4, 8, 12, and 16 days), followed by their transfer on a growth regulator-free medium. The highest response in terms of percent regeneration (90%), average number of shoots/explant (23.17 ± 2.15), and maximum shoot length (5.3 ± 0.83 cm) was achieved by pretreating the nodal explants with 50 μ mol/L TDZ for 8 days. On increasing the concentration of TDZ, the number of shoots per explant was reduced. Similarly, at lower concentration, the percentages of regeneration as well as the number of shoots were drastically reduced.

22.2.11 *Kigelia pinnata*

Kigelia pinnata is a fast-growing, multipurpose tree used for ornamental and wayside planting belonging to the family Bignoniaceae. Various parts of the plant are employed for medicinal purposes by certain indigenous people. Traditional healers in India have used various parts of this plant to treat a wide range of skin ailments, from relatively mild complaints, such as fungal infections, boils, and psoriasis, to the more serious diseases like leprosy, syphilis, and skin cancer. Other medicinal applications include the treatment of dysentery, ringworm, tapeworm, postpartum hemorrhaging, malaria, diabetes, pneumonia, and toothache. An antimalarial compound known as lapachol has been extracted from the root of *K. pinnata.* Another compound obtained from the wood, quinone, shows antimalarial activity against drug-resistant strains of *Plasmodium falciparum* superior to chloroquine and quinine. Conventionally, *K. pinnata* reproduces via viable seeds, but the low percentage of seed viability limits its natural propagation. Thus in vitro propagation of the plant was established (Thomas and Puthur 2004). The surface-sterilized nodal segments were cultured on MS (Murashige and Skoog 1962) medium supplemented with various concentrations (1–7 mM) of 2,4-D in a test for callus induction. Callus was subcultured onto fresh medium (MS + 3 mM 2,4-D) every 45 days. For multiple shoot induction, calli were transferred to MS medium supplemented with various concentrations (0.5–9 mM) of TDZ alone or in combination with NAA (0.5 and 1 mM). The optimum response in terms of percentage of explants producing shoots and the highest number of shoot buds per explant were recorded on MS medium supplemented with TDZ (3 mM) and NAA (0.5 mM). On this medium 100% cultures responded with an average 28 shoots per culture. The regenerated shoots attained a height of about 2 cm in about 45 days of callus culture.

22.2.12 *Solanum tuberosum*

Potato is one of the most important widely grown crops and is an integral part of diet in the entire world. It produces more protein (524 kg/ha) as compared to wheat (254 kg/ha). It also supplies at least 12 essential minerals including Vitamin C. Apical shoot explants of *Solanum tuberosum* L. cvs. Desiree and Cardinal were grown on MS (Murashige and Skoog 1962) medium containing three different concentrations of TDZ (10^{-8}, 10^{-9} or 10^{-10} M). The maximum number of shoots (2.66 and 2.96) was obtained on MS + TDZ (10^{-8}) in cvs. Cardinal and Desiree, respectively (Sajid and Aftab 2009). The highest number of roots (12.60 and 14.90) and nodes (7.90 and 7.20) was observed on MS medium in the two cultivars. The maximum fresh and dry weight of the plantlets (0.543 g and 0.0524 g) in cv. Cardinal was obtained on MS medium containing10^{-9} M TDZ. In Desiree, the highest fresh and dry weights (1.0560 and 0.0965 g, respectively) were observed on MS medium containing 10^{-10} M TDZ.

22.2.13 *Morus alba*

Morus sp. is an invaluable tree for the sericulture industry as it is the only source of food for mori silkworms. Three cultivars of mulberry S-36, S-1, and K-2 were selected for in vitro regeneration experiments (Thomas 2003). Cotyledonary explants were cultured 7, 14, and 21 days after embryo culture. Individual cotyledons were excised from seedlings about 1 mm below the cotyledonary node and cultured on MS medium supplemented with 2–9 μM TDZ and BAP. The embryos cultured on MS medium supplemented with 5 μM BAP produced well-developed cotyledons, hypocotyl, and radicle 7 days after culture. The induction of multiple shoots varied with the age of the cotyledons as well as the concentration and type of growth regulator used. A significantly greater number of shoots were formed from cotyledons of 14-day-old embryos cultured on MS medium containing TDZ and BAP compared to 7- and 21-day-old explants. TDZ at different concentration was found to induce more shoots per explant as compared to BAP at the same concentration and explant age. A maximum number of shoots were produced on MS medium fortified with 7 μM TDZ. On this medium the 14-days-old explants produced an average number of 20.3 shoots per explant in S-36 cultivar. Whereas 7.3 shoots were produced in cultivar K-2 followed by 5.6 shoots in cultivar S-1. The shoots produced were isolated individually and elongated on MS medium augmented with 5 μM BAP.

22.2.14 *Salvia officinalis*

Salvia officinalis L., common sage (family, Lamiaceae), one of the important medicinal plant species, is cultivated in several countries mostly to obtain the dried leaves to be used as raw material in medicine and perfumery industries. Recent research has shown that sage essential oil can recover the memory and has shown promise in the treatment of Alzheimer's disease. It is also used in treating bronchial asthma, inflammatory affection, atherosclerosis, cataracts, ischemic heart disease, cancer, hepatotoxicity, and insufficient sperm mobility. Conventionally, *S. officinalis* is propagated through seeds; however, in nature, seeds germinate slowly and remain dormant for a long time. Alternatively, cutting can be used but low population size hampers the process. Therefore, in vitro methods for large-scale multiplication would be a viable option (Jafari et al. 2017). This research was conducted to develop an indirect organogenesis regeneration protocol for *Salvia officinalis* L. via callus which was obtained from leaf and internode explants. Among these explants internode explant gave best callus induction on MS medium supplemented with 0.5 mg/l 6-benzylaminopurine (BAP), 2.0 mg/l α-naphthalene acetic acid (NAA). The calli formed were subcultured on MS medium fortified with 0.5 mg/l thidiazuron (TDZ). A maximum of 70% shooting was observed, and a maximum of 2.5 shoots were produced on TDZ-augmented medium. The elongated shoots were transferred to MS/2 medium fortified with different concentrations of NAA and IBA for root induction.

22.2.15 *Lavandula angustifolia*

Lavandula angustifolia "Munstead" (English lavender) family Lamiaceae (Labiatae) is a hardy perennial shrub rich in aromatic essential oils and is valuable for its pharmaceutical, aromatic, and culinary properties. This genus is relatively rich in phenolic constituents, with 19 flavones and 8 anthocyanins. *Lavandula angustifolia* is stated to be a carminative, spasmolytic, tonic, and antidepressant and is also used for treating nervous headache, neuralgia, rheumatism, depression, insomnia, windy colic, fainting, toothache, sprains, sinusitis, stress, and migraine. Aromatherapy involves massage using a much diluted essential oil or mixture of essential oil to the bath or a basin of hot water or using burners. The present investigation was carried out to study the in vitro shoot proliferation, root formation, and ex vitro acclimatization of English lavender (Munstead) (Hamza et al. 2011). Nodal explant showed a good response for producing the highest survival percentage and shoot and leaf numbers compared with the shoot tip one. Among the tested cytokinins, TDZ at 0.20 mg/L recorded the highest shoot number (30.55 shoots), followed by BAP at 0.80 mg/L (16.50 shoots). The weakest effects on shoot number were recorded for media supplemented with all KIN concentrations; however, it tabulated the tallest shoot length.

22.3 TDZ-Mediated Embryogenesis

22.3.1 *Tylophora indica*

Tylophora indica (Burm. f.) Merrill, previously called as *Tylophora asthematica*, a member of Asclepiadaceae, is an important indigenous medicinal plant found in restricted localities in the Indian subcontinent. The roots have a sweetish taste turning acrid, an aromatic odor, and a brittle fracture. They possess stimulant, emetic, cathartic, expectorant, stomachic, and diaphoretic properties and are used for the treatment of asthma, bronchitis, whooping cough, dysentery, diarrhea, and rheumatic gouty pains. Apparently due to non-availability of sufficient quality planting materials, commercial plantations of this important aromatic and medicinal species have not been widely attempted, and presently only the wild population is exploited for extraction purposes. Due to overexploitation and lack of organized cultivation, the wild populations have declined fast. An efficient procedure has been developed for inducing somatic embryogenesis from mature leaves of *Tylophora indica* was established (Chandrasekhar et al. 2006). Leaf bits were cultured facing the adaxial and abaxial side toward the medium. They were inoculated on MS medium containing 2,4-D (1.0, 1.5 and 2.5 μM), TDZ (0.25, 0.5, and 0.75 μM), or BA (0.5 and 1.0 μM) alone or in combinations. Leaf sections were initially cultured on Murashige and Skoog's (MS) medium supplemented with thidiazuron (TDZ) in addition with 2,4-dichlorophenoxy acetic acid (2,4-D); particularly 0.5 μM TDZ along with 1.5 μM 2,4-D was very effective in inducing somatic embryos (71.6%). Plants were regenerated from in vitro somatic embryos plated on semisolid medium devoid of growth regulators.

22.3.2 *Vitis vinifera*

Grapes (*Vitis vinifera*) are one of the most commonly consumed fruits in the world. Grapes can be used in numerous forms, viz., raisins, jam, jelly, and beverages like juice and wine, and also added for other culinary purposes. Being an inexpensive fruit, the demand for grapes is increasing day by day due to its immense potential toward improving the health of the humans. The biological activities of grapes has been studied widely which showed that it is rich in phenolic compounds with approximately two-thirds of grape polyphenols present in skin and seeds. These grapes owing to its phenolic compounds provide wholesome health benefits including cardioprotective, anti-inflammatory, anticarcinogenic, antimicrobial, and antioxidant properties. In vitro protocols for callus induction, somatic embryogenesis, and plant regeneration using leaf explants of three varieties of grapes (Thompson seedless, Sonaka, and Tas-e-Ganesh) were developed (Malabadi et al. 2010). Surface-sterilized leaf explants were cultured on Nitsch and Nitsch basal medium (Nitsch and Nitsch 1969) supplemented with a range of TDZ concentrations (0.45–11.35 μM) and 2,4-D at a concentration of 4.52 μM singly and in combination. The leaf explants responded well, and callus was induced after 2–4 weeks of culture on NN medium supplemented with 4.52 μM 2,4-D and 4.54 μM TDZ in all the three varieties. The highest percentage of somatic embryogenesis (Thompson seedless, 78%; Sonaka, 56.2%; and Tas-e-Ganesh, 48%) was observed. Somatic embryos recovered per gram fresh weight of embryogenic tissue were 55.0 in Thompson seedless, and the number of seedlings recovered per gram fresh weight of embryogenic tissue was 43.0. This study has opened possibilities for large-scale clonal propagation of grapes.

22.3.3 *Theobroma cacao*

Cacao trees (*Theobroma cacao*) are grown principally in rainforest areas in the tropical regions of the world. Cacao seeds are the sources of cacao powder and butter, which are important ingredients in chocolate and confectionary products. Cocoa butter is also used in a number of pharmaceutical and cosmetic products. Vegetative propagation of cacao is limited because of the low propagation rate, intensive labor, and associated costs. Simultaneously cacao has proven to be recalcitrant to in vitro shoot regeneration and organogenesis. Plant regeneration via somatic embryogenesis provides an alternative approach for clonal propagation of cacao since somatic embryos are produced through bipolar development of somatic cells; plants derived from somatic embryos are genetically identical to their parental donor cells and have the growth characteristics of seed-derived plants. A procedure for the regeneration of cacao plants from staminode explants via somatic embryogenesis was developed (Li et al. 1998). Rapidly growing calli were induced by culturing staminode explants on primary callus growth medium supplemented with 20 g l^{-1} sucrose, 9 μM 2,4-D, and various concentration of TDZ (22.7–454.4 nM). Calli were further subcultured, and somatic embryos were formed from embryogenic calli. A TDZ concentration of 22.7 nM was found to be the optimal concentration for effective induction of somatic embryos from cacao. Two types of somatic embryos were identified on the basis of their visual appearance and growth behavior. Plants raised via somatic embryos showed morphological and growth characteristics similar to those of seed-derived plants.

22.3.4 *Myrica rubra*

Myrica rubra is a dioecious species and its progeny is highly heterozygous. Conventional vegetative propagation methods such as air layering and grafting are not rapid to meet the need of elite varieties. Somatic embryogenesis provides great promise of mass propagation and could be used as genetic engineering vehicle to develop non-chimeric transgenic plants. This present study accomplishes plant regeneration through direct somatic embryogenesis from cotyledon explants of *Myrica rubra*. Somatic embryogenesis was induced on woody plant medium (WB) (Sugawara et al. 1994) supplemented with thidiazuron (TDZ) alone or in combination with 2,4-D from mature cotyledon explants of *Myrica rubra* (Asghar et al. 2013). All concentrations of TDZ except 1.0 mgL^{-1} induced somatic embryos and adventitious shoots simultaneously within 2 months of culture. Addition of 2,4-D in the medium significantly improved induction of somatic embryos. Frequency of embryogenesis was only 3.34% with 7.00 embryos per explants when TDZ was fortified as a single growth regulator which was improved to 22.00% with the addition of 0.1 mgL^{-1} 2,4-D in the media. Repetitive embryogenesis was induced on optimized concentrations (0.5 mgL^{-1} BA and 0.05 mgL^{-1} TDZ) of two cytokinins in combination with various concentrations of 2,4-D. Continuous culture of the explants with cluster of embryos on the induction media did not induce repetitive embryogenesis. On repetitive embryogenesis induction media, most of the embryos induced were smaller in size than those of the primary embryos during their induction stage. TDZ in combination with IBA induced adventitious shoots on the surface of somatic embryo explants. TDZ (0.2 mgL^{-1}) plus IBA (1.0 mgL^{-1}) was the most effective combination with maximum number (8.5) of shoots per explant. Shoot elongation was achieved on the media supplemented with 0.5 mgL^{-1} BA concentration plus 0.1 mgL^{-1} NAA. The plants were rooted and successfully hardened.

22.3.5 *Murraya koenigii*

Murraya koenigii (L.) Spreng, popularly known as curry leaf plant, is a small aromatic tree belonging to the family Rutaceae that grows widely in Southeast Asia. Its leaves are slightly pungent, bitter, and acidulous in taste. Fresh and dried leaves are used extensively as a flavoring agent in many Indian culinary practices. The aromatic components of this tree are widely utilized in the medicinal field. A reproducible protocol for direct and indirect somatic embryogenesis was established (Paul et al. 2011). Embryogenic callus was obtained from 90% zygotic embryonic axis

(ZE) and 70% cotyledon (COT) explants in Murashige and Skoog (MS) basal medium supplemented with 8.88 μM 6-benzyladenine (BA) and 2.675 μM α-naphthalene acetic acid (NAA). Globular somatic embryos were induced and further matured from such embryogenic callus by subsequent culture on the same basal media containing thidiazuron (TDZ) (2.27–9.08 μM). The highest frequency of somatic embryos (14.58 ± 0.42) was recovered from ZE-derived callus after 6 weeks. The age and type of explant and concentration of TDZ played an important role in the development of somatic embryos. Explants excised from 60-day-old seed differentiated from 96.67% of ZE explants and 86.67% from COT explants when cultured on MS basal medium supplemented with 4.54 and 9.08 μM TDZ, respectively, after 4 weeks. The best result obtained for the average frequency of somatic embryos (11.28 ± 0.32) was from ZE explants, which was significantly higher than COT explants (7.34 ± 0.97). Most of the somatic embryos (above 95%), irrespective of their origin, germinated after 4 weeks in 1/2 MS basal media.

22.3.6 *Phalaenopsis aphrodite*

Phalaenopsis aphrodite subspecies *formosana* is qualified to be a model plant in the recent orchid research which has been intensively studied in the past 10 years, including in vitro protocols, flowering and photosynthetic physiology, chloroplast genomic analysis, global analysis of transcriptome, and modified ABCDE model of flowering. In the global horticultural trade, *Phalaenopsis* (i.e., moth orchids) is one of the most popular plants in the production of pot plants and cut flowers. It is mainly due to their beautiful flowers, ease of cultivation in the artificial conditions, and a long vase life. An alternative in vitro protocol for embryo induction directly from intact living seedlings of *Phalaenopsis aphrodite* subspecies *formosana* was established (Feng and Chen 2014). Without the supplementation of plant growth regulators (PGRs), no embryos were obtained from all the seedlings when cultured on the solid medium. In contrast, embryos formed from the seedlings on the two-layer medium and the two-step culture system without the use of PGRs (Fig. 22.4). It was found that the age of the seedlings affected embryo induction. The 2-month-old seedlings typically had higher embryogenic responses when compared with the 4-month-old seedlings in the two-layer medium or two-step system. For the 2-month-old seedlings, 1 mg/L TDZ resulted in the highest number of embryos at the distal site of the shoot. However, on the leaves' surface, 0.5 mg/L TDZ induced the highest number of embryos. When the 2-month-old seedlings were cultured using the two-step method at 1 mg/L of TDZ, the highest embryogenic response was obtained, with an average of 44 embryos formed on each seedling. These adventitious embryos were able to convert into plantlets in a PGR-free 1/2 MS medium, and the plantlets had normal morphology and growth.

Fig. 22.4 *Phalaenopsis aphrodite* subsp. *formosana*, direct somatic embryogenesis from intact seedlings. (**a**) A flowering potted plant, (**b**) in vitro seed germination, (**c**) cluster of embryos, (**d**) embryos turned into protocorm-like bodies, (**e**) embryogenesis from the leaf surface, (**f**) foliar embryos formed, (**g**) foliar embryos of developing leaves, (**h**) numerous embryos initiated, (**i**) plantlet conversion, (**j**) rooted plantlets (*Source*: Feng and Chen 2014; Article DOI: https://doi.org/10.1155/2014/263642)

22.3.7 *Coffea arabica*

Coffea is an extremely important perennial agricultural crop in tropical areas with more than 6.5 million tons of green beans being produced every year on about 11 million hectares. The genus *Coffea* (Rubiaceae) consists of about 80 species in which only *Coffea arabica* (arabica) and *Coffea canephora* (robusta) are important for the production of *Coffea* beans. *C. arabica* contributes nearly 70% of the coffee consumed worldwide due to its superior quality, and *C. canephora* accounts for the rest 30%. Conventional breeding of coffee is difficult because of the long duration of cultivation before the seeds are set. Plant regeneration via various tissue culture methods could be very effective for propagation and improvement of coffee plants. Somatic embryogenesis is a highly useful method for the large-scale propagation of species of economic interest. Somatic embryos are widely considered to be of single cell origin; hence this is advantageous for transformation studies. The rapid direct and repetitive somatic embryogenesis in *Coffea arabica* and *C. canephora* genotypes was tested on Murashige and Skoog medium containing thidiazuron (TDZ) in concentrations of 2.27–11.35 μM (Giridhar et al. 2004). Segments taken from cotyledon leaf, first leaf, and stalk of regenerated plantlets produced clusters of somatic embryos directly from cut portions of explants on TDZ (9.08 μM)-containing medium within a period of 2 months. Subculturing of these embryo clusters produced more secondary embryos on reduced TDZ (0.045–0.91 μM)-containing medium, and these subsequently developed into plantlets (80–85%) on development medium followed by rooting on MS basal medium. This direct somatic embryogenesis from leaf and hypocotyl explants in *Coffea* sp. is a strong evidence of cell totipotency. The rapid somatic embryo induction protocol would be useful for the mass propagation, direct regeneration, and genetic transformation of selected elite lines.

22.3.8 *Vigna umbellata*

Rice bean, an under exploited tropical legume, is a native of Southeast Asia. In India, the crop is mainly found in the Western and Eastern Ghats and the NE Himalayas but is also grown in the sub-temperate Western Himalaya in the Uttaranchal and Himachal Pradesh hills. It is grown for its nutritious seeds and green pods as vegetable and also as a leguminous fodder crop in Kerala, Orissa, and West Bengal because of its higher fodder production potential. Rice bean has a comparatively higher content of proteins than other crops. Amino acid content is well suited for human digestion. But despite the great advantages, there are some biotic and abiotic constraints which affect its potential. Strategies to overcome these yield-limiting factors by conventional breeding have been slow due to the lack of desirable level of genetic variability in germplasm. An efficient in vitro regeneration protocol for Indian cultivar (RBL-50) of rice bean *Vigna umbellata* (Thunb.).

Fig. 22.5 Somatic embryogenesis and plant regeneration in rice bean. (**a**) Callus induction from cotyledonary node explant. (**b**) Formation of shoot buds on regenerative callus. (**c–d**) Induction of somatic embryos. (**e–i**) Various stages of somatic embryos: (**f**) globular, (**g**) heart, (**h**) torpedo, and (**i**) cotyledonary. (**j**) Germination of somatic embryos. (**k**) Abnormal germination of somatic embryos. (**l**) Potted plantlets raised from germination of somatic embryos (*Source*: Saini and Chopra 2012)

Ohwi and Ohashi via somatic embryogenesis has been developed (Saini and Chopra 2012). Highly proliferating (98%) calli cultures were initiated from the cotyledonary node containing both half cotyledons on semisolid MSB medium (MS salts and B5 vitamins) supplemented with 12.5 μM thidiazuron (TDZ) alone. Type and concentrations of growth regulators influenced the frequency of somatic embryogenesis. TDZ was found responsive for somatic embryogenesis than BAP, 2,4-D, and picloram, and the best result (18 somatic embryos per explant) was obtained with 12.5 μM TDZ in combination with 2.5 μM 6-benzylaminopurine (BAP). Sustained cell division resulted in the formation of cell aggregates, which progressed to the globular and heart-shaped somatic embryos and then, if they differentiated properly, to the torpedo shape and cotyledonary stages. The transfer of embryos onto MS basal medium enabled the embryos to achieve complete maturation and germination (Fig. 22.5). The percentage of germinating embryos increased significantly from 20 to 50 when 1.5 μM BAP and 2.0 μM gibberellic acid (GA_3) were supplemented to the MS basal media. In vitro-raised plantlets with well-developed roots were successfully hardened in a greenhouse and established in soil.

22.3.9 *Crocus sativus*

Saffron (*Crocus sativus* L.) is one of the most valuable industrial crops which is particularly important in exploration and income revenue. In vitro propagation of saffron either through somatic embryogenesis or cormogenesis is considered as an efficient alternative method for large-scale propagation of pathogen-free corms. In order to develop an efficient protocol for in vitro propagation of saffron, a factorial experiment was carried out based on completely randomized design to investigate the effects of various concentrations of TDZ (0, 0.1, 0.25, and 0.5 mg l^{-1}) on somatic embryogenesis induction from five different types of corm explants (terminal or axillary buds, upper or lower parts of the corm tissue, and terminal buds from pre-treated corms at 4 °C for 2 weeks) (Sheibani et al. 2007). The results revealed that TDZ concentration affected the induction of somatic embryogenesis significantly, while different types of corm explants showed no significant effect on this process. Among TDZ concentrations used, 0.5 mg l^{-1} was the most effective treatment for embryogenesis induction. Embryogenic calli proliferated well when subcultured into MS medium supplemented with 0.25 mg l^{-1} TDZ before transferring to hormone-free MS medium containing 6% sucrose for maturation. Matured embryos were transferred to half-strength MS medium without growth regulators for further development, from which microcorms were produced at the basal part after 3 months.

22.3.10 *Psoralea corylifolia*

Psoralea corylifolia L. (Fabaceae), commonly known as "babchi," is an endangered medicinal plant distributed in the tropical region of the world. The plant is used in indigenous system of medicine as a laxative, aphrodisiac, anthelmintic, diuretic, and diaphoretic in febrile conditions. It is specially recommended in the treatments of leucoderma, leprosy psoriasis, and inflammatory diseases. It is a seed-propagated species; however, the germination percentage is very low (5–7%). The low percentage of seed germination coupled with non-judicious wild collection for pharmaceuticals pose a serious threat to its existence in the nature. Tissue culture and in vitro plant regeneration system provide an alternative means for mass proliferation and ex situ conservation of endangered plant species. The development of efficient in vitro regeneration systems is needed to facilitate the application of recombinant DNA technology to the improvement of crop germplasm. In the present study, a simple, rapid, and effective system to regenerate *Psoralea corylifolia* plants via direct somatic embryogenesis from nodal segments has been established (Faisal et al. 2008). The embryogenic cells proliferated, formed somatic embryos, and were subsequently converted into normal plantlets under optimized culture conditions. The frequency of somatic embryogenesis was strongly influenced by the concentration of thidiazuron (TDZ) (0.0, 10.0, 11.0, 12.0, 13.0, 14.0 15.0, 16.0, 17.0, 18.0, 19.0, 20.0 μM) in the medium. The highest frequency (82%) of somatic embryogenesis was observed on Murashige and Skoog medium containing 16.0 μM TDZ. The somatic embryos, when transferred to plant growth regulator-free MS basal medium, developed further to heart-shaped, torpedo, and cotyledonary stages within 2 weeks. Conversion of somatic embryos into plantlets was achieved by isolating somatic embryos with distinct cotyledons and transferring them onto half-strength MS medium containing 1.0 μM gibberellic acid (GA_3). Subsequently, the regenerated plantlets were successfully established in ex vitro condition with 90% survival.

22.3.11 *Pimpinella tirupatiensis*

Pimpinella tirupatiensis Balk and Subr., locally known as "adavikothimeera" (forest coriander), is a herbaceous medicinal plant, distributed on Tirumala Hills (1000 m above MSL) of Chittoor District, Andhra Pradesh. It is a narrow endemic species (Umbelliferae) of seasonal occurrence with underground tuberous root system. Dried roots of *P. tirupatiensis* are administered along with few other ingredients to cure colic and rheumatic ailments in cattle. The local Yanadi tribal community uses the tuberous roots of *P. tirupatiensis* to cure severe ulcers of the stomach, throat, and genital organs and also as aphrodisiac and as abortifacient agents. Fruits are used to cure asthma and are considered as an effective remedy for "flatulent colic." Conventional propagation methods through seed and root tubers for cultivation of *P. tirupatiensis* are beset with limited planting material and poor fruit setting. The availability of the seed is also very less due to its dispersal by wind, on attaining maturity. In the present investigation, a regeneration protocol through somatic embryogenesis is attempted to conserve this rare species of Umbelliferae for posterity (Prakash et al. 2001). Hypocotyl segments were excised from 4-week-old aseptic seedlings of *Pimpinella tirupatiensis* and were cultured on MS medium with TDZ (1 mg/l) and NAA (0.5 mg/l), which gave rise to friable, pink callus after 4 weeks of culture. Embryogenic callus on transfer to MS medium containing TDZ (1 mg/l) produced somatic embryos after 8 weeks having dark green shoots and white hairy roots. On MS + TDZ (1 mg/l) + BA (1 mg/l), somatic embryo formation was enhanced. Embryos isolated and germinated in the presence of MS + TDZ (1.0 mg/l) and GA3 (1.0 mg/l) showed normal flowering without any morphological variation on transplantation to soil.

22.4 Conclusion

TDZ is widely applied in plant in vitro or in vivo that influences a number of parameters in plants. It was firstly used as a defoliant for cotton. A miscellaneous range of responses with a high grade of efficacy is induced via TDZ application. Exploitation of TDZ in plant cell culture systems in the early 1980s for induction of adventitious shoot regeneration produced a considerable interest in understanding the plant morphogenesis and different physiological parameters. The complex nature of the biochemical and morphological responses that have been reported for plant tissues exposed to TDZ has provided some indication of the cascade of physiological reactions within the plant tissues. It was reported that TDZ induces shoot regeneration in many plant species. TDZ fraction plays a very important role in morphogenesis like lower concentration which induces axillary shoot proliferation, whereas higher fraction causes adventitious shoot development. An array of complex physiological mechanisms like functions of an intact molecule in both alone and in engaged system are involved in TDZ-treated somatic embryogenesis, and also TDZ-treated tissues maintain and enhance the accumulation and transport of auxin. All these results suggest that TDZ has a keen role in the induction of stimulation of plant growth regulator processes and physiological maintenance of plant tissues during culture process. TDZ is believed to be the best synthetic cytokinin present for the regeneration of numerous plant species. TDZ improved greatly the ex vivo generation and multiplication of species recalcitrant to propagation. In several cases, growth of explants accelerated when transferred from amino purine cytokinin-cultured medium onto TDZ-fortified medium. TDZ is much effective in concentrations 10–1000 times less than the other phytohormones. It has been observed that an expanded range of concentrations to be effective ex vivo, dependent on the species, explant status, and objective. In certain procedures, a twofold culture system is conducted with pronounced success, where TDZ-fortified initial medium induces shoot multiplication which is followed by secondary medium containing low level of TDZ or other phytohormones to enhance shoot organogenesis. Plant's response to TDZ may be attributed to the change of oxidative stresses in plant cell, especially during the shoot regeneration or embryo formation.

References

Ahmad N, Anis M (2007) Rapid plant regeneration protocol for cluster bean (*Cyamopsis tetragonoloba* L. Taub.) J Hortic Sci Biotechnol 82:585–589

Ahmed MR, Anis M (2012) Role of TDZ in the quick regeneration of multiple shoots from nodal explant of *Vitex trifolia* L. an important medicinal plant. Appl Biochem Biotechnol 168:957–966

Alatar AA (2015) Thidiazuron induced efficient in vitro multiplication and ex vitro conservation of *Rauvolfia serpentina* – a potent antihypertensive drug producing plant. Biotechnol Biotechnol Equip 29(3):489–497

Arndt FR, Rusch R, Stillfried HV, Hanisch B, Martin WC (1976) A new cotton defoliant. Plant Physiol 57:S–99

Asghar S, Abbas SJ, Wahab F, Khan NH, Ahmad N, Chen L, He X, Qin Y (2013) Direct induction of somatic embryogenesis and plant regeneration from cotyledon explants of *Myrica rubra* Sieb. & Zucc. Afr J Agric Res 8(2):216–223

Babaei N, Abdullah NAP, Saleh G, Abdullah TL (2014) An efficient in vitro plantlet regeneration from shoot tip cultures of *Curculigo latifolia*, a medicinal plant. Sci World J 2014:1–9

Baghel S, Bansal YK (2015) Thidiazuron promotes *in vitro* plant regeneration and phytochemical screening of *Guizotia abyssinica* Cass. – a multipurpose oil crop. World J Pharm Pharm Sci 4(1):1193–1217

Baskakov YA, Shapovalov AA, Zhirmunskaya NM, Ovsyannikov TV (1981) Interrelationship of growth-regulating activity and phytotoxicity of synthetic cytokinins. Dokl Akad Nauk SSSR 267:1514–1517

Burkhanova EA, Fedina AB, Baskakov YA, Yu A, Kulaeva ON (1984) Comparative study of the action of 6-benzylaminopurine, thidiazuron, and cartolin on growth of intact pumpkin seedlings. Sov Plant Physiol 31:8–12

Capelle SC, Mok DWS, Kirchner SC, Mok MC (1983) Effects of thidiazuron on cytokinin autonomy and the metabolism of N6-(Y2-Isopentyl) [8-14c] adenosine in callus tissues of *Phaseolus lunatus* L. J Plant Physiol 73(3):796–802

Caramori LPC, Favaro S, Vieira LGE (2001) Thidiazuron as a promoter of multiple shoots in cotton explants (*Gossypium hirsutum* L.) Acta Sci 23(5):1195–1197

Carl AH, John EP (1993) Thidiazuron: a potent cytokinin for woody plant tissue culture. Plant Cell Tissue Organ Cult 33:105–119

Cassells A, Curry R (2001) Oxidative stress and physiological, epigenetic and genetic variability in plant tissue culture: implications for micro-propagators and genetic engineers. Plant Cell Tissue Org Cult 64(2–3):145–157

Chandrasekhar T, Mohammad Hussain T, Rama Gopal G, Srinivasa Rao V (2006) Somatic embryogenesis of *Tylophora indica* (Burm.f.) Merril., an important medicinal plant. Int J Appl Sci Eng 4(1):33–40

Davies PJ (1995) The plant hormones: their nature, occurrence, and functions. In: Davies PJ (ed) Plant hormones: physiology, biochemistry and molecular biology. Kluwer, Boston, pp 3–38

Donna IL, John EP (2004) Thidiazuron stimulates adventitious shoot production from *Hydrangea quercifolia* Bartr., leaf explants. Sci Hortic 101:121–126

Faisal M, Anis M (2006) Thidiazuron induced high frequency axillary shoot multiplication in *Psoralea corylifolia* L. Biol Plant 50:437–440

Faisal M, Shahzad A, Anis M (2008) Somatic embryogenesis and plant regeneration from nodal explants in *Psoralea corylifolia* L. Int J Plant Dev Biol 2(2):111–113

Faisal M, Alatar AA, Hegazy AK, Alharbi SA, El-Sheikh M, Okla MK (2014) Thidiazuron induced in vitro multiplication of *Mentha arvensis* and evaluation of genetic stability by flow cytometry and molecular markers. Ind Crop Prod 62:100–106

Fathima N, Anis M (2011) Thidiazuron induced high frequency axillary shoot multiplication in *Withania somnifera* L. Dunal J Med Plants Res 5(30):6681–6687

Feng JH, Chen JT (2014) A novel in vitro protocol for inducing direct somatic embryogenesis in *Phalaenopsis aphrodite* without taking explants. Sci World J 2014:1–7

Gamborg OL, Phillips GC (1995) Plant cell, tissue and organ culture–fundamental methods. Springer – Verlag, Berlin

George EF (2008) Plant tissue culture procedure – background. In: George EF, Hall MA, De Kler KGJ (eds) Plant propagation by tissue culture, 3rd edn. Springer, Dordrecht, pp 1–28

Giridhar P, Kumar V, Indu EP, Ravishankar GA, Chandrasekar A (2004) Thidiazuron induced somatic embryogenesis in *Coffea Arabica* L. and *Coffea canephora* P ex Fr. Acta Bot Croat 63(1):25–33

Grabkowska R, Sitarek P, Wysokińska H (2014) Influence of thidiazuron (TDZ) pretreatment of shoot tips on shoot multiplication and ex vitro acclimatization of *Harpagophytum procumbens*. Acta Physiol Plant 36:1661–1672

Grossmann K (1991) Induction of leaf abscission in cotton is a common effect of urea and adenine-type cytokinins. Plant Physiol 95:234–237

Guo B, Abbasi BH, Zeb A, Xu LL, Wei YH (2011) Thidiazuron: a multi-dimensional plant growth regulator. Afr J Biotechnol 10(45):8984–9000

Hamza AM, Omaima M, Kafie AE, Kasem MM (2011) Direct micropropagation of English Lavender (*Lavandula angustifolia* Munstead) plant. J Plant Prod 2(1):81–96

Huettman CA, Preece JE (1993) Thidiazuron: a potent cytokinin for woody plant tissue culture. Plant Cell Tissue Organ Cult 33:105–119

Jafari S, Daneshvar MH, Salmi MS, Abadi LJ (2017) Indirect organogenesis and plant regeneration in common sage (*Salvia officinalis* L.): an important medicinal plant of Iran. Mod Appl Sci 11(5):22–29

Jariteh M, Ebrahimzadeh H, Niknam V, Mirmasoumi M, Vahdati K (2015) Developmental changes in protein, proline and some antioxidant enzymes activities in somatic and zygotic embryos of Persian walnut (*Juglans regia* L.) Plant Cell Tissue Org Cult 122:101–115

Ji ZL, Wang SY (1988) Reduction of abscisic acid content and induction of sprouting in potato, *Solanum tuberosum* L., by thidiazuron. J Plant Growth Regul 7:37–44

Jones MPA, Yi ZJ, Murch SJ, Saxena PK (2007) Thidiazuron induced regeneration of *Echinacea purpurea* L.: micropropagation in solid and liquid culture systems. Plant Cell Rep 26:13–19

Kane ME (2005) Shoot culture procedures. In: Trigiano RN, Gray DJ (eds) Plant development and biotechnology. CRC Press, Washington, DC

Kher MM, Joshi D, Nekkala S, Nataraj M, Raykundaliya DP (2014) Micropropagation of *Pluchea lanceolata* (Oliver & Hiern.) using nodal explant. J Hortic Res 22(1):35–39

Kokotkiewicz A, Luczkiewicz M, Hering A, Ochocka R, Gorynski K, Bucinski A, Sowinski P (2012) Micropropagation of *Cyclopia genistoides*, an endemic South African plant of economic importance. Zurich Nat Sci 67:65–76

Kozai T, Xiao Y (2006) A commercialized photoautotrophic micropropagation system. In: Gupta SD, Ibaraki Y (eds) Plant tissue culture engineering. Springer, Dordrecht, pp 355–371

Lata H, Chandra S, Wang YH, Raman V, Khan IA (2013) TDZ-induced high frequency plant regeneration through direct shoot organogenesis in *Stevia rebaudiana* Bertoni: an important medicinal plant and a natural sweetener. Am J Plant Sci 4:117–128

Lee JM (2004) Plant cell culture and its applications. In: Goodman RM (ed) Encyclopedia of plant and crop sciences. Marcel Dekker, New York, pp 931–933

Li Z, Traore A, Maximova S, Guiltinan MJ (1998) Somatic embryogenesis and plant regeneration from floral explants of cacao (*Theobroma cacao* L.) using thidiazuron. In: In Vitro Cell Dev Biol Plant, vol 34, pp 293–299

Lin CH, Lee LY, Tseng MJ (1994) The effect of stratification and thidiazuron treatment on germination and protein synthesis of *Pyrus serotina* Rehd. cv. Niauli. Ann Bot 73:515–523

Malabadi RB, Vijayakumar S, Nataraj K, Mulgund GS (2010) Induction of somatic embryogenesis and plant regeneration in grapes (*Vitis vinifera* L.) Bot Res Int 3(2):48–55

Mok MC, Mok DWS (1985) The metabolism of [^{14}C]-thidiazuron in callus cultures of *Phaseolus lunatus*. Physiol Plant 65:427–432

Moyo M, Aremu AO, Vanstaden J (2015) Insights into the multifaceted application of microscopic techniques in plant tissue culture systems. Planta 242(4):773–790

Murashige T, Skoog F (1962) A revised medium for rapid growth and bioassays with tobacco tissue cultures. Physiol Plant 15:473–497

Murthy BNS, Murch SJ, andSaxena PK (1998) Thidiazuron: a potent regulator of in vitro plant morphogenesis. In: In Vitro Cell Dev Plant, vol 34, pp 267–275

Nitsch JP, Nitsch C (1969) Haploid plants from pollen grains. Science 163:85–87

Paul S, Dam A, Bhattacharya A, Bandyopadhyay TK (2011) An efficient regeneration system via direct and indirect somatic embryogenesis for the medicinal tree *Murraya koenigii*. Plant Cell Tissue Organ Cult 105:271–283

Phillips GC (2004) In vitro morphogenesis in plants – recent advances (invited review). In: In Vitro Cell Dev Biol Plant, vol 40, pp 342–345

Pospisilova J, Ticha I, Kadlecek S et al (1999) Acclimatization of micropropagated plants in ex vitro conditions. Biol Plant 42:481–497

Prakash E, Khan SV, Meru E, Rao KR (2001) Somatic embryogenesis in *Pimpinella tirupatiensis* Bal. and Subr., an endangered medicinal plant of Tirumala hills. Curr Sci 81(9):1239–1242

Reynolds AG, Wardle DA, Zurowski C, Looney NE (1992) Phenylureas CPPU and thidiazuron affect yield components, fruit composition, and storage potential of four seedless grape selections. J Am Soc Hortic Sci 1(17):85–89

Saini R, Chopra AR (2012) In vitro plant regeneration via somatic embryogenesis in rice bean *Vigna umbellata* (Thunb.) Ohwi and Ohashi: an underutilized and recalcitrant grain legume. J Environ Res Dev 6(3):452–457

Sajid ZA, Aftab F (2009) Effect of thidiazuron (TDZ) on in vitro micropropagation of *Solanum tuberosum* L. cvs. Desiree and Cardinal. Pak J Bot 41(4):1811–1815

Sheibani M, Azghandi AV, Hemati SH (2007) Induction of somatic embryogenesis in Saffron using Thidiazuron (TDZ). Pak J Biol Sci 10(20):3564–3570

Siddique I, Anis M (2006) Thidiazuron induced high frequency shoot bud formation and plant regeneration from cotyledonary node explants of *Capsicum annuum* L. Indian J Biotechnol 5:303–308

Sivanesan I, Hwang SJ, Jeong BR (2010) Influence of plant growth regulators on axillary shoot multiplication and iron source on growth of *Scrophularia takesimensis* Nakai-a rare endemic medicinal plant. Afr J Biotechnol 7:4484–4490

Skoog F, Miller CO (1957) Chemical regulation of growth and organ formation in plants tissue cultured in vitro. Symp Soc Exp Biol 11:118–131

Sugawara F, Yamamoto N, Tanaka O (1994) Plant regeneration in in vitro culture of leaf, stem and petiole segments of *Actinidia polygama* Miq. Plant Tissue Cult Lett 11:14–18

Sujatha G, Ranjitha Kumari BD (2007) High-frequency shoot multiplication in *Artemisia vulgaris* L. using thidiazuron. Plant Biotechnol Rep 1:149–154

Thomas TD (2003) Thidiazuron induced multiple shoot induction and plant regeneration from cotyledonary explants of mulberry. Biol Plant 46(4):529–533

Thomas TD, Puthur JT (2004) Thidiazuron induced high frequency shoot organogenesis in callus from *Kigelia pinnata* L. Bot Bull Acad Sin 45:307–313

Venglat SP, Sawhney VK (1994) Ectopic formation of trichomes and stomata in floral organs of *Arabidopsis thaliana* induced by thidiazuron. Can J Bot 72:671–677

Vogel G (2005) Deriving 'controversy-free' ES cells is controversial. Science 310:416–417

Wang SY, Steffens GL, Faust M (1986) Breaking bud dormancy in apple with a plant bioregulator, thidiazuron. Phytochemistry 25:311–317

Wang R, He LS, Xia B, Tong JF, Li N, Peng F (2009) A micropropagation system for cloning of hemp (*Cannabis sativa* L.) by shoot tip culture. Pak J Bot 41(2):603–608

Zubkova NF, Bukashkina ZV, Markina LG, Belova NA (1991) Defoliating effect of fusicoccin. Agrokhimiya 12:86–92

23 TDZ-Induced Plant Regeneration in *Jatropha curcas*: A Promising Biofuel Plant

Nitish Kumar, Vacha D. Bhatt, Shaik G. Mastan, and Muppala P. Reddy

Abstract

In recent years, *Jatropha curcas* has pronounced attention due to its capacity of production of biodiesel. Uniform large-scale propagation of *J.* curcas is one of the significant keys that will eventually decide victory. Direct regeneration is one of the methods which help in the production of uniform and homogenous plant, and TDZ plays an important role in the production of plantlets by direct organogenesis in several number of plant species including *J. curcas.* Measuring the economical importance of *J. curcas* and the role of TDZ in shoot regeneration, the present book chapter briefly reviews the impact of TDZ on shoot bud induction from various explants of *J. curcas.*

Keywords

Jatropha curcas · Regeneration · Somatic Embryogenesis · TDZ

N. Kumar (✉)
Centre for Biological Sciences (Biotechnology), Central University of South Bihar, Patna, Bihar, India
e-mail: nitish@cub.ac.in

V. D. Bhatt
Department of Botany, SavitribaiPhule Pune University, Pune, India

S. G. Mastan
Aditya Degree and PG College, Kakinada, Andhra Pradesh, India

M. P. Reddy
Center for Desert Agriculture, King Abdullah University of Science and Technology, Thuwal, Saudi Arabia

N. Ahmad, M. Faisal (eds.), *Thidiazuron: From Urea Derivative to Plant Growth Regulator*, https://doi.org/10.1007/978-981-10-8004-3_23

23.1 Introduction

Jatropha curcas (family, Euphorbiaceae), one of the potent biofuel plant, has conquered pronounced attention in recent years due to its potential of biodiesel production. *J. curcas* arises as the most gifted oilseed tree due to high oil content and wider range of adaptability. *J. curcas* is a drought-resistant, perennial, multipurpose plant of Latin American origin, but it is now distributed throughout the tropical regions of the world. *J. curcas* is a cross-pollinated plant which produced seeds of heterozygous in nature and high degree of variation in population. Biodiesel manufactured from *J. curcas* has been effectively experienced in both stationary and mobile engines without alteration in any parts of the engine. The crude oil of *Jatropha* meets the biofuel characters of quality standards of rapeseed and can be simply transformed into biodiesel, meeting European and US standards (Tiwari et al. 2007; Azam et al. 2005). *J. curcas* is conventionally propagated through seeds and vegetative cuttings. Propagation through vegetative cuttings is not suitable to meet the demand of high planting material. This creates it imperative for exploration of methods for its quick propagation. Since TDZ is potent a plant growth regulator and efforts have been carried out to regenerate *J. curcas* through direct shoot morphogenesis as well as callus-mediated regeneration using TDZ (Sujatha et al. 2005; Deore and Johnson 2008; Kumar and Reddy 2010; Khurana-Kaul et al. 2010; Kumar et al. 2010a, b, c; Kumar et al. 2011a, b; Sharma et al. 2011; Kumar and Reddy 2012; Li et al. 2012; Chiangmai et al. 2015; Gopale and Zunjarrao 2013; Zhang et al. 2013; Aishwariya et al. 2015; Liu et al. 2015, 2016).

TDZ was found to have an effect to plant similar to auxin and cytokinin (Murthy et al. 1998; Guo et al. 2011), although it had the chemical structure different from these two plant growth regulators (Murthy et al. 1998; Guo et al. 2011). TDZ may involve in modulating endogenous plant growth regulators, resulting to the modification of plant cell membranes and uptake and assimilation of plant nutrient (Murthy et al. 1998). For these reasons, TDZ was used to study in various plant species and explant types (Murthy et al. 1998; Ahmed and Anis 2012; Sharma et al. 2011). Pan et al. (2016) observed that 75 μM and 225 μM TDZ treatments stimulated development of pistil, which enhanced the female flower numbers along with the inflorescence meristem development. TDZ application enhanced the branch orders of the dichasia on the inflorescence. The total number of flowers was reduced, but TDZ enhanced the total number of fruit of *J. curcas* by stimulating development of pistil. The exploitation of TDZ for regeneration has been reported vastly superior over adenine-based cytokinin for a number of plant species (Siddique and Anis 2007; Husain et al. 2007; Gopale et al. 2013; Zhang et al. 2013; Aishwariya et al. 2015; Liu et al. 2015; 2016). Ferrante et al. (2002, 2003) have reported that TDZ decreased degradation of chlorophyll in leaves of cut flowers of chrysanthemums, tulips, and alstroemeria. Mutui et al. (2005) demonstrated that TDZ markedly delay the onset of leaf senescence in *Pelargonium* and decrease flower abscission and the senescence of flowers and leaves in cut inflorescences of lupins and phlox (Sankhla et al. 2003, 2005). Although the exact working principle of TDZ is not well known, some evidence proposes that TDZ can modulate biosynthesis of cytokinin and/or its metabolism and may mimic the auxin activity (Murthy et al. 1998; Mok et al. 2000). It has been assumed that the long-lived cytokinin effects provided by TDZ application not only stops the yellowing of leaf but also minimizes the sensitivity of ethylene (Dinesh et al. 1996; Ferrante et al. 2002; Sankhla et al. 2005). Upsurges in endogenous cytokinin, auxin, and ethylene have been observed in response to TDZ application (Murthy et al. 1998). As a consequence, TDZ has been publicized to be suitable for speedy regeneration of plant in numerous recalcitrant plant species through organogenesis (Malik and Saxena 1992). So, considering the importance of TDZ in shoot regeneration of plant, the present chapter is reviewed on shoot regeneration of *J. curcas* using TDZ.

23.2 Regeneration of *J. curcas* Plantlets Using TDZ

23.2.1 Direct Organogenesis

Direct organogenesis from explants, without intervening the callus induction phase, was required particularly for the development of uniform plant. It also show high genomic homogeneity than those produced through callus-intervening organogenesis steps. There are several research groups observed on direct regeneration of plantlets using TDZ from various explants, viz., leaf, shoot tip, cotyledonary leaf, petiole, hypocotyl, and epicotyl explants (Sujatha et al. 2005; Deore and Johnson 2008; Kumar and Reddy 2010; Khurana-Kaul et al. 2010; Kumar et al. 2010a, b, c, 2011a, b; Sharma et al. 2011; Kumar and Reddy 2012; Gopale et al. 2013; Zhang et al. 2013; Aishwariya et al. 2015; Liu et al. 2015; 2016). An example of complete direct organogenesis protocol from cotyledonary petiole and leaf explant (Fig. 23.2) using TDZ is shown (Figs. 23.1 and 23.2).

Sujatha et al. (2005) found shoot bud regeneration from axillary nodes and leaf tissue of nontoxic variety of *J. curcas* on MS medium containing supplemented with 2.3–46.5 mM Kn, 2.2–44.4 mM BAP, and 2.3–45.4 mM TDZ separately. Axillary bud culture for 12 weeks resulted in production of 24, 10, and 4 shoot buds per node from explants initially cultured on MS medium supplemented with TDZ, BAP, and Kn, respectively, and concluded that TDZ stimulated greater percentage of shoot bud induction from axillaries as compared to BAP.

Deore and Johnson (2008) reported that the best percentage of regeneration of shoot buds from leaf using TDZ. They observed that TDZ in the culture medium has greater role on the shoot bud regeneration as compared to BAP, and it was also found that only callus has been produced, if BAP only was added in culture MS medium. Deore and Johnson (2008) first observed and reported that TDZ plays an important role in regeneration of direct shoot buds, and successive studies carried and proved that TDZ significantly influenced the regeneration of shoot buds.

Kumar et al. (2010a, b, c) observed that regeneration of shoot buds along with callus were induced from leaves of 2-year-old plants cultured on MS medium

Fig. 23.1 Direct shoot bud induction from cotyledonary petiole explants of nontoxic cultivar of *J. curcas*. Direct shoot bud induction from (**a**) in vitro cotyledonary petiole in horizontal position, (**b**) in vivo cotyledonary petiole in horizontal position, (**c**) in vitro cotyledonary petiole in vertical position, and (**d**) in vivo cotyledonary petiole in vertical position on MS medium with 2.27 μM TDZ. (**e**) Shoot proliferation of induced shoot buds on MS medium with 10 μM kinetin (Kn), 4.5 μM 6-benzyl aminopurine (BAP), and 5.5 μM α-naphthaleneacetic acid (NAA). (**f**) Elongation of proliferated shoot on MS medium with 2.25 μM BAP and 1.5 8.5 μM IAA. (**g**) Development of roots on half strength of MS medium with 15 μM IBA, 5.7 μM IAA, and 16.5 μM NAA + 0.25 mg l^{-1} activated charcoal after 4 weeks. (**h**) Regenerated plants in polybags (Source: Kumar et al. 2010b; License No. 4117420633543)

comprised with 2 μM TDZ, 2 μM BAP, and 1 μM IBA, wherein 83.3% leaf explants responded. A maximum of 83.3% of explants formed adventitious shoot buds in MS medium having 2 μM BAP, 2 μM TDZ, and 1 μM IBA which produced to a mean of 9.7 shoots per explant after 8 weeks of culture on multiplication medium. The presence of TDZ, BAP, and IBA in the initiation medium significantly enhanced the formation of shoot bud at lower concentrations. However, higher concentration of cytokinins in the medium had an inhibitory effect on shoot bud initiation, reducing the number of shoots buds to 4.3 per explant.

Kumar and Reddy (2010) found maximum percentage of shoot bud induction (58.35%), and the number of shoots per explant (10.10) was achieved after 6 weeks when in vitro petiole explants were placed horizontally on MS medium having 2.27 μ M TDZ. Further they reported that little TDZ concentrations regenerated relatively lower number of shoot buds, but these elongated quickly into shoots in subsequent culture. In contrast, MS media comprising high TDZ concentration had more visible shoot buds or primordia, but only a few were able to develop into shoots.

Fig. 23.2 Shoot regeneration from cotyledonary leaf explants of *J. curcas*. Direct organogenesis from (**a**) in vitro cotyledonary leaf explant (bar 5 mm), (**b**) in vivo cotyledonary leaf explant (bar 5 mm) on MS medium with 2.27 μM thidiazuron (TDZ) after 6 weeks. (**c**) Shoot proliferation of induced shoot buds on MS medium with 10 μM kinetin (Kn) + 4.5 μM 6-benzyl aminopurine (BAP) + 5.4 μM α-naphthaleneacetic acid (NAA) after 4 weeks (bar 100 mm). (**d**) Elongation of proliferated shoot on MS medium with 2.25 μM BAP + 8.5 μM indole-3-acetic acid (IAA) after 6 weeks (bar 5 mm). (**e**) Elongated shoot cultured on half-strength basal MS liquid medium supplemented with 15 μM indole-3-butyric acid (IBA) + 5.7 μM IAA + 5.5 μM NAA for root induction (bar 5 mm). (**f**) Development of roots at the base of auxins treated elongated shoot on half-strength basal MS medium with 0.25 mg/L activated charcoal after 4 weeks (bar 1 mm). (**g**) Regenerated plant in polybag after 4 weeks (bar 150 mm). (**h**) Regenerated plant in pot soil after 6 months under natural condition (100 mm) (Source: Kumar et al. 2010a; License No. 4117420944428)

Khurana-Kaul et al. (2010) fund that TDZ exhibited double shoot bud induction potential as compared to BAP. A maximum number of shoot bud induction (18 shoot buds per leaf explant) was observed on MS medium having 0.98 μM IBA and 0.90 μM TDZ and concluded that TDZ is best plant growth regulator in elicitation of morphogenic response.

Kumar et al. (2010a) found that the regeneration efficiency from cotyledonary leaf explants was more in TDZ comprising medium as compared to BAP. Percentage of shoot bud induction varied from 23.6% to 93.4%, and the number of regenerated shoots per explant varied from 8.6 to 26.2 on TDZ containing MS medium, whereas on BAP comprising MS medium, percentage of shoot bud induction varied from 14.1% to 56.1%, and the number of shoots per explant varied from 2.8 to 6.9. Kumar et al. (2010b) observed the frequency of shoot bud induction (59.11%), and the number of shoots (5.01) per explant of nontoxic variety of *J.* curcas was achieved on MS medium having 2.27 μM TDZ.

Kumar et al. (2011a) develop an efficient, reproducible, and simple plant regeneration method through direct regeneration from petiole explants of nontoxic *J. curcas*. They used MS medium which is supplemented with various concentrations of TDZ for shoot regeneration. The best percentage of shoot bud regeneration (57.61%) and a number of regenerated shoots (4.98) per explant were achieved when in vitro petiole explants were positioned horizontally on MS medium having 2.27 μM TDZ. Kumar et al. (2011b) discover the possessions of various plant hormones, viz., TDZ or BAP, independently and in mixture with IBA, on direct organogenesis from field-grown mature and in vitro leaf explants and glasshouse and in vitro-grown seedling cotyledonary leaf explants of nontoxic *J. curcas*. In the entire tested medium having various concentrations of TDZ and BAP, the maximum percentage of shoot bud induction (81.07%) and the number of regenerated shoots per explants (20.17) were found on 9.08 μ M TDZ comprising MS medium from in vitro-grown cotyledonary leaf explants.

Sharma et al. (2011) observed the best results on regenerations of shoot buds irrespective of germplasm or genotype were achieved on the MS medium comprising 0.5 mg/LTDZ. They observed 88.8% regeneration efficiency and 16.9 buds per explant on MS medium containing 0.5 mg/LTDZ.

Kumar and Reddy (2012) reported that TDZ in the MS medium greatly control the initiation of shoot bud regeneration irrespective of germplasm or genotype. The frequency of initiation of shoot bud regeneration and the number of regenerated shoot buds per explant were directly proportional to the TDZ concentration. Of the various tested concentrations (0–9.08 μM) of TDZ, the maximum shoot bud regeneration percentage (66.97%) and maximum regenerated shoot bud number (13.76) per explant were found in the occurrence of 9.08 M TDZ, among the genotypes studied.

Gopale et al. (2013) reported that TDZ in the shoot regeneration medium has better response on the regeneration of adventitious shoot buds, whereas callus induction was observed on BAP without TDZ. 2.27 μM TDZ helped in the induction of adventitious shoot buds in 55% leaf explants, whereas only 35% shoot bud regeneration was observed when 4.55 μM TDZ was added in MS medium.

Zhang et al. (2013) reported that the maximum regeneration of shoots was found on MS medium comprising 0.5 mg/L Kn, 1 .0 mg/L TDZ, and 0.5 mg/L GA_3. TDZ was the main plant growth regulators which initiated shoot bud induction, while $GA_3$and Kn played an important role in the elongation of regenerated shoot buds and in the enhancement of the number of plantlets per leaf, respectively. The highest regeneration of shoot buds (>90%) was achieved when the 1–2 mg/l TDZ was added in MS medium which was significantly higher than for MS medium with 0.5 mg/L TDZ. 2.0 mg/L TDZ concentrations significantly enhanced regeneration of compact shoot buds which are not able to proliferate and elongate due to the compact nature of regenerated shoot buds.

Aishwariya et al. (2015) reported that TDZ is most potent plant growth regulators like substance which initiates high-frequency in vitro shoot bud regeneration. Earlier shoot bud regeneration was observed when *J. curcas* (CJC – 19) calli were treated with 3 mg/L TDZ for 3–5 days.

Liu et al. (2015) found that treating the petiole explants with 5–120 mg/L of TDZ for time length of 5–80 min increased the shoot bud regeneration percentage and enhanced the quality of the originated shoot buds significantly. The maximum regeneration of shoot buds (65.78%) and shoot bud number (6.77) per explant was observed in the second petiole explants of germplasm M-1 which is treated with 20 mg/L TDZ solution for 20 min.

Liu et al. (2016) reported that the maximum frequency (63.68%) of regeneration of shoot buds and the maximum number (6.58) of regenerated buds per explant were observed on MS medium containing 0.3 mg/L TDZ. Explants are also treated with 20 mg/L TDZ for different time lengths. For the purpose of studying the influence of time length of TDZ solution treatment on the induction of shoot buds, petiole explants were treated with 20 mg/L TDZ for different durations before inoculation of explants onto the plant growth regulator fee MS medium. Observation results indicated that treatment with 20 mg/L TDZ for 20 min was the most appropriate, by which the maximum shoot bud regeneration frequency (91.36%) and the maximum number of regenerated shoot buds per explant (12.01) were obtained. Nevertheless, when the explants were treated with 20 mg/L TDZ for 40 min, the regeneration percentage of adventitious buds was reduced.

23.2.2 Indirect Organogenesis

Indirect organogenesis is a method in which de novo development whole plant arises, and it includes two different stages: dedifferentiation and redifferentiation. Dedifferentiation occurs when formation of bulk of undifferentiated cells, i.e., called callus, was formed due to the hastening of cell division in isolated plant tissue. After dedifferentiation, i.e., after the formation of callus from single plant cell, redifferentiation starts, i.e., formation of shoot and root primordia originates from a single or a cluster of callus cells, and the organ primordia give rise to whole plantlets. Indirect organogenesis, i.e., formation of whole plants from callus in *J. curcas*, is also controlled by the media composition comprising different concentration and combination of plant growth regulators. There are limited studies that were reported on indirect organogenesis of *J. curcas* using TDZ via intervening callus (Li et al. 2012; Chiangmai et al. 2013).

Li et al. (2012) observed that the MS media comprising TDZ induce the induction of adventitious shoot buds from callus and no callusing without TDZ. It was also observed that the frequency of shoot bud induction from callus declined by increasing the concentration of TDZ. MS medium containing 1 mg/L Kn and 0.1 mg/L IBA with 1 mg/L TDZ increased the number of induction of shoot buds from callus and also improved the adventitious shoot regeneration among several tested regeneration mediums. These observations recommended that in vitro plant shoot induction media comprising TDZ in combination with other plant growth regulators could induce the adventitious shoots formation, and the combination of 1 mg/L Kin, 0.1 mg/L IBA, and 1 mg/L TDZ was the most promising medium for adventitious shoot formation from callus.

Chiangmai et al. (2013) observed that MS medium comprising only TDZ in the range of 0.1 to 1.0 mg/L was the greatly responding MS medium in the induction of morphogenic callus. The results showed that highest yellow/green compact callus frequency was observed in the MS mediums comprising either 0.1 or 1.0 mg/L TDZ treatment. MS medium having both 1.5 mg/L IBA and 0.1 mg/L TDZ showed the highest value of differentiated callus (96.7%).

23.2.3 Somatic Embryogenesis

Somatic embryogenesis is also one of the important processes of plant regeneration in which single inherently somatic cell is converted to bipolar structure like gametic embryo. Somatic embryogenesis not only assists in in vitro plant regeneration but also plays a significant role in studies of genetic and metabolic engineering. There were very few studies reported on somatic embryogenesis of *J. curcas* using TDZ (Mahalakshmi et al. 2014; Mweu et al. 2016).

Mahalakshmi et al. (2014) showed that 0.2 mgl^{-1} Kn and 1 mgl^{-1} NAA showed highest percentage in the induction of callus from the first petiole. Frequency of formation of somatic embryo was more (35.16%) from the first petiole on MS medium having 0.5 mg/L TDZ and 0.4 mg/L gibberellic acid. Highest germination percentage of somatic embryo (66.85%) was recorded on MS medium comprising 0.25 mg/LKn, 1 mg/L BAP, and 0.5 mg/L IAA.

Mweu et al. (2016) reported that the maximum induction of callus (85%) was observed in combination of 0.1 mg/L TDZ, 0.5 mg/L IAA, 0.6 mg/ L Kn, and 1.5 mg/L BAP within 8 weeks in all germplasm. The combination 0.1 mg/L TDZ, 0.5 mg/L IAA, 0.6 mg/ L Kn, and 1.5 mg/L BAP is a revolution in induction of high callus and morphogenetic development.

23.3 Conclusion

Considerable improvement has been carried out concerning shoot regeneration of *J. curcas* by different explants using TDZ. Interestingly, much data has accumulated during the last decades indicating that TDZ shows powerful cytokinin-like activity. TDZ have an important role in promoting in vitro response from (I) direct organogenesis of shoot buds from explants, (II) regeneration of shoot buds from induced callus from various explants, (III) development of somatic embryo from callus, and (IV) enhancement of the percentage of regeneration of explants as well as the number of shoot buds per explants. Nevertheless, also for highly potent plant growth regulators, conditions have to be optimized concerning combination with other growth regulators since several reports indicate the highest organogenic capability, if TDZ is applied as the sole growth regulators, while other studies describe the need to couple with a certain auxin.

References

Ahmed MR, Anis M (2012) Role of TDZ in the quick regeneration of multiple shoots from nodal explant of *Vitex trifolia* L.-an important medicinal plant. Appl Biochem Biotechnol 168:957–966

Aishwariya V, Ramrao RK, Kokila DE, Arul L, Sudhakar D, Kumar KK, Balasubramanian P (2015) Impact of TDZ (thidiazuron) pulse treatment in single and multiple shoot formation in calli of *Jatropha curcas* L. Int J Adv Res 3:879–884

Azam MM, Waris A, Nahar NM (2005) Prospects and potential of fatty acid methyl esters of some non-traditional seed oils for use as biodiesel in India. Biomass Bioenergy 29:293–302

Chiangmai PN, Pootaeng-on Y, Meetum P, Kamkajon K, Yuiam W, Rungphan N, Ninkaew P (2015) Regeneration of adventitious shoots from callus and leaf explants in *Jatropha curcas* L. 'Phetchaburi'. Silpakorn U Sci Tech J 9(1):28–39

Deore AC, Johnson TS (2008) High frequency plant regeneration from leaf-disc cultures of *Jatropha curcas* L.: an important biodiesel plant. Plant Biotechnol Rep 2:7–11

Dinesh R, Sankhla N, Sankhla D, Kachhwaha S, Upadhyaya A (1996) Effect of thidiazuron on growth, antioxidants and lipid peroxidation in chickpea seedlings. Proc Plant Growth Regul Soc Am 23:211–216

Ferrante A, Hunter DA, Hackett WP, Reid MS (2002) Thidiazuron-a potent inhibitor of leaf senescence in Alstroemeria. Postharvest Biol Technol 25:333–338

Ferrante A, Mensuali-Sodi A, Serra G, Tognoni F (2003) Treatment with thidiazuron for preventing leaf yellowing in cut tulips and *Chrysanthemum*. Acta Horticul 624:357–363

Gopale KD, Zunjarrao RS (2013) In vitro culture of *Jatropha curcas* L: a biofuel plant. Int J Pure Appl Sci Technol 16(2):46–54

Guo B, Abbasi BH, Zeb A, Xu LL, Wei YH (2011) Thidiazuron: a multi-dimensional plant growth regulator. Afr J Biotechnol 10:8984–9000

Husain MK, Anis M, Shahzad A (2007) In vitro propagation of Indian kino (*Pterocarpus marsupium* Roxb.) using thidiazuron. In Vitro Cell Dev Biol Plant 43:59–64

Khurana-Kaul V, Kachhwaha S, Kothari SL (2010) Direct shoot regeneration from leaf explants of *Jatropha curcas* in response to thidiazuron and high copper contents in the medium. Biol Plant 54:369–372

Kumar N, Reddy MP (2010) Plant regeneration through the direct induction of shoot buds from petiole explants of *Jatropha curcas*: a biofuel plant. Ann Appl Biol 156:367–375

Kumar N, Vijayanand KG, Reddy MP (2010a) Shoot regeneration from cotyledonary leaf explants of *Jatropha curcas*: a biodiesel plant. Acta Physiol Plant 32:917–924

Kumar S, Kumaria S, Tandon P (2010b) Efficient in vitro plant regeneration protocol from leaf explant of *Jatropha curcas* L – a promising biofuel plant. J Plant Biochem Biotechnol 19(2):275–277

Kumar N, Vijayanand KG, Reddy MP (2010c) In vitro plant regeneration of non-toxic *Jatropha curcas* L: direct shoot organogenesis from cotyledonary petiole explants. J Crop Sci Biotechnol 13:189–194

Kumar N, Vijayanand KG, Reddy MP (2011a) In vitro regeneration from petiole explants of non-toxic *Jatropha curcas*. Ind Crop Prod 33:146–151

Kumar N, Vijayanand KG, Reddy MP (2011b) Plant regeneration of non-toxic *Jatropha curcas*-impacts of plant growth regulators, source and type of explants. J Plant Biochem Biotechnol 20:125–133

Kumar N, Reddy MP (2012) Thidiazuron (TDZ) induced plant regeneration from cotyledonary petiole explants of elite genotypes of *Jatropha curcas*: a candidate biodiesel plant. Ind Crop Prod 39:62–68

Li ZG, Gong M, Yang SZ, Long WB (2012) Efficient callus induction and indirect plant regeneration from various tissues of *Jatropha curcas*. Afr J Biotechnol 11(31):7843–7849

Liu Y, Tong X, Hui WK, Liu T, Chen X, Li J, Zhuang CX, Yang YS, Liu ZL (2015) Efficient culture protocol for plant regeneration from petiole explants of physiologically mature trees of *Jatropha curcas* L. Biotechnol Biotecnological Equip 29:479–488

Liu Y, Yin XG, Lu JN, Zhu HB, Li LF, Shi YZ, Yang YS (2016) An efficient protocol for inducing regeneration in physic nut (*Jatropha curcas* L.) Bangladesh J Bot 45:827–833

Malik KA, Saxena PK (1992) Thidiazuron induces high- frequency shoot regeneration in intact seedlings of pea (*Pisumsativum*), chickpea (*Cicer arietinum*) and lentil (*Lens culinaris*). Aust J Plant Physiol 19:731–740

Mahalakshmi R, Eganathan P, Parida A (2014) In vitro regeneration from different ages of petioles of physic nut (Jatropha curcas L.) Afr J Biotechnol 13:265–273

Mok MC, Martin RC, Mok DVS (2000) Cytokinins: biosynthesis, metabolism and perception. In Vitro Cell Dev Biol Plant 36:102–107

Murthy BNS, Murch SJ, Saxena PK (1998) Review: Thidiazuron: a potential regulator of in vitro plant morphogenesis. In Vitro Cell Dev Biol Plant 34:267–275

Mutui TM, Mibus, Serek M (2005) Effects of thidiazuron, ethylene, abscisic acid and dark storage on leaf yellowing and rooting of *Pelargonium* cutting. J Hortic Sci Biotechnol 80:543–550

Mweu CM, Nyende A, Onguso J (2016) Efficient somatic embryogenesis of *Jatropha curcas* L. from petiole and leaf discs. IntJ Biotechnol Mol Biol Res 7:29–35

Pan BZ, Luo Y, Song L, Chena MS, Li JL, Xu ZF (2016) Thidiazuron increases fruit number in the biofuel plant *Jatropha curcas* by promoting pistil development. Ind Crop Prod 81:202–210

Sankhla N, Mackay WA, Davis TD (2003) Reduction of flower abscission and leaf senescence in cut *Phlox* inflorescence by thidiazuron. Acta Horticult 628:837–841

Sankhla N, Mackay WA, Davis TD (2005) Effect of thidiazuron on senescence of flowers in cut inflorescences of *Lupinus densiflorus* Benth. Acta Hortic 669:239–243

Sharma S, Kumar N, Reddy MP (2011) Regeneration in *Jatropha curcas*-factors affecting the efficiency of in vitro regeneration. Ind Crop Prod 34:943–951

Siddique I, Anis M (2007) In vitro shoot multiplication and plantlet regeneration from nodal explants of *Cassia angustifolia* (Vahl.)—a medicinal plant. Acta Physiol Plant 29:333–338

Sujatha M, Makkar HPS, Becker K (2005) Shoot bud proliferation from axillary nodes and leaf sections of non-toxic *Jatropha curcas* L. Plant Growth Regul 47:83–90

Tiwari AK, Kumar A, Raheman H (2007) Biodiesel production from Jatropha cur- cas with high free fatty acids: an optimized process. Biomass Bioenergy 31:569–575

Zhang C, Fu S, Tang G, Hu X, Guo J (2013) Factors influencing direct shoot regeneration from mature leaves of *Jatropha curcas*, an important biofuel plant. In Vitro Cell Dev Biol Plant 49:529–540

Role of Thidiazuron in Modulation of Shoot Multiplication Rate in Micropropagation of *Rauvolfia* Species

24

Sheikh Altaf Hussain, Naseem Ahmad, Mohammad Anis, Abdulrahman Abdullah Alatar, and Mohammad Faisal

Abstract

Thidiazuron (TDZ) is a light yellow crystalline phenylurea compound with tremendous morphogenic potential on shoot regeneration ranging from small grasses to large tree species. This morphoregulatory potential of TDZ has led to well-established micropropagation systems in various woody plant species where other cytokinins are least effective. Short exposure of TDZ is sufficient to stimulate the best regeneration system, while the prolonged or continuous exposure has various negative effects on growing cultures. The present communication reviewes the morphogenic response of TDZ in *Rauvolfia* species. A differential growth response was observed on explants when inoculated on Murashige and Skoog (MS), 1962 medium supplemented with different concentrations of TDZ. Nodal explants of *R. tetraphylla* cultured on MS + 5 μM TDZ gave optimal (90%) regeneration response for maximum (9.2 ± 1.20) shoot production, while in the case of *R. serpentina*, MS + 0.8 μM TDZ proved to be best for producing highest mean shoot number (5.7 ± 0.28) with 77% regeneration rate after 4 weeks of culture. However, for shoot tip explants of *R. hookeri*, 0.45 μM TDZ on ½ MS proved best and gave 70% regeneration, with a maximum shoot number of 1.50 ± 0.22 and a mean shoot length of 1.18 ± 0.04 cm after 45 days of culture. Responsive nodal explants of *R. tetraphylla* and *R. serpentina* when cultured continuously on the same TDZ-supplemented media lead to adverse effects like shoot distortion, fasciation, or hyperhydricity in growing shoots and resulted in stunted growth. These negative effects of prolonged TDZ

S. A. Hussain · N. Ahmad (✉) · M. Anis
Plant Biotechnology Laboratory, Department of Botany, Aligarh Muslim University, Aligarh, UP, India

A. A. Alatar · M. Faisal
Department of Botany and Microbiology, Faculty of Science, King Saud University, Riyadh, Kingdom of Saudi Arabia

N. Ahmad, M. Faisal (eds.), *Thidiazuron: From Urea Derivative to Plant Growth Regulator*, https://doi.org/10.1007/978-981-10-8004-3_24

exposure were apprehended when the cultures were transferred to MS basal medium devoid of TDZ. The transference of cultures to secondary medium not only showed the positive effect on growing cultures but also increased shoot proliferation and shoot multiplication rates in both the species. After 4 weeks of transfer to secondary medium, shoot number increased up to 23.1 ± 0.4 and 18.5 ± 1.25 per explant in *R. tetraphylla* and *R. serpentina*, respectively. Regenerated shootlets of ≥4 cm were excised and transferred to various rooting medium supplemented with different concentrations of auxins like IAA, IBA, and NAA. IBA in all the three species proved best for in vitro rooting with maximum root mean number of 6.9 ± 0.34 and mean root length of 5.2 ± 0.4 cm at 1.0 μM IBA in *R. serpentina* after 4 weeks of incubation. Rooted plantlets were acclimatized in culture room and finally transferred to garden soil with 90% survival rate without any genetic or morphogenic abnormality.

Keywords
Nodal segment · Micropropagation · *Rauvolfia serpentina* · *Rauvolfia tetraphylla* · *Rauvolfia hookeri* · Clonal fidelity

Abbreviations

BA	Benzyl adenine
IBA	Indole-3-butyric acid
MS	Murashige and Skoog's medium
PGRs	Plant growth regulators
RAPD	Random amplified polymorphic DNA
TDZ	N-Phenyl-N′-(1,2,3-thiadiazol-5-yl)urea

24.1 Introduction

Rauvolfia belongs to family Apocynaceae (dogbane) which is represented by evergreen perennial shrubs or undershrubs, widely distributed in tropical Asia, Africa, and America. There are 76 *Rauvolfia* species (https://en.wikipedia.org/wiki/Rauvolfia) or 85 species (Kumar et al. 2011), 5 of them, namely, *R. densiflora*, *R. hookeri*, *R. micrantha*, *R. serpentina*, and *R. tetraphylla*, are found in India (Bhattacharjee 1998) which are introduced and now became naturalized (Anonymous 1973). These plant species are pharmacologically the most important due to the presence of more than 50 potent alkaloids like ajmaline, ajmalicine, reserpine, serpentine, tetraphyllincine, etc. Reserpine is globally used as an antihypertensive, which depresses central nervous system and lowers blood pressure (Anonymous 2003). These medicinal herbs are also used to cure and relief the diseases like insomnia, anxiety, cholera, diarrhea, dysentery, fever, anxiety, epilepsy, and other disorders of the central nervous system (Kirtikar and Basu 1993; Ghani 1998) from prehistoric times. Root extract of these species is used as laxative, thermogenic and diuretic and is highly recommended in difficult childbirth, as it increases the uterine contraction cycles. It is also used as antidote to cure snake bite (Ghani 1998; Tona et al. 1999).

From last few decades, natural population of these species is rapidly declining because of overexploitation from the wild to meet the growing needs of increasing population and to fulfill the traditional demand from pharmaceutical sectors (Nayar and Sastry 1987). Root, stem, bark, and leaves are used in medicine; thus, sometimes the whole plant is uprooted. Therefore, to ensure its natural growth and conservation, the Government of India has restricted the export of its drugs to reduce its exploitation, which will result in the shortage of these alkaloids in the global markets (Anonymous 2003; Faisal et al. 2005). This led to an active search and research for these alkaloids in other related species of *Rauvolfia*. Further, these species have poor seed germination and viability with insufficient and problematic conventional propagation methods which may in turn lead to further destruction of these species. To meet the basic and increasing demand, there is an urgent need to develop non-conventional methods for propagation and conservation of these plant species.

Biotechnology offers new tools, techniques, technologies, and strategies for easy sustainable development and utilization of natural resources which are facing extinction threat due to overexploitation, unsustainable utilization, and indiscriminate harvesting because of their industrial and pharmacological importance. Plant tissue culture offers valuable ways to overcome all these problems that are found in natural methods of propagation. In vitro techniques offer a powerful tool for mass multiplication by increasing the number of propagules through either direct or indirect regeneration methods by using various explants like cotyledonary node (CN) , internodes, leaves, nodal and shoot tips on growing medium supplemented with different cytokinins alone or in combination with various auxins. The modern biotechnological tools are facilitating mass multiplication for population enhancement and ecorestoration via ex situ conservation of aromatic, endangered, and vulnerable medicinal plants where traditional methods are helpless and inadequate (Krishnan et al. 2011; Pence 2011).

24.2 *Rauvolfia tetraphylla* (L.)

Rauvolfia tetraphylla commonly known as milk bush or bee still plant is an endangered tropical medicinal plant. It is a small evergreen woody perennial shrub growing up to the height of 1.5 m. In India, it is found in Uttar Pradesh, Bihar, Orissa, Madhya Pradesh, Andhra Pradesh, West Bengal, and Kerala (Anonymous 2003). It is a rich source of bioactive chemicals with the presence of about 30 indole alkaloids. Reserpine, the major and potent one that depresses the central nervous system, produces sedation and lowers blood pressure. The other main alkaloids present in this plant are ajmaline, alstonine, canescine, corynanthine, deserpidine, isoreserpiline, isoreserpine, serpagine, serpentine, tetraphyllicine, yohimbine, and pseudoyohimbine (Anonymous 2003; Anitha and Kumari 2006). Since the beginning of human civilization, this medicinal herb is used as medicine to cure and control the diseases like central nervous system disorders, diarrhea, dysentery, and other intestinal infections. It is also used as anthelmintic and antidote against snake bite. Root extract is used to stimulate uterine contraction and is recommended in difficult childbirth cases. Due to its high medicinal properties, it is facing indiscriminate and extensive collection from the wild. Further, it has low seed viability and poor germination percentage with lack of commercial plantation; thus, for its conversation, the Government of India has constrained the export of its drugs to lessen its overexploitation, which in turn may result in a shortage of its alkaloids in the world market (Anonymous 2003; Faisal et al. 2005). Under these conditions, its micropropagation on large scale is the only way to meet the growing and upcoming challenges of this high value medicinal plant. Need of the hour is to devise an efficient micropropagation system for *R. tetraphylla*.

24.3 Role of TDZ on Nodal Explants in *Rauvolfia tetraphylla*

Faisal et al. (2005) studied the regeneration potential of nodal explants collected from a 2-year-old *Rauvolfia tetraphylla* plant maintained at the botanical garden of the Department of Botany, Aligarh Muslim University, Aligarh. The explants were inoculated on MS medium alone or supplemented with different concentrations (0.1–10 μM) of TDZ. MS medium devoid of TDZ did not respond even after 6 weeks of incubation and thus served as the control. All the tested concentrations of MS with TDZ facilitated shoot bud induction. Among the tested treatments, MS + 5 μM TDZ proved to be optimal for maximum 90% regeneration with shoot mean number of 9.2 + 1.20 after 4 weeks of incubation (Table 24.1). However, when the growing cultures were transferred onto the similar fresh TDZ-supplemented medium, some deleterious effects like fasciation, distortion, and necrosis in growing shoots was observed. Further, premature but normal flowering was also observed on MS + 5.0 μM TDZ.

To avoid the ill effects of prolonged TDZ exposure, the cultures were transferred to secondary hormone-free medium, and positive morphogenic effects were seen on growing cultures with increased shoot multiplication in all the treatments. Further, to obtain the best optimum period for maximum regeneration and shoot multiplication from nodal explants, a differential TDZ exposure of 2, 4, and 6 weeks was given before transfer to hormone-free medium. The 4-week treatment of 5.0 μM TDZ proved to be ideal for maximum shoot regeneration (18.5 ± 1.25) prior to transfer to their hormone-free medium. Healthy shootlets ~4 cm were excised and transferred to MS rooting medium fortified with different (0.1–2.0 μM) concentrations of IBA and IAA. 0.5 μM IBA proved best for a maximum (4.80 ± 0.58) number of roots with 100% rooting response after 4 weeks of incubation (Table 24.2). Plantlets having 4–6 fully expanded leaves with healthy root shoot system were transferred to sterile vermiculate in pots. These plantlets were acclimatized, hardened, and transferred to earthen pots and finally maintained in greenhouse with 90%

Table 24.1 In vitro shoot bud induction and proliferation in *Rauvolfia* species using TDZ

Species	Explant used	TDZ concentration Used (μM)	Optimum Treatment (μM)	Response	No. of shoot buds induced	No. of proliferated Shoots on secondary medium	References
R. tetraphylla	Nodal	1.0, 2.5, 5.0, 7.5, or 10	MS + 5.0	Shoot bud induction in TDZ-supplemented MS medium, multiplication in secondary medium devoid of PGR	09	19	Faisal et al. (2005)
R. serpentina	Nodal	0.1, 0.3, 0.5, 0.8. 1.0, or 2.5	MS + 0.8	Pretreatment of explants, shoot bud induction on TDZ-supplemented medium, and multiplication in PGR-free medium	07	23	Alatar (2015)
R. hookeri	Shoot tip	0.41, 0.45, or 1.36	1/2 MS + 0.45	Shoot bud induction and multiplication in TDZ-supplemented medium	02	No PGR-free medium used	Ranjusha and Gangaprasad (2014)

Table 24.2 Effect of various auxins on in vitro rooting in *Rauvolfia species*

Species	Auxin used (μM)			Optimum	No. of roots induced	References
	IAA	IBA	NAA	Concentration		
R. tetraphylla	0.1, 0.5, 1.0, or 2.0	0.1, 0.5, 1.0, or 2.0	–	0.5 μM IBA	5	Faisal et al. (2005)
R. serpentina	–	0.1, 0.5, 1.0, or 2.0	–	1.0 μM IBA	7	Alatar (2015)
R. hookeri	–	2.46, 4.93, 7.38, 9.84, or 14.76	2.68, 5.37, 8.04, 10.72, or 16.08	7.38 μM IBA	4	Ranjusha and Gangaprasad (2014)

survival rate. All the regenerated plants were morphologically similar without any detection of somaclonal variation.

24.4 *Rauvolfia serpentina* (L.)

Rauvolfia serpentina (L.), commonly known as sarpagandha, is an evergreen woody perennial, erect shrub, distributed in Bangladesh, Bhutan, China, Indonesia, India, Malaysia, Nepal, Pakistan, Sri Lanka, Vietnam, and India (Anonymous 2003; Dey and De 2010). The plant is the source of many prized and therapeutically important bioactive chemicals like serpentine, reserpine, deserpidine, ajmaline, rescinnamine, yohimbine, etc. (Anonymous 2003), which are used from prehistoric times to cure and control the diseases like high blood pressure, insomnia, anxiety, insomnia, cholera, diarrhea, dysentery, fever, anxiety, and other disorders of the central nervous system and epilepsy (Ghani 1998). Reserpine is globally used as an antihypertensive (Anonymous 2003). Root extract is used as laxative and also possesses thermogenic and diuretic properties. It is also used as antidote to snake venom (Ghani 1998; Tona et al. 1999). Poor seed viability, low germination rate, industrial demands, and indiscriminate collection from their natural habitat lead to its fast disappearance; thus, every factor contributed much to make it endangered and has been enlisted in Red Data Book (Ravikumar et al. 2000) of IUCN. Thus, there is an immediate need for its conservation, mass propagation, sustainable utilization, and ecorestoration. Conventional propagation can be achieved from seed, which results in a high degree of genetic and phenotypic variation. Micropropagation can provide an opportunity to obtain large number of homogeneous plants.

24.5 Effect of TDZ on Nodal Explants in *Rauvolfia serpentina*

An efficient and rapid in vitro propagation of *R. serpentina* using nodal explants has been standardized and described by Alatar (2015). Healthy nodal explants (15 mm) were excised, washed thoroughly, and inoculated on MS medium supplemented with different concentrations (0.1–2.5 μM) of TDZ. MS + 0.8 μM TDZ proved best for maximum (77%) shoot regeneration with 5.7 ± 0.28 mean shoot number and shoot length of 3.4 ± 0.43 cm after 4 weeks of culture (Table 24.1). However, when the cultures were grown continuously on the same fresh medium beyond 4 weeks, some negative effects due to prolonged TDZ exposure were observed. So to minimize these negative effects, the responsive explants were transferred to a secondary medium devoid of any PGR for better growth and proliferation. Normal growth was observed in all cultures in secondary medium with successive increase in growth parameters like shoot number and shoot length up to the fifth subculture, and beyond which a decrease in shoot number was observed. In the second phase of the experiment, a pulse treatment of 4, 8, 12, or 16 days of different (5, 25, 50, 75, or 100 μM) concentrations of TDZ was given to nodal explants in MS liquid medium before inoculation to MS basal medium. After 8 weeks of incubation on MS basal medium, 90% regeneration was observed in 8-day treated (50 μM) TDZ explants with 23.1 ± 2.5 shoot mean number and mean shoot length of 5.3 ± 0.83 cm. Thus, short-term exposure of 50 μM TDZ for 8 days adds more effectiveness to the protocol with increase in regeneration frequency from 77% to 90% as compared to the untreated TDZ-exposed nodal segments. Microshoots (5 cm) were excised and transferred to MS rooting medium supplemented with different concentrations (0.1, 0.5, 1.0, 1.5, or 2.0 μM) of IBA. After 4 weeks of incubation, 87% rooting response was observed on 1.0 μM IBA with a maximum root mean number of 6.9 ± 0.34 and mean root length of 5.2 ± 0.4 cm (Table 24.2). Rooted plantlets with six or more fully expanded leaves were transferred to pots filled with soilrite and were covered with polybags to ensure proper humidity. After 1 month of acclimatization and hardening, these plants were transferred to pots containing garden soil and were maintained in the greenhouse with 90% survival rate. Further, ex vitro stability parameters like chlorophyll (chl. a, chl. b), and carotenoid content, net photosynthetic rate, and other biochemical parameters like SOD and CAT were studied on 0, 7, 14, 21, and 28 days of acclimatization. The chl. a and chl. b content increased from 0 to 28 days of acclimatization, while carotenoid content first decreased from 0 to 14 days and then started increasing after 14 days of acclimatization. Similarly net photosynthetic rate first got decreased from day 1 to day 7 and thereafter started increasing up to 28th day due to formation of new leaves. Maximum activity of SOD enzyme (23 unit, mg^{-1} protein) was recorded at the 7th day of transfer where after gradual decrease in the activity was recorded upto the 28th day at which the enzymatic activity became conparable to the naturally grown plant. CAT activity of acclimatized plants increased continuously from 2.5 to 5.1 μ mol min^{-1} from day 1 to day 28. All these observations showed the regulation of plant protective mechanisms against oxidative stress and cleansing of H_2O_2 by converting it into O_2 and H_2O and physiological stability during acclimatization period (Alatar 2015).

24.6 *Rauvolfia hookeri* (L)

Rauvolfia hookeri, a rare and endemic medicinal shrub of southern Western Ghats of India, is an evergreen flowering shrub. Its roots possess many phytochemical alkaloids like ajmalicine, sarpagine, and serpentine (Anonymous 1973), which are similar to its other family members like *R. serpentina* and *R. tetraphylla*. Thus, its root extract is mixed with the root extract of other Indian *Rauvolfia* species. Due to its narrow geographical distribution (Gopalan and Henry 2000), poor seed viability, low germination rate, increasing market demand, inefficient vegetative propagation methods, commercial overexploitation (Mohanan and Sivadasan 2002), large-scale root collection (Arora 1983), and whole plant uprooting had led to its great decrease in natural population in Western Ghats. Thus, there is an immediate need for its conservation, and micropropagation is the only way to meet its conservation standards.

24.7 Effect of TDZ on Shoot Tip Explants in *Rauvolfia hookeri*

Ranjusha and Gangaprasad (2014) reported an efficient and successful micropropagation protocol for *Rauvolfia hookeri* which is described here. Shoot tip explants collected from 2-year-old *R. hookeri* plant maintained in the greenhouse were inoculated on full or half-strength MS medium, enriched with various (0.41, 0.45, or 1.36 μM) concentrations of TDZ. Among tested concentrations, 1/2 MS + 0.45 μM TDZ proved to be optimal for maximum 70% regeneration, with mean shoot number 1.50 ± 0.22 and 1.18 ± 0.04 cm mean shoot length per explant after 45 days of culture (Table 24.1). Full-strength MS medium did not respond in bud breaking and failed in induction of shoots. Shootlets of size 4 cm were excised and transferred to full- or half-strength MS rooting medium supplemented with different concentrations of IBA or NAA. 1/2 MS medium supplemented with 7.38 μM IBA gave 86% rooting response with 3.66 ± 0.21 mean root number and root length of 2.16 ± 0.17 cm per shootlet after 4 weeks of incubation (Table 24.2). Healthy plantlets with well-developed root and shoot system were transferred to pots containing soilrite. These plantlets were hardened and acclimatized properly and transferred to field conditions with 70% survival rate with no detectable variations as proved by RAPD analysis of regenerated plants. The genetic stability of the regenerated plants was evaluated by 10 RAPD decamer primers, with an average of 4.8 reproducible bands per primer. Primer (OPX – 17) gave a minimum number of three bands, while OPI – 12 produced a maximum number of seven bands. This micropropagation protocol developed from shoot tip explants by Ranjusha and Gangaprasad appears to be the first established protocol for clonal propagation with 100% genetic stability of *R. hookeri.*

24.8 Conclusion

The present communication describes efficiency of in vitro regeneration protocols of selected *Rauvolfia* species through nodal (*R. tetraphylla* and *R. serpentina*) or shoot tip explants (*R. hookeri*) without any morphological abnormalities or genetic changes. Summing up it can be said that TDZ is more effective in bud breaking and shoot multiplication at lower concentrations. Further, short-term exposure of TDZ up to 4 weeks is more effective than continuous exposure of TDZ that causes certain abnormalities like shoot distortion, fascination, and hyperhydricity in growing cultures. To overcome and apprehend these negative effects of prolonged TDZ exposure, the responsive cultures should be transferred to a secondary medium devoid of any PGRs. A positive effect on growth parameters were observed on this medium with better shoot multiplication and proliferation rates than other cytokinins. To increase the regeneration, induction, and shoot multiplication rates, a unique pretreatment of TDZ may be given to the explants in MS liquid medium before inoculation to MS medium supplemented with different concentrations of TDZ. These well-established plant tissue culture protocols can be utilized for true-to-type production of elite germplasm of other rare and endangered medicinal plants for conservation and ecorestoration. It is suggested that these effective approaches of plant biotechnology can be employed throughout the year for high-volume production of planting material to challenge growing human greed and industrial lust.

References

Alatar AA (2015) Thidiazuron induced efficient in vitro multiplication and ex vitro conservation of *Rauvolfia serpentina* – a potent antihypertensive drug producing plant. J Biotechnol Biotechnol Equip 29:489–497

Anitha S, Kumari BDR (2006) Reserpine accumulation in NaCl treated Calli of *Rauvolfia tetraphylla* L. J Sci Asia 32:417–419

Anonymous (1973) The wealth of India, a dictionary of Indian raw materials and industrial products, vol VIII. CSIR, New Delhi, p 376

Anonymous (2003) The wealth of India: a dictionary of Indian raw materials and industrial products. CSIR, New Delhi. 2003

Arora RK (1983) Threatened plants of India-Some considerations on native genetic resources, pp 296–302. In: Jain SK, RR Rao (eds) An assessment of the threatened plants of India. Botanical Survey of India, Howrah, 334 p

Bhattacharjee SK (1998) Handbook of medicinal plants. Pointer Publishers, India, pp 343–345

Dey A, De JN (2010) *Rauvolfia serpentina* (L). Benth. Ex Kurz. – a review. Asian J Plant Sci 9:285–298

Faisal M, Ahmad N, Anis M (2005) Shoot multiplication in *Rauvolfia tetraphylla* L. using thidiazuron. Plant Cell Tissue Organ Cult 80:187–190

Ghani A (1998) Monographs in medicinal plants of Bangladesh. Chemical constituents and uses., 2nd ed. Asiat Soc Bangladesh 2:276

Gopalan R, Henry A N. (2000). Endemic plants of India. Bishen Singh Mahendrapal, 1 p. https://en.wikipedia.org/wiki/Rauvolfia

Kirtikar KR, Basu BD (1993) Indian medicinal plants, vol II, II edn. Dehra Dun Publishers, Culcutta. 9, pp 285–298

Krishnan PN, Decruse SW, Radha RK (2011) Conservation of medicinal plants of Western Ghats, India and its sustainable utilization through in vitro technology. In Vitro Cell Dev Biol Plants 47:110–122

Kumar A, Bhardwaj MK, Upadhyay AK, Tiwari A, Bikram DO (2011) Quantitative determination of Yohimbine alkaloids in the different part of the *Rauvolfia tetraphylla*. J Chem Pharm Res 3(2):907–910

Mohanan N, Sivadasan M (2002) Flora of Agasthyamala. Bishen Singh Mahendrapal Singh, Dehradun, p 442

Murashige T, Skoog F (1962) A revised medium for rapid growth and bioassays with tobacco tissue cultures. Physiol Plant 15:473–497

Nayar MP, Sastry AR K (1987) Red data book of Indian Plants vol I, Botanical Survey of Calcutta India

Pence CV (2011) Evaluating costs for the in vitro propagation and preservation of endangered plants. In Vitro Cell Dev Biol Plant 47:176–187

Ranjusha AP, Gangaprasad (2014) An efficient micropropagation protocol for *Rauvolfia hookeri* Srinivas and Chithra and assessment of clonal fidelity by RAPD analysis. Int J Agric Environ Biotech 7:205–212

Ravikumar K, Ved DK, Vijaya Sankar R, Udavan PS (2000). 100 red listed medicinal plants of conservation concern in Southern India. Foundation for Revitalisation of Local Health Traditions, Bangalore 467 p

Tona L, Ngimbi NP, Tsakala M, Mesia K, Cimanga K (1999) Antimalarial activity of 20 crude extracts from nine African medicinal plants used in Kinshasa, Congo. J Ethno pharmacol 68:193–203

Effect of TDZ on Various Plant Cultures

25

Sandeep R. Pai and Neetin S. Desai

Abstract

This chapter provides the effect of thidiazuron (TDZ) on various plant cultures. Plant cell cultures still remain to be of great benefit to many disciplines including studies, viz., physiology, mechanism, etc. Apart from plant potency, this supremacy can be attributed to the increase in number of plant growth regulators (PGRs). Growth regulators are the mile stones in plant tissue culture history. Plant growth regulators depict some interesting functions; they singly, in synergy or antagonistically, function in growth of plant. Also, their concentrations play pivotal role in plant response. These PGRs are categorized in one of the five classes of plant hormones: auxins, gibberellins (GAs), cytokinins (CKs), ethylene (C_2H_4), and abscisic acid (ABA). In recent years apart from natural PGRs available, different synthetic PGRs are made available. The use of thidiazuron (N-phenyl-N′-1,2,3-thiadiazol-5-ylurea) has been successfully demonstrated to promote axillary shoot proliferation and to encourage shoot formation in plants. Recalcitrant woody species have been great responders to TDZ, reason being its high cytokinin-like activity and better response. It facilitates initiation of multiple shoots in many recalcitrant woody tree species. It has been observed that lower concentrations (<1 μM) of TDZ show greater axillary proliferation compared to other cytokinins. Besides, it has many adverse effects on culture, viz., higher concentration of TDZ causes no shoot elongation. Thus, the present chapter reveals the effect of TDZ on various plant cultures.

Keywords

Thidiazuron · Cytokinin · Auxin · Plant growth regulator · Recalcitrant · Woody

S. R. Pai (✉) · N. S. Desai
Amity Institute of Biotechnology (AIB), Amity University, Mumbai - Pune Expressway, Bhatan, Post – Somathne, Panvel, Mumbai 410206, Maharashtra, India
e-mail: srpai@mum.amity.edu

N. Ahmad, M. Faisal (eds.), *Thidiazuron: From Urea Derivative to Plant Growth Regulator*, https://doi.org/10.1007/978-981-10-8004-3_25

25.1 Introduction

Humans have a great deal of allure toward plants, due to obvious inclusions of many of them in food, clothing, medicine, and other purposes. Researchers have been studying on various aspects of the plants to understand their system. Optimization on availability of the plants throughout the season has been one among the many other challenges which has been answered efficiently. With the emergence of biotechnology, the things have been easy, especially plant tissue culture has played a pivotal role in optimizing the yield of plants. This was only possible by learning that plant growth can be regulated by adding some chemicals. Initial studies demonstrated that substances isolated from human urine can regulate the plant growth, which was later identified as indole acetic acid, a substance that had been known for decades (Mitchell and Rice 1942).

Cytokinins comprise of a separate class of growth promoters; they stimulate synthesis of proteins and actively take part in cell cycle control. Perhaps due to which they thought to promote chloroplast maturation and delay senescence in plants. Cytokinins when applied to plant tissue, biochemically, causes the treated part to act as a sink for amino acids, which then migrates to the nearby sites (George et al. 2008), thus causing the most noticeable effect of cytokinin in the tissue. They are always used in ratio with auxin to encourage cell division and manage morphogenesis. They are known to overcome apical dominance and proclaim lateral buds from dormant tissues.

25.2 Thidiazuron as Cytokinin

Naturally occurring cytokinin compounds and their derivatives are available in market. Kinetin was the first cytokinin to be discovered at professor Skoog's laboratory in the University of Wisconsin. This came out as a result out of the experiemnts to promote continuing growth of the callus which formed on tobacco stem sections in vitro. Further, as George et al. (2008) mentions that chromatography of alcohol-soluble yeast extract fractions proved to be purine, researches for supplementary sources of purines were evaluated to observe the potency toward callus growth. Until then, extracts of herring sperm DNA showed a molecule with similar spectrum and chemical behavior. Its isolation and crystallization from DNA samples under acidic conditions lead to a new growth factor "kinetin." This molecule stimulated cell division in cells which otherwise might have become multinuclear (Miller et al. 1955a, b; Miller 1961a, b) and was later identified as 6-furfuryl aminopurine. Skoog et al. (1965) proposed the general term "cytokinin" to envelop all molecules that show such similar activity.

Kinetin is not accepted as natural cytokinin and has arisen as structural rearrangements in original isolates (Hecht 1980). There are many cytokinins naturally available and identified which structurally resemble to kinetin. They are structurally either freebases, glucosides, ribosides, or nucleotides of kinetin (Entsch et al. 1980).

Their utility in plant tissue culture work has been widely studied, viz., trans-zeatin (4-hydroxy-3-methyltrans-2-butenylaminopurine), iP (N6-Δ2isopentenyladenine), and dihydrozeatin (6-(4-hydroxy-3-methyl-trans-2-butenyl)aminopurine).

Higher costs of natural cytokinins (iP and zeatin) make them an unpopular choice for commercial routine laboratory practices; however, they are still a popular choice in research laboratories. Substituted purines and phenyl urea are two largely vouched groups of synthetic cytokinins, prevalent due to their potent cytokinin-like properties. Synthetic derivatives of natural occurring cytokinins are extremely potent; they chiefly include N6-substituted adenine derivatives. There are few other molecules to some extent structurally similar but possess such activity, such as 4-alkylaminopteridines (Iwamura et al. 1980a, b) and 6-benzyloxypurines. Some of these analogues are reported to be more active than kinetin or benzyladenine (BA) and are particularly effective in promoting morphogenesis (Wilcox et al. 1978, 1981). The 1-deaza analogue of zeatin riboside (Rogozinska et al. 1973; Rodaway and Lutz 1985; Kaminek et al. 1987) also has cytokinin activity. Current-day literature suggests that BA and its derivatives are widely accepted forms of cytokinins. Topolin is one such derived group which is aromatic and naturally occurring cytokinins.

Natural cytokinin, viz., yeast extract or coconut milk, is used in media as organic supplements. Shantz and Steward (1955) demonstrated that they contain physiologically active substances including natural cytokinin zeatin and 1,3-diphenylurea. The two molecules in the same series are 2Cl-4PU (N-(2-chloro-4-pyridyl)-N′-phenylurea) and 2,6Cl-4PU (N-(2,6-dichloro-4-pyridyl)-N′-phenylurea), which is supposed to be the most active.

A thiadiazole-substituted phenylurea, thidiazuron (TDZ) (N-phenyl-N′-1,2,3-thiadiazol-5-ylurea) which was earlier registered as a cotton defoliant (Arndt et al. 1976)with the product name "Dropp," has demonstrated high cytokinin activity (Mok et al. 1982). Diphenylurea is a rather weakly active cytokinin (Bottomley et al. 1963; Miller 1960; Strong 1956), but particular derivatives of N-phenyl-N′-4-pyridylurea exhibit cytokinin activity equal to or exceeding that of zeatin in the tobacco callus bioassay (Takahashi et al. 1978).

Mok and coworkers (1982) studied cytokinin property of TDZ on *Phaseolus* species. The experiment was part of their study on cytokinin metabolism in species (Mok et al. 1978, 1979, 1980; Armstrong et al. 1981); they examined effects of TDZ and other substituted urea compounds on the growth of cytokinin-dependent callus tissues of *Phaseolus lunatus* cv. Kingston. It was understood that TDZ has more potent cytokinin activity compared to N-phenyl-N′-4-pyridylureaderivatives and most other active cytokinins of adenine type. It was also concluded that there are two derived classes of urea, (i) pyridylurea and (ii) thiadiazolyl urea, comprising of various compounds with cytokinin activity. It was also mentioned that these activities were equivalent or exceeding to that of the most active cytokinins of the adenine type (Mok et al. 1982).

25.3 Chemistry

In recent years, TDZ appeared as a very effectual PGR in plant tissue culture experiments for a wide range of species like herbs, shrubs, climbers, crop plants, and majorly trees (Murthy et al. 1998). Thidiazuron had a commercial impact for its strongest use in defoliating leaves to facilitate collection of bolls from cotton plant (Arndt et al. 1976). It has also proved to protect chlorophyll from degradation in detached leaves. TDZ exhibits a high level of activity at concentrations as low as 10 pM for a relatively short period (Bakulev et al.). Chemically it is 1-phenyl-3-(1,2,3-thiadiazol-5-yl)urea. It has been demonstrated to have herbicidal properties. TDZ has a chemical formula $C_9H_8N_4OS$, molar mass of 220.251 g/mol, and density of 1.51 g/cm^3. Structurally it is different from both auxins and adenine-type cytokinins. It possesses two functional groups, phenyl and thiadiazole. It has been demonstrated that replacement or modifications in these groups will result in reduction of kinin activity of TDZ.

Thomas and Katterman (1986) demonstrated dose-dependent effect of thidiazuron in radish and soybean. They concluded that the general growth and cell division stimulation become saturated at low levels of TDZ. That is, TDZ was 10^2 times more effective in the radish and 10^4 times more effective in soybean compared to purine cytokinins (Thomas and Katterman 1986).

As per physical chemistry guidelines, TDZ is found as colorless, odorless crystals, white to yellow in color. It has a melting point of 210.5 °C and a vapor pressure of 4×10^{-6} mPa (25 °C). It is soluble in water at a rate of 31 mg/L in a neutral condition (pH 7) with temperature of 25 °C and also soluble in organic solvents (hexane, 0.002; methanol, 4.20; dichloromethane, 0.003; toluene, 0.400; acetone, 6.67; ethyl acetate, 1.1 g/L 20 °C). Interestingly it has been written that TDZ rapidly gets converted to photoisomer, 1-phenyl-3-(1,2.5-thiadiazol-3-yl)urea in the presence of light (λ >290 nm). TDZ is hydrolytically stable at room temperatures, at pH 5.9.

25.4 Mode of Action

TDZ has been used singly or in synergy with other PGRs, mainly auxins. Positive effects on culture of *Geranium* by replacing auxin and cytokinin with TDZ have been successfully been demonstrated (Visser et al. 1992). Further studies elucidated role of TDZ in induction and regeneration in many species (Murthy and Saxena 1998). Not only this but the recalcitrant cultures successfully responded and regenerated to TDZ (Malik and Saxena 1992b; Murthy and Saxena 1998).

The flow of actions on treatment of TDZ involves reprogramming and expression of morphological and genetic cell competent to undergo development which further leads to morphogenesis. Studies have showed association of TDZ in metabolism of PGRs. TDZ earlier was categorized as cytokinin; this was because of its natural cytokinin-like response. Metabolism of endogenous plant growth regulators has a direct relation to the presence of TDZ during morphogenesis and regulates endogenous growth. The role of TDZ in morphogenesis is intimately related to the

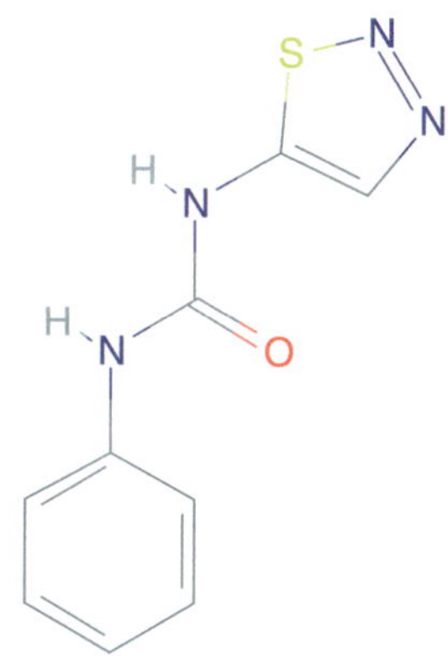

Fig. 25.1 Chemical structure of TDZ (Source: https://pubchem.ncbi.nlm.nih.gov/compound/thidiazuron#section=Top)

metabolism of endogenous growth regulators. In some independent experiments, increased levels of endogenous auxin, ethylene, and ABA were recorded in response to TDZ treatment (Murthy et al. 1995; Yip and Yang 1986; Capelle et al. 1983; Thomas and Katterman 1986; Mok et al. 1987; Ji and Wang 1988; Hutchinson and Saxena 1996b) (Fig. 25.1).

Actions of TDZ have been discussed by many workers in details; referring to most of the articles, it feels like TDZ can be compared in following ways to the classical PGRs to understand the exact mechanisms (Fig. 25.2).

25.4.1 Cytokinin-Like Action

According to Mok et al. (1982), TDZ holds cytokinin-like activity, and also in several other bioassays, the application of TDZ elicited effects in association with cytokinins (Thomas and Katterman 1986; Visser et al. 1995). There are reports suggesting TDZ promoted synthesis and/or accumulation of endogenous cytokinins (Thomas and Katterman 1986; Murthy et al. 1995; Hutchinson and Saxena 1996b). Kefford et al. (1968) reported that this may potentially occur due to (a) increase in synthesis, (b) decrease in catabolism, or (c) release of biologically active cytokinin molecules from non-active storage forms. Interconversion of labeled cytokinin ribonucleosides and ribonucleotides in the presence of TDZ and zeatin was studied by Mok et al. (1987)). Here, TDZ inhibited formation of nucleotides, whereas zeatin sustained in production of nucleotides from ribonucleosides. This indicated involvement of TDZ in regulating metabolism for production of endogenous cytokinins (Mok et al. 1987). Its cytokinin-like action was also evident when tissue of *Phaseolus lunatus* callus proliferated on medium with TDZ and on a cytokinin-free medium. This indicated positive alteration and involvement of TDZ in pathways for production of cytokinin-active adenine derivatives bearing N6-isoprenoid side chains (Capelle et al. 1983).

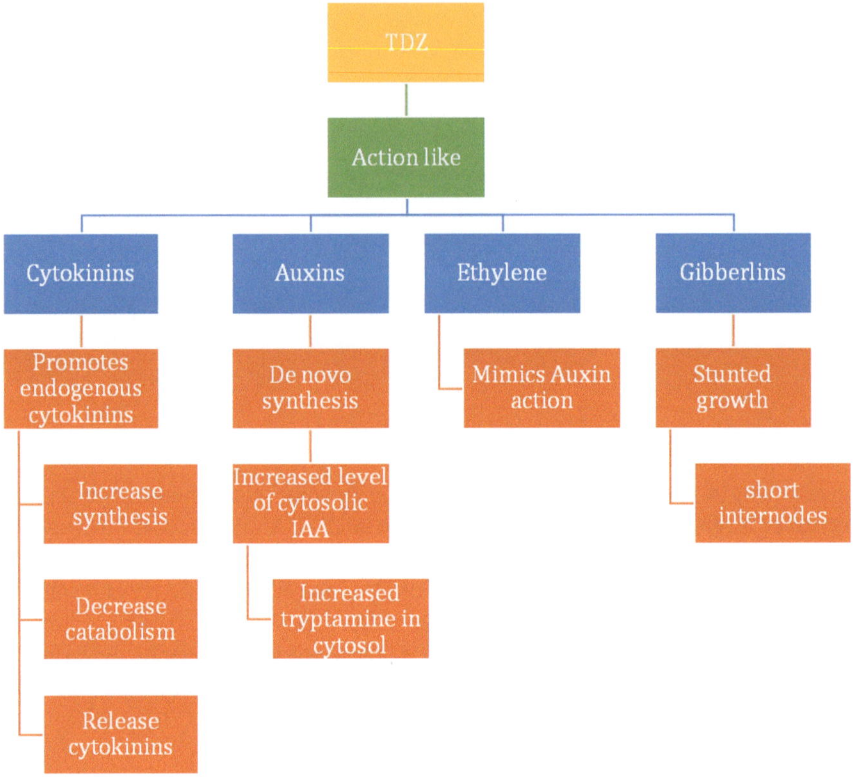

Fig. 25.2 Various PGR-like actions of TDZ

Thidiazuron is structurally dissimilar to any other naturally occurring cytokinins especially purine based. Further, its action to induce somatic embryogenesis (SE) makes it different from any other purine-based cytokinins, wherein the later alone has never been reported to induce SE. Therefore, properties of TDZ might be apparent via different mechanisms. The biological responses induced with purine-based cytokinins and phenylureas in the presence of competitive inhibitors and the measurement of a relationship between activity and structure suggested a common site of action for these two groups of growth regulators (Kefford et al. 1968; Iwamura et al. 1980b). Recently, Nagata et al. (1993) isolated a cytokinin-specific binding protein (CSBP) from etiolated mung bean seedlings and showed that the association constant of CSBP for CPPU, a phenylurea derivative, was higher than that for BA. This finding clearly demonstrated the existence of a common cytokinin-specific binding protein for both types of compounds.

25.4.2 Auxin-Like Action

Auxins have a vital role in differentiation of cell aggregates which are preliminary requirement for regeneration. TDZ has proved to induce SE in *Arachis hypogea* and *Azadirachta indica* (Murthy and Saxena 1998). In an independent study, Murthy et al. (1995) observed de novo synthesis of auxins in peanut seedlings grown on TDZ-containing medium. It was observed that there was an increased level of IAA and other monoamine alkaloid compound tryptamine in cytoplasm. It was further observed that the use of PCIB (2-(p-chlorophenoxy)-2-methylpropionic acid) popularly known as clofibric acid which is a herbicide functioning against auxin biosynthesis, for reduction of TDZ-induced somatic embryogenesis in both *Geranium* (Hutchinson et al. 1996) and *Arachis hypogea*. Similarly, TIBA (2, 3, 5-triiodobenzoic acid), an inhibitor of polar auxin transport (Thomson et al. 1973), reduced SE but failed to decrease auxin levels in TDZ-treated plant tissues (Hutchinson et al. 1996).

25.4.3 Ethylene-Like Action

Promotive and inhibitory effects of ethylene in somatic embryogenesis have been reported (Biddington 1992). Supplementation of media with TDZ for induction of *Geranium* somatic embryogenesis results in elevated levels of ethylene in the space of culture vessel (Hutchinson et al. 1997a). On the other hand, decrease in ethylene level by its inhibitor AVG (aminoethoxyvinylglycine) improved the embryogenic response in geranium hypocotyls. Thus, indicating ethylene can be produced as a negative by-product of TDZ-mediated metabolic cascade (Hutchinson et al. 1997a). Utilization of exogenous ethylene or 1-aminocyclopropane-l-carboxylic acid, an ethylene forerunner, diminished the embryogenic result to an undistinguished level as observed in the TDZ-actuated culture (Hutchinson et al. 1997a). Auxin-like metabolic response of TDZ was based on the observation of Suttle (1985, 1986) on 2-(p-chlorophenoxy)-2-methylpropionic acid (PCIB) that inhibited TDZ-mediated ethylene production in cotton. Since auxin treatment additionally brought about expanded ethylene generation (Suttle 1984; Yip and Yang 1986), it is conceivable to infer that the leaf abscission may not be an immediate aftereffect of TDZ treatment yet rather a consequence of the auxin reaction instigated by TDZ.

25.4.4 Gibberellin-Like Action

There are not really any reports observed on coordinated balance of endogenous gibberellins by TDZ. In any case, a few reports propose change of TDZ-prompted somatic embryogenesis in geraniums (Hutchinson et al. 1997b) by GA synthesis inhibitors (triazoles and ancymidol), supporting that gibberellins are affected by TDZ. Legume seedlings growing on TDZ medium exhibited stunted growth habit (Murthy et al. 1995, 1998); also foliar spray or soil drenching of TDZ significantly affected stem elongation making them stunted in ginseng and geranium (Sanago et al. 1995; Proctor et al. 1996). Woody plants regenerating into adventitious shoots on TDZ medium will be dwarf with short internodes (Lu 1993; Pai et al. 2017), which indicates changed metabolism in gibberellins. Although to what extent it explains connection of TDZ with gibberellins remains to be the area to be examined.

25.5 Thidiazuron and Its Applications

European Union banned the use of thidiazuron in agriculture. It was one of the agricultural chemicals in the framework of the European Pesticides Directive 91/414/EEC that must have an environmental and health assessment to obtain a new authorization (Wikipedia). In spite of the fact being known that TDZ is modestly poisonous to fish and aquatic organisms and also that it is not easily biodegradable, yet it is utilized in different parts of the world, including United States, Australia, and Mexico. Since it is not extremely toxic to birds or bees, the World Health Organization has made substance classification as: "It was rated unlikely to present acute hazard in normal use."

25.5.1 Plant-Based Applications

Each living cell of a plant body has a capacity to grow into a whole plant (totipotency) by means of de novo formation of organs or somatic embryos. Plant growth substances (viz., auxins and cytokinins) mainly regulate the whole process of acquisition of competency, dedifferentiation, and redifferentiation. Further, the above processes in tissue cell cultures are effectively governed by TDZ singly or in combination with other PGRs.

25.5.1.1 Callus Formation

The presence of auxins in cell culture media helps in the induction of cell proliferation and callus growth. Weedicides such as 2,4-dichlorophenoxyacetic acid along with other synthetic auxins [naphthaleneacetic acid (NAA) and others] have successfully and extensively been utilized in tissue culture. The importance of another synthetic compound "TDZ" has been observed in the induction of callus formation in assorted plant cultures, which has also exhibited higher rate of cell proliferation compared to other PGRs. For example, TDZ induced a 30-fold increase in the growth of callus cultures over other plant growth regulators (Capelle et al. 1983). In addition, the callus absorbed less TDZ than other plant growth regulators, thereby indicating a relatively high intrinsic activity of TDZ (Capelle et al. 1983). Apart from its cytokinin-like activity, TDZ is also used in the formation of callus. Table 25.1 enlists few such recent examples of plants in which TDZ has been used in combination with other PGRs from production or maintenance of callus.

Table 25.1 Use of TDZ in formation of callus

Plant name	TDZ (mg/l)	Other PGR (mg/l)	References
Phalaenopsis spp.	0.5	2,4 D (0.5)	Chen et al. (2000)
Crocus spp.	4.0	NAA (4.0)	Verma et al. (2016)
Aconitum balfourii	0.5	NAA (1.0)	Gondval et al. (2016)
Fragaria × ananassa	0.2 and 0.5	2, 4 D (0.02)	Cappelletti et al. (2016)
Hypericum triquetrifolium	0.4	IAA (0.5)	Azeez et al. (2017)
Curcuma soloensis	2.5 uM	BA (1.2 uM), 2, 4 D (1.2 uM)	Zhang et al. (2011)
Cymbidium spp.	0.01	NAA (0.1)	Huan et al. (2004)
Artemisia absinthium	2.0	NAA (1.0)	Tariq et al. (2014)
Digitalis spp.	0.5	IAA (0.25)	Cingoz et al. (2014)
Mangosteen	2.25 uM	BA (2.22)	Te-chato and Lim (2000)

25.5.1.2 Shoot Formation

There is generous confirmation that TDZ not only helps in the induction of bud (axillary) break and production of adventitious buds but also plays a major role in shoot production of diverse crops ranging from tropical fruit trees to roots and tuber crops. Medium fortified with high levels of cytokinins used for culturing explants stimulates multiple shoots or bud formation. A comprehensive review of plants that have been micropropagated using TDZ as growth regulator has been published (Huetteman and Preece 1993; Lu 1993). TDZ has been effectively utilized in regeneration of wood plant species (Briggs et al. 1988; Preece and Imel 1991; Baker and Bhatia 1993), but without the help of high concentration of adenine type of cytokinins, organogenesis is not possible. Also, TDZ has been utilized adequately in species in which purine-type cytokinins were incapable. In any case, there have been reports of issues with transformation of TDZ-induced shoots into complete plantlets, poor shoot elongation, and inadequate rooting (Huetteman and Preece 1993; Lu 1993). This debilitated development of TDZ-induced regenerants may come about because of utilization of supraoptimal levels of TDZ in the media or the presence of the compound in cultured tissues. Table 25.2 provides an insight in the published recent literature, wherein TDZ has been used in shoot formation.

25.5.1.3 Somatic Embryogenesis

Exogenous auxin to cytokinin ratio plays an important role in production of embryogenetic tissues especially somatic embryos. TDZ singly has been found to substitute for both the auxin and cytokinin prerequisite of substantial embryogenesis in numerous species (Saxena et al. 1992; Visser et al. 1992; Gill et al. 1993). Addition of TDZ in culture media invigorated in vitro somatic embryogenesis in *Nicotiana* (Gill et al. 1993), *Arachis hypogea* (Saxena et al. 1992; Murthy et al. 1995), geranium (Visser et al. 1992), chickpea (Murthy et al. 1996), neem (Murthy and Saxena 1998), and St. John's wort (Murch et al. unpublished information), at a significantly higher rate compared to the known phytohormones. In different cases, concurrent

Table 25.2 Use of TDZ in shoot formation in different plant species

Plant name	TDZ (mg/l)	Other PGRs (mg/l)	References
Phalaenopsis spp.	0.5	2,4 D (0.5)	Chen et al. (2000)
Aconitum balfourii	0.5	–	Gondval et al. (2016)
Solanum tuberosum	2.0	–	Sherkar and Chavan
Oryza sativa	0.5	BAP (0.5), Kn (1.5), NAA (0.5)	Dina et al. (2016)
Brassica oleracea	0.33 and 0.088	Adenine (79.70) and IAA (0.22)	Gambhir et al. (2017)
Curcuma soloensis	2.5 uM	BA (9.0 μM), NAA (1.2 μM)	Zhang et al. (2011)
Swertia lawii	3.0	IBA (0.3)	Kshirsagar et al. (2015)
Ancistrocladus heyneanus	6.81 uM	BAP (13.31 uM)	Pai et al. (2008)
Achyranthes aspera	0.1	BAP (3.0)	Pai et al. (2017)
Strawberry	0.5	2, 4 D (0.02)	Cappelletti et al. (2016)
Blueberry	0.5	2iP (0.2)	Cappelletti et al. (2016)
Agapanthus praecox	4.5 uM	BA (22.2 uM), IAA (2.9 uM)	Baskaran and Van Staden (2013)

production of shoots and somatic embryos has additionally been recorded (Bates et al. 1992). Despite the fact that the activity of TDZ as a cytokinin-like compound is all around archived, the previously mentioned reports of somatic embryogenesis give proof to a part of TDZ in regulation of auxin metabolism, as the enlistment of somatic embryogenesis is a response usually connected with auxins.

A novel arrangement of TDZ-induced regeneration is the improvement of somatic embryos on intact seedlings (Malik and Saxena 1992b). Somatic embryos developed at various sites on intact pea, peanut, and chickpea seedlings germinated on TDZ-fortified media. In purine cytokinin BA (N 6-benzyladenine) utilization as a part of a similar procedure, de novo shoots appeared at the regenerative area, demonstrating that the TDZ-initiated somatic embryogenesis is not exclusively a cytokinin-dependent response. Despite the fact that the seedling culture system was initially produced for recovering substantial seeded legumes, viz., pea, bean, and peanut, this system has been therefore utilized as a part of a wide assortment of different plants including geranium (Quresbi and Saxena 1992) and neem (Murthy and Saxena 2015). The following table depicts the use of different concentrations of TDZ along with other PGRs to achieve somatic embryos (Table 25.3).

25.5.1.4 Protoplast Culture

Plant regeneration from protoplasts more often than not continues through a callus stage; in any case, somatic embryogenesis might be started from the protoplast-derived cells (Song et al. 1990). The significance of the nearness of both auxin and cytokinin in the cultures to stimulate protoplast division and development is well studied (Cook and Meyer 1981), and henceforth, most protoplast culture media contain mixture of auxins and cytokinins. TDZ (in combination with auxins like NAA,

Table 25.3 Use of TDZ in induction of somatic embryos in various plant species

Plant name	TDZ (mg/l)	Other PGRs (mg/l)	References
Crocus spp.	2.0	IAA (2.0) + BAP (2.0)	Verma et al. (2016)
Mirabilis jalapa	0.5	BAP (2.0)	Rohela et al. (2016)
Lachenalia montana	1 uM	2, 4 D (0.5 uM)	Baskaran and Van Staden (2017)
Crocus olivieri	2.0	IAA (2.0)	Verma et al. (2016)
Malaxis densiflora	6.80	2, 4 D (3.39)	Mahendran and Bai (2016)
Crocus sativus	2.5 uM	Picloram (2.00 uM)	Devi et al. (2014)
Lachenalia viridiflora	1.0 uM	Picloram (2.50 uM)	Kumar et al. (2016)
Digitalis trojana	1.0	IAA (0.5)	Verma et al. (2012)
Cajanus cajan	10.0 uM	–	Singh et al. (2003)
Azadirachta sp.	1.0 uM	–	Gairi and Rashid (2004)

naphthoxyacetic acid (NOA), or 2,4-D) at various levels (0.001–20 btM) has been utilized amid introductory periods of cell wall formation around protoplasts, initiation of cell division (Chupeau et al. 1993; Reustle et al. 1995), and in later stages to complete the recovery from protoplast-derived callus (Lenzner et al. 1995). Wallin and Johansson (1989) showed that TDZ supported division of leaf protoplasts of apple at a superior rate than either BA or zeatin. Likewise, TDZ was more viable than kinetin and zeatin for development of willow (*Salix viminalis* L.) protoplast cultures (Vahala and Eriksson 1991). Chupeau et al. (1993) reported that TDZ was more successful at lower levels compared to adenine-type cytokinins. The role of TDZ in induction of regeneration from protoplast cultures is by all accounts to a great extent a cytokinin-like response, yet the correct mechanism for this impact stays undetermined.

25.5.1.5 In Vivo Regeneration

The effectiveness of TDZ as an inductive molecule for morphogenesis is not restricted to tissue culture systems. Sanago et al. (1995) found that TDZ actuated the arrangement of regenerative outgrowths on root tissues and adventitious shoots at the crown region of greenhouse-grown geraniums. Also, pot-developed *Spathiphyllum* plants splashed with TDZ created countless shoots, both at and beneath soil levels (Henny 1995). Proctor et al. (1996) observed the arrangement of adventitious buds on the shoulders of ginseng tap roots when TDZ was applied either as foliar shower or soil soaked. This solid relationship among TDZ and morphogenic forms instigated in vitro and in vivo is exclusive to the molecule and gives a few experimental systems for surveying the biochemical responses to TDZ and additionally researching the variables that manage plant morphogenesis.

Thidiazuron does not degrade by cytokinin oxidase and is stable (Mok et al. 1987). It is considered more active compared to BAP or zeatin, and a lower concentration of it is effective in tissue culture experiments. TDZ is more efficient in most of the species yet had made a noteworthy accomplishment in woody species. A portion of the impediments revealed in a few species include hyperhydricity of shoots

(Debergh et al. 1992; Briggs et al. 1988; Cousineau and Donnelly 1991), anomalous leaf morphology (van Nieuwkerk et al. 1986; Cambecedes et al. 1991), shorter internodes and smaller shoots (Fasolo et al. 1989; Meyer and van Staden 1988; Desai et al. 2016), and trouble in prolongation and establishing of recovered shoots (Meyer and Kerns 1986; Meyer and van Staden 1988). In such cases, hyperhydricity can be overwhelmed by utilizing unlocked profound petri dishes in the shoot initiation stage, vented caps in the shoot lengthening stage, or higher levels of gelling agent. Shoot quality can be enhanced by utilizing a mixture of TDZ and purine cytokinin (Briggs et al. 1988). Preece and Imel (1991) accounted that the majority of shoots regenerated on TDZ medium were short, yet elongated after shifting them to medium containing IBA and 2iP. To overcome other issues mentioned above, explants ought to be prompted with the most minimal successful TDZ concentration and continued TDZ medium for the slightest time that is required for each species (Lu 1993). In another investigation, Pawar et al. (2015) discovered unfavorable impact of TDZ on nature of callus and content of proline and glutamine in rice.

Conclusively, in years of its usage since its revelation, TDZ has demonstrated to advance in its application from basic cytokinin to every single other feature of plant tissue culture. Impact of any PGR rely upon many factors as fixation, levels of different endogenous PGR, ecological conditions, signaling factors, and for the most part affectability of plant species. In this manner, it is hard to anticipate the activity of exogenous utilization of any PGR including TDZ for new plant system. A scrutiny of writing uncovers that it has effectively been utilized to induce axillary or adventitious shoot multiplication in various plant species including herbaceous and perennials. Less has been comprehended about the activity of TDZ on morphogenesis and furthermore its biochemical and physiological basis. In any case, studies done on different species and tissue culture response of TDZ have laid more extensive application in plant tissue culture.

Acknowledgement Authors are indebted to the Head of Amity University, Mumbai.

References

Armstrong DJ, Kim SG, Mok MC, Mok DWS (1981) Genetic regulation of cytokinin metabolism in *Phaseolus* tissue cultures. In: Caud-Lenoel C, Guern J (eds) Metabolism and molecular activities of Cytokinins. Springer-Verlag, Berlin, p 97

Arndt F, Rusch R, Stilfried HV (1976) SN 49537, a new cotton defoliant. Plant Physiol 57:99

Azeez H, Ibrahim K, Pop R, Pamfil D, Hârţa M, Bobiş O (2017) Changes induced by gamma ray irradiation on biomass production and secondary metabolites accumulation in *Hypericum triquetrifolium* Turra callus cultures. Ind Crop Prod 108:183–189

Baker BS, Bhatia SK (1993) Factors effecting adventitious shoot regeneration from leaf explants of quince (*Cydonia oblmga*). Plant Cell Tissue Organ Cult 35:273–277

Baskaran P, Van Staden J (2013) Rapid in vitro micropropagation of *Agapanthus praecox* South Afr. Aust J Bot 86:46–50

Baskaran P, Van Staden J (2017) Ultrastructure of somatic embryo development and plant propagation for *Lachenalia Montana*. South Afr J Bot 109:269–274

Bates S, Preece JE, Navarrete NE, Sarnbeek JW, van Gafbey GR, Van Sambeek JW (1992) Thidianiron stimulates shoot organogenesis and somatic embryogenesis in white ash (*Frmims aimericana* L). Plant Cell Tiss Organ Cult 31:21–29

Biddington NL (1992) The influence of ethylene in plant tissue culture. Pl Growth Regul 11:173–187

Bottomley W, Kefford NP, Zwar JA, Goldacre PL (1963) Kinin activity from plant extracts. I. Biological assays and sources of activity. Aust J Biol Sci 16:395

Briggs BA, McCulloch SM, Edick LA (1988) Micropropagation of azaleas using thidiazuron. Acta Hortic 226:205–208

Cambecedes J, Duron M, Decourtye L (1991) Adventitious bud regeneration from leaf explants of the shrubby ornamental honeysuckle, *Lonicera nitida* Wils. cv. 'Maigrun': effects of thidiazuron and 2,3,5-triiodobenzoic acid. Plant Cell Rep 10:471–474

Capelle SC, Mok DWS, Kirchner SC, Mok MC (1983) Effects of TDZ on cytokinin autonomy and the metabolism of N6-(DELTA2-isopentenyl) [8-14C] adenosine in callus tissues of *Phaseolus lunatus* L. Plant Physiol 73:796–802

Cappelletti R, Sabbadini S, Mezzetti B (2016) The use of TDZ for the efficient in vitro regeneration and organogenesis of strawberry and blueberry cultivars. Sci Hortic 207:117–124

Chen Y, Chang C, Chang W (2000) A reliable protocol for plant regeneration from callus culture of *Phalaenopsis*. In Vitro Cell Dev Bio Plant 36(5):420–423

Chupeau MC, Lemoine M, Chupeau Y (1993) Requirement of thidianiron for healthy protoplast development to efficient tree regeneration of a hybrid poplar (*Poplus tremziia x P. alba*). J Plant Physiol 141:601–609

Cingoz GS, Verma SK, Gurel E (2014) Hydrogen peroxide-induced antioxidant activities and cardiotonic glycoside accumulation in callus cultures of endemic *Digitalis* species. Plant Physiol Biochemist 82:89–94

Cousineau JC, Donnelly DJ (1991) Adventitious shoot regeneration from leaf explants of tissue cultured and greenhouse-grown raspberry. Plant Cell Tissue Organ Cult 27:249–255

Debergh P, Aitken-Christie J, Cohen D, Grout B, Arnold S, von Zimmerman R, Ziv M (1992) Reconsideration of the term 'vitrification' as used in micropropagation. Plant Cell Tissue Organ Cult 30:135–140

Desai M, Pramod HJ, Upadhya V, Sailo L, Hegde HV, Pai SR (2016) In vitro rapid multiplication protocol for ex situ conservation of the rare, endemic medicinal plant *Achyranthes coynei*. Planta Med Lett 3(04):e87–e90

Devi K, Sharma M, Ahuja PS (2014) Direct somatic embryogenesis with high frequency plantlet regeneration and successive cormlet production in saffron (*Crocus sativus* L.) South Afr J Bot 93:207–216

Dina ARJM, Ahmad FI, Wagiran A, Samad AA, Rahmat Z, Sarmidi MR (2016) Improvement of efficient in vitro regeneration potential of mature callus induced from Malaysian upland rice seed (*Oryza sativa* cv. Panderas). Saudi J Biol Sci 23(1):S69–S77

Entsch B, Letham DS, Parker CW, Summons RE & Gollnow BI (1980) Metabolites of cytokinins (Skoog, ed), pp 109–118

Fasolo F, Zimmerman RH, Fordham I (1989) Adventitious shoot formation on excised leaves of in vitro grown shoots of apple cultivars. Plant Cell Tissue Organ Cult 16:75–87

Gairi A, Rashid A (2004) Direct differentiation of somatic embryos on different regions of intact seedlings of *Azadirachta* in response to thidiazuron. J Plant Physiol 161(9):1073–1077

Gambhir G, Kumar P, Srivastava DK (2017) High frequency regeneration of plants from cotyledon and hypocotyl cultures in *Brassica oleracea* cv. Pride of India. Biotech Rep 15:107–113

George EF, Hall MA, Klerk GJD (2008) Plant growth regulators II: cytokinins, their analogues and antagonists. In: George EF, Hall MA, Klerk GJD (eds) Plant propagation by tissue culture. Springer, Dordrecht

Gill R, Gerrath JM, Saxena PK (1993) High-frequency direct somatic embryogenesis in thin layer cultures of hybrid seed geranium (*Pelargonium X hortorum*). Can J Bot 71:408–413

Gondval M, Chaturvedi P, Gaur AK (2016) Thidiazuron – induced high frequency establishment of callus cultures and plantlet regeneration in *Aconitum balfourii* Stapf.: an endangered medicinal herb of North-West Himalayas. Indian J Biotechnol 15:251–255

Hecht SM (1980) Probing the cytokinin receptor site(s) (Skoog F, ed), pp 144–160

Henny RI (1995) Thidiazuron increases basal bud and shoot development in *Spathiphyllum* 'petite'. Plant Growth Reg Soc Ame Quart 23:13–16

Huan LVT, Takamura T, Tanaka M (2004) Callus formation and plant regeneration from callus through somatic embryo structures in Cymbidium orchid. Plant Sci 166(6):1443–1449

Huetteman CA, Preece JE (1993) Thidiazuron: a potent cytokinin for woody plant tissue culture. Plant Cell Tissue Organ Cult 33:105–119

Hutchinson MJ, Saxena PK (1996b) Role of purine metabolism in TDZ-induced somatic embryogenesis of geranium (*Pelargonium X hortorum*) hypocotyls cultures. Physiol Plant 98:517–522

Hutchinson MJ, Murr DP, Krishnaraj S, Senaratna T, Saxena PK (1997a) Does ethylene play a role in TDZ-regulated somatic embryogenesis of geranium (*Pelargonium X hortorum*) hypocotyl cultures. In Vitro Cell Dev Biol 33P:136–141

Hutchinson MJ, Krishnaraj S, Saxena PK (1997b) Inhibitory effect of GA z on the development of TDZ-induced somatic embryogenesis of geranium (*Pelargonium X hortorum*) hypocotyl cultures. Plant Cell Rep 16:435–438

Iwamura H, Masuda N, Koshimizu K, Matsubara S (1980a) Effects of 4-alkylaminopteridines on tobacco callus growth. Plant Sci Lett 20:15–18

Iwamura H, Fujita T, Koyama S, Koshimizu K, Kumazawa Z (1980b) Quantitative structure-activity relationship of cytokinin-active adenine and urea derivatives. Phytochemistry 19:1309–1319

Ji ZL, Wang SY (1988) Reduction of abscisic acid content and induction of sprouting in potato, *Solanum tuberosum* L., by TDZ. J Plant Growth Regul 7:37–44

Kaminek M, Vanek T, Motyka V (1987) Cytokinin activities of N6 -benzyladenosine derivatives hydroxylated on the side-chain phenyl ring. J Plant Growth Regul 6:113–120

Kefford NP, Zwar JA, Bruce MI (1968) Antagonism of purine and urea cytokinin activities by derivatives of benzylurea. In: Wightman F, Setterfield G (eds) Biochemistry and physiology of plant growth substances. Runge Press, Ottawa, pp 61–69

Kshirsagar PR, Chavan JJ, Umdale SD, Nimbalkar MS, Dixit GB, Gaikwad NB (2015) Highly efficient in vitro regeneration, establishment of callus and cell suspension cultures and RAPD analysis of regenerants of *Swertia lawii* Burkill. Biotech Rep 6:79–84

Kumar V, Moyo M, Van Staden J (2016) Enhancing plant regeneration of *Lachenalia viridiflora*, a critically endangered ornamental geophyte with high floricultural potential. Sci Hortic 211:263–268

Lenzner S, Zoglauer K, Schieder O (1995) Plant regeneration from protoplasts of sugar beet (*Beta vulgaris*). Physiol Plant 94:342–350

Lu C (1993) The use of thidiazuron in tissue culture. In Vitro Cell Dev Biol 29:92–96

Mahendran G, Bai VN (2016) Direct somatic embryogenesis of *Malaxis densiflora* (A. Rich.) Kuntze J Genet Eng Biotechnol 14(1):77–81

Malik KA, Saxena PK (1992b) TDZ induces high-frequency shoot regeneration in intact seedlings of pea (*Pisum sativum*), chickpea (*Cicer arietinum*), and lentil (*Lens culinaris*). Aust J Plant Physiol 19:731–740

Meyer MM, Kerns HR (1986) Thidiazuron and in vitro shoot proliferation of Celtis occidentalis L. Abst. in Proceedings of the VI International Congress Plant Tissue & Cell Culture, Minneapolis, 149

Meyer HJ, van Staden J (1988) In vitro multiplication of *Ixia flexuosa*. Hortscience 23(6):1070–1071

Miller CO (1960) An assay for kinetin-like materials. Plant Physiol 35(Suppl. XXVI):26

Miller CO (1961a) A kinetin-like compound in maize. Proc Nat Acad Sci USA 47:170–174

Miller CO (1961b) Kinetin related compounds in plant growth. Annu Rev Plant Physiol 12:395–408

Miller CO, Skoog F, Von Saltza M, Strong FM (1955a) Kinetin, a cell division factor from deoxyribonucleic acid. J Am Chem Soc 77:1392

Miller CO, Skoog F, Okumura FS, Von Saltza MH, Strong FM (1955b) Structure and synthesis of kinetin. J Am Chem Soc 77:2662–2663

Mitchell JW, Rice RR (1942) Plant growth regulators, Publisher Washington, D.C.: U.S. Dept. Agriculture Volume no.495

Mok MC, Mok DWS, Armstrong DJ (1978) Differential cytokinin structure- activity relationships in *Phaseolus*. Plant Physiol 61:72

Mok MC, Kim SG, Armstrong DJ, Mok DWS (1979) Induction of cytokinin autonomy by N,N-diphenylurea in tissue cultures of *Phaseolus lunatus* L. Proc Natl Acad Sci USA 76:3880–3884

Mok MC, Mok DWS, Armstrong DJ, Rabakoarihanta A, Kim SG (1980) Cytokinin autonomy in tissue cultures of *Phaseolus*: a genotype-specific and heritable trait. Genetics 94:675

Mok MC, Mok DWS, Armstrong DJ et al (1982) Cytokinin activity of Nphenyl-N′-1,2,3-thidiazol-5-ylurea (TDZ). Phytochemistry 21:1509–1511

Mok MC, Mok DWS, Turner JE et al (1987) Biological and biochemical effects of cytokinin-active phenylurea derivatives in tissue culture systems. Hortscience 22:1194–1197

Murthy BNS, Saxena PK (1998) Somatic embryogenesis and plant regeneration of Neem (*Azadirachta indica* A. Juss). Plant Cell Rep 17:469–475

Murthy BNS, Murch SJ, Saxena PK (1995) TDZ-induced somatic embryogenesis in intact seedlings of peanut (*Arachis hypogaea*): endogenous growth regulator levels and significance of cotyledons. Physiol Plant 94:268–276

Murthy BNS, Victor J, Singh R et al (1996) In vitro regeneration of chickpea (*Cicer arietinum* L.): stimulation of direct organogenesis and somatic embryogenesis by TDZ. J. Plant Growth Regul 19:233–240

Murthy BNS, Murch SJ, Saxena PK (1998) Thidiazuron: a potent regulator of in vitro plant morphogenesis. In Vitro Cell Dev Biol Plant 34:267

Nagata R, Kawachi E, Hashimoto Y et al (1993) Cytokinin-specific binding protein in etiolated mung bean seedlings. Biochem Biophys Res Commun 19:543–549

van Nieuwkerk JP, Zimmerman RH, Fordham I (1986) Thidiazuron stimulation of apple shoot proliferation in vitro. Hort Science 21:516–518

Pai SR, Nimbalkar MS, Pawar NV, Kedage VV, Dixit GB (2008) In vitro embryo culture and ex situ regeneration studies in *Ancistrocladus heyneanus* Wall. ex Grah. Plant Cell Biotechnol Mol Biol 9(3&4):1–6

Pai SR, Upadhya V, Hedge HV, Joshi RK, Kholkute SD (2017) In vitro rapid multiplication and determination of triterpenoids in callus cultures of *Achyranthes aspera* Linn. Indian J Biotech (In Press)

Pawar B, Kale P, Bahurupe J, Jadhav A, Kale A, Pawar S (2015) Proline and glutamine improve in vitro callus induction and subsequent shooting in rice. Rice Sci 22(6):283–289

Preece JE, Imel MR (1991) Plant regeneration from leaf explants of *Rhododendron* 'P. J. M. hybrids'. Sci Hortic 48:159–170

Proctor JTA, Slimmon T, Saxena PK (1996) Modulation of root growth and organogenesis in TDZ-treated ginseng (*Panax quinquefolium* L.) J Plant Growth Regul 20:201–208

Quresbi JA, Saxena PX (1992) Adventitious shoot induction and somaticembryogenesis with intact seedlings of several hybrid seed geranium (Pelragonium X hortorum bailey) varieties. Plant Cell Rep 11:443–448

Reustle G, Harst M, Alleweldt G (1995) Plant regeneration of grape (Vitis sp.) protoplasts isolated from embryogenic tissue. Plant Cell Rep 15:238–241

Rodaway S, Lutz AW (1985) Nitroguanidines: a new class of synthetic cytokinins. Plant Physiol 77(Suppl. 21) (Abst. 109)

Rogozinska JH, Kroon C, Salemink CA (1973) Influence of alterations in the purine ring on biological activity of cytokinins. Phytochemistry 12:2087–2092

Rohela GK, Damera S, Bylla P, Korra R, Pendili S, Thammidala C (2016) Somatic embryogenesis and indirect regeneration in *Mirabilis jalapa* Linn. Mater Today Proc 3((10) B):3882–3891

Sanago MHM, Murch SJ, Slimmon TY et al (1995) Morphoregulatory role of TDZ: morphogenesis of root outgrowths in TDZ-treated geranium (*Pelargonium X hortorum* bailey). Plant Cell Rep 15:205–211

Saxena PK, Malik KA, Gill R (1992) Induction by TDZ of somatic embryogenesis in intact seedlings of peanut. Man Ther 187:421–424

Shantz EM, Steward FC (1955) The identification of compound A from coconut milk as 1,3-diphenylurea. J Am Chem Soc 77:6351–6353
Singh ND, Sahoo L, Sarin NB, Jaiwal PK (2003) The effect of TDZ on organogenesis and somatic embryogenesis in pigeonpea (*Cajanus cajan* L. Millsp). Plant Sci 164(3):341–347
Skoog F, Strong FM, Miller CO (1965) Cytokinins. Science 148:532–533
Song J, Sorensen EL, Liang GH (1990) Direct embryogenesis from single mesophyll protoplasts in alfalfa (*Medicugo sativa* L). Plant Cell Rep 9(2):1–25
Strong FM (1956) Topics in microbial chemistry. Wiley, New York, p 98
Suttle JC (1984) Effects of the defoliant TDZ on leaf abscission and ethylene evolution from cotton seedlings. In: Fuchs Y, Chalutz E (eds) Ethylene. Biochemical, physiological and applied aspects. Martinus Nijhoff/Dr. W. Junk Publishers, The Hague, pp 277–278
Suttle JC (1985) Involvement of ethylene in the action of the cotton defoliant TDZ. Plant Physiol 78:272–276
Suttle JC (1986) Disruption of the polar auxin transport system in cotton seedlings following treatment with the defoliant TDZ. Plant Physiol 86:241–245
Takahashi S, Shudo K, Okamoto T, Yamada K, Isogai Y (1978) Cytokinin activities of N-phenyl-N′-(4-pyridyl)urea derivatives. Phytochemistry 17:1201–1207
Tariq U, Ali M, Abbasi BA (2014) Morphogenic and biochemical variations under different spectral lights in callus cultures of *Artemisia absinthium* L. J Photochem Photobiol B Biol 130:264–271
Te-chato S, Lim M (2000) Improvement of mangosteen micropropagation through meristematic nodular callus formation from in vitro-derived leaf explants. Sci Hortic 86(4):291–298
Thomas JC, Katterman ER (1986) Cytokinin activity induced by TDZ. Plant Physiol 81:681–683
Thomson KS, Hertel R, Muller S et al (1973) 1-N-naphthylphthalamic acid and 2,3,5-triiodobenzoic acid. In vitro biding to particulate cell fractions and action on auxin transport in corn coleoptiles. Planta 109:337–352
Vahala T, Eriksson T (1991) Callus production from willow (*Salix viminalis* L.) protoplam. Plant Cell Tissue Organ Cult 27:243–248
Verma SK, Sahin G, Yucesan B, Ekera I, Sahbaza N, Gurel S, Gurela E (2012) Direct somatic embryogenesis from hypocotyl segments of *Digitalis trojana* Ivan and subsequent plant regeneration. Ind Crops Prod 40:76–80
Verma SK, Das AK, Cingoz GS, Uslu E, Gurela E (2016) Influence of nutrient media on callus induction, somatic embryogenesis and plant regeneration in selected Turkish crocus species. Biotechnol Rep (Amst) 10(66–74)
Visser C, Qureshi JA, Gill R et al (1992) Morphoregulatory role of TDZ. Substitution of auxin and cytokinin requirement for the induction of somatic embryogenesis in *Geranium* hypocotyl cultures. Plant Physiol 99:1704–1707
Visser C, Fletcher RA, Saxena PK (1995) TDZ stimulates expansion and greening in cucumber cotyledons. Physiol Mol Biol Plants 1:21–26
Wallin A, Johansson L (1989) Plant regeneration nom leaf mesophyll protoplasts of in vitro cultured shoots of a columnar apple. J Plant Physiol 135:565–570
Wilcox EJ, Selby C, Wain RL (1978) Studies on plant growth-regulating substances. L. The cytokinin activity of some substituted benzyloxypurines. Ann Appl Biol 88:439–444
Wilcox EJ, Selby C, Wain RL (1981) The cytokinin activities of 6-α-alkylbenzyloxy-purines. Ann Appl Biol 97:221–226
Yip WK, Yang SF (1986) Effect of TDZ, a cytokinin-active urea derivative, in cytokinin-dependent ethylene production systems. Plant Physiol 80:515–519
Zhang S, Liu N, Sheng A, Ma G, Wu G (2011) Direct and callus mediated regeneration of *Curcuma soloensis* Valeton (Zingiberaceae) and ex vitro performance of regenerated plants. Sci Hortic 130(4):899–905

Check for updates

Role of Thidiazuron in Tissue Culture of Orchids

26

Remya Mohanraj

Abstract

Orchids are among the elite of flowering plants, fetching a very high price in the international market. The orchid industry of today is dependent on micropropagation for planting material, since orchid seeds lack a functional endosperm and do not germinate naturally without a fungal association. Proper growth and maintenance of orchid plantlets in tissue culture require an external supply of growth regulators. Recently, thidiazuron (TDZ) has surfaced as an effective bioregulant in plant cell and tissue cultures by virtue of its morpho-regulatory potential. It is reported to be the matchless synthetic cytokinin available and has significantly improved the in vitro regeneration and multiplication of species recalcitrant to propagation. This chapter summarizes the various developments and recent advancements in the application of TDZ for tissue culture of orchids.

Keywords

Orchids · Thidiazuron · Micropropagation

26.1 Introduction

Orchids occupy the pinnacle in the world of ornamental plants because of their colossal horticultural importance and fetch a very high price in the international market. They are a group of extremely interesting plants possessing floricultural, commercial and therapeutic values that have been used in traditional healing system of medicine as well as in the treatment of a number of diseases in many parts of the world (Bastin and Jeyachandran 2015).

R. Mohanraj (✉)
Houston Community College, Houston, TX, USA

N. Ahmad, M. Faisal (eds.), *Thidiazuron: From Urea Derivative to Plant Growth Regulator*, https://doi.org/10.1007/978-981-10-8004-3_26

In nature, seedling orchids are found only rarely, usually at the bases of mature orchid plants. Though the orchid pods contain thousands of powdery seeds, they rarely germinate and are difficult to use directly in the field for mass propagation. This could be attributed to the fact that orchid seeds lack a metabolic machinery and functional endosperm and require a fungal association for germination.

The primary method used by orchid breeders for orchid seed germination is the asymbiotic method in which seeds are cultured aseptically on a nutrient medium supplemented with a simple carbon source such as sucrose (Tan et al. 1998). Successful asymbiotic germination of orchid seeds has been reported for many species. During seed germination, the embryo first forms tuberose PLBs (Protocorm like bodies), that eventually develops into a complete plant.

Formulation of the growth medium which includes a right combination of growth hormones/growth adjuvants is a very essential component in the tissue culture of orchids. In the above context, thidiazuron (TDZ) a cytokinin-like substance is gaining importance in the tissue culture of orchids.

This chapter focuses on the application of thidiazuron in the micropropagation of orchids and attempts to provide an account of its mechanism of action and recent advances.

26.2 General Outline of Orchid Tissue Culture

26.2.1 Media Formulation

The very first step is the formulation of an appropriate medium. In general, a tissue culture medium contains micro- and macronutrients that the plant needs. The medium also contains organic compounds such as hormones, carbohydrates, and vitamins. The appropriate ratio of auxin/cytokinin varies for different species. Sucrose is used as a common source of carbohydrate in concentrations of 2–4%, but glucose can also be used if necessary. Vitamins are added as thiamin and nicotinic acid at 0.5 ppm concentration. In some cases citric acid is added to the media because it is beneficial in preventing browning and cell death during the subculturing steps. Usually agar, or an agar substitute called Gelrite, is used as the gelling agent and is added at 0.5–0.6%. The medium is autoclaved at about 120 °C and 15 psi for 20 min to make it sterile (Liam 2014).

26.2.2 Selection of Explant

Only the healthiest and ideal plant material is seleted for use as an explant. Orchid tissue culture could be done using wide variety of explants including meristems, shoot tips, immature embryos, etc. (Liam 2014).

26.2.3 Sterilization of Explant

The selected explant is surface sterilized and inoculated on a sterile medium under aseptic conditions.

The glassware containing the culture is placed into a growth chamber at a temperature of 25 °C and at an appropriate photoperiodicity. This step is called "establishment" since during this phase the cells multiply and become established to their environment (Liam 2014).

26.2.4 Subculturing

Subculturing is done after a few weeks of culture. During this stage, the explants are removed from their culture tubes and reinoculated on to fresh media. This is done so as to maintain a stable culture. When shoots have developed, they are excised and placed into tubes with a tissue culture medium that supports rooting. The medium for rooting is generally supplemented with a slightly higher auxin concentration (Liam 2014).

26.2.5 Acclimatization

The acclimatization phase prepares the young plantlets to adapt themselves to the ex vitro conditions. Plantlets are removed from their culture tubes and are washed to remove any remaining culture medium. The plantlets are now placed in a potting medium, typically containing sphagnum moss and peat. They are kept under a mist to prevent desiccation and brought gradually to common greenhouse conditions (70% humidity, 21–24 °C, and a 16 light to 8 h dark photoperiod) (Liam 2014). The various steps involved in the tissue culture of orchids are showin the following diagram (Fig. 26.1).

26.3 Thidiazuron in Orchid Tissue Culture

Thidiazuron has been used in the recent past to achieve success in orchid micropropagation. This section attempts to provide an account of such cases.

Thidiazuron (TDZ) was used to obtain shoot regeneration of the epiphytic orchid *Xenikophyton smeeanum* (Reichb. f.) from shoot tip sections and protocorm-like bodies (PLBs) by Gangadhar et al. (2011).

High-frequency direct shoot proliferation was induced from the shoot segments of three epiphytic orchids, *Cymbidium aloifolium*, *Dendrobium aphyllum* and *Dendrobium moschatum* on Murashige and Skoog's medium supplemented with N6-benzyladenine(BA) or thidiazuron (TDZ). Thidiazuron was found to be more effective in inducing shoot proliferation at 2.2–4.5 PM (Nihar Ranjan Nayak 1997).

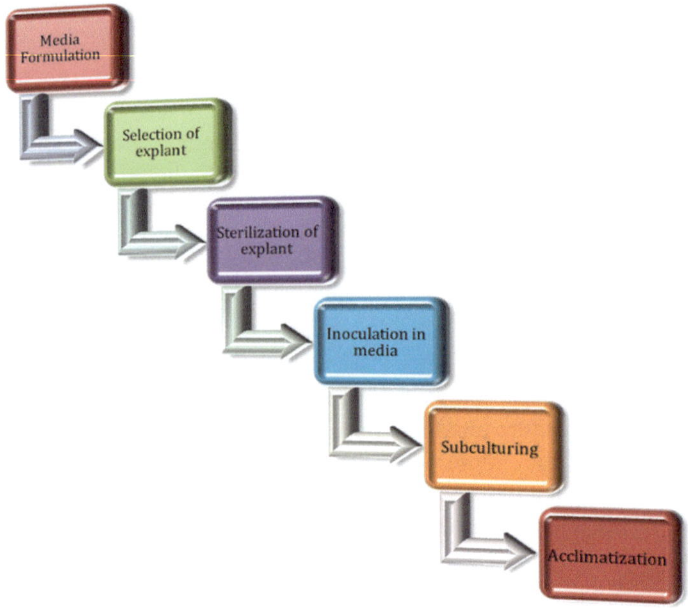

Fig. 26.1 Process diagram for the general steps of orchid tissue culture

Yung-i Lee and Nean Lee (2003) reported that, in *Cypripedium formosanum*, an endangered slipper orchid species, totipotent callus, could be induced from seed-derived protocorm segments on a quarter-strength Murashige and Skoog medium containing 4.52 μM 2,4-dichlorophenoxyacetic acid and 4.54 μM thidiazuron.

It has been reported by Malabadi et al. 2004 that Protocorm-like bodies could be observed in *Vanda coerulea* when thin shoot tip sections were cultured on Vacin and Went's (1949) basal medium supplemented with 11.35 μM thidiazuron. Treatment with thidiazuron for four weeks resulted in the highest percentage of protocorm-like bodies that survived and ultimately produced healthy shoots with two to three leaves. However, prolonged culture for more than 8 weeks with thidiazuron resulted in the development of fasciated or distorted shoots (Malabadi et al. 2004).

Natarajan Kannan 2009, formulated shoot regeneration media for *Cymbidium* with MS mineral salts and vitamins along with 0.05–0.15 mg dm-3 thidiazuron and 0.1 mg dm-3 NAA (Natarajan Kannan 2009).

Seed-derived protocorm sections of *Phalaenopsis gigantea* were cultured on solidified New Dogashima Medium basal medium containing 0–5.0 mg l-1 BA or 0–1.0 mg l-1 thidiazuron. It was observed that sections cultured on the media containing low concentrations of BA or thidiazuron multiplied faster as compared to those cultured on other media with higher concentrations of BA and the control medium. Also, thidiazuron concentration of 0.1–0.3 mg l-1 increased the percentage of protocorm sections that proliferated (Latip et al. 2010).

Micropropagation of the endangered orchid, *Rhynchostylis retusa*, was conducted by culturing explants on half-strength Murashige and Skoog medium containing thidiazuron alone or in combination with naphthalene acetic acid and activated charcoal. It was observed that higher concentrations of NAA and lower concentrations of TDZ produced superior shoots and roots (Aung Htay Naing et al. 2010).

When the effect of N6-benzyladenine and thidiazuron on the in vitro induction of protocorm like bodies was studied in *Dendrobium* it was found that thidiazuron was an effective inducer of protocorm like bodies and their proliferation than BA (Panjan and Kamnoon 2011).

A study on the effect of plant growth regulators on in vitro propagation of *Cymbidium faberi* demonstrated that the optimal medium for adventitious shoot induction was 1/2 MS + 1.0 mg/L thidiazuron (TDZ) + 0.5 mg/L NAA (Tao et al. 2011).

A protocol for the induction of callus from hybrid Cymbidium was developed by substituting KIN with thidiazuron, Teixeira da Silva (2012).

The effects of culture media, sucrose concentrations, and plant growth regulators on adventitious shoot regeneration from shoot tip culture of *V. coerulea* were studied by Jitsopakul et al. (2013). It was reported that the combination of 0.5 mg/l NAA and 2 mg/l TDZ gave the optimal number of roots per explant and plantlet height after 3 months of culture.

An efficient mass propagation protocol through direct shoot bud formation from axenic nodal segments of *D. chrysanthum* was reported by Hajong et al. (2013). The maximum frequency of explants forming buds, highest number of shoots/explant, the bud forming capacity index of 14.33, and the maximum length of shoots were obtained in MS medium supplemented with 5 μM each of thidiazuron and BAP. The authors note that the protocol developed will not only help to alleviate the pressure on the natural population under stress but will also help in meeting its demands in pharmaceutical and ornamental industries and also form the basis for conservation (Hajong et al. 2013).

Effects of chitosan and thidiazuron combinations on multiplication, differentiation, and genetic stability of *Phalaenopsis gigantea* protocorm like bodies (PLBs) were investigated using different media combinations by Samarfard (2014). PLBs were cultured in solid New Dogashima medium and Vacin and Went medium supplemented with different concentrations of chitosan and thidiazuron. The study revealed that ND medium with 10 $mg{\cdot}L^{-1}$ chitosan and 0.1 $mg{\cdot}L^{-1}$ thidiazuron is most effective in the production of PLBs (Samarfard 2014).

Protocorm-like bodies (PLBs) were induced from the pseudostem segments of *Dendrobium nobile*, using thidiazuron, by Bhattacharyya et al. (2014). It was reported that, although PLB induction was observed at higher concentrations of thidiazuron, plantlet regeneration from those PLBs was affected adversely. A combination of 1.5 mg/l TDZ and 0.25% activated charcoal was found to be the best for rooting (Bhattacharyya et al. 2014).

Asymbiotic in vitro propagation of the mature seeds obtained from undehisced green pod of *Vanda testacea*, was carried out by transferring the PLBs to medium supplemented with 0.20 mg/l thidiazuron (Sebastinraj et al. 2014).

Balilashaki et al. (2014) attempted to micropropagate *Phalaenopsis amabilis* using flower stalk nodes and sterile leaves obtained from node cultures. Results of the study indicated that the production of protocorms increased with an augmented concentration of thidiazuron.

Bastin and Jeyachandran (2015) used ten different media for orchid seed germination in *Trichoglottis tenera*. One of the media tried was thidiazuron (TDZ 0.2–2.0 mg/l), coconut water (15% ml/l), and banana powder (0.4% mg/l), and another one was full- and half-strength Murashige and Skoog (MS) medium (1962) fortified with different combinations of plant growth regulators, i.e., 6-benzylaminopurine (BAP 0.2–2.0 mg/l and 0.1–1.0 mg/l), α-naphthalene acetic acid (NAA 2.0–0.2 mg/l and 0.2–2.5 mg/l), kinetin (1.0–1.0 mg/l), thidiazuron (0.2–2.0 mg/l), coconut water (15% ml/l), and tomato powder (4.0% and 0.4%) (Bastin and Jeyachandran 2015).

A mass propagation protocol for *Dendrobium* Indonesia Raya "Ina" via somatic embryogenesis pathway in producing high-qualified seedlings was successfully established by Rachmawati et al. (2015). It was observed that shoot tips derived from in vitro plantlets cultured on half-strength MS semisolid medium supplemented with 1.5 mg/L thidiazuron and 0.5 mg/L BAP resulted in the shortest embryonic callus initiation time.

Kou et al. (2016) reported that thidiazuron, which has the potential to affect the level of endogenous auxins and cytokinins, triggers PLB formation and plantlet regeneration in *Rosa canina*. Their results demonstrate that different cells in rhizoid tips acquired regeneration competence after induction by thidiazuron.

Dini and Iriawati (2016) conducted a study to examine the best medium and explants for regeneration of *Phalaenopsis* "Join Angle x Sogo Musadian." In vitro and in vivo roots and leaves were cultured in half-strength Murashige and Skoog medium supplemented with various combinations of plant growth regulators (TDZ, 2,4-D, NAA, BAP, and IAA). A medium containing 1 ppm thidiazuron and 2 ppm 2,4-D was found to be beneficial for the regeneration of PLBs from in vitro leaf explants.

A study aimed at determining the effect of coconut water on seed germination and embryo development, as well as the effect of NAA on shoots development of *Dendrobium antennatum*, was conducted by Utami and Hariyanto (2016). MS medium supplemented with 1 mg/L thidiazuron +1 mg/L NAA was found to be the most suitable for shoot development.

Deepak and Shashi 2016, compared the germination of *Herminium lanceum* seeds at different stages of embryo development on four diiferent media. The maximum number of shoots per culture from PLBs were found to develop on BM + 4 μM TDZ and the best rhizogenic response was obtained using BM + 4 μ M TDZ + 0.1 μM IBA (Deepak and Shashi 2016).

26.4 Mode of Action of Thidiazuron

Originally developed for mechanized harvesting of cotton bolls, thidiazuron (*N*-phenyl-*N′*-1,2,3-thiadiazol-5-yl urea), a substituted phenylurea compound, has manifested itself as a highly potent bioregulant of morphogenesis in the tissue culture of a wide variery of plant species. It has been reported that thidiazuron in culture induces a diverse array of responses ranging from induction of callus to formation of somatic embryos. Though structurally different from either auxins or purine-based cytokinins, thidiazuron mimics both auxin and cytokinin effects on growth and differentiation of cultured explants. Studies have pointed out that thidiazuron might exhibit its action directly through modulation of the endogenous plant growth regulators, or as a result of induced stress. Thidiazuron could also act via modifications in cell membranes, energy levels, nutrient uptake, or nutrient assimilation. While a low concentration of thidiazuron induces axillary shoot proliferation, a higher fraction induces adventitious shoot development (Murthy 1998).

It has been hypothesized that thidiazuron induces regeneration through a metabolic cascade that begins with a signaling event, followed by accumulation, and transport of endogenous plant signals such as auxin and melatonin, a system of secondary messengers, and a concurrent stress response (Jones et al. 2007).

26.5 Conclusion

The findings listed in this chapter lead us to conclude that thidiazuron is one of the most effective synthetic cytokinins for regeneration of orchid seedlings. It has been observed in various instances that the use of thidiazuron in combination with other growth regulators improves regeneration. Also, it should be noted that many of the studies on orchid tissue culture suggest that a low concentration of thidiazuron is highly beneficial for regeneration. In spite of the numerous instances of success using thidiazuron, further research on the exact mechanism of action is warranted.

References

Aung HN, Park IS, Hwang YJ, Chung JD, Lim KB (2010) In vitro micropropagation and conservation of *Rhynchostylis retusa* BL. Hortic Environ Biotechnol 51:440–444

Balilashaki K, Ruhangiz N, Siamak K, Aboozar S (2014) Micropropagation of *Phalaenopsis amabilis* cv. Cool ‘Breeze’ with using of flower stalk nodes and leaves of sterile obtained from node cultures. Int J Farm Allied Sci 3:823–829

Bastin M, Jeyachandran R (2015) Ex situ conservation of trichoglottis tenera (Lindl.) a threatened, and endangered orchid of western ghats using asymbiotic seed germination technique. Int J Recent Sci Res 6:3488–3496

Bhattacharyya P, Suman K, Reemavareen D, Pramod T (2014) Genetic stability and phytochemical analysis of the in vitro regenerated plants of *Dendrobium nobile* Lindl., an endangered medicinal orchid. Meta Gene 2:489–504

Deepak KS, Shashi BB (2016) In vitro propagation and chemical profiling of *Herminium lanceum* (Thunb. ex Sw.) Vuijk, a medicinally important orchid, for therapeutically important phenolic acids. Plant Biotechnol 33:153–160

Dini M, Iriawati (2016) Regeneration of plantlets through PLB (protocorm-like body) formation in *Phalaenopsis* 'Join Angle X Sogo Musadian'. J Math Fund Sci 48:204–212

Gangadhar S, Mulgund K, Nataraja RB, Malabadi S, Kumar V (2011) TDZ induced in vitro propagation of an epiphytic orchid *Xenikophyton smeeanum* (Reichb. f.) Res Plant Biol 4:07–15

Hajong S, Kumaria S, Tandon PJ (2013) Effect of plant growth regulators on regeneration potential of axenic nodal segments of *Dendrobium chrysanthum* wall. Ex Lindl Agr Sci Tech 15:1425–1435. http://sites.psu.edu/orchidtissueculture/2014/03/22/dendrobium-kingianum-tissue-culture-propagation/

Jitsopakul N, Kanchit T, Keiko I (2013) Efficient adventitious shoot regeneration from shoot tip culture of *Vanda coerulea*, a Thai orchid. Sci Asia 39:449–455

Jones MP, Cao J, O'Brien R, Murch SJ, Saxena PK (2007) The mode of action of thidiazuron: auxins, indoleamines, and ion channels in the regeneration of *Echinacea purpurea* L. Plant Cell Rep 26:1481–1490

Kou Y, Cunquan Y, Qingcui Z, Guoqin L, Jing N, Zhimin M, Chenxia C, Teixeira da Silva JA, Liangjun Z (2016) Thidiazuron triggers morphogenesis in *Rosa canina* L. Protocorm-like bodies by changing incipient cell fate. Front Plant Sci 7:557

Latip MA, Rosmah M, Zaleha AA, Lai HT, Lilly MG, Rimi R (2010) Effects of N6 -Benzyladenine and Thidiazuron on proliferation of *Phalaenopsis gigantea* Protocorms. Asia Pac J Mol Biol Biotechnol 18:215–218

Lee Y-I, Lee N (2003) Plant regeneration from protocorm-derived callus of *Cypripedium* Formosanum. In Vitro Cell Dev Biol Plant 39:475–479

Liam Farrell (2014) http://sites.psu.edu/horticulture202blueberrypropagation/2014/03/22/dendrobium-kingianum-tissue-culture-propagation/#respond

Malabadi BR, Mulgund GS, Nataraja K (2004) Efficient Regeneration of *Vanda coerulea*, an Endangered Orchid Using Thidiazuron. Plant Cell Tissue Org Cult 76:289–293

Murthy BNS, Murch SJ, Saxena PK (1998) Thidiazuron: a potent regulator of in vitro plant morphogenesis. In Vitro Cell Dev Biol Plant 34:267–275

Natarajan K (2009) An in vitro study on micropropagation of *Cymbidium* orchids. Curr Biotica 3:244–250

Nihar RN, Shiba PR, Satyanarayan P, Nihar RN, Shiba PR, Satyanarayan P (1997) In vitro propagation of three epiphytic orchids, *Cymbidium aloifolium* (L.) Sw., *Dendrobium aphyllum* (Roxb.) Fisch. and *Dendrobium moschatum* (Buch-Ham) Sw. through thidiazuron-induced high frequency shoot proliferation. Sci Hortic 71:243–250

Panjan S, Kamnoon K (2011) Efficient direct protocorm-like bodies induction of dwarf *Dendrobium* using Thidiazuron. Not Sci Biol 3:88–92

Rachmawati BW, Nurhajati AM, Ni Made AW, Agus P (2015) Shoot tips derived-somatic embryogenesis in mass propagation of *Dendrobium* Indonesia Raya 'Ina' Fitri. Emirates J Food Agric 27:1–10

Samarfard S, Mihdzar AK, Saleh BK, Halimi MS, Seyed AR (2014) In vitro propagation and detection of somaclonal variation in *Phalaenopsis gigantea* as affected by chitosan and thidiazuron combinations. Hortscience 49:82–88

Sebastinraj J, John Britto S, Vinoth Kumar D, Philip Robinson J, Thangavel P (2014) Rapid propagation of *Vanda testacea* (Lindl.) Rchb.F. – a highly medicinal value epiphytic orchid of India. World J Agric Sci 10:223–230

Tan TK, Loon WS, Khor E, Loh CS (1998) Infection of *Spathoglottis plicata* (Orchidaceae) seeds by mycorrhizal fungus. Plant Cell Rep 18(1-2):14–19

Tao Jun LYFK, Zhao D (2011) Effects of plant growth regulators on in vitro propagation of *Cymbidium faberi* Rolfe. Afr J Biotechnol 10:15639–15646

Teixeira da Silva JA (2012) New basal media for protocorm-like body and callus induction of hybrid *Cymbidium*. J Fruit Ornamental Plant Res 20:127–133

Utami ESW, Hariyanto S (2016) The effect of organic nutrient and growth regulators on seed germination, embryo and shoots development of *Dendrobium antennatum* Lindl. Orchid by in vitro. Biosaintifika J Biol Biol Educ 8:165–171

Vacin EF, Went FW (1949) Some pH Changes in Nutrient Solutions. Bot Gaz 110(4):605–613

Thidiazuron as an Elicitor in the Production of Secondary Metabolite

27

Bengu Turkyilmaz Unal

Abstract

The secondary metabolites are known to play a major role in the adaptation of plants to their environment. They are also used by humans as food additives and as pharmaceuticals. Various strategies have been developed to improve the production of secondary metabolites in plant cell and tissue culture. Strain improvement, methods for the selection of high-producing cell lines, medium optimizations, and treatment of the undifferentiated cells with abiotic or biotic elicitors such as heavy metals, chitosan, methyljasmonate, salicylic acid, and thidiazuron (TDZ) can lead to an enhancement in secondary metabolite production. TDZ, a substituted phenylurea (N-phenyl-1,2,3-thiadiazol-5-yl urea), has gained a considerable attention during the past decades due to its efficient role in plant cell and tissue culture. TDZ has shown both auxin- and cytokinin-like effects, although, chemically, it is totally different from commonly used auxins and cytokinins. A number of physiological and biochemical events in cells are induced or enhanced by TDZ. To attain continuous and quick industrial production, when used in the appropriate concentration, the addition of stimulants such as TDZ to the growing medium will contribute to production of secondary metabolites.

Keywords

Biotechnology · Cytokinin · Natural products · Pharmaceutical plants · TDZ

B. Turkyilmaz Unal (✉)
Arts and Sciences Faculty, Biotechnology Department, Nigde Omer Halisdemir University, Nigde, Turkey
e-mail: bturkyilmaz@ohu.edu.tr

N. Ahmad, M. Faisal (eds.), *Thidiazuron: From Urea Derivative to Plant Growth Regulator*, https://doi.org/10.1007/978-981-10-8004-3_27

Plants are an important part of our daily diet. In addition to essential primary metabolites (e.g., amino acids, carbohydrates, and lipids), bioactive compounds produced by plants have been used for centuries as dyes, food additives, pigments, pesticides, cosmetic products and perfumes, and various chemical substances (Balandrin and Klocke 1988). Since the early days of mankind, plants with natural compounds also have been used by humans to health disorders and illness (Wyk and Wink 2004). Even today, the World Health Organization is predicting that up to 80% of people still based on traditional remedies for their medicines (Katzung 1995; Pezzuto 1996; Roberts 1988). They are the traditional source of many chemicals used as pharmaceuticals (Barz and Ellis 1981; Deus and Zenk 1982). Medicinal plants that are the most important source of lifesaving drugs are mainly used as crude drugs and extracts. Many effective and active substances are used in the form of isolated compounds, including many alkaloids such as ajmaline (antiarrhythmic), berberine (psoriasis), caffeine (stimulant), capsaicin (rheumatic pains), codeine (antitussive), colchicines (gout), ephedrine (stimulant), galantamine (acetylcholinesterase inhibitor), morphine (pain killer), papaverine (phosphodiesterase inhibitor), pilocarpine (glaucoma), quinine (antimalarial), reserpine (antihypertensive), scopolamine (travel sickness), yohimbine (aphrodisiac), and various types of cardiac glycosides (heart insufficiency) (Wink et al. 2005), glycosides, flavonoids, volatile oils, tannins, resins, etc. (Karuppusamy 2009; Namdeo 2007). In recent years, natural products have been replaced by synthetic drugs.

Secondary metabolites play an important role in the adaptation of plants to their environment and at the same time constitute an important pharmaceutical source (Rao and Ravishankar 2002). Biotechnological technologies will extend and enhance the usefulness of plants as renewable resources of valuable chemicals. There has been considerable interest in plant cell cultures as an alternative to traditional agriculture for the industrial production of secondary metabolites (Dicosmo and Misawa 1985). The advantage of this method is that it can ultimately provide a continuous, reliable source of natural products. Plant cell and tissue cultures can be generated in artificial nutrient media under aseptic conditions from meristematic cells, callus cells, and explants such as plant leaves or stems or organs. The major advantages of cell cultures include (1) synthesis of bioactive secondary metabolites is running in controlled environment, independently from climatic and soil conditions; (2) negative biological influences that affect secondary metabolites production in the nature are eliminated (microorganisms and insects); (3) it is possible to select cultivars with higher production of secondary metabolites; and (4) with automatization of cell growth control and metabolic processes' regulation, cost price can decrease and production increase (Mulabagal and Tusay 2004).

Most of these secondary metabolites are isolated from wild or cultivated plants because their chemical synthesis is extremely difficult and not economical (Namdeo 2007). The production of these compounds is often less than 1% of the dry weight and depends on the developmental stage and physiological condition of the plant (Oksman-Caldenteyl and Inze 2004; Dixon 2001). Biotechnological approaches, particularly plant tissue culture, play a crucial role in increasing the production of desired medical compounds from plants (Rao and Ravishankar 2002). All plants also utilize other metabolic pathways producing compounds that are not used in growth and development: these are secondary metabolites. These pathways are usually species specific and can only be activated at certain stages of growth and development or at abiotic and biotic stress conditions. The dividing line between primary and secondary metabolism is indistinct because many of the intermediates in primary metabolism are also intermediates in secondary metabolism (Yeoman and Yeoman 1996).

The production of secondary metabolites in plant cell suspension cultures has been reported from various medicinal plants. The production of solasodine from calli of *Solanum elaeagnifolium* and pyrrolizidine alkaloids from root cultures of *Senecio* sp. are examples (Nigra et al. 1987; Toppel et al. 1987). Cephaeline and emetine were isolated from callus cultures of *Cephaelis ipecacuanha* (Jha et al. 1988). Scragg et al. isolated quinoline alkaloids in significant quantities from globular cell suspension cultures of *Cinchona ledgeriana*. Capsaicin is obtained from *Capsicum annuum* L. (Islek et al. 2016). Enhanced indole alkaloid biosynthesis in the suspension culture of *Catharanthus roseus* has also been reported (Zhao et al. 2001).

The production of large-scale plant secondary metabolites can be technically performed in the bioreactors (Tripati and Tripati 2003). The use of this method for large-scale production of secondary metabolites has been limited since the low and unreliable yields of products to date (Rao and Ravinkhasar 2002). It can be applied for some expensive products, but some of the most interesting products are produced in very small quantities, while others cannot be produced in plant cell or tissue cultures. However, there are a few examples: shikonin production by cell suspension cultures of *Lithospermum erythrorhizon*, berberine production by cell cultures of *Coptis japonica* (Fujita and Tabata 1987), rosmarinic acid production by cell cultures of *Coleus blumei*, sanguinarine production by cell cultures of *Papaver somniferum* (Ulbrich et al. 1985; Eilert et al. 1985) of using plant cells as factories successfully to produce high-value secondary metabolites (Namdeo 2007).

Various strategies such as strain improvement, methods for the selection of high-producing cell lines, medium optimizations that may lead to 20–30-fold increase (Verpoorte et al. 2002), and elicitors have been developed to improve the production of secondary metabolites. This means that the maximization of the production and accumulation of secondary metabolites (Mulabagal and Tsay 2004).

Recently used elicitation is the induction of secondary metabolite production by either biotic or abiotic treatments (Johnson et al. 1991). In vitro plants and/or plant cell responses to biotic (microbial) or abiotic (physical or chemical) factors are known as "elicitors." Elicitation is a process by which plants induce or enhance the synthesis of secondary metabolites and ensures their survival, persistence, and competitive power (Namdeo 2007). The production of secondary metabolites can be enhanced by the treatment of the undifferentiated cells with elicitors such as methyljasmonate, salicylic acid, chitosan, thidiazuron, and heavy metals (Alatar 2015; Guo et al. 2011; Islek et al. 2016). Metabolites are induced or increased biosynthesis when the elicitor is added in trace amounts (Radman et al. 2003).

Elicitors can be classified on the basis of their “nature” like abiotic elicitors or biotic elicitors or on the basis their “origin” like exogenous elicitors and endogenous elicitors. Biotic and abiotic elicitors allow the reduction of process time to achieve high product concentration (Barz et al. 1988; Eilert 1987; Dicosmo and Tallev 1985).

TDZ, also used as an elicitor, has gained importance for decades due to its active role in plant cell and tissue culture (Guo et al. 2011). TDZ was synthesized by German Schering Corporation for defoliation of cotton (*Gossypium hirsutum*) (Arndt et al. 1976). TDZ, a modified phenylurea (N-phenyl-1,2,3-thiadiazol-5-yl urea), is a potent plant growth regulator with oxine and cytokinin-like activity in various culture systems (Khan et al. 2006). However, it is totally different from chemically widely used oxine and cytokinins. TDZ activity may exceed even that of most active adenine-type cytokinins (Mok et al. 1982). It is a light yellow-crystalline chemical that is slightly soluble in water but soluble in higher levels in organic solvents such as ethanol, acetone, benzene, DMSO, etc. (Murthy 1997). TDZ emerged as an effective bioregulant in cell and tissue cultures in a wide array of plant species (Hosseini-Nasr and Rashid 2000; Li et al. 2000; Matand and Prakash 2007; Svetla et al. 2003). Although the mode of action is not yet fully understood, some physiological and biochemical events in cells are induced or increased by TDZ (Guo et al. 2011; Vijaya et al. 2010). It causes a variety of morphogenic responses ranging from tissue proliferation to the induction of shoot buds and somatic embryos (Alatar 2015; Siddique and Anis 2007). TDZ is also known to stimulate strong shoot growth activity in many plant species (Dhavala and Rathore 2010; Kumar and Srivastava 2015; San et al. 2015).

Phyllanthin and hypophyllanthin amounts of callus from node and leaf explants of *Phyllanthus amarus* used in traditional folk medicine in India due to many pharmaceutical applications [including use in the treatment of fever, jaundice, ascites, hemorrhoids, frequent menstruation, skin ulcer, and diabetes mellitus (Rai and Mehrotra 2007) and hypoglycemic, antibacterial, antifungal, and antiviral (Sharma et al. 2001), antinociceptive (Santos et al. 2000), antitumor (Islam et al. 2008), antimutagenic (Kumar and Kuttani 2005), and anti-inflammatory (Kiemer et al. 2003) properties] doubled when TDZ was applied to their culture (Unander et al. 1995; Sharma et al. 2001).

Merwilla plumbea, sold at herbal markets in South Africa, is a popular but threatened traditional medicinal plant. Due to high demand, commercial production is being done. When growth regulators such as TDZ were added to the growing medium, the total phenolic, flavonoid, gallotannin, and condensed tannin contents were found to be 3–16 times higher than in naturally grown plants (Baskaran et al. 2012).

The *Scutellaria* (Lamiaceae) genus have been used for centuries in traditional Chinese medicine for the treatment of hyperlipidemia, arteriosclerosis, allergy, antibacterial, and inflammatory diseases. The pharmacological effect is due to the presence of flavonoids such as baicalein, baicalin, wogonin, and wogonoside (Shang et al. 2010). The highest increase was detected in polyphenol production in TDZ treatment (Grzegorczyk-Karolak et al. 2015).

Capsaicin obtained from pepper species is considered to be a potential chemical for pharmaceutical industries. Islek et al. (2016) found that TDZ as an elicitor applied at different concentrations and times increased the amount of capsaicin of *Capsicum annuum* L. cell suspension cultures to the control.

27.1 Conclusion

The secondary metabolite production is very important for the food, perfumery, and pharmaceutical industries. When these compounds are produced by conventional methods, their amounts are small and time is lost. In order to achieve continuous and rapid industrial production, the addition of stimulants to the growth medium will contribute to the production of secondary metabolites. TDZ is an effective elicitor in the production of secondary metabolites as well as functions in plant tissue culture.

References

Alatar AA (2015) Thidiazuron induced efficient in vitro multiplication and ex vitro conservation of *Rauvolfia serpentina* – a potent antihypertensive drug producing plant. Biotechnol Biotechnol Equip 29(3):489–497

Arndt FR, Rusch R, Stillfried HV, Hanisch B, Martin WC (1976.) SN 49537) A new cotton defoliant. Plant Physiol 57(5):99

Balandrin MF, Klocke JA (1988) Medicinal, aromatic and industrial materials from plants. In: Bajaj YPS (ed) Biotechnology in agriculture and forestry, vol 40. Springer, Berlin, pp 1–35

Barz W, Ellis B (1981) Potential of plant cell cultures for pharmaceutical production. In: Beal JL, Reinhard E (eds) Natural products as medicinal agents. Hippokrates, Stuttgart, pp 471–507

Barz W, Daniel S, Hinderer W, Jaques U, Kessmann H, Koster J, Tiemann K (1988) Elicitation and metabolism of phytoalexins in plant cell cultures. In: Pais M, Deus B, Zenk MH (eds) Exploitation of plant cells for the production of natural compounds. Biotechnol Bioeng 1982, 24:1965–1974

Baskaran P, Ncube B, Van Staden J (2012) In vitro propagation and secondary product production by *Merwilla plumbea* (Lindl.) Spet. Plant Growth Regul 67:235

Dhavala A, Rathore TS (2010) Micropropagation of *Embelia ribes* Burm f. through proliferation of adult plant axillary shoots. In: Vitro Cell Dev Bio Plant 46, pp 180–191

Deus B, Zenk MH (1982) Exploitation of plant cells for the production of natural compounds. Biotechnol Bioeng 24:1965–1974

Dicosmo F, Misawa M (1985) Eliciting secondary metabolism in plant cell cultures. Trends Biotechnol 3:318–322

Dicosmo F, Tallevi SG (1985) Plant cell cultures and microbial insult: interactions with biotechnological potential. Trends Biotechnol 3:110–111

Dixon RA (2001) Natural products and plant disease resistance. Nature 411:843–847

Eilert U (1987) Elicitation: methodology and aspects of application. In: Constabel F, Vasil I (eds) Cell culture and somatic cell genetics of plants, vol 4. Academic, San Diego, pp 153–196

Eilert U, Kurz WGW, Constabel F (1985) Stimulation of sanguinarine accumulation in *Papaver somniferum* cell cultures by fungal elicitors. J Plant Physiol 119:65–76

Fujita Y, Tabata M (1987) Secondary metabolites from plant cells – pharmaceutical applications and progress in commercial production. In: Green CE et al (eds) Plant tissue and cell culture. Alan R. Liss, New York, pp 169–185

Grzegorczyk-Karolak I, Kuzma L, Wysokinska H (2015) The effect of cytokinins on shoot proliferation, secondary metabolite production and antioxidant potential in shoot cultures of *Scutellaria alpina*. Plant Cell Tissue Organ Cult 122:699–708

Guo B, Abbasi BH, Zeb A, Xu LL, Wei YH (2011) TDZ: a multi-dimensional plant growth regulator. Afr J Biotechnol 10:8984–9000

Hosseini-Nasr M, Rashid A (2000) Thidiazuron-induced shoot-bud formation on root segments of *Albizia julibrissin* is an apex-controlled, light-independent and calcium-mediated response. Plant Growth Regul 36:81–85

Islam A, Selvan T, Mazumdar UK, Gupta M, Ghosal S (2008) Antitumor effect of phyllanthin and hypophyllanthin from *Phyllanthus amarus* against ehrlich ascites carcinoma in mice. Pharmacol Online 2:796–807

Islek C, Turkyilmaz UB, Koc E, Kaya D (2016) Effect of thidiazuron used as an elicitor in the production of capsaicin on total protein and phenolic amounts, antioxidant enzyme activities of pepper plants. Int J Environ Agric Biotechnol (IJEAB) 1(3):603–609

Jha S, Sahu NP, Mahato SB (1988) Production of the alkaloids emetine and cephaeline in callus cultures of *Cephaelis ipecacuanha*. Planta Med 54:504–506

Johnson TS, Ravishankar GA, Venkataraman LV (1991) Elicitation of capsaicin production in freely suspended cells and immobilized cell cultures of *Capsicum frutescens* Mill. Food Biotechnol 5:197–205

Karuppusamy S (2009) A review on trends in production of secondary metabolites from higher plants by in vitro tissue, organ and cell cultures. J Med Plants Res 3(13):1222–1239

Katzung BG (1995) Basic and clinical pharmacology, 6th edn. Prentice Hall International (UK) Limited, London

Khan H, Siddique I, Anis M (2006) Thidiazuron induced somatic embryogenesis and plant regeneration in *Capsicum annuum*. Biol Plant 50:789–792

Kiemer AK, Hartung T, Huber C, Vollmar AM (2003) *Phyllanthus amarus* has anti-inflammatory potential by inhibition of iNOS, COX2, and cytokines via the NF-jB pathway. J Hepatol 38(3):289–297

Kumar KBH, Kuttan R (2005) Chemoprotective activity of an extract of *Phyllanthus amarus* against cyclophosphamide induced toxicity in mice. Phytomedicine 12(6–7):494–500

Kumar P, Srivastava DK (2015) Effect of potent cytokinin thidiazuron on in vitro morphogenic potential of broccoli (*Brassica oleracea* L. var. italica), an important vegetable crop. Indian J Plant Physiol 20(4):317–323

Li H, Murch SJ, Saxena PK (2000) Thidiazuron-induced de novo shoot organogenesis on seedlings, etiolated hypocotyls and stem segments of Huang-qin. Plant Cell Tissue Organ Cult 62:169–173

Matand K, Prakash CS (2007) Evaluation of peanut genotypes for in vitro plant regeneration using thidiazuron. J Biotechnol 130:202–207

Mok MC, Mok DWS, Armstrong DJ, Shudo K, Isogai Y, Okamoto T (1982) Cytokinin activity of N-phenyl-N1-1,2,3-thiadiazol-5-ylurea (Thidiazuron). Phytochemistry 21:1509–1511

Mulabagal V, Tsay HS (2004) Plant cell cultures – an alternative and efficient source for the production of biologically important secondary metabolites. Int J Appl Sci Eng 2:29–48

Murthy S (1997) Morpho-physiological role of thidiazuron in plants. PhD thesis, University of Guelph, Canada

Namdeo AG (2007) Plant cell elicitation for production of secondary metabolites: a review. Pharmacogn Rev 1(1):69–79

Nigra HM, Caso OH, Guilietti AM (1987) Production of solasodine by calli form different parts of *Solanum elaeagnifolium* Cav plants. Plant Cell Rep 6:135–137

Oksman-Caldenteyl KM, Inze D (2004) Plant cell factories in the post-genomic era: new ways to produce designer secondary metabolites. Trends Plant Sci 9:9

Pezzuto J (1996) Taxol® production in plant cell culture comes of age. Nat Biotechnol 14:1083

Radman R, Saez T, Bucke C, Keshavarz T (2003) Elicitation of plant and microbial systems. Biotechnol Appl Biochem 37:91–102

Rai V, Mehrotra S (2007) Chromium-induced changes in ultramorphology and secondary metabolites of *Phyllanthus amarus* Schum & Thonn.— an hepatoprotective plant. Environ Monit Assess 147:307–315

Rao RS, Ravishankar GA (2002) Plant tissue cultures; chemical factories of secondary metabolites. Biotechnol Adv 20:101–153

Roberts MF (1988) Medicinal products through plant biotechnology. In: Robins RJ, Rhodes MJC (eds) Manipulating secondary metabolism in culture. University Press, Cambridge, pp 201–216

San B, Karakurt Y, Donmez F (2015) Effects of Thidiazuron and activated charcoal on in vitro shoot proliferation and rooting of Myrtle (*Myrtus communis* L.) J Agric Sci 21(2):177–183

Santos A, Campos R, Miguel O, Filho V, Siani A, Yunes R, Calixto J (2000) Antinociceptive properties of extracts of new species of plants of the genus *Phyllanthus* (Euphorbiaceae). J Ethnopharmacol 72:229–238

Shang XF, He XR, He XY, Li MX, Zhang RX, Fan PC, Zhang QL, Jia ZP (2010) The genus *Scutellaria* an ethnopharmacological and phytochemical review. J Ethnopharmacol 128:279–313

Sharma PC, Yelne MB, Dennis TJ (2001) Database on medicinal plants used in ayurveda, vol 3. Central Council for Research in Ayurveda and Siddha, Delhi, pp 512–536

Siddique I, Anis A (2007) Rapid micropropagation of *Ocimum Basilicum* using shoot tip explants precultured in thidiazuron supplemented liquid medium. Biol Plant 51:787–790

Svetla DY, Sara G, Ervin F, Simcha LY, Moshe AF (2003) Auxin type and timing of application determine the activation of the developmental program during in vitro organogenesis in apple. Plant Sci 165:299–309

Toppel G, Witte L, Riebesehl B, Von Borstel K, Hartman T (1987) Alkaloid patterns and biosynthetic capacity of root cultures from some pyrrolizidine alkaloid producing Senecio spp. Plant Cell Rep 6:466–469

Tripathi F, Tripathi JN (2003) Role of biotechnology in medicinal plants. Trop J Pharm Res 2(2):243–253

Ulbrich B, Wiesner W, Arens H (1985) Large scale production of rosmarinic acid from plant cell cultures of *Coleus blumei* Benth. In: Neumann KH et al (eds) Primary and secondary metabolism of plant cell cultures. Springer, Berlin, pp 293–303

Unander DW, Webster GL, Blumberg BS (1995) Usage and bioassays in *Phyllanthus* (Euphorbiaceae). IV. Clustering of antiviral uses and other effects. J Ethnopharmacol 45:1–18

Verpoorte R, Contin A, Memelink J (2002) Phytochem Rev 1:13. https://doi.org/10.1023/A:1015871916833

Vijaya SN, Udayasri PVV, Aswani KY, Ravi BB, Phani PY, Vijay VM (2010) Advancements in the production of secondary metabolites. J Nat Prod 3:112–123

Wink M, Alfermann AW, Franke R, Wetterauer B, Distl M, Windhovel J, Krohn O, Fuss E, Garden H, Mohagheghzaden A, Wildi E, Ripplinger P (2005) Sustainable bioproduction of phytochemicals by plant in vitro cultures: anticancer agents. Plant Genet Resour 3:90–100

Wyk BEV, Wink M (2004) Medicinal plants of the world. Briza, Pretoria

Yeoman MM, Yeoman CL (1996) Tansley review no. 90 manipulating secondary metabolism in cultured plant cells. New Phytol 134:553–569

Zhao J, Zhu WH, Hu Q, Guo YQ (2001) Compact callus cluster suspension cultures of *Catharanthus roseus* with enhanced indole alkaloid biosynthesis. In-vitro Cell Dev Biol-Plant 37:68–72

28 Thidiazuron: An Effective Plant Growth Regulator on In Vitro Cloning of Slow-Growing Economic Rattan Palms in Southern Western Ghats for Eco-restoration and Consistent Utilization

T. S. Preetha, A. S. Hemanthakumar, and S. Seeni

Abstract

Rattans, the spiny climbing palms in Southeast Asia, are the fast depleting high-value non-wood forest produce primarily used for making furniture, baskets and handicraft items contributing significantly to the rural economy in the provinces where they occur. The most seriously affected species in the Western Ghats are *Calamus thwaitesii, C. rotang*, *C. travancoricus, C. nagabettai*, etc. which despites its wide distribution is indiscriminately exploited to the extent that mature useful canes are no more available. Since seed availability is scarce and traditional propagation methods are insufficient to meet the demand, in vitro cloning holds great promise for large-scale production and consistent utilization of planting material through eco-restoration activities. However, in vitro cloning of rattan palms is not thoroughly worked out, and there are only few isolated reports on micropropagation of rattan palms wherein the effectiveness of purine-based cytokinins such as BA or kinetin has been ruled out with very slow and little regeneration efficiency. The original work presented here reports the efficiency of thidiazuron (TDZ), a potent multidimensional cytokinin capable of fulfilling both the cytokinin and auxin requirements of various regeneration responses, viz. new meristem formation, and promotes multiple shoot induction from pre-existing meristems of elite rattan palms with special reference to *C. rotang*. TDZ has played a significant role to speed up inducing multiple shoot formation from offshoot/sucker-derived shoot tips with greater efficiency than other cytokinins (BA, 2-iP) in micro-cloning of rattan palms. The TDZ-induced

T. S. Preetha
Department of Botany, University College, Thiruvananthapuram, Kerala, India

A. S. Hemanthakumar (✉) · S. Seeni
Biotechnology and Bioinformatics Division, Jawaharlal Nehru Tropical Botanic Garden and Research Institute, Palode, Thiruvananthapuram, Kerala, India

N. Ahmad, M. Faisal (eds.), *Thidiazuron: From Urea Derivative to Plant Growth Regulator*, https://doi.org/10.1007/978-981-10-8004-3_28

in vitro multiplication scheme, developed for the first time in rattan palm tissue culture, can be gainfully employed for the focused production and supply of elite female rattan palms for open seeding and commercial plantation activities.

Keywords
Rattan palm · *Calamus* · In vitro cloning · Shoot tip · Thidiazuron · Reintroduction

28.1 Introduction

Rattans or canes are a unique and versatile group of prickly climbing or trailing palms with characteristic scaly fruits and solid stem classified under the Lepidocaryoid major group of the palm family. They include about 14 genera and 600 species distributed in the forests of Indo-Malayan and African regions. In India, they are represented by 5 genera with a total of 51 species mostly distributed in the tropical evergreen, semievergreen and moist deciduous forests of the Eastern Himalayas, Western Ghats and Andaman and Nicobar group of islands. They are variously described as the most important non-timber forest product, monocotyledonous wood, nonconventional wood, poor man's wood, fashion-proof forest product and environment-friendly renewable resource. They have also remained very much a part of the culture and village life of the provinces where they occur. Humans have used rattan for livelihood and subsistence for many centuries throughout the documented history of mankind (Anonymous 1983; Sarma 1989; Manokaran 1990; Tombac and Lapis 1993; Johnson et al. 1997; Sastry 2000). They carry remarkable social and economic importance due to their unique characteristics such as strength, durability, elasticity, lightness, lustrous brown colour look and bending ability which make them true 'green gold' (Mohan Ram and Tandon 1997). However, despite their unlimited aesthetic and utility value, these 'poor cousins' of bamboos have largely remained underdeveloped, much maligned, little appreciated and totally ignored resource of any tropical forestry development.

Canes are inexpensive rare materials for making traditional furniture and baskets by the rural dwellers. They find a variety of traditional and nontraditional uses on account of their remarkable beauty, pliability, light weight, high strength to weight ratio, lengths in which they can be obtained and bending ability, all of which enhance their aesthetic value. As a high-value fashion-proof commodity of booming 'multi-million dollar industry', it is much sought after for designer trade goods to grace the homes of urbanites the world over. The world demand for well-processed good quality canes is so high that even trebling the present supplies will not meet the requirements (Menon 1980). Unfortunately, natural forests are the only source of canes even today. During the last 20 years, over-exploitation of the rattans from the Western Ghats region has led to the disappearance of most of the species from all accessible areas. Since reproduction of canes is dependent on seeds, poor seed set contributes significantly to poor regeneration of the species in the forests where they occur. Species like *C. thwaitesii*, *C. rotang*, *C. travancoricus*, *C. nagabetttai*,

C. dransfieldi and *C. vattayila* are now seen in very restricted localities with miniscule number in their populations. However, many species are disappearing from their natural habitats due to (a) unscientific harvesting, (b) indiscriminate exploitation, (c) biotic interference and (d) habitat destruction/fragmentation and consequent microclimatic changes. It is widely recognized that in cases where seed availability is scarce or traditional vegetative propagation methods are impractical, micropropagation through tissue culture is an attractive alternative (Wochok 1981) for rapid in vitro multiplication of highly useful canes so that supply of planting materials as part of a revegetation programme and sustained deliveries of quality raw materials to the industry would be a reality at least in the future.

As far as rattan palm is concerned, shoot induction through conventional propagation methods is much slow, and similar response was also noticed by the use of commonly used cytokinins, viz. BA, 2-iP, kinetin, etc. in nonconventional propagation methods. However in the present study, a highly appreciable rapid response with tenfold multiplication was observed in TDZ-supplemented MS medium. Therefore this multidimensional plant growth regulator is used to achieve in vitro cloning, eco-restoration and consistent utilization of the rattan palms particularly in *C. rotang*.

28.2 In Vitro Cloning of Rattan Palms

28.2.1 Plant Materials

Three economically important species of rattan palms, viz. *Calamus thwaitesii*, *C. rotang*, and *C. travancoricus* collected from different forest segments in the Western Ghats of Kerala (India), were used as the study materials for the present investigation. Among these species, *C. rotang*, a good quality cane distributed to the plain along the backwaters and coasts was chosen for the detailed demonstration. *C. rotang* is mainly used for making furniture and basket industry, and its fruits are used in the treatment for cancer (Chakraberti and Mukerji 1968). The powdered seeds are useful for treating abdominal ulcer (Hartwell 1970), while the roots are useful for treating snake bites, antidysenteric, antibilious, tonic, febrifuge and depuration purposes.

28.2.2 Collection of Suckers/Offshoots

Suckers or offshoots (3–6 years old) were the main source of explants for rattan palm tissue culture (Fig. 28.1a). The suckers of the candidate species *C. rotang* were collected from estuaries of Ashramam (Kollam district) of Kerala State Forest Department. The suckers were collected with maximum number of roots with a ball of earth and immediately defoliated to avoid evapotranspiration. They were packed in gunny bags and without exposure to direct sun brought to the institute, planted in pure river sand medium in 30 cm diameter pots and reared in the mist house for

Fig. 28.1 In vitro cloning of *Calamus rotang*. (**a**) Offshoots of *C. rotang*. (**b**) Dissected out shoot tip. (**c**) Axillary shoot initiation in MS medium supplemented with 5 mgl^{-1} BA. (**d**, **e**) Shoot bud initiation from the pre-existing buds in the base of rhizomatous portion of shoot tip cultured in MS medium augmented with 0.2 mgl^{-1} TDZ. (**f**) Shoot bud elongation through subculture passage at 6-week interval in MS medium augmented with 0.2 mgl^{-1} TDZ. (**g**) Abnormal shoots formed around the shoot tip in MS medium augmented with higher concentration (>1 mgl^{-1}) of TDZ.

2–4 weeks with frequent irrigation. These offshoots served as the source of explants for in vitro cloning. It is a different matter that unlike other palms, the rattans are extremely thorny making it difficult to extract the shoot tips from the growing palm or offset. Again in an actively growing offshoot, the shoot tip with the basal rhizomatous part is so deeply located amidst thorny leaves that it has to be fished out, making the decontamination procedures rigorous and difficult to obtain meaningful number of surface decontaminated, responding shoot tips often collected from accessible forests.

28.2.3 Surface Sterilization

The rigorous and stepwise surface decontamination procedure followed during different stages of shoot tip explant preparation from the intact offshoots helped to obtain up to 70% contamination-free explants that responded to culture conditions. This multistep procedure essentially consisted of:

(a) Agitating the downsized offshoots (having only three leaves) in a solution of 5% sodium hypochlorite and 0.5% Teepol for 20 min
(b) Immersing and decontaminating the further trimmed and downsized offshoot in 5% sodium hypochlorite solution for 10 min
(c) Treating the exposed shoot tip (~0.5 cm) having basal ~1.5 cm rhizomatous base in 0.1% (w/v) streptomycin sulphate solution for 40 min
(d) Final passage through 0.1% $HgCl_2$ (w/v) for 5 min

Each step was accompanied by liberal washing in sterile distilled water or sterile double distilled water. Further each step was found important, and deviation from it resulted in either increased percentage contamination or lethality of the tissue. Browning of the explant during the course of downsizing was common. Bacterial contamination and severe browning were the major problems, which cause up to 30% loss of the shoot tip explants. In some explants, browning and release of exudates occurred together, while in others browning of the tissue was independent of exudation.

Fig. 28.1 (continued) (**h**) Abnormal shoots turned brown and necrosed after 10 weeks in media supplemented with 2 mgl^{-1} TDZ. (**i**) Shoot multiplication observed in MS medium supplemented with 0.4 mgl^{-1} BA and 0.1 mgl^{-1} each of TDZ and NAA. (**j**) Proliferative mass of marved shoots observed after 6 months through continuous subculture of shoot tip at 6-week intervals in MS media supplemented with 0.2 mgl^{-1} TDZ. (**k**) Mixture of normal and callogenic shoots formed from the base of sucker after 6 months through regular subculture at 6-week intervals in MS medium supplemented with 0.2 mgl^{-1} TDZ. (**l, m**) Shoot elongation in normal shoots followed by rhizogenesis in half MS medium supplemented with 0.2 mgl^{-1} each of BA, 2-iP and 0.1 mgl^{-1} NAA. (**n**) 12-month-old rooted plants ready for deflasking. (**o**) Deflasked plants for hardening. (**p**) Hardened and established plants ready for field transfer. (**q**) Two-year-old field established plant

The shoot tip with basal rhizome part obtained and subjected to vigorous decontamination had to be thinned with the removal of young/primordial leaves and leaf sheath one by one during the disinfection process. This had to be done due to the underground, soiled nature of the explants which are expectedly loaded with microorganisms of all sorts. Underground plant parts harbouring different microorganisms and the need to surface decontaminate them using rigorous procedures are reported in anthurium (Teng 1997), taro (Yongwei Li et al. 2002) and *Zantedeschia aethiopica* (Kritzinger et al. 1998). In such cases, multistep disinfection process is usually followed. Even then, only 70–80% of the shoot tips could be saved from infection. Based on the results obtained, it is safe to infer that at least the shoot tip explants of the rattan palms are free from endophytic bacteria and fungi. The contamination-free explants obtained compare well with the reported success rate of decontamination in anthurium (Geier 1986; Teng 1997), taro (Tim et al. 1990), mango (Thomas and Ravindra 1997) and rattan palms (Hemanthakumar 2011). However, the observed rate of bacterial contamination higher than fungal ones indicates certain level of bacterial colonization in the outer parts of the rhizome is still possible, even if such associations are not strictly endophytic. If some of these microbes have co-evolved along with the plants, their decontamination process per se may be difficult (George 1993). Whatever may be the nature of the plant-microbe interaction in rattan palms, it is evident that microbes associated with the explants grow faster than the plant cells, and if allowed to continue, the tissue may not grow and eventually get plagued with bacterial colonies, phenolic oxidants and necrotic responses. The two-step (5% sodium hypochlorite with 0.5% Teepol) treatment with bleaching agent, fairly long treatment (40–80 min) of the downsized explant with 0.1% streptomycin sulphate and the last treatment with 0.1% $HgCl_2$ for 4–5 min together offered a reasonable decontamination protocol for different species, the differences in size of the explants between the species notwithstanding. The shoot tip explants of *C. rotang* were always larger than those of *C. travancoricus*. To be successful with the decontamination procedure, the explants had to be collected during summer months (February–March), and those collected during rainy season obviously showed higher percentage contamination due to possible penetration of microbes along with rainwater and subsequent colonization of parts including crown and leaf sheath making the decontamination process difficult and cumbersome.

28.3 Culture Initiation

For culture initiation, the suckers with exposed shoot tips (~0.5–0.7 cm) and rhizome part (~1.5 cm) (Fig. 28.1b) were inoculated to full-strength Murashige and Skoog (MS) medium (Murashige and Skoog 1962) augmented with required concentrations and combinations of plant growth regulators, viz. 0.2–0.4 mgl^{-1} N6-benzylaminopurine (BA), 0.05–2.0 mgl^{-1} thidiazuron (TDZ) and 0.1 mgl^{-1} α-naphthaleneacetic acid (NAA); and the pH was adjusted to 5.8 with 1N NaOH or HCl. The gelling of the medium was done by boiling with 0.5%

(w/v) agar. Then 60 ml^{-1} aliquot media were dispensed into 250 ml Erlenmeyer flasks/bottles and closed tightly with nonabsorbent cotton plugs or polypropylene caps as the case may be. The media was sterilized by autoclaving at 121 °C and 1.1 $kgcm^{-2}$ pressure for 20 min. Subculturing was made every 6 weeks. In certain cases 2–3 transfers each of 2-week duration had to be followed to minimize exudation and oxidative damage. Unless otherwise mentioned, all the cultures were incubated at 24 ± 2 °C under 12 h photoperiod with a photon flux intensity of 50–60 μEm^{-2} s^{-1} provided by cool white fluorescent tubes (Philips India Ltd., Mumbai). All the cultures were observed at weekly intervals and data were recorded after 4–16 weeks.

28.4 Comparative Effect of TDZ with Other Weak Cytokinins on Culture Initiation

The less affected shoot tip explants cultured in MS agar nutrient medium responded with elongation or emergence of new axillary buds or both, depending on the growth regulator supplementation. Seventy percent of the explants cultured in basal medium showed shoot tip elongation to attain 1.0–3.0 cm length in 6 weeks followed by decline and degeneration of the entire shoot in 12 weeks. Addition of cytokinin was essential to sustain the growth of the apical bud and also to induce axillary bud formation, but the degree of these responses varied between types and concentrations of the cytokinins. The widely used cytokinins, BA and 2-iP, reversed the elongation of shoot tip only marginally, and the percentage of axillary shoot/bud release even at the optimal concentration of these cytokinins (5.0 mgl^{-1}) was only 10–23% (Fig. 28.1c) (Table 28.1). Elongation of the apical bud in the basal medium or in medium supplemented with low concentrations (0.5–1.0 mgl^{-1}) of BA or 2-iP occurred from the fourth day onwards. The inverse relationship between apical bud growth and axillary bud release became evident when TDZ was used. This cytokinin at concentrations ranging from 0.2 to 2.0 mgl^{-1} suppressed apical bud elongation and stimulated maximum percentage (66.67%) and number (3.0 ± 0.20) of axillary shoot formation at 0.2 mgl^{-1} concentration. Shoot bud initiation was axillary in nature as it occurred from the pre-existing buds in the base of the shoot tip, more specifically in the rhizomatous portion (Fig. 28.1d, e). The TDZ-induced emergence of small number of axillary buds was observed after 6 weeks of culture, and the buds grew into shoots of 1.6–2.5 cm only during the subculture period (Fig. 28.1f). The growth of apical bud was arrested altogether. The shoots formed in the presence of higher concentrations (>0.5 mgl^{-1}) of TDZ were also axillary in nature though they looked abnormal at concentrations exceeding 1.0 mgl^{-1} (Fig. 28.1g). Such shoots induced by 2.0 mgl^{-1} TDZ turned brown and necrosed after 10 weeks of culture (Fig. 28.1h). Combinations of TDZ, with other weak cytokinins and NAA at the concentrations tested, neither improved the axillary shoot formation nor the quality of the shoots significantly.

Exogenous supply of cytokinins was essential to arrest longitudinal growth of shoot tip but also to keep the apical bud alive and active in different species of *Calamus*. Cytokinins other than TDZ (BA, 2-iP, kinetin) were just sufficient to

Table 28.1 Multiple shoot bud/shoot formation in shoot tip explants of *C. rotang*

PGRs (mgl^{-1})				Percentage response				Number of axillary shoots/sucker	Length of shoots (cm)
TDZ	BA	2-iP	NAA	Contamination	Browning/ necrosed	Longitudinal growth of the shoot tip	Axillary shoot formation		
0.0	–	–	–	23.33	6.67	73.33	0.00	0.00 ± 0.00	0.00 ± 0.00
0.02	–	–	–	20.00	3.33	63.33	13.33[f]	2.00 ± 0.00[a, b]	3.60 ± 0.03
0.05	–	–	–	16.67	10.00	50.00	23.33[e]	1.50 ± 0.16[b]	3.36 ± 0.03
0.10	–	–	–	16.67	16.67	23.33	43.33[c]	2.00 ± 0.18[a]	2.62 ± 0.09
0.20	–	–	–	13.33	10.00	10.00	66.67[a]	3.00 ± 0.20[a]	2.26 ± 0.11
0.30	–	–	–	23.33	13.33	6.67	56.67[b]	2.16 ± 0.17[a, b]	1.92 ± 0.13
0.50	–	–	–	10.00	20.00	3.33	66.67[a]	1.86 ± 0.15[b]	1.60 ± 0.17
1.0	–	–	–	16.67	10.00	3.33	70.00[a]	1.71 ± 0.11[b]	1.26 ± 0.13
2.0	–	–	–	23.33	26.67	3.33	46.67[c]	1.80 ± 0.19[b]	1.02 ± 0.07
–	0.5	–	–	20.00	3.33	76.67	0.00	0.00 ± 0.00	0.00 ± 0.00
–	1.0	–	–	16.67	10.00	73.33	0.00	0.00 ± 0.00	0.00 ± 0.00
–	3.0	–	–	13.33	16.67	63.33	6.67[g]	2.00 ± 0.00[a, b]	3.90 ± 0.03
–	5.0	–	–	23.33	20.00	33.33	23.33[e]	2.00 ± 0.31[a, b]	3.32 ± 0.16
–	7.0	–	–	20.00	36.67	33.33	10.00[f, g]	2.00 ± 0.00[a, b]	3.15 ± 0.11
–	–	0.5	–	13.33	6.67	80.00	0.00	0.00 ± 0.00	0.00 ± 0.00
–	–	1.0	–	16.67	6.67	76.67	0.00	0.00 ± 0.00	0.00 ± 0.00
–	–	3.0	–	13.33	10.00	76.66	0.00	0.00 ± 0.00	0.00 ± 0.00
–	–	5.0	–	23.33	23.33	43.33	10.00[f, g]	2.00 ± 0.00[a, b]	3.75 ± 0.08
–	–	7.0	–	23.33	20.00	46.67	10.00[f, g]	1.00 ± 0.00[b]	3.70 ± 0.00
–	0.2	0.2	–	20.00	20.00	63.33	0.00	0.00 ± 0.00	0.00 ± 0.00
–	0.5	0.5	–	16.67	13.33	63.33	10.00[f, g]	3.00 ± 0.00[a]	3.13 ± 0.09
–	1.0	1.0	–	20.00	16.67	53.33	10.00[f, g]	2.00 ± 0.00[a, b]	3.15 ± 0.17
–	0.2	0.2	0.1	16.67	13.33	43.33	26.67[e]	1.67 ± 0.13[b]	3.12 ± 0.10
–	0.5	0.5	0.1	30.00	13.33	40.00	16.67[f]	2.00 ± 0.31[a]	2.65 ± 0.22

0.1	0.4	–	0.1	13.33	10.00	33.33	43.33^{c}	2.75 ± 0.21^{a}	2.81 ± 0.09
0.1	0.4	–	0.2	30.00	20.00	13.33	36.67^{d}	$2.00 \pm 0.18^{a,b}$	2.64 ± 0.11
0.2	0.4	–	0.1	20.00	26.67	13.33	$40.00^{c,d}$	1.80 ± 0.09^{b}	2.39 ± 0.09
–	0.5	–	0.2	13.33	13.33	73.33	0.00	0.00 ± 0.00	0.00 ± 0.00
–	1.0	–	0.5	26.67	20.00	53.33	0.00	0.00 ± 0.00	0.00 ± 0.00
–	3.0	–	1.0	16.67	23.33	40.00	$20.00^{e,f}$	$2.00 \pm 0.00^{a,b}$	2.65 ± 0.14
–	5.0	–	1.0	30.00	33.33	23.33	13.33^{f}	3.00 ± 0.00^{a}	2.80 ± 0.07
–	3.0	–	2.0	30.00	26.67	30.00	13.33^{f}	$2.00 \pm 0.00^{a,b}$	2.60 ± 0.03
–	5.0	–	2.0	33.33	33.33	23.33	$10.00^{f,g}$	3.00 ± 0.00^{a}	2.63 ± 0.07

Data represents mean ± SE of 20 replicates repeated thrice, recorded after 12 weeks of culture. Values followed by the same letter in the superscript do not differ significantly at $p \leq 0.05$ based on ANOVA and LSD t-test reverse the longitudinal growth of the apical bud (apical dominance) but were short of inducing high percentage axillary bud proliferation. In *C. rotang*, BA and 2-iP could not reverse apical dominance completely, and in *C. thwaitesii*, high concentration (3.0 mgl^{-1}) of BA was required to induce minimal additional bud formation (Hemanthakumar et al. 2014). Poor caulogenic response was noticed in combinations of cytokinins without TDZ. The inability of conventional cytokinins (BA, 2-iP) in isolation or in combination with auxin to induce multiple shoot formation at reasonable levels made it compulsory to try TDZ individually and in combinations. TDZ was very effective to induce multiple shoot formation though its concentration required for optimum bud release and frequencies varied between species. Lowest concentration of TDZ (0.5 mgl^{-1}) was required to induce the formation of maximum shoots in *C. rotang* within 12 weeks. The differential frequencies, number of buds formed and the period required for shoot bud initiation may indicate the genotypic differences between the species (Mullins et al. 1997; Lyyra et al. 2006). However, in all the species, in the presence of TDZ, apical dominance was completely arrested, and the shoots initiated did not elongate beyond 1.5 cm. Additional shoot bud initiation was axillary as it occurred from the base of the shoot tip particularly in the rhizomatous portion. The inverse relationship between apical bud growth and axillary bud release as observed and stunted growth of even the newly formed axillary shoots could be only related to the known inhibitory influence of TDZ on the longitudinal growth of existing and newly formed shoots (Huetteman and Preece 1993).

Thidiazuron is capable of fulfilling both the cytokinin and auxin requirements of various regeneration responses (Jones et al. 2007). It can induce new meristem formation, promote shoot development from pre-existing meristems and induce adventitious bud regeneration in a number of species including recalcitrant woody plants (Murthy et al. 1998; Cuenca et al. 2000; Bunn et al. 2005). One of the most effective uses of TDZ has been in the regeneration of woody plant species in which organogenesis was only possible with high concentrations of adenine-type cytokinins or in those species in which these compounds were ineffective (Murthy et al. 1998). The results presented in this study are consistent with these observations, as TDZ has played an important role in inducing multiple shoot formation from shoot tips derived from offshoots/suckers with greater efficiency than other cytokinins (BA, 2-iP).

The indispensability of TDZ for shoot initiation was again confirmed when combinations of cytokinins on shoot initiation were tested. Only combinations involving TDZ proved good for bud initiation in 43.33% of the shoot tip in *C. rotang*. Among the various concentrations and combinations of PGRs tested in this species, maximum of 2.75 ± 0.21 shoots with somewhat elongated shoots (2.81 ± 0.09 cm) were initiated after 12 weeks of culture in the presence of 0.4 mgl^{-1} BA and 0.1 mgl^{-1} each of TDZ and NAA. The beneficial effect of NAA for obtaining somewhat elongated shoots even in the presence of the cytokinins was evident. In certain species, combinations involving cytokinins and auxins act synergistically to obtain normal lengthy shoots during culture initiation/multiplication. This kind of observation was noticed in *Robinia pseudoacacia* L., *Sorbus aucuparia* L., *Tilia cordata* Mill.

(Chalupa 1987) and *Yucca aloifolia* (Atta-Alla and Van Staden 1997) and *Citrus grandis* (Paudyal and Haq 2000). Overall analyses of the results on shoot initiation reveal that meaningful initiations of shoot buds in culture initiation experiments were possible only with TDZ in isolation or in combination with other PGRs. Shoots induced by higher (2.0 mgl^{-1}) concentrations of TDZ were always distorted or fasciated as observed in *C. rotang*. Other cytokinins are too weak to initiate buds in shoot tip cultures of *Calamus*. TDZ has been used with great success in such species as *Malus domestica* (Elobeidy and Korban 1988; Fasolo et al. 1989), *Pyrus* sp. (Chevreau et al. 1989), *Rhododendron* (Preece and Imel 1991), *Rubus* (Cousineau and Donnelly 1991) and *Populus* sp. (Russell and Mc Cown 1988) wherein the shoots initiated remained stunted. Since the number of buds formed even in the presence of TDZ was relatively small, that too, after prolonged period of incubation, the rattans form a specialized group of palms, where even repeated trials with strong cytokinins like TDZ do not yield good number of shoot initiation required of a mass multiplication scheme. There is no wonder why previous workers have not met with great success in vitro mass multiplication of shoots in *Calamus* species (Yusoff et al. 1985). This is also true with other workers including Patena et al. (1984), Umali-Garcia (1985), Barba et al. (1985), Gunawan and Yani (1986), Yusoff (1989), Dekkers and Rao (1989) and Padmanabhan and Ilangovan (1989, 1993) who worked on different species of rattans. This work has established that it is possible to clone rattan palm through tissue culture using TDZ as the principal cytokinin; however, initiation of large number of shoot buds required of mass multiplication schemes is a distant reality.

28.4.1 Shoot Multiplication Through Subculture Passages

To study the influence of hormonal regimes on shoot multiplication, the emergent axillary shoot buds obtained after 12 weeks of culture were dissected out of the basal rhizomatous part and transferred to the same nutrient medium (full-strength MS medium) with varied hormonal regimes [TDZ (0.05–1.0 mgl^{-1}); BA (0.2–0.5 mgl^{-1})+2-iP (0.2–0.5 mgl^{-1})+NAA (0.1 mgl^{-1}), BA (0.4 mgl^{-1})+TDZ (0.1–0.2 mgl^{-1})+NAA (0.1–0.2 mgl^{-1})]. The rhizomatous parts free of axillary buds were again transferred to the culture initiation medium or medium supplemented with changed hormonal concentrations for additional bud emergence. Buds proliferated were isolated and subcultured at regular intervals through passages each of 6-week duration.

All the axillary shoots/buds dissected out of the primary explant cultures (offshoot tip/sucker tip) and subcultured in the presence of majority of the selected hormone supplements were not morphogenetically active (Table 28.2). The percentage of the explants taking part in multiplication varied from 20 to 93.33 with relatively higher percentage response recorded in the second subculture passage compared to the first. In none of the treatments, shoot elongation was impressive, and the presence of the auxin in the combination did not seem to contribute to significant shoot elongation. Invariably all the shoots cultured in the presence of BA

Table 28.2 Multiplication of shoots through subculture of axillary shoots/buds proliferated upon the primary explants of *C. rotang*

PGRs (mgl^{-1})				1st subculture passage		2nd subculture passage		Total no. of shoots/explant after two subcultures	Length of shoot after two subculture passages
TDZ	BA	2-iP	NAA	% response	No. of shoots	% response	No. of shoots		
0.1	–	–	–	46.67^{c}	2.00 ± 0.16^{b}	73.33^{c}	2.14 ± 0.15^{b}	4.14^{c}	1.43 ± 0.14
0.2	–	–	–	63.33^{a}	3.00 ± 0.20^{a}	93.33^{a}	3.11 ± 0.23^{a}	6.11^{a}	1.06 ± 0.10
0.2	0.4	–	0.1	53.33^{b}	1.60 ± 0.11^{c}	83.33^{b}	1.67 ± 0.15^{c}	3.27^{d}	1.36 ± 0.13
0.1	0.4	–	0.2	56.67^{b}	2.33 ± 0.16^{b}	66.67^{d}	2.71 ± 0.28^{b}	5.04^{b}	1.60 ± 0.13
0.1	0.4	–	0.1	63.33^{a}	3.16 ± 0.17^{a}	80.00^{b}	3.75 ± 0.20^{a}	6.91^{a}	2.15 ± 0.13
–	0.2	0.2	0.1	33.33^{d}	2.25 ± 0.21^{b}	43.33^{e}	2.50 ± 0.13^{b}	4.75^{c}	2.30 ± 0.13
–	0.5	0.5	0.1	20.00^{e}	1.50 ± 0.11^{c}	26.67^{f}	2.00 ± 0.18^{b}	3.50^{d}	1.93 ± 0.17

Data represents mean ± SE of 20 replicates repeated thrice, recorded after every 6 weeks of culture. Values followed by the same letter in the superscript do not differ significantly at $p \leq 0.05$ based on ANOVA and LSD t-test alone showed signs of decline in 4 weeks and necrosis in 8–12 weeks. Presence of TDZ (0.1–0.2 mgl^{-1}) was essential for shoot multiplication, but 2–3 shoots developed from the base of each shoot tip during each of the subculture passage remained stunted and distorted. The number of shoots proliferated upon the subcultured shoots/buds in the presence of both TDZ and BA was relatively few, short and healthy. A combination of 0.1 mgl^{-1} TDZ, 0.4 mgl^{-1} BA and 0.1 mgl^{-1} NAA was adjudged the best to induce multiplication of 3–4 shoots (2.15 ± 0.13 cm length) from each subcultured shoot through the two subculture passages each of 6-week duration (Fig. 28.1i). Even in this optimized combination of PGRs, subculture of the isolated shoots/buds beyond three cycles resulted in multiplication of shoots of which 20–30% were abnormal having rosette-type wrinkled leaves, unexpanded threadlike leaves and colloid protrusions as the case may be. However, the shoot buds raised in the presence of 0.2 mgl^{-1} TDZ and subcultured as such at 6-week intervals in the same medium produced a proliferative mass of fasciated shoots (Fig. 28.1j) as well as a mixture of normal and callogenic shoots from the base of the suckers (Fig. 28.1k).

Combinations of hormones tried in *C. rotang* were more desirable as the shoots obtained were somewhat elongated and negative influence of TDZ in inducing fasciated shoot development and poor rooting could be avoided to some extent. Invariably, the multiplication responses in terms of percent shoot multiplication and the number of shoots formed were less in the first subculture passage compared to the second subculture. This might be due to the better acclimatization and consequent increase in the caulogenic responses of the shoots in the second subculture. Since repeated subculture in the presence of 0.01–0.2 mgl^{-1} TDZ beyond three to four subcultures led to formation of abnormal/distorted shoots, combinations of hormones including TDZ were better desired than the individual concentration of TDZ for the in vitro clonal multiplication of this species. The inhibitory influence of TDZ on shoot elongation was further confirmed when its replacement by 2-iP led to significant increase in shoot length and not so in shoot number. Both percentage response and number of multiplied shoots were reduced if TDZ was replaced by 2-iP in a combination of BA-2-iP-NAA (instead of BA-TDZ-NAA).

28.4.2 Shoot Elongation

For shoot elongation, shoot cultures of 2–3 cm length obtained after different subculture passages were isolated either individually or as groups (3–5 shoots) and transferred to full and half strength MS agar/liquid medium free of hormones or medium containing a combination of 0.2–0.5 mgl^{-1} each of BA and 2-iP, 0.1 mgl^{-1} NAA or 0.4 mgl^{-1} BA and 0.1 mgl^{-1} each of TDZ and NAA. Presence of TDZ in the medium containing full- or half-strength salts promoted shoot proliferation but not shoot elongation. Under selected combinations of other cytokinins (BA, 2-iP) and NAA, shoots cultured in half-strength medium showed better elongation than those cultured in full-strength basal and half-strength basal media (Table 28.3). However, the salt drained, yet PGR supplemented half-strength elongation medium did not

Table 28.3 Shoot elongation of *C. rotang* through two subculture passages

Media used	PGRs (mgl^{-1})			1st subculture passage		2nd subculture passage	
	BA	2-iP	NAA	No. of additional shoots/explant	Length of shoot (cm)	No. of additional shoots/explant	Length of shoot (cm)
Full MS	–	–	–	1.40 ± 0.12^{b}	3.95 ± 0.50	1.33 ± 0.13^{b}	5.28 ± 0.64
	0.5	0.5	0.1	2.50 ± 0.12^{a}	2.77 ± 0.43	2.33 ± 0.18^{a}	3.02 ± 0.48
	0.2	0.2	0.1	1.60 ± 0.26^{b}	3.77 ± 0.46	2.00 ± 10.8^{a}	3.57 ± 0.54
½ MS	–	–	–	1.50 ± 0.13^{b}	4.33 ± 0.62	1.33 ± 0.13^{b}	5.33 ± 0.87
	0.5	0.5	0.1	1.75 ± 0.21^{b}	4.25 ± 0.41	1.50 ± 0.22^{b}	4.56 ± 0.71
	0.2	0.2	0.1	1.50 ± 0.16^{b}	4.90 ± 0.48	1.50 ± 0.16^{b}	5.56 ± 0.72

Data represents mean ± SE of 20 replicates repeated thrice, recorded after every 6 weeks of culture. Values followed by the same letter in the superscript do not differ significantly at $p \leq 0.05$ based on ANOVA and LSD t-test

support additional shoot formation from the base of the transferred shoots as much as the full-strength medium. The same trend was observed during the second subculture at the end of which the mean number and length of additional shoots formed per shoot respectively were 1.50 ± 0.16 and 5.56 ± 0.72 cm in half-strength and 2.00 ± 1.08 and 3.57 ± 0.54 cm in full-strength media both containing 0.2 mgl^{-1} each of BA, 2-iP and 0.1 mgl^{-1} NAA. Apart from shoot elongation, these hormonal combinations in half-strength medium induced rhizogenesis in 40% of the shoot cultures (Fig. 28.11, m) while in the full-strength medium and half-strength medium containing higher concentrations of cytokinin, the same was reduced to 10–20%.

However, shoot buds multiplied through subculture two to four times had to be necessarily transferred to full-/half-strength MS medium with/without hormones for shoot elongation. Although half-strength medium devoid of hormones also favoured shoot elongation, the number of shoots harvested was relatively less presumably due to the dependence of the young shoots on hormones for multiplication. The inhibitory effect of high salts present in full-strength MS medium on shoot elongation and consequent choice of half-strength nutrient medium for shoot elongation is also reported by earlier workers in different systems, viz. *Syzygium alternifolium* (Sha Valli Khan et al. 1999), *Eucalyptus* F_1 hybrids (Joshi et al. 2003) and *Calophyllum apetalum* (Nair and Seeni 2002). For shoot elongation, the medium has to be devoid of TDZ (Russell and MC Cown 1986; Singha and Bhatia 1988; Fasolo et al. 1989; Preece and Imel 1991; Neuman et al. 1993; Huetteman and Preece 1993; Tiwari et al. 2001; Thengane et al. 2006) but supplemented with relatively higher concentrations of other cytokinins (0.5 mgl^{-1} each of BA and 2-iP) in half-strength medium. Though reduced concentrations of relatively weaker cytokinins (BA and 2-iP) were indispensable for shoot elongation, use of full-strength MS medium contributed to increase in number of shoots each of reduced length compared to half-strength salts which promoted lengthy, but reduced number of shoots as in *C. rotang*. Again, at least two transfers of the shoots (6 + 6 = 12 weeks) to fresh medium were required for optimal shoot elongation. Invariably in the cultures, mainly due to the spillover effect from the multiplication stage, a marginal number

Table 28.4 Rooting of shoots in *C. rotang*

Auxins (mgl^{-1})	Rooting response percentage				No. of roots per shoot	Length of roots (cm)
	4 weeks	6 weeks	8 weeks	12 weeks		
IAA						
0.0	0.00	0.00	0.00	0.00	0.00 ± 0.00	0.00 ± 0.00
1.0	0.00	0.00	0.00	3.33^{g}	0.00 ± 0.00	0.00 ± 0.00
2.0	0.00	0.00	3.33^{g}	13.33^{f}	0.17 ± 0.41^{d}	6.10 ± 0.41^{a}
3.0	0.00	3.33^{e}	13.33^{f}	26.67^{e}	0.83 ± 0.75^{c}	6.08 ± 0.24^{a}
4.0	0.00	3.33^{e}	16.67^{e}	50.00^{d}	1.29 ± 0.49^{a}	6.05 ± 0.16^{a}
5.0	0.00	6.67^{d}	20.00^{d}	50.00^{d}	1.43 ± 0.53^{a}	6.08 ± 0.22^{a}
IBA						
1.0	0.00	0.00	3.33^{g}	20.00^{e}	1.00 ± 0.00^{b}	5.10 ± 0.09^{b}
2.0	0.00	3.33^{e}	10.00^{f}	26.67^{e}	1.00 ± 0.00^{b}	6.73 ± 0.36^{a}
3.0	0.00	10.00^{c}	23.33^{d}	63.33^{c}	1.25 ± 0.46^{a}	6.16 ± 0.40^{a}
4.0	0.00	6.67^{d}	43.33^{c}	83.33^{a}	1.38 ± 0.52^{a}	5.90 ± 0.41^{b}
5.0	3.33^{c}	6.67^{d}	40.00^{c}	50.00^{d}	1.51 ± 0.53^{a}	4.84 ± 0.38^{c}
NAA						
1.0	3.33^{c}	3.33^{e}	26.67^{d}	56.67^{d}	1.00 ± 0.00^{a}	4.75 ± 0.18^{c}
2.0	20.00^{b}	30.00^{b}	$60.00^{a,b}$	70.00^{b}	1.14 ± 0.38^{a}	5.84 ± 0.55^{b}
3.0	30.00^{a}	46.67^{a}	66.67^{a}	83.33^{a}	1.25 ± 0.46^{a}	5.92 ± 0.39^{b}
4.0	26.67^{b}	36.67^{b}	56.67^{b}	70.00^{b}	1.43 ± 0.53^{a}	6.42 ± 0.46^{a}
5.0	26.67^{b}	$40.00^{a,b}$	53.33^{b}	56.67^{d}	1.50 ± 0.55^{a}	5.84 ± 0.42^{b}

Data represents mean ± SE of 20 replicates repeated thrice, recorded after every 6 weeks of culture. Values followed by the same letter in the superscript do not differ significantly at $p \leq 0.05$ based on ANOVA and LSD t-test

of shoots were also added during the elongation phase. Such observations have been recorded mostly in *Vitis rotundifolia*, *Rhododendron* (Huetteman and Preece 1993) and bamboo (Singh et al. 2001).

28.4.3 Rooting of Shoots

The individual shoots/shoot clumps (3–5 shoots) of 1.5–7.0 cm length were dissected out carefully and implanted into full- and half-strength MS agar or liquid MS media augmented with various concentrations of auxins, viz. IAA, IBA and NAA (1.0–5.0 mgl^{-1}). None of the condensed (1.5–3.0 cm) shoots raised in culture media containing 2.0 mgl^{-1} TDZ and directly transferred to rooting media containing auxins were rooted even after 12 weeks of transfer. By contrast, the unrooted shoots of 3.5–5.5 cm length harvested from the full-strength or half-strength shoot elongation medium were readily rooted in the presence of the auxins (Table 28.4). Rhizogenic response of the shoots varied between the auxins, with 30% of the shoots rooted within 4 weeks in the presence of 3.0 mgl^{-1} NAA followed by 10% rooting recorded after 6 weeks in IBA and 13.33% after 8 weeks in IAA. In the order of merit, after 12 weeks, 83.33% of the shoots were rooted in 3.0 mgl^{-1} NAA and 4.0 mgl^{-1} IBA respectively and only 50% in IAA (4.0–5.0 mgl^{-1}). In all the shoots, irrespective of the auxins tried, only 1–2 roots were formed and the roots so formed were 4.6–8.3 cm long. Growth of the shoots observed in all the treatments during the rooting period was marginal (0.4–0.6 cm).

Though the supplementation of the medium with auxins was essential to get rooting of the shoots, despite the presence of auxins, shoots that were multiplied in higher concentration of TDZ (2.0 mgl^{-1}) never rooted presumably due to the anti-rhizogenic influence of this synthetic cytokinin. Percentage of rooting was significant if the shoots were treated with NAA (3.0 mgl^{-1}) or IBA (4.0 mgl^{-1}) in two transfers of 12 weeks in *C. rotang*. The remarkable similarity of 1–2 root formations in all the species tested indicated that root formation may be a predetermined character in the presence of an inducer auxin like NAA. By and large, NAA was more desirable than other auxins for safe and efficient root induction in all the species particularly in *C. rotang*. In such other species of *Calamus*, as *C. simplicifolius* (Zhang Fangqiu 1993) and *C. egregious* (Zeng Bingshan 1997), also the in vitro shoots formed only 1–2 roots which are in agreement with our observation.

28.4.4 Hardening and Nursery Establishment

In order to harden and establish the micropropagated plants, rooted plantlets of 5–10 cm length obtained after 12 months were weaned from the bottles, washed thoroughly in running tap water to remove traces of nutrient media adhering to the plantlets and treated with 0.1% Dithane M-45 for 5 min to avoid fungal contamination. The fungicide-treated plants were transferred to 10 cm diameter earthen pots and also poly bags filled with pure river sand medium and maintained in the mist house under constant irrigation at 28 ± 2 °C and 80 ± 5% RH for hardening. The clones were observed periodically and data on establishment recorded at regular intervals of time. The established 4–6-month-old clones were transferred to the nursery where they were maintained under diffused light and regular watering. After 6–12 months, nursery established plants were transferred to selected forest segments and their field performance was assessed.

Out of the 225 rooted plants (Fig. 28.1n) weaned from the flasks (Fig. 28.1o), washed and treated with Dithane M-45 fungicide, 88% (198) got established in poly bags in the mist house in 4 months (Table 28.5). Same plants transplanted in potting medium and maintained under regular irrigation in the 50% shade house showed

Table 28.5 Hardening and establishment of 12-month-old clonal plants of *C. rotang*

Deflasked plants			Plants grown to 4 months		
Reared place	No. of plantlets transferred	Length of shoot (cm)	Length of shoot (cm)	No. of new leaves	% survival
Mist house	225	7.3 ± 0.41	10.62 ± 0.26	0.6 ± 0.12	198 (88.00)
50% shade house	30	7.52 ± 0.33	10.26 ± 0.36	0.4 ± 0.12	18 (60.00)

low percentage (60%) establishment. The percentage establishment came down further, if the plants were transferred to the shade house for a long period. Some of the root-free shoots transplanted and maintained under both the conditions showed signs of survival initially but were subsequently dried up in 6–8 weeks. During the 6-month period of hardening in the mist house and subsequent 2 months in shade house, the successfully established plants produced a maximum of 1–2 leaves (Fig. 28.1p).

28.4.5 Reintroduction

The microclones already established in the nursery for 6–8 months showed conspicuous growth as evidenced from emergence of new leaves similar to the reintroduced species of *C. thwaitesii* (Hemanthakumar et al. 2013). In exceptional cases, emergence of new leaves may occur from second month onwards. The plants so raised and reintroduced into the natural forest segment of the Western Ghats during the southwest monsoon period recorded a maximum of 76.19% establishment after 2 years (Fig. 28.1q) (Table 28.6). Based on the results obtained, a protocol for in vitro clonal propagation of *C. rotang* was developed and presented in Fig. 28.2. Care should be taken to introduce the plants during the pre-monsoon showers in late May or monsoon rains in early June which facilitated easy establishment of the plants back in nature without getting desiccated. The uniformly high establishment of the plants recorded in different experimental sites further strengthened the possible utility of the methods followed for reintroduction in other parts of India as well where again rattan forms a natural resource. It should be noted that hitherto no successful studies on in vitro cloning and reintroduction of the propagated rattan plants have ever been attempted. The methods described make certain for the first time successful cloning and restoration of rattan palms. The observation that the cloned plants reintroduced into the forest segments showed uniform growth and were free from morphological and growth abnormalities further confirmed the utility of the methods for conservation and revegetation purposes. Hopefully, this kind of reintroduction practices is not a cause for objection from the point of environment, as tissue culture-mediated propagation of plants is widely accepted in horticulture and micropropagation itself is a specialized form of vegetative propagation.

Table 28.6 Reintroduction of clonal plants of *C. rotang* into selected forest segments of Western Ghats

Forest segments	Total no. of plants introduced/transferred	Plants established	Percentage establishment	Mean no. of new leaves/plant
Palode	42	32	76.19	7.7 ± 0.33
Aryankavu	23	17	73.91	6.9 ± 0.16

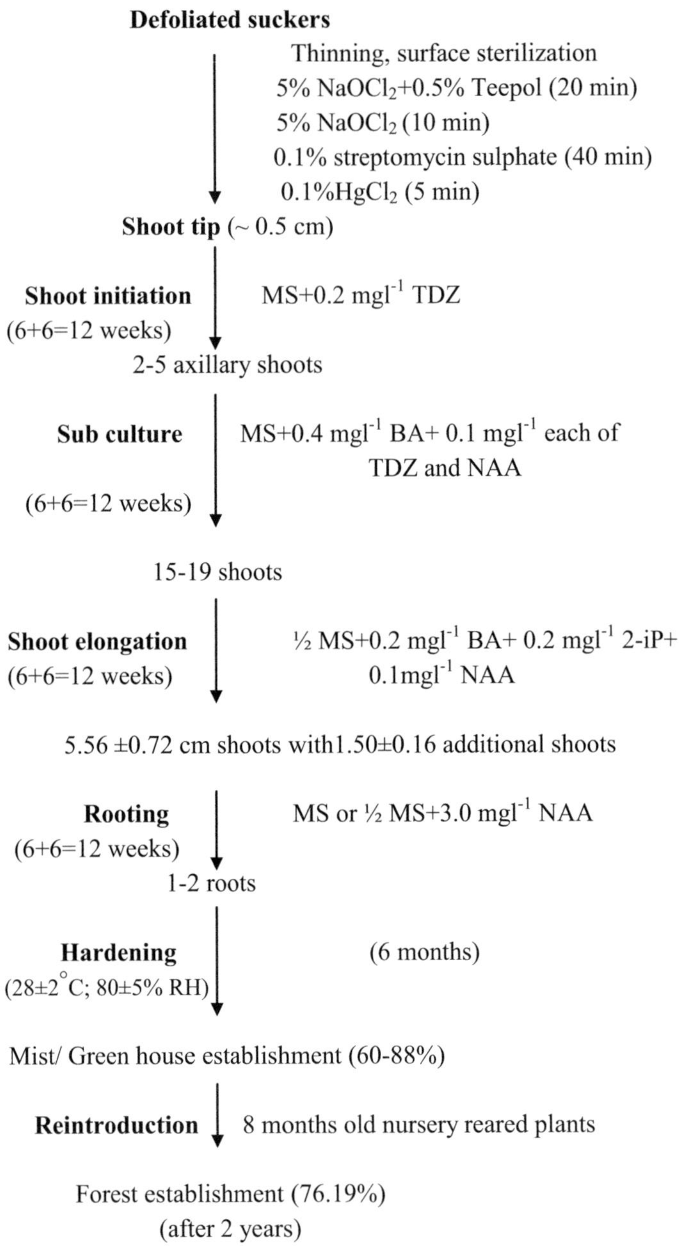

Fig. 28.2 Schematic representation of in vitro clonal multiplication of *C. rotang*

28.5 Conclusion

The viable in vitro cloning system established from wild suckers/offshoot culture demonstrated here can be used for the consistent production of elite female clones for afforestation activities and sustained delivery of high-quality raw materials to cane processing units for strengthening cane industry. The present investigation also reveals that in the in vitro cloning of rattan palms, use of TDZ is indispensable for achieving direct organogenesis in the shoot tip cultures and is the first report as far as rattan palm is concerned. Prolonged subculture of the initiated shoot buds in the presence of TDZ should be avoided as it interferes with shoot elongation and rooting of the multiplied shoots. The results also suggest that rattans are very slow-growing palms and not easily amenable to in vitro manipulation. Hence for the early response and rate of shoot multiplication, a strong cytokinin TDZ was inevitable in culture media than other cytokinins tried, viz. BA, 2-iP, kinetin, etc. Thus the present investigation constitutes the platform for other researchers to improve the in vitro cloning system with innovation to meet the challenges faced by the cane-based cottage industry as it is the livelihood for millions of poor in the developing countries that are otherwise rich in natural resources.

References

Anonymous (1983) Diary, Govt. of India Press, Govt. of Kerala, Ernakulam, India

Atta-Alla H, Van Staden J (1997) Micropropagation and establishment of *Yucca aloifolia*. Plant Cell Tissue Organ Cult 48:209–212

Barba RC, Patena LJ, Mercado MM, Lorico L (1985) Tissue culture of rattan (*Calamus manillensis* Wendl.) In: Proceedings of the second national symposium on tissue culture of rattan, University Partanian Malaysia

Bingshan Z (1997) Tissue culture of *Calamus egregius*. J Central South Forestry University 17(4):563–569

Bunn E, Senaratna T, Sivasithamparam K, Dixon KW (2005) In vitro propagation of *Eucalyptus phylacis* L. Johnson and K. Hill., a critically endangered relict from Western Australia. In Vitro Cell Dev Biol Plant 41:812–815

Chakraberti SK, Mukerji B (1968) Search for anticancer drugs from Indian medicinal plants. J Res on Indian Medicine 3:1–122

Chalupa V (1987) Effect of benzylaminopurine and thidiazuron on in vitro shoot proliferation of *Tilia cordata* Mill., *Sorbus aucuparia* L. and *Robinia pseudoacacia* L. Biol Plant (Praha) 29:425–429

Chevreau E, Skirvin RM, Abu-Qaoud HA, Korban SS, Sullivan JG (1989) Adventitious shoot regeneration from leaf tissue of three pear (*Pyrus* sp.) cultivars in vitro. Plant Cell Rep 7:688–691

Cousineau JC, Donnelly DJ (1991) Adventitious shoot regeneration from leaf explants of tissue cultured and greenhouse-grown raspberry. Plant Cell Tissue Organ Cult 27:249–255

Cuenca B, Ballester A, Vieitez AM (2000) In vitro adventitious bud regeneration from internode segments of beech. Plant Cell Tissue Organ Cult 60:213–220

Dekkers AJ, Rao AN (1989) Some observations on in vitro culture of *Calamus trachycoleus*. In: Rao AN, Yusoff AM (eds) Proceedings of the seminar on tissue culture of forest species. Forest Research Institute Malaysia and International Development Research Centre, Singapore, pp 63–68

Elobeidy A, Korban SS (1988) The effect of thidiazuron on shoot regeneration from apple leaf discs. Hort Sci 23:755

Fangqiu Z (1993) A study on rattan tissue culture. For Res 6(5):486–492

Fasolo F, Zimmerman RH, Fordham I (1989) Adventitious shoot formation on excised leaves of in vitro grown shoots of apple cultivars. Plant Cell Tissue Organ Cult 16:75–87

Geier T (1986) Factors affecting plant regeneration from leaf segments of *Anthurium scherzerianum* Schott cultured in vitro. Plant Cell Tissue Organ Cult 6:115–125

George EF (1993) Plant propagation by tissue culture, part I, the technology, 2nd edn. Exegetics Limited, Edington

Gunawan LW, Yani SA (1986) In vitro propagation of rattan (*Calamus manan* Miq) for agroforestry plantations. VIth International Congress of Plant Tissue and Cell Culture, Minneapolis, p 282

Hartwell JL (1970) Plants used against cancer. Lloydia 33:315–318

Hemanthakumar AS (2011) Studies on embryo and tissue cultures of three economically important rattans (*Calamus* spp.). PhD thesis, University of Kerala, Kerala, India

Hemanthakumar AS, Preetha TS, Krishnan PN, Seeni S (2013) Utilization of zygotic embryos of an economic rattan palm *Calamus thwaitesii* Becc. (Arecaceae) for somaplant regeneration and cryobanking. 3Biotech 3:195–203

Hemanthakumar AS, Preetha TS, Padmesh P, Krishnan PN, Seeni S (2014) Micro-cloning of an economic rattan palm *Calamus thwaitesii* for eco-restoration programme. Biologia 69(5):618–624

Huetteman CA, Preece JE (1993) Thidiazuron: a potent cytokinin for woody plant tissue culture. Plant Cell Tissue Organ Cult 33:105–119

Johnson TS, Narayan SB, Narayana DBA (1997) Rapid in vitro propagation of *Saussurea lappa*, an endangered medicinal plant through multiple shoot cultures. In Vitro Cell Dev Biol Plant 33:128–130

Jones MPA, Yi Z, Murch SJ, Saxena PK (2007) Thidiazuron induced regeneration of *Echinaceae purpurea* L.: micropropagation in solid and liquid culture systems. Plant Cell Rep 26:13–19

Joshi I, Bisht P, Sharma VK, Uniyal DP (2003) In vitro clonal propagation of mature *Eucalyptus* F1 Hybrid (*Eucalyptus tereticornis* SM. X *E. grandis* Hill ex. Maiden). Silvae Genetica 52:110–113

Kritzinger EM, Jansen Van Vuuren R, Woodward B, Rong IH, Spreeth MH, Slabbert (1998) Elimination of external and internal contaminants in rhizomes of *Zantedeschia aethiopica* with commercial fungicides and antibiotics. Plant Cell Tissue Organ Cult 52:61–65

Li Y, Xu C, Chen J (2002) Establishment of virus-free taro (*Colocasia esculenta* cv. *Fenghuayunaitou*) by meristem-tip culture combined with thermotherapy. Pakistan J Plant Pathology 1(2–4):40–43

Lyyra S, Lima A, Merkle S (2006) In vitro regeneration of *Salix nigra* from adventitious shoots. Tree Physiol 26:969–975

Manokaran N (1990) The state of rattan and bamboo trade. RIC Paper No. 7. FRIM, Kepong

Menon KD (1980) Rattan: a state of the art review. In: Rattan – a report of a workshop held in Singapore. International Research centre, Canada, pp 13–74

Mohan Ram HY, Tandon R (1997) Bamboos and rattans: from riches to rags. Proc Indian National Sci Acad 63:245–267

Mullins KV, Llewellyn DJ, Hartney VJ, Strauss S, Dennis ES (1997) Regeneration and transformation of *Eucalyptus camaldulensis*. Plant Cell Rep 16:787–791

Murashige T, Skoog F (1962) A revised medium for rapid growth and bioassays with tobacco tissue cultures. Physiol Plant 15:473–497

Murthy BNS, Murch SJ, Saxena PK (1998) Thidiazuron: a potent regulator of in vitro morphogenesis. In Vitro Cell Dev Biol Plant 34:267–275

Nair LG, Seeni S (2002) Rapid clonal multiplication of *Morinda umbellata* Linn. (Rubiaceae), a medicinal liana, through cultures of nodes and shoot tips from mature plant. Phytomorphology 52:77–81

Neuman MC, Preece JE, Van Sabeek JW, Gaffney GR (1993) Somatic embryogenesis and callus production from cotyledon explants of eastern black walnut (*Juglans nigra* L.) Plant Cell Tissue Organ Cult 32:9–18

Padmanabhan D, Ilangovan R (1989) Studies on embryo culture in *Calamus rotang* Linn. RIC Bulletin 8(1/4.):1):6–9

Padmanabhan D, Ilangovan R (1993) Surgical induction of multiple shoots in embryo cultures of *Calamus gamblei* Becc. RIC Bulletin 12:8–12

Patena LF, Mercado MMS, Barba RC (1984) Rapid propagation of rattan (*Calamus manillensis* H. A. Wendl.) by tissue culture. Philip J Crop Sci 9:217–218

Paudyal KP, Haq N (2000) In vitro propagation of Pummelo (*Citrus grandis* L. Osbeck). In Vitro Cell Dev Biol Plant 36:511–516

Preece JE, Imel MR (1991) Plant regeneration from leaf explants of *Rhododendron* 'PJM Hybrids'. Scientia Hort 48:159–170

Russell JA, Mc Cown (1986) Thidiazuron-stimulated shoot differentiation from protoplast-derived calli of *Populus*. VIth International Congress Plant Tissue Cell Culture Abstracts, 49

Russell JA, Mc Cown (1988) Recovery of plants from leaf protoplasts of hybrid poplar and aspen clones. Plant Cell Rep 7:59–62

Sarma SS (1989) Plants in Yajurveda, Tirupathi (India). KS Vidya Peetha 286

Sastry C (2000) Bamboo in the new millennium: opportunities and challenges. In XXI IUFRO WORLD Congress, Kuala Lumpur, Malaysia

Sha Valli Khan PS, Hausman JF, Rao KR (1999) Clonal multiplication of *Syzygium alternifolium* (Wight.) Walp., through mature nodal segments. Silvae Genetics 48(1):45–50

Singh M, Jaiswal U, Jaiswal VS (2001) Thidiazuron-induced shoot multiplication and plant regeneration in bamboo (*Dendrocalamus strictus* Nees). J Plant Biochem Biotechnol 10:133–137

Singha S, Bhatia SK (1988) Shoot proliferation of pear cultures on medium containing thidiazuron and benzylamino purine. Hortic Sci 23:803

Teng WL (1997) Regeneration of *Anthurium* adventitious shoots using liquid or raft culture. Plant Cell Tissue Organ Cult 49(2):153–156

Thengane SR, Bhosle SV, Deodhar SR, Pawar KD, Kulkarni DK (2006) Micropropagation of Indian laurel (*Calophyllum inophyllum*), a source of anti-HIV compounds. Curr Sci 90:1393–1397

Thomas P, Ravindra MB (1997) Shoot tip culture in mango: influence of medium, genotype, explant factors, season and decontamination treatments on phenolic exudation, explant survival and axenic culture establishment. J Hort Sci 72(5):713–722

Tim W, Yam JLP, Young I, Kap PL, Fan I, Arditti J (1990) Induction of callus from axillary buds of taro (*Colocasia esculenta* Var. Esculenta, Araceae) and subsequent plantlet regeneration. Plant Cell Rep 9:459–462

Tiwari V, Tiwari KN, Singh BD (2001) Comparative studies of cytokinins on in vitro propagation of *Bacopa monniera*. Plant Cell Tissue Organ Cult 66:9–16

Tombac CC, Lapis AB (1993) Indigenous people and rattan. FORSPA Publication 5:12

Umali-Garcia M (1985) Tissue culture of some rattan species. In: Wong KM, Manokaran N (eds) Proceedings of the Rattan seminar. The Rattan Information Centre and Forest Research Institute, Malaysia, pp 23–32

Wochok ZS (1981) The role of tissue culture in preserving threatened and endangered plant species. Biol Conserv 20:83–89

Yusoff AM (1989) Shoot formation in *Calamus manan* under in vitro. In: Rao AN, Yusoff AM (eds) Proc of the seminar on tissue culture of forest species. Forest Research Institute Malaysia and International Development Research Centre, Singapore, pp 45–49

Yusoff AM, Jung SU, Paranjothy K (1985) Tissue culture of *Calamus manan*. Sympossium Tisu Tumbuhan Kebangsaan Kell dan Bengkel Kultur Tisu Getah Antarabangsa. Universiti Pertanian Malaysia, Serdang

MIX
Papier aus verantwortungsvollen Quellen
Paper from responsible sources
FSC® C105338

If you have any concerns about our products,
you can contact us on
ProductSafety@springernature.com

In case Publisher is established outside the EU,
the EU authorized representative is:
Springer Nature Customer Service Center GmbH
Europaplatz 3, 69115 Heidelberg, Germany

Printed by Libri Plureos GmbH
in Hamburg, Germany